# Introductory Statistics for the Behavioral Sciences

# Introductory Statistics for the Behavioral Sciences

*Eighth Edition*

**R. Brooke Lea**
**Barry H. Cohen**

WILEY

Published by John Wiley & Sons, Inc., Hoboken, New Jersey.
Published simultaneously in Canada.



***Library of Congress Cataloging-in-Publication Data***

Names: Lea, R. Brooke author | Cohen, Barry H.,
1949- author
Title: Introductory statistics for the behavioral sciences /
R. Brooke Lea, Barry H. Cohen.
Description: Eighth edition. | Hoboken, New Jersey : Wiley, [2026] |
Series: Introductory statistics for the behavioral sciences | Includes
index.
Identifiers: LCCN 2025030350 (print) | LCCN 2025030351 (ebook) | ISBN
9781394234738 paperback | ISBN 9781394234752 adobe pdf | ISBN
9781394234745 epub
Subjects: LCSH: Social sciences–Statistical methods | Psychometrics |
Sociology–Statistical methods | Educational statistics
Classification: LCC HA29 .W445 2025 (print) | LCC HA29 (ebook)
LC record available at https://lccn.loc.gov/2025030350
LC ebook record available at https://lccn.loc.gov/2025030351

Cover design: Wiley
Cover image: Courtesy of Tom Navratil and Dan Navratil

Set in 10/12pt SlimbachStd by Straive, Pondicherry, India

*Brooke dedicates this book to Emily and Jackson,*
*the two parameters that keep him Normal,*
*and Barry dedicates the book to Leona and the newest additions*
*to our family: Maya, Leo, Jackson, Liv, and Roman*

# Contents

# Preface

The eighth edition of this textbook has been a delight and honor to create. The book is one of the longest-running statistics texts in the behavioral sciences. (This title has been in print continuously since 1971.) In creating this revision, we tried to maintain the original purpose of the text as defined by Joan Welkowitz and Jacob "Jack" Cohen, and expressed in the preface to the second edition: "to introduce and explain statistical concepts and principles clearly and in a highly readable fashion, assuming minimal mathematical sophistication, but avoiding a 'cookbook'; approach to methodology."

At the same time that we worked to uphold the original mission of this classic text, we have made substantial improvements to nearly every chapter. In addition, we have added a new chapter that presents an accessible introduction to the evolving and controverial role that null hypothesis testing plays in behavioral research, along with newer methods (e.g., bootstrapping and Bayesian statistics) that have gained popularity and importance since the last edition. In this new chapter—and throughout the book—we have adopted a more "comprehension-based" approach that emphasizes how an understanding of statistical concepts informs students' comprehension of primary sources of research. Borrowing the comprehension vs. production distinction from psycholinguistics, students of behavioral science use their statistical knowledge to comprehend (e.g., read and understand) research at least as much as they use it to produce statistical output (e.g., perform statistical tests). Our approach, especially for more advanced topics, is to ensure that readers of this book understand statistical techniques at a level that will be useful in a variety of educational and experiential contexts.

We also took a hard look at the examples that are used in the book and updated them to provide a more diverse and inclusive perspective on behavioral science. We have made substantial changes to the data set used for exercises by replacing variables that some students might find triggering (e.g., math phobia) with those that are more approachable.

We listened to feedback from students and adopters of the previous edition, who commented that once students have mastered the fundamentals of computing means and standard deviations by hand, the practice becomes needlessly tedious when exercises focus on more complex computations such as $t$ tests, Pearson $r$, and ANOVA. Accordingly, we now supply summary statistics (means and SDs) so that students can focus their efforts on statistical concepts they are working to master. Of course, we also supply the raw data in case instructors would prefer students to continue to drill on computing descriptive statistics step by step, or use statistical calculators or spreadsheets.

As always, we emphasize the through-lines that link newer topics with those that came before them in the text. We view this conceptual connective tissue as an important facilitator of learning content that, while divided into discrete topics and chapters, is deeply interconnected. Accordingly, we now provide specific page numbers when referring to formulas, tables, or figures presented earlier in the book to encourage students to refer back to familiar content that reinforces these connections.

Naturally, we took this opportunity to correct any mistakes we and others had noticed in previous editions, and to add some explanatory sentences for the concepts that we know our students have struggled with in the past.

As with the seventh edition of this text, the supplementary materials of this edition will overlap with those for the fourth edition of Barry Cohen's graduate-level statistics text, *Explaining Psychological Statistics* (https://www.wiley.com/en-us/Explaining+Psychological+Statistics%2C+4th+Edition-p-9781118652145), also published by John Wiley & Sons.

There are several structural changes in this edition that are worth emphasizing, as described next.

## 1. Adding Page Numbers When Referring to Prior Material in the Text

One of our chief strategies for making statistics understandable is to point out the connections between related concepts; mastering behavioral statistics is easier when students can use prior knowledge to help scaffold new material. We have facilitated this process by providing both the name of the prior element and the page on which it appears (e.g., Formula 3.2, page 51). We hope this change will encourage students to appreciate the extent to which statistical knowledge is cumulative.

## 2. Reduced Emphasis on Rarely Used Procedures and Formulas

As the field advances and makes room for new techniques (see the new Chapter 12), older, rarely used procedures can be relegated to more specialized texts. Examples include linear interpolation procedures for percentiles and cumulative frequency distributions, and computational formulas that are only useful for processing a large quantity of numbers. Learning to compute the sum of squares by hand, for example, is mainly useful as a way for students to understand how examining deviations from the mean can lead to a useful descriptive measure of dispersion. The computational formulas for *SS*, by contrast, hold no pedagogical value and only exist to facilitate arriving at the correct value of *SS* with large amounts of data. The ubiquity of automated computation, however, obviates the need for such formulas and techniques.

## 3. Updating the Computer Exercises, SPSS Sections, and New Bridge to R

We revised Ihno's data set to create Jackson's data set for this edition by changing the variables so that they lead to more interesting exercises. We also took this opportunity to update our Bridge to SPSS sections to reflect changes in the most recent versions of SPSS. (The version current during the writing of this edition is 29.) In several chapters we have now included screenshots to illustrate the main SPSS dialog boxes commonly used by researchers to perform the analyses described in this text. In some cases, we have also included results boxes from the output that SPSS produces for particular analyses, in order to draw connections between the terms used by SPSS to label its output and the corresponding (sometimes different) terms used in this text. Furthermore, we now include "Bridge to R" sections as part of the online Supplementary Materials, which will help students using RStudio to successfully complete our Computer Exercises.

## 4. New Chapter on Newer, More Advanced Statistical Techniques

The new Chapter 12, titled "Beyond Null Hypothesis Testing," addresses some important topics in behavioral statistics that have emerged since our last edition. We begin by reviewing some of the more trenchant critiques of null hypothesis testing, along with some cogent rebuttals that are sometimes omitted in treatments of this debate. A concrete analogy involving disease detection illustrates what a $p$ value can and cannot tell you. This discussion leads to an introduction to "Robust Statistics" that includes a review of bootstrapping techniques along with the important role that nonparametric statistics can play in behavioral science. Every student should learn about the problems introduced by $p$ Hacking, HARKing, and the file drawer problem. We explain the dangers associated with these practices, and how easily researchers can inadvertently find themselves falling into such traps. That discussion naturally leads into a treatment of the replication crisis in behavioral research, along with some proposed solutions. The last topic in this new chapter introduces alternatives to null hypothesis testing, including the "new" statistics, which emphasizes the interpretation of effect sizes and confidence intervals. Finally, we include an appendix to this chapter that serves as a primer to the use of Bayesian statistics. Overall, this chapter presents these advanced topics and debates in an accessible, friendly manner.

## 5. Updated Ancillaries

We have updated the Student Companion page on the Wiley website for the new edition, which can be found at: www.wiley.com/go/ISBS8e

The Student Companion Site includes the following items:

- **Study Guide:** A lively chapter-by-chapter review of the text with additional exercises and answers. Created by graduate students who recently served as teaching assistants for statistics, it provides another perspective on the material presented in this text.
- **PowerPoint Slides:** Expanded and updated for this new edition, these slides provide convenient summaries of the important points of each chapter, and can help instructors to organize their lectures around the key concepts for each statistical topic.

We have also completely updated the Wiley Instructor Companion Site for the eighth edition, which can also be found at: www.wiley.com/go/ISBS8e

In addition to the two items on the student site, the Instructor Website includes the following items:

- **Instructor's Manual:** Step-by-step answers to all of the computational exercises in the text.
- **Test Bank:** Multiple choice questions, both conceptual and computational, that can be used to create quizzes to assess the students' mastery of each chapter in the eighth edition. The test bank includes questions for Chapter 12 that can be used for testing or to spur discussion of issues concerning the use of different statistical approaches.

R. Brooke Lea
Barry H. Cohen

# Acknowledgments

Thanks are due to our many encouraging friends and relatives, to all the colleagues and adopters who made many useful comments on previous editions, and especially to Columbia University Professor Katherine Fox-Glassman. Professors Steve Guglielmo and Annie Pezzala at Macalester College have been inspirational co-conspirators in convincing psychology majors that learning statistics can be a (surprisingly) exciting and rewarding voyage. Once again, we want to take this opportunity to thank our many students who, throughout the years, have provided invaluable feedback on our teaching of statistics, as well as on earlier editions of this text and its accompanying materials. Special thanks go to the recent survivors of the "RIP" sequence at Macalester College for their trenchant suggestions for improving the text.

This latest edition surely owes its very existence to Patricia Rossi (former Executive Editor), Darren Lalonde (current Acquisitions Editor), and Nathanael Jude Mcgavin (Managing Editor) at John Wiley & Sons, whose support and prodding have ensured that this text will now begin its 40th year in continuous print. This new edition has also benefited greatly from the able assistance of Christina Weyrauch (Senior Editorial Assistant), as well as Sundhar Karuthudiyan (Content Refinement Specialist). We are also grateful to those responsible for the attractive look and design of this text's cover—especially, Tom and Daniel Navratil. We remain ever grateful to Grace Jackson and Samantha Gaies for creating the lively and engaging Study Guide for the previous edition, most of which has survived in the supplementary web material for this eighth edition.

We are more than delighted to acknowledge a substantial debt of gratitude to Jackson Pearlman for numerous invaluable contributions to this edition. In addition to revising the data set (now named after him) that we use for the computer exercises, Jackson was instrumental in helping us update many examples in the text that were of questionable cultural relevance in the mid-2020s. He also updated and contributed importantly to the material we provide in the Instructor and Student companion sites for this text. Simply put, this edition would not have been possible without Jackson Pearlman.

Finally, we would like to express our gratitude to those who have supported us in all the small ways that matter during our work on this project. Brooke shouts out thanks and love to Emily Fields and Jackson Lea, and Barry does the same for Leona Gizzi.

R. Brooke Lea
Barry H. Cohen

## Postscript

We cannot close this Acknowledgments section without sadly paying homage to the two departed professors who, along with a colleague now living in Florida, wrote and edited the original text and kept it fresh and alive over the course of several decades. BHC had the pleasure of working closely with the senior author, Dr. Joan Welkowitz, on the previous edition of this text, and it is not possible to overestimate the continued influence of both her and her former co-author and friend, Jacob Cohen, on this text and on our approach to the teaching of statistics. The one living author of the original text, Robert Ewen, wrote nearly all of the exercises and thought questions you see in this edition. Though many of the exercises and examples in the text have been updated, Dr. Ewen's careful craftsmanship and editorial skills are still visible in this edition.

R. Brooke Lea
Barry H. Cohen

# Glossary of Symbols

*Numbers in parentheses indicate the chapter in which the symbol first appears.*

| | |
|---|---|
| $a_{YX}$ | $Y$-intercept of linear regression line for predicting $Y$ from $X$ (10) |
| $\alpha$ | criterion (or level) of significance; probability of Type I error (5) |
| $\alpha_{EW}$ | experiment-wise alpha (14) |
| $\alpha_{FW}$ | family-wise alpha (15) |
| $\alpha_{pc}$ | alpha per comparison (14) |
| $b_{YX}$ | slope of the linear regression line for predicting $Y$ from $X$ (10) |
| $\beta$ | probability of Type II error (5) |
| $1 - \beta$ | power (11) |
| Cf | cumulative frequency (2) |
| $\chi^2$ | statistic following the chi square distribution (13) |
| $D$ | difference between two scores (7) |
| $\overline{D}$ | mean of the *D*s (7) |
| $d$ | effect size involving two samples (7) |
| $\mathbf{d}$ | effect size involving two populations (11) |
| $Df$ | degrees of freedom (6) |
| $df_{\mathrm{Bet}}$ | degrees of freedom between groups (13) |
| $df_{\mathrm{W}}$ | degrees of freedom within groups (13) |
| $df_1$ | degrees of freedom for factor 1 (15) |
| $df_2$ | degrees of freedom for factor 2 (15) |
| $df_{1\times 2}$ | degrees of freedom for interaction (15) |
| $\delta$ | delta (11) |
| $\eta^2$ | eta squared; effect size in multiple samples (13) |
| $\mathbf{f}$ | effect size involving multiple populations (13) |
| $f$ | frequency (2) |
| $f_e$ | expected frequency (18) |
| $f_o$ | observed frequency (18) |
| $F$ | statistic following the *F* distribution (13) |
| $g$ | Hedge's adjusted effect size involving two samples (7) |
| $H$ | statistic for the Kruskal–Wallis test (13) |
| $H_0$ | null hypothesis (5) |
| $H_A$ | alternative hypothesis (5) |
| HSD | Tukey's Honestly Significant Difference (14) |
| $i$ | case number (1) |
| $k$ | a constant (1) |
| $k$ | number of groups (13) |
| LSD | Fisher's Least Significant Difference (14) |
| Mdn | median (3) |
| $MS$ | mean square (13) |
| $MS_{\mathrm{Bet}}$ | mean square between groups (13) |
| $MS_W$ | mean square within groups (13) |
| $MS_1$ | mean square for factor 1 (15) |
| $MS_2$ | mean square for factor 2 (15) |
| $MS_{1\times 2}$ | mean square for interaction (15) |
| $\mu$ | population mean (3) |
| $n$ | number of observations in one of two or more equal-sized samples (7) |
| $N_{\mathrm{T}}$ | total number of subjects or observations (1) |

| | |
|---|---|
| $N_i$ | number of observations or subjects in group $i$ (13) |
| $\omega^2$ | omega squared; proportion of variance accounted for in a population (10) |
| $\pi$ | hypothetical population proportion (6) |
| $p$ | probability of attaining results as extreme as yours when the null hypothesis is true (5) |
| $P$ | observed sample proportion (6) |
| $P(A)$ | probability of event $A$ (17) |
| PR | percentile rank (2) |
| $\phi$ | phi coefficient (18) |
| $\phi_C$ | Cramér's $\phi$ (18) |
| $q$ | studentized range statistic (14) |
| $r_C$ | matched pairs rank biserial correlation coefficient (8) |
| $r_G$ | Glass rank biserial correlation coefficient (8) |
| $r_{pb}$ | point-biserial correlation coefficient (10) |
| $r_s$ | Spearman rank-order correlation coefficient (9) |
| $r_{XY}$ | sample Pearson correlation coefficient between $X$ and $Y$ (9) |
| $\overline{R}$ | mean of a set of ranks (8) |
| $\rho_{XY}$ | population correlation coefficient between $X$ and $Y$ (9) |
| $s$ | sample standard deviation (5) |
| $s^2$ | population variance estimate (5) |
| $s^2_D$ | variance of the $D$s (7) |
| $s^2_{pooled}$ | pooled variance (7) |
| $s_{\overline{X}}$ | standard error of the mean (6) |
| $s_{\overline{X}_1 - \overline{X}_2}$ | standard error of the difference (7) |
| $s_{est}$ | estimate of $\sigma_y'$ obtained from a sample (10) |
| SIQR | semi-interquartile range (3) |
| $SS$ | sum of squares (3) |
| $SS_T$ | total sum of squares (13) |
| $SS_{Bet}$ | sum of squares between groups (13) |
| $SS_W$ | sum of squares within groups (13) |
| $SS_1$ | sum of squares for factor 1 (15) |
| $SS_2$ | sum of squares for factor 2 (15) |
| $SS_{1 \times 2}$ | sum of squares for interaction (15) |
| $\sum$ | summation sign (1) |
| $\sigma$ | population standard deviation (3) |
| $\sigma^2$ | population variance (3) |
| $\sigma_p$ | standard error of a sample proportion (6) |
| $\sigma_T$ | standard error of the ranks of independent samples (8) |
| $\sigma_{T_M}$ | standard error of the ranks of matched samples (8) |
| $\sigma_{\overline{X}}$ | standard error of the mean when $\sigma$ is known (5) |
| $\sigma_{est}$ | standard error of estimate for predicting $Y$ (10) |
| $t$ | statistic following the $t$ distribution (6) |
| $T$ | T score (4) |
| $T_E$ | expected sum of the ranks (8) |
| $T_i$ | sum of ranks in group $i$ (8) |
| $X'$ | predicted $X$ score (10) |
| $\overline{X}$ | sample mean (3) |
| $\overline{X}_i$ | mean of group $i$ (13) |
| $\overline{X}_G$ | grand mean (13) |
| $Y'$ | predicted $Y$ score (10) |
| $z$ | standard score (4) |

# About the Companion Website

This book is accompanied by a companion website:

**www.wiley.com/go/ISBS8e**

The website includes:

- Power Point
- Study Guide
- Test Bank
- Instructor Manual

# Part I
# Descriptive Statistics

# Chapter 1
# Introduction

## PREVIEW

### Why Study Statistics?

What are four important reasons why knowledge of statistics is essential for anyone majoring in psychology, sociology, or education?

### Descriptive and Inferential Statistics

What is the difference between descriptive and inferential statistics?
Why must behavioral science researchers use inferential statistics?

### Populations, Samples, Parameters, and Statistics

What is the difference between a population and a sample?
Why is it important to specify clearly the population from which a sample is drawn?
What is the difference between a parameter and a statistic?

### Measurement Scales

What types of scales are used to measure variables in the behavioral sciences?

### Independent and Dependent Variables

What is the difference between observational and experimental studies?

### Summation Notation

Why is summation notation used by statisticians?
What are the eight rules involving summation notation?

### Jackson's Study

An example that provides a common thread tying together all of the subsequent chapters.

### Summary

### Exercises

### Thought Questions

### Computer Exercises

### Bridge to SPSS

*Introductory Statistics for the Behavioral Sciences*, Eighth Edition. R. Brooke Lea and Barry H. Cohen.

Companion website: www.wiley.com/go/ISBS8e

## Why Study Statistics?

This book is written primarily for undergraduates majoring in psychology, or one of the other behavioral sciences. There are four reasons why a knowledge of statistics is essential for those who wish to conduct or consume behavioral science research:

1. ***To understand the professional literature.*** Most professional literature in the behavioral sciences includes results that are based on statistical analyses. Therefore, you will be unable to understand important articles in scientific journals and books unless you understand statistics. It is possible to seek out secondhand reports that are designed for the statistically uninformed, but those who prefer this alternative to obtain firsthand information perhaps should not be majoring in the field of behavioral science.
2. ***To understand and evaluate statistical claims made in the popular media.*** "This has been an important reason to acquire quantitative reasoning skills for decades, as advances in technology have led to an explosion of quantitative claims in a variety of media outlets." Unfortunately, the difference between claims that are statistically sound and those that merely appear that way can be difficult to detect without some formal training. You will make many important life decisions over the next decade that will require weighing probabilities under conditions of uncertainty. A competence with basic statistics will empower you to make maximal use of the available information and help protect you from those who may wish to mislead you with pretty graphs and numbers.
3. ***To understand the rationale underlying research in the behavioral sciences.*** Statistics is not just a catalog of procedures and formulas. It offers the rationale on which much of behavioral science research is based—namely drawing inferences about a population based on data obtained from a sample. Those familiar with statistics understand that research consists of a series of educated guesses and fallible decisions, not right or wrong answers. Those without knowledge of statistics, by contrast, cannot understand the strengths and weaknesses of the techniques used by behavioral scientists to collect information and draw conclusions.
4. ***To carry out behavioral science research.*** In order to contribute competent research to the behavioral sciences, it is necessary to design the statistical analysis *before* the data are collected. Otherwise, the research procedures may be so poorly planned that not even an expert statistician can make any sense out of the results. To be sure, it is possible (and often advisable) to consult someone more experienced in statistics for assistance. Without some statistical knowledge of your own, however, you will find it difficult or impossible to convey your needs to someone else and to understand the replies.

Save for these introductory remarks, we do not see it as our task to persuade you that statistics is important in psychology and other behavioral sciences. If you are seriously interested in any of these fields, you will find this out for yourself. Accordingly, this book does not rely on documented examples selected from the professional literature to prove to you that statistics really is used in these fields. Instead, we have devised several artificial, but realistic, examples with numerical values that reveal the processes and issues involved in statistical analyses as clearly as possible.

One example we use throughout this book is based on a hypothetical study performed by a new student named Mia on the relative friendliness of four dormitory halls on her campus. We present the data for this study in the first exercise at the end of this chapter and return to it in many of the subsequent chapters.

We have tried to avoid a "cookbook" approach that places excessive emphasis on computational recipes. Instead, the various statistical procedures and the essential underlying concepts have been explained at length, and insofar as possible in standard English, so that you will know not only what to do but *why* you are doing it. Do not, however, expect to learn the material in this book from a single reading; the concepts involved in statistics, especially inferential statistics, are sufficiently challenging that it is often said that the only way to completely understand the material is to teach it (or write a book about it). Having said that, however, there is no reason to approach statistics with fear and trembling. You certainly do not need any advanced understanding of mathematics to obtain a good working knowledge of basic statistics. What *is* needed is mathematical comprehension sufficient to cope with beginning high school algebra and a willingness to work at new concepts until they are understood, which requires, in turn, a willingness to spend some time working through at least half of the exercises at the end of each chapter (and, perhaps, some additional exercises from the online Study Guide).

## Descriptive and Inferential Statistics

One purpose of statistics is to summarize or describe the characteristics of a set of data in a clear and convenient fashion. This is accomplished by what is called **descriptive statistics**. For example, your grade point average (GPA) serves as a convenient summary of all of the grades that you have received in college. Part I of this book is devoted to descriptive statistics.

A second function of statistics is to make possible the solution of an extremely important problem. Behavioral scientists can never measure *all* of the cases in which they are interested. For example, a clinical psychologist studying the effects of various kinds of therapies cannot obtain data on every single mental health patient in the world; a developmental psychologist studying age differences in attitudes cannot measure all of the millions of children and adults in the United States; a cognitive psychologist cannot observe the reading behavior of all literate adults. Behavioral scientists want to know what is happening in a given **population**—a large group (theoretically an infinitely large group) of people, animals, objects, or responses that are alike in at least one respect (e.g., all college students in the United States). They cannot measure the entire population, however, because it is so large that it would be too time-consuming and expensive to do so. What to do?

Turns out there's a reasonably simple solution: Measure just a relatively small number of cases drawn from the population (i.e., a **sample**), and use inferential statistics to make educated guesses about the population. **Inferential statistics** makes it possible to draw inferences about what is happening in the population based on what is observed in a sample from that population. (This point is discussed at greater length in Chapter 5.) The subsequent parts of this book are devoted to inferential statistics, which makes frequent use of some of the descriptive statistics discussed in Part I.

## Populations, Samples, Parameters, and Statistics

As the above discussion indicates, the term **population** as used in statistics does not necessarily refer to people. For example, the population of interest may be that of all white rats of a given genetic strain or all responses of a single participant's eyelid in a conditioning experiment.

Whereas the population consists of all of the cases of interest, a **sample** consists of any subgroup drawn from the specified population. It is important that the population be clearly specified. For example, a group of 100 Macalester College freshmen might be a well-drawn sample from the population of all Macalester freshmen or a poorly drawn sample from the population of all undergraduates in the United States (poorly drawn because it probably will not be representative of all US undergraduates). It is strictly proper to apply (i.e., *generalize*) the research results only to the specified population from which the sample was drawn. (A researcher *may* justifiably argue that her results are more widely generalizable, but she is on her own if she does so because the rules of statistical inference do not justify this.)

A **statistic** is a numerical quantity (such as an average) that summarizes some characteristic of a sample. A **parameter** is the corresponding value of that characteristic in the population. For example, if the average studying time of a sample of 100 New York University sophomores is 7.4 hours per week, then 7.4 is a statistic. If the average studying time of the population of all NYU freshmen is 9.6 hours per week, then 9.6 is the corresponding population parameter. Usually, the values of population parameters are unknown because the population is too large to measure in its entirety, and appropriate techniques of inferential statistics are therefore used to estimate the values of population parameters from sample statistics. If the sample is properly selected, the sample statistics will often give good estimates of the parameters of the population from which the sample was drawn; if the sample is poorly chosen, erroneous conclusions are likely to occur. Whether you are doing your own research or reading about that produced by someone else, you should always check to be sure that the population to which the results are generalized is proper in light of the sample from which the results were obtained.

## Measurement Scales

You may have noticed that we have used the term *data* several times without talking about where the data come from. It should come as no surprise that in the behavioral sciences, the data generally come from measuring some aspect of the behavior of a human or animal. Unlike physics, in which there are quite a few important **constants** (values that are always the same, such as the speed of light or the mass of an electron), the behavioral sciences deal mainly with the measurement of **variables**, which can take on a range of different values. An additional complication faced by the behavioral scientist is that some of the variables of most interest can be difficult to measure (e.g., self-esteem). In this section, we discuss the measurement scales most commonly used in the behavioral sciences. It is important to know which scale you are using because that choice often determines which statistical technique is appropriate. The scales differ with respect to how finely they can distinguish differences among instances. For example, the nominal scale can only distinguish whether an item is in one category or another, but the categories have no inherent order (e.g., the color of your iPhone). The ordinal scale involves measurements that can distinguish order on a set of values (e.g., bigger, smaller). Interval scales add the capability to measure quantities on a scale that has equal intervals between units (e.g., inches; temperature), and ratio scales are interval scales for which a value of zero means that *absolutely none* of the variables being measured is present. We describe them below in order of complexity.

### Nominal Scales

The crudest form of measurement is to classify items by assigning names to them (categorization), which does not involve any numerical precision at all. Such a scale is called a **nominal scale**. For example, a person's occupation can only be "measured" on a nominal scale (e.g., accountant, lawyer, politician, computer programmer). We can count the number of people who fall into each category, but (unlike ordinal, interval, or ratio scales) there is no obvious order to the categories, and certainly no regular intervals between them. We refer to such categorical data as being **qualitative**, as distinguished from data measured on one of the **quantitative** scales described next.

### Ordinal Scales

Sometimes it is possible to order your categories, even though the intervals are not precise. The most common example of this in psychological research is called a Likert scale (after its creator, Rensis Likert, pronounced "Like-ert"), on which respondents rate their agreement with some statement by choosing, for instance, among "strongly agree," "agree," "uncertain," "disagree," and "strongly disagree." The order of the categories is clear, but because there is no way to be sure that they are equally spaced (is the psychological distance between "strongly agree" and "agree" the same as between "agree" and "uncertain?"), this type of scale lacks the interval property and is therefore called an **ordinal scale**. Although it is a somewhat controversial practice, many behavioral researchers simply assign numbers to the categories (e.g., strongly agree is 1, agree is 2, etc.) and then treat the data as though they came from an interval scale.

Another, less common, way that an ordinal scale can be created is by rank ordering. It may not be possible to measure, in a precise way, the creativity of paintings produced by students in an art class, but a panel of judges could rank them from most to least creative, with perhaps a few paintings tied at the same rank.

It is important to distinguish between the *characteristics of the variable* we are measuring, on one hand, and the particular *scale with which we choose to measure that variable*, on the other. For example, suppose a teacher wishes to measure how often middle school students raise their hands in class. Hand-raising frequency is a variable that can be measured quantitatively by just counting how often each student raises a hand. However, when the teacher uses these data to understand hand-raising behavior in his class, his purposes may be best served by organizing the students into three broad, but ordinal, groups: (1) those who never raised their hand (most of the group); (2) those who did so just a few times; and (3) those who did so many times. In that case, he will want to create an ordinal scale to present data that was originally measured on an interval scale (described next). In other words, just because you are using a variable that has the *potential* to be measured on a particular scale does not necessarily mean that the data you collect will possess or display the characteristics of that scale. Sometimes researchers have quantitative measurements that vary in such an odd way that it becomes more useful just to rank them all and use the ranks in place of the original measurements. We explain this somewhat unusual practice when we describe statistical tests based on ordinal data in Chapter 8.

### Interval/Ratio Scales

The most precise scales are the kinds that are used for physical measurement. For instance, the temperature of the skin at your fingertips can be related to the amount of stress that you experience (high stress can cause the constriction of peripheral blood vessels, resulting in a decrease in skin temperature). Using either the Fahrenheit or Celsius temperature scale allows a precise measurement of skin temperature. Because degrees on either scale are fixed units that are always the same size, you can be sure that the interval between, say, 32 and 33 degrees Celsius is exactly the same as the interval between 18 and 19 degrees Celsius. These two temperature scales are therefore called *interval scales*.

Another desirable characteristic that scales may have is the ratio property, which requires that the scale have a true zero point, that is, a measurement of zero on the scale indicates that there is really nothing left of what is being measured. For example, a measurement of zero pounds indicates the total absence of weight. If the scale has a true zero point in addition to the interval property, a measurement of 6 units, for instance, will actually indicate twice as much of what is being measured as compared to only 3 units. Thus, it is appropriate to call these scales *ratio scales*. A measurement of zero degrees on the Kelvin temperature scale means that there is no temperature at all (i.e., absolute zero), which is not the case for the Fahrenheit or Celsius scales. Therefore, Kelvin is a ratio scale, whereas the other two are just interval scales (zero degrees Celsius, e.g., does not indicate the complete absence of temperature—it just means that water will change form into ice).

It is only the interval property that is needed for precise measurement. So, for our purposes, it is common to make no distinction between interval and ratio scales, referring instead to interval/ratio data, or simply *quantitative data*, as opposed to qualitative data.

It's worth noting that the quantitative/qualitative distinction may best be thought of as a dimension on which the clearest cases are at the extremes, with fuzzier cases falling toward the middle. For example, when members of one category can be distinguished from members of another based only on *qualitative features of the items* (e.g., shape or color), not on the quantity of such features, then you have categorical (nominal) data; it makes sense that data of this sort are called *qualitative*. By contrast, variables that assign values based on the *quantity of the variable measured* (e.g., weight, length), rather than the type measured, are reasonably called *quantitative*. But what if you have ranked paintings on the basis of creativity, as in the example from the section on ordinal variables above? Is the difference in creativity between the #1 ranked painting and the #2 ranked painting only qualitative (one painting is in a different category of creativity from the other), or quantitative (one exhibits more creativity than the other)? It may not be clear. Arguably both qualitative *and* quantitative differences create the separation between the two paintings. What about survey data based on participants' answers to questions on a 10-point scale? As we stated earlier, social scientists often behave as though these data possess interval-scale properties and therefore analyze them quantitatively. So, are ordinal data qualitative, quantitative, or both? As our examples illustrate, the quantitative/qualitative distinction may not be as useful for characterizing ordinal data as it is for categorical and interval/ratio data. In any case, in this book, we describe specific methods for dealing with ordinal data, and at some point, you can decide for yourself when to treat your ordinal data as though it came from an interval or ratio scale (looking, of course, at what your colleagues are doing with similar data).

## Independent and Dependent Variables

Most behavioral research can be classified into one of two categories: **observational** or **experimental**.

In the simplest experiment, a researcher creates two conditions. The participants assigned to one of the conditions get some form of treatment, such as a pill intended to reduce depression. Those assigned to the other condition get something that superficially resembles the treatment, such as an identical-looking pill that is totally ineffective (i.e., a placebo); they are part of a control group. These two conditions are the two different levels of an **independent variable**, or one that is created by the experimenter. Commonly, an independent variable is one whose levels are qualitative (e.g., a pill that contains medicine versus a placebo that does not). The **dependent variable**, or the variable that is measured by the experimenter and is expected to change from one level of the independent variable to another, is usually quantitative (such as a self-rating of depression). We will begin to describe such experiments in Part II of this book when we turn to inferential statistics.

True independent variables are under the control of the experimenter, as in the example above. **Quasi-independent variables** are those that are *not* under the control of the experimenter. Quasi-independent variables are also referred to as subject variables, selected variables, or grouping variables. For example, imagine a psychologist is interested in whether owning a dog is associated with greater empathy for humans. She would want to study one group of people who own dogs, and a separate group of people who do not. Dog ownership is the independent variable in this study; but is it a "true-independent variable" or a "quasi-independent variable?" The question you should ask yourself is "Did the experimenter randomly assign participants to the two groups, or was dog ownership decided by some process not controlled by the experimenter?" Clearly, experimenters cannot (easily) randomly assign people to be either a dog owner or not a dog owner—group membership in this case was decided by the people themselves. Therefore, the variable "dog ownership" is a subject variable—it is a characteristic of the participant not influenced by the experimenter. Such variables are considered quasi-independent variables. Other examples of quasi-independent variables are religion, marital status, and age. In these cases, the experimenter cannot decide whether you will be Lutheran or Jewish or whether you will be single or married; participants' group memberships have already been determined before they arrive at the study.

Why does it matter whether an independent variable is "true" or "quasi?" Inferences concerning causality depend on this distinction. True independent variables allow experimenters to draw a causal link between the independent variable and the dependent variable (assuming the rest of the experiment was conducted properly). Establishing causality is important to progress in science. In the hypothetical antidepressant experiment described previously, the randomized trial design allows the researcher to establish whether the active ingredient in the pill really *caused* people to become less depressed. With a quasi-independent variable, such causal conclusions cannot be made (e.g., if married people are less depressed than people who are not married, is it because being married is causing a reduction in depression, or because those who are less depressed are more likely to get married?).

Some independent variables can be truly experimental or quasi-experimental depending on the design of the study. For example, suppose you wanted to know whether exercising leads to better health (as measured by the number of sick days taken at work). You could ask a sample of workers about their exercise habits and number of sick days over a six-month period. If the data reveal that those who exercise the most have the fewest sick days, the researcher is *not* entitled to draw a causal relationship between the two (i.e., that more exercise causes better health). In this example, exercise habit is a subject variable; the experimenter did not manipulate how many hours per week each person spends in the gym. That lack of control leaves open other possible explanations. Maybe the causal arrow points the other way and people who are healthy tend to (and are able to) work out more. Perhaps some third variable like optimism drives both health and exercise. You just cannot be sure which conclusion is correct with a quasi-experimental design.

Now imagine that the researcher had assigned people at random to work out 0, 5, 10, or 15 hours per week at a gym and that after six months she got the same inverse association between hours spent exercising and number of sick days. In this situation, the researcher *can* claim that exercise causes fewer sick days. Other possibilities such as those mentioned earlier have been controlled for. For example, optimism cannot explain the results because random assignment would have distributed the optimistic people evenly among the four workout groups. Note also that true experiments are usually more difficult to conduct compared to quasi-experimental versions of the same studies. Therefore, a trade-off often exists between effort and expense, on one hand, and the sort of conclusions that can be drawn, on the other. In fact, behavioral researchers often study the relationships among variables that are not convenient, or even possible, to ever manipulate. If one simply measures the relationship between two dependent variables (e.g., self-esteem in teenagers and their family's annual income to see if those from more affluent families tend to have higher—or lower—self-esteem), this is an observational (i.e., quasi-experimental) study. Such a study can reveal that self-esteem and family income are somehow related, but behavioral researchers are often concerned with understanding *how* and *why* they are related.

Methods for dealing with studies in which both the independent and dependent variables (or two dependent variables) are quantitative will be presented in Chapter 9. Research in which both variables being related are qualitative will be discussed in the last chapter of this book. Simple studies, in which one variable has only one or two qualitative levels, while the other is continuous and therefore has many (theoretically, infinite) possible levels, will be analyzed in Chapters 6 and 7. For now, it is important to begin developing the tools you will need to work with quantitative data. One of the most basic of such tools is the summation procedure, to which we turn next.

### Understanding Why

In the previous section, we touched on the importance of understanding the "why" of how variables such as exercise and sick days are related. In this book, we invite you to understand the "why" of the statistical procedures you will be learning, because (a) concepts are often easier to remember if you know something about what they are made up of; and (b) concepts often become more interesting once you can understand why they are the way they are. For example, soon you will learn that the mean of a set of $z$-scores is always zero and the standard deviation is always 1.0. You can take our word for it and memorize these factoids, but if you understand *why* these properties of $z$-scores must be true you are more likely to remember that it is the mean that is zero, not the standard deviation. Plus you might even allow yourself to think "that's pretty cool" when you learn why these properties are true. The eight summation rules described below will allow you to unpack the "why" of many fundamental statistical concepts you will learn in this book. Before we get to the rules, let us make sure we are clear about summation notation.

## Summation Notation

Mathematical formulas and symbols can appear forbidding if you are not used to working with them. Once you get to know them, however, you see that they actually simplify matters. They are just convenient ways to clearly and concisely convey information that would be much more awkward to express in words, much in the same way that "LOL" and "OMG" simplify expression in email and text messages. In statistics, a particularly important symbol is the one used to represent the *sum* of a set of numbers—that is, the value obtained by adding up all of the numbers.

To illustrate the use of summation notation, let us suppose that eight students take a 10-point quiz. Letting $X$ stand for the variables in question (quiz scores), let us further suppose that the results are as follows:

$$X_1 = 7 \quad X_2 = 9 \quad X_3 = 6 \quad X_4 = 10$$
$$X_5 = 6 \quad X_6 = 5 \quad X_7 = 3 \quad X_8 = X_N = 4$$

Notice that $X_1$ represents the score of the first student on $X$; $X_2$ stands for the score of the second student; and so on. Also, the *number of scores* is denoted by $N$; in this example, $N = 8$. The last score may be represented by either $X_8$ or $X_N$. The *sum of all the X scores* is represented by

$$\sum_{i=1}^{N} X_i$$

where $\sum$, the Greek capital letter sigma, stands for "the sum of" and is called the *summation sign*. The subscript below the summation sign indicates that the sum begins with the first score ($X_i$ where $i = 1$), and the superscript above the summation sign indicates that the sum continues up to and including the last score ($X_i$ where $i = N$ or, in this example, 8). Thus,

$$\begin{aligned}\sum_{i=1}^{N} X_i &= X_1 + X_2 + X_3 + X_4 + X_5 + X_6 + X_7 + X_8 \\ &= 7 + 9 + 6 + 10 + 6 + 5 + 3 + 4 \\ &= 50\end{aligned}$$

Most of the time, however, the sum of *all* the scores is needed in the statistical analysis. In such situations, it is customary to omit the indices $i$ and $N$ from the notation, as follows:

$$\sum X = \text{sum of all the } X \text{ scores}$$

The absence of written indication as to where to begin and end the summation is taken to mean that all the $X$ scores are to be summed.

### Summation Rules

Certain rules involving summation notation will prove useful in subsequent chapters. But first, we would like to flash back to junior high when you learned the order in which mathematical operations should be computed. Do you recall learning the order of operations, PEMDAS, together with a phrase to help you remember it like "Please Excuse My Dear Aunt Sally?" Or perhaps "Pandas Eat: Mustard on Dumplings, and Apples with Spice?" "PEMDAS" stands for "Parentheses, Exponents, Multiplication and Division, and Addition and Subtraction." It describes the order in which you should perform mathematical operations: Parentheses are taken care of before exponents, followed by multiplication and division (which are at the same rank), followed by addition and subtraction (also at the same rank). As you will see, these summation rules often involve operations beyond addition, so Please Excuse My Dear Aunt Sally. Let us suppose that the eight students previously mentioned take a second quiz, denoted by $Y$. The results of both quizzes can be summarized conveniently as follows:

| **Students (*S*)** | **Quiz1 (*X*)** | **Quiz 2 (*Y*)** |
|---|---|---|
| 1 | 7 | 8 |
| 2 | 9 | 6 |
| 3 | 6 | 4 |
| 4 | 10 | 10 |
| 5 | 6 | 5 |
| 6 | 5 | 10 |
| 7 | 3 | 9 |
| 8 | 4 | 8 |

We have already seen that $\sum X = 50$. The sum of the scores on the second quiz is equal to

$$\begin{aligned}\sum Y &= Y_1 + Y_2 + Y_3 + Y_4 + Y_5 + Y_6 + Y_7 + Y_8 \\ &= 8 + 6 + 4 + 10 + 5 + 10 + 9 + 8 \\ &= 60\end{aligned}$$

The following rules are illustrated using the small set of data just shown, and you should verify each one carefully.

**Rule 1.** $\sum(X + Y) = \sum X + \sum Y$

| **Illustration** | ***S*** | ***X*** | ***Y*** | ***X + Y*** |
|---|---|---|---|---|
| | 1 | 7 | 8 | 15 |
| | 2 | 9 | 6 | 15 |
| | 3 | 6 | 4 | 10 |
| | 4 | 10 | 10 | 20 |
| | 5 | 6 | 5 | 11 |
| | 6 | 5 | 10 | 15 |
| | 7 | 3 | 9 | 12 |
| | 8 | 4 | 8 | 12 |
| | | $\sum X = 50$ | $\sum Y = 60$ | $\sum(X + Y) = 110$ |
| | | $\sum X + \sum Y = 110$ | | |

If you remember your Aunt Sally, this rule should be intuitively obvious; the same total should be reached regardless of the order in which the scores are added.

**Rule 2.** $\sum(X - Y) = \sum X - \sum Y$

| Illustration | $S$ | $X$ | $Y$ | $X - Y$ |
|---|---|---|---|---|
| | 1 | 7 | 8 | −1 |
| | 2 | 9 | 6 | 3 |
| | 3 | 6 | 4 | 2 |
| | 4 | 10 | 10 | 0 |
| | 5 | 6 | 5 | 1 |
| | 6 | 5 | 10 | −5 |
| | 7 | 3 | 9 | −6 |
| | 8 | 4 | 8 | −4 |
| | | $\sum X = 50$ | $\sum Y = 60$ | $\sum(X - Y) = -10$ |
| | | $\sum X - \sum Y = -10$ | | |

As with the first rule, it makes no difference whether you subtract first and then sum $[\sum(X - Y)]$ or obtain the sums of $X$ and $Y$ first and then subtract $(\sum X - \sum Y)$.

Unfortunately, matters are not so simple when multiplication and squaring are involved.

**Rule 3.** $\sum XY \neq \sum X \sum Y$

That is, first multiplying each $X$ score by the corresponding $Y$ score and then summing $(\sum XY)$ is *not* equal to summing the $X$ scores $(\sum X)$ and summing the $Y$ scores $(\sum Y)$ first and then multiplying once $(\sum X \sum Y)$.

| Illustration | $S$ | $X$ | $Y$ | $XY$ |
|---|---|---|---|---|
| | 1 | 7 | 8 | 56 |
| | 2 | 9 | 6 | 54 |
| | 3 | 6 | 4 | 24 |
| | 4 | 10 | 10 | 100 |
| | 5 | 6 | 5 | 30 |
| | 6 | 5 | 10 | 50 |
| | 7 | 3 | 9 | 27 |
| | 8 | 4 | 8 | 32 |
| | | $\sum X = 50$ | $\sum Y = 60$ | $\sum(X - Y) = 373$ |
| | | $\sum X \sum Y = (50)(60) = 3{,}000$ | | |

**Rule 4.** $\sum X^2 \neq (\sum X)^2$

That is, first squaring all of the $X$ values and then summing $(\sum X^2)$ is *not* equal to summing first and then squaring that single quantity, $[(\sum X)^2]$

| Illustration | $S$ | $X$ | $X^2$ |
|---|---|---|---|
| | 1 | 7 | 49 |
| | 2 | 9 | 81 |
| | 3 | 6 | 36 |
| | 4 | 10 | 100 |
| | 5 | 6 | 36 |
| | 6 | 5 | 25 |
| | 7 | 3 | 9 |
| | 8 | 4 | 16 |
| | | $\sum X = 50$ | $\sum X^2 = 352$ |
| | $(\Sigma X)^2 = (50)^2 = 2{,}500$ | | |

**Rule 5.** **If $k$ is a *constant* (a fixed numerical value), then**

$$\sum k = Nk$$

**Suppose that $k$ = 3. Then,**

| **Illustration** | $S$ | $k$ |
|---|---|---|
| | 1 | 3 |
| | 2 | 3 |
| | 3 | 3 |
| | 4 | 3 |
| | 5 | 3 |
| | 6 | 3 |
| | 7 | 3 |
| | 8 | 3 |
| | | $\sum k = 24$ |
| | $Nk = (8)(3) = 24$ | |

**Rule 6.** **If $k$ is a constant,**

$$\sum(X + k) = \sum X + Nk$$

**Suppose that $k$ = 5. Then,**

| **Illustration** | $S$ | $X$ | $K$ | $X + k$ |
|---|---|---|---|---|
| | 1 | 7 | 5 | 12 |
| | 2 | 9 | 5 | 14 |
| | 3 | 6 | 5 | 11 |
| | 4 | 10 | 5 | 15 |
| | 5 | 6 | 5 | 11 |
| | 6 | 5 | 5 | 10 |
| | 7 | 3 | 5 | 8 |
| | 8 | 4 | 5 | 9 |
| | | $\sum X = 50$ | $\sum k = Nk = 40$ | $\sum(X + k) = 90$ |
| | | $\sum X + Nk = 50 + (8)(5) = 90$ | | |

This rule follows directly from Rules 1 and 5.

**Rule 7.** **If $k$ is a constant,**

$$\sum(X - k) = \sum X - Nk$$

The illustration of this rule is similar to that of Rule 6 and is left to the reader as an exercise.

**Rule 8.** **If $k$ is a constant,**

$$\sum kX = k\sum X$$

**Suppose that $k$ = 2. Then,**

| **Illustration** | $S$ | $X$ | $k$ | $kX$ |
|---|---|---|---|---|
| | 1 | 7 | 2 | 14 |
| | 2 | 9 | 2 | 18 |
| | 3 | 6 | 2 | 12 |
| | 4 | 10 | 2 | 20 |
| | 5 | 6 | 2 | 12 |
| | 6 | 5 | 2 | 10 |
| | 7 | 3 | 2 | 6 |
| | 8 | 4 | 2 | 8 |
| | | $\sum X = 50$ | | $\sum kX = 100$ |
| | | $k\sum X = (2)(50) = 100$ | | |

These eight summation rules will come up in this book from time to time and will help us understand why certain principles of key statistical concepts work the way they do. As noted earlier, summation notation is akin to handy abbreviations, such as IMHO, FYI, and LMAO. Summation rules operate like bits of grammar. If you want to modify a verb so that it expresses past tense in English, you add "ed." If you want to modify scores by multiplying each one by a constant and know their sum, just add up the numbers and multiply the sum by the constant. We encourage you to acquire these rules as a way of understanding what statistical expressions mean, not just when to use them.

## Jackson's Study

The process of summation can be considered the "workhorse" of statistics; you will see it used in one form or another in all of the statistical procedures described later in this book. Specifically, the summation sign is an important part of most statistical formulas, and we think it is vital that you understand how statistical formulas work. Although we expect, and certainly hope, that all statistics students will eventually become proficient at performing statistical analyses by computer, we believe that there is important educational value in asking students to apply basic statistical formulas directly to small sets of numbers to see how the formulas work and to thus gain a greater conceptual understanding of the statistical results being generated by computer programs. At the same time, we want to help you become familiar with the use of statistical software, and the statistical package we chose to describe in this textbook is SPSS, because it is very popular among behavioral researchers, and is relatively easy to learn for beginners. For analyzing by computer, we created a fairly large data set (100 participants and more than a dozen variables), which is included in a table in the Appendix of this book and is also available in several electronic formats online. In addition to exercises at the end of each chapter, which ask the reader to apply the formulas of that chapter to small sets of data, we have also included a set of computer exercises in each chapter, and each of those exercises refers to the same large data set in the Appendix, which is described next.

The data in the Appendix comes from a hypothetical study performed by Jackson, an advanced doctoral student, who is the teaching assistant (TA) for several sections of a statistics course. The 100 participants are the students who were enrolled in Jackson's sections and who attended one of two review classes that they conducted each week as the TA for the course. (Of course, all of their students voluntarily signed properly informed consent forms, and their study was approved by the appropriate review board at their hypothetical school.) Their data were collected on two different days. On the first day of classes, the students who came to either Jackson's morning or afternoon review session filled in a brief background questionnaire on which they provided contact information, some qualitative data (undergrad major, why they had enrolled in statistics, whether they regularly drink coffee, and whether they exercise regularly) and some quantitative data (number of previous teachers they identified as outstanding, a measurement of their general academic interest on a scale of 9–50 with a higher score correlating to higher interest in their academic pursuits, and a rating of their tendency to procrastinate on a scale from 0 to 10).

The rest of Jackson's data were collected as part of an experiment that they conducted on another day in the middle of the semester. The combined results of the two class sessions on that day add up to a total of 100 students who participated in the experiment. (Due to late registration and other factors, not all of Jackson's students answered the questionnaire that measured their overall academic interest.) One week prior to the experiment, Jackson gave a 10 multiple-choice question quiz on recently covered material during a review session, and these scores were collected and included in the data set. A week later, at the beginning of the experimental session, Jackson explained how each student could take their own pulse, after which a one-minute interval was provided during which they counted the number of beats and wrote down that number as their (baseline) heart rate in beats per minute (bpm). Then, each student reported how many cups of coffee they had drunk since waking up that morning and filled out an anxiety questionnaire consisting of 10 items, each rated (0–4) on a 5-point Likert scale. The questionnaire items inquired about anxiety and how the student was feeling at the present time (e.g., "Would you say that you are now feeling tense and restless? Circle one: Not at all; Somewhat; Moderately; Quite a bit; Extremely"). Total scores could range from 0 to 40, and provided a measure of baseline anxiety.

Next, Jackson handed out a one-page quiz consisting of 11 multiple-choice statistics questions on material covered during the preceding two weeks; for the sake of the experiment, the students were led to believe that this quiz was a required part of the course and the scores would count toward their final grades. The students were asked to keep the quiz face down while taking and recording their (prequiz) pulse and filling out the anxiety questionnaire for a second time (i.e., prequiz anxiety). Then Jackson told the students they had 15 minutes to take the fairly difficult quiz. They also told them that the first 10 questions were worth 1 point each but that the 11th question was worth 3 points of extra credit. Jackson's experimental manipulation consisted of varying the difficulty of the 11th question. Twenty-five quizzes were randomly distributed at each level of difficulty of the final question: easy, moderate, difficult, and impossible to solve. After the quizzes were collected at the end of the 15 minutes, Jackson asked the students to provide heart rate and anxiety data (postquiz) one more time. Finally, Jackson explained the experiment, adding that the 11th quiz question would not be scored and that, although the students would get back their quizzes with their score for the first 10 items, that score would not influence their grade for the statistics course and, instead, participation in this experiment rewarded 1 point of extra credit toward their final. All of the data from Jackson's experiment is printed out in the Appendix at the end of this book.

The computer exercises in this book can be solved by any major statistical package, and because the data for Jackson's study is also available on the Web as both a tab-delimited text file and an Excel file (go to www.wiley.com/go/ISBS8e), most packages will be able to read in the data directly from the electronic file. However, because we will be teaching the basics of SPSS in this textbook in the Bridge to SPSS sections at the end of each chapter, we have also included Jackson's data as an SPSS file on that web page. The Bridge to SPSS sections at the end of each chapter can help you solve the computer exercises in each chapter, unless, of course, your statistics instructor has assigned different statistical software.

## Summary

*Descriptive statistics* are used to summarize and make large quantities of data understandable. *Inferential statistics* are used to draw inferences about numerical quantities (called *parameters*) concerning *populations* based on numerical quantities (called *statistics*) obtained from *samples*. Some behavioral variables cannot be measured quantitatively (e.g., choice of religion) but can only be measured qualitatively using categories (e.g., Catholic, Jewish, Buddhist), which comprise a *nominal* scale. If the categories can be placed in order (e.g., the belts awarded for different levels of skill in the martial arts—black belt, brown belt, etc.), an *ordinal* scale has been created. If the scale involves precise measurement resulting in units of equal size, the data are considered to be *quantitative*, whether the scale has a true zero point (*ratio* scale) or not (*interval* scale). (Ordinal scales are often treated as quantitative in behavioral research; nominal scales always produce *qualitative* data.) Experiments involve measuring dependent variables that are expected to vary somewhat as a function of the different levels of one or more independent variables created by the researcher. *Observational* research involves comparing dependent variables to each other because no variables are being manipulated, nor are participants being assigned to experimental conditions.

The summation sign, $\sum$, is used to indicate "the sum of" and occurs frequently in statistical work. Remember that $\sum X$ is a shorthand version of

$$\sum_{i=1}^{N} X_i$$

(where $N$ = number of subjects or cases).

Summation Rules:

1. $\sum(X + Y) = \sum X + \sum Y$
2. $\sum(X - Y) = \sum X - \sum Y$
3. $\sum XY$ (multiply first, then add) $\neq \sum X \sum Y$ (add first, then multiply)
4. $\sum X^2$ (square first, then add) $\neq (X)^2$ (add first, then square)
5. If $k$ is a constant, $\sum k = Nk$
6. $\sum(X + k) = \sum X + Nk$
7. $\sum(X - k) = \sum X - Nk$
8. $\sum kX = k\sum X$

## Exercises

A first-year student from Oklahoma named Mia is trying to decide which of four dormitories to list as her first choice. She wants to live in the dorm with the friendliest students. So, she makes up a friendliness questionnaire, on which a score of 20 indicates the friendliest a student can be and zero, the least friendly. She manages to get 15 students each from two of the dorms to fill out her questionnaire, 10 from a third dorm, and 5 from a fourth. The individual friendliness scores are shown in Table 1.1. Exercise 1 in this chapter refers to the set of data just described. We will return to this data set several times in later chapters.

1. Compute the following (these values will not by themselves help Mia to make a decision about the dormitories, but we will use these values as steps in future exercises to answer Mia's concern):

   (a) For Turck Hall:

   $$\sum X = \_\_\_ \quad \sum X^2 = \_\_\_ \quad \left(\sum X\right)^2 = \_\_\_$$

   (b) For Kirk Hall:

   $$\sum X = \_\_\_ \quad \sum X^2 = \_\_\_ \quad \left(\sum X\right)^2 = \_\_\_$$

   (c) For Dupre Hall:

   $$\sum X = \_\_\_ \quad \sum X^2 = \_\_\_ \quad \left(\sum X\right)^2 = \_\_\_$$

**Table 1.1 Hypothetical Scores on a 20-Point Friendliness Measure for Students from Four Different Dormitories at One Midwestern College**

| | | | | | |
|---|---|---|---|---|---|
| **Turck Hall ($N$ = 15)** | 17 | 18 | 6 | 13 | 9 |
| | 17 | 11 | 16 | 5 | 15 |
| | 11 | 13 | 10 | 20 | 14 |
| **Kirk Hall ($N$ = 15)** | 17 | 8 | 12 | 12 | 3 |
| | 12 | 7 | 14 | 1 | 11 |
| | 9 | 8 | 6 | 11 | 7 |
| **Dupre Hall ($N$ = 10)** | 9 | 11 | 6 | 5 | 4 |
| | 9 | 10 | 4 | 5 | 7 |
| **Doty Hall ($N$ = 5)** | 14 | 8 | 17 | 6 | 10 |

(d) For Doty Hall:

$$\sum X = \_\_\_ \sum X^2 = \_\_\_ (\sum X)^2 = \_\_\_$$

2. Express the following words in symbols.
   (a) Add up all the scores on test $X$, then add up all the scores on test $Y$, and then add the two sums together.
   (b) Add up all the scores on test $G$. To this, add the following: the sum obtained by squaring all the scores on test $P$ and then adding them up.
   (c) Square all the scores on test $X$. Add them up. From this, subtract 6 times the sum you get when you multiply each score on $X$ by the corresponding score on $Y$ and add them up. To this, add 4 times the quantity obtained by adding up all the scores on test $X$ and squaring the result. To this, add twice the sum obtained by squaring each $Y$ score and then adding them up. (Compare the amount of space needed to express this equation in words with the amount of space needed to express it in symbols. Do you see why summation notation is necessary?)
3. Five students are enrolled in an advanced course in psychology. Two quizzes are given early in the semester, each worth a total of 10 points. The results are as follows:

| Student | Quiz 1 ($X$) | Quiz 2 ($Y$) |
|---|---|---|
| 1 | 0 | 2 |
| 2 | 2 | 6 |
| 3 | 1 | 7 |
| 4 | 3 | 6 |
| 5 | 4 | 9 |

(a) Compute each of the following:

$$\sum X = \_\_\_ \qquad \left(\sum X\right)^2 = \_\_\_ \qquad \sum(X - Y) = \_\_\_$$

$$\sum Y = \_\_\_ \qquad \left(\sum Y\right)^2 = \_\_\_ \qquad \sum X - \sum Y = \_\_\_$$

$$\sum X^2 = \_\_\_ \qquad \sum(X + Y) = \_\_\_ \qquad \sum XY = \_\_\_$$

$$\sum Y^2 = \_\_\_ \qquad \sum X + \sum Y = \_\_\_ \qquad \sum X \sum Y = \_\_\_$$

(b) Using the results of part (a), show that each of the following rules listed in this chapter is true:

Rule 1: ___ $=$ ___
Rule 2: ___ $=$ ___
Rule 3: ___ $\neq$ ___
Rule 4: ___ $\neq$ ___ ($X$ data)
___ $\neq$ ___ ($Y$ data)

(c) After some consideration, the instructor decides that Quiz 1 is excessively difficult and also decides to add 4 points to each student's score. This can be represented in symbols by using $k$ to stand for the constant amount in question, 4 points.

Using Rule 6, compute $\sum(X + k) =$ ___ $+$ ___ $=$ ___.
Compute $\sum X + k =$ ___ $+$ ___ $=$ ___. (Note that this result is different from the preceding one.)
Now add 4 points to each student's score on Quiz 1 and obtain the sum of these new scores.

(d) Had the instructor been particularly uncharitable, he might have decided that Quiz 2 was too easy and subtracted 3 points from each student's score on that quiz. Although this is a new problem, the letter $k$ can again be used to represent the constant; here, $k = 3$.

Using Rule 7, compute $\sum(Y - k)$ = _____ − _____ = _____.
Compute $\sum Y - k$ = _____ − _____ = _____. (Note that this result is different from the preceding one.)
Now subtract 3 points from each student's score on Quiz 2 and obtain the sum of these new scores.

(e) Suppose that the instructor decides to double all of the original scores on Quiz 1.

Using Rule 8, compute $\sum kX$ = _____ × _____ = _____.
Now double each student's score on Quiz 1 and obtain the sum of these new scores.

4. For each of the following (separate) sets of data, compute the values needed in order to fill in the answer spaces. Then answer the additional questions that follow.

*Data set 1:*

| $S$ | $X$ | $Y$ | | |
|---|---|---|---|---|
| | | | $N$ = _____ | |
| 1 | 1 | 2 | $\sum X$ = _____ | $\sum Y$ = _____ |
| 2 | 3 | 5 | $\sum X^2$ = _____ | $\sum Y^2$ = _____ |
| 3 | 1 | 0 | $(\sum X)^2$ = _____ | $(\sum Y)^2$ = _____ |
| 4 | 0 | 1 | $\sum XY$ = _____ | $\sum X \sum Y$ = _____ |
| 5 | 2 | 3 | $\sum(X + Y)$ = _____ | $\sum(X - Y)$ = _____ |

*Data set 2:*

| $S$ | $X$ | $Y$ | | |
|---|---|---|---|---|
| | | | $N$ = _____ | |
| 1 | 7.14 | 0 | $\sum X$ = _____ | $\sum Y$ = _____ |
| 2 | 8.00 | 2.60 | $\sum X^2$ = _____ | $\sum Y^2$ = _____ |
| 3 | 0 | 4.32 | $(\sum X)^2$ = _____ | $(\sum Y)^2$ = _____ |
| 4 | 4.00 | 2.00 | $\sum XY$ = _____ | $\sum X \sum Y$ = _____ |
| 5 | 4.00 | 6.00 | $\sum(X + Y)$ = _____ | $\sum(X - Y)$ = _____ |
| 6 | 1.00 | 1.15 | | |
| 7 | 2.25 | 1.00 | | |
| 8 | 10.00 | 3.00 | | |

| | *set 1* | *set 2* |
|---|---|---|
| If every $X$ score is multiplied by 3.2, what is the new $\sum X$ in each set? | _____ | _____ |
| If 7 is subtracted from every $Y$ score, what is the new $\sum Y$ in each set? | _____ | _____ |
| If 1.8 is added to every $X$ score, what is the new $\sum X$ in each set? | _____ | _____ |
| If every $Y$ score is divided by 4, what is the new $\sum Y$ in each set? | _____ | _____ |

(Hint: Use the appropriate summation rule in each case so as to make the calculations easier.)

5. Compute the values needed to fill in the blanks.

*Data set 3:*

| S | X | Y |
|---|---|---|
| 1 | 97 | 89 |
| 2 | 68 | 57 |
| 3 | 85 | 87 |
| 4 | 74 | 76 |
| 5 | 92 | 97 |
| 6 | 92 | 79 |
| 7 | 100 | 91 |
| 8 | 63 | 50 |
| 9 | 85 | 85 |
| 10 | 87 | 84 |
| 11 | 81 | 91 |
| 12 | 93 | 91 |
| 13 | 77 | 75 |
| 14 | 82 | 77 |

| | |
|---|---|
| $N =$ ______ | |
| $\sum X =$ ______ | $\sum Y =$ ______ |
| $\sum X^2 =$ ______ | $\sum Y^2 =$ ______ |
| $(\sum X)^2 =$ ______ | $(\sum Y)^2 =$ ______ |
| $\sum XY =$ ______ | $\sum X \sum Y =$ ______ |
| $\sum(X + Y) =$ ______ | $\sum(X - Y) =$ ______ |

## Thought Questions

1. What is the difference between (a) a population and a sample? (b) a parameter and a statistic? (c) descriptive statistics and inferential statistics?
2. What is the difference between a constant and a variable?
3. What is the difference between a ratio scale and an interval scale?
4. What important property do interval scales have that ordinal and nominal scales do *not* have?
5. If we are classifying psychiatric patients as having one or another of five different psychological disorders—Major Depressive Disorder, Bipolar Disorder, Generalized Anxiety Disorder, Obsessive Compulsive Disorder, and Phobic Disorder—what kind of measurement scale are we using?
6. A poll of sportswriters ranks the 25 best college football teams in the country, where #1 is the best team, #2 is the second-best team, and so on. What kind of measurement scale is this?
7. What kind of measurement scales are used for each of the following variables in Jackson's study? (a) Whether or not the student regularly reads for enjoyment in their leisure time. (b) The undergraduate major of each student. (c) The number of previous outstanding teachers each student reported having. (d) Score on the statistics quiz given before the experiment.
8. What is the difference between an independent variable and a dependent variable?

## Computer Exercises

1. Read Jackson's data into your statistical software package. For those not using the Statistical Package for the Social Sciences (SPSS), we have provided the data in two convenient formats: a tab-delimited text file and an Excel spreadsheet (Microsoft Office 2021). For the convenience of SPSS users, we have also included the data as an SPSS.sav file, though your instructor may want you to know how to read text or Excel files into SPSS (go to www.wiley.com/go/ISBS8e).
2. Label the values of the categorical (i.e., qualitative) variables according to the following codes: For regular exercise, 0 = Not regularly and 1 = Regularly; for

undergrad major, 1 = Psychology, 2 = Geology, 3 = Biology, 4 = Sociology, and 5 = Economics. Your instructor may ask you to fill in missing value codes for any data that are missing (e.g., blank cell in the Excel spreadsheet).

3. A good many statistical functions can be performed in Excel. As a first step, use the Sum function to add up the scores for each of the quantitative variables in the Excel file of Jackson's data.
4. Create a new variable that adds 10 points to everyone's academic interest score. How does the sum of this variable compare to the sum of the original variable? What general rule is being illustrated by this comparison?
5. Create a new variable that is 10 times the statistics quiz score. How does the sum of this variable compare to the sum of the original variable? What general rule is being illustrated by this comparison?

## Bridge to SPSS

SPSS is certainly not the only statistical package (*statpack*) available, and your instructor may prefer to teach you how to use SAS, Stata, Minitab, *R*, or other statistical software. However, because there are commonalities among disparate statistical packages, it is likely that, by reading the Bridge sections in each chapter, your understanding of the statistical procedures covered in that chapter will be enhanced, regardless of which statistical software you are using. For those students who will be using SPSS in their course, these sections should help you not only obtain the statistics you will need to complete the computer exercises but to translate the sometimes idiosyncratic terminology used by SPSS into the more conventional language that we are using to describe the same topics in this book. We also show you a few tricks and shortcuts, but note that our Bridge to SPSS sections are not meant to replace a basic guide to the SPSS package (one advantage of using such a popular statpack is that there are many good guides available at a variety of different levels and prices). Moreover, we understand that you may not be using the latest SPSS version, which will probably be 30.0 by the time this book is printed. Therefore, we only describe aspects of SPSS that have changed very little in recent years.

SPSS uses three distinct windows: a Data window; an Output window; and a Syntax window. Each window can be stored as a file, with different extensions to identify the type of window (.sav, .spv, and .sps, respectively). We will begin by describing the Data window in this section and introduce the other two windows in later chapters. The Data window has two tabs. The Data View tab is a spreadsheet in which you can enter your data; it is initially set so that the rows represent cases (often different participants) and columns represent different variables. The other tab is the Variable View, and in this tab, the rows represent your variables and the columns represent different ways to describe your data.

The first column in Variable View is **Name**. Unlike in Excel, you cannot have spaces in your variable names. The next column is **Type**. There are quite a few types to choose from, but the most commonly used are *Numeric* (only numbers) and *String* (allows letters and some symbols, as well as numbers). One of the rightmost columns is **Measure**, and it allows you to classify each of your variables as being measured by one of the following three types of scales: nominal, ordinal, or scale. The first two terms are used the same way by SPSS that we have defined in this chapter. *Scale* is SPSS's term for interval/ratio data. In general, numerical data are set to **scale** by default, whereas string data—which contain letters instead of, or in addition to, numbers—are set to **nominal**. In practice, these scale designations are usually not important for statistical calculations, because SPSS uses them mainly to determine the way some charts are displayed.

For simplicity, Jackson's data set is presented entirely in terms of numbers, even for the categorical variables of regular exercise and undergrad major. To assign meaningful labels to the arbitrary numbers we have used to represent the different levels of the categorical variables, go down the **Values** column of Variable View until you reach the row for a categorical variable. Then click on the right side of that cell to open the *Value Labels* box. For exercise, you would type **0** for Value, and then tab to Value Label, where you can type **not**

**regularly**. Click the **Add** button, and provide a label (e.g., regularly) for value **1**, then **Add** again, and **OK**. The process is similar for undergrad major. Note that if you use the **Missing** column to define a particular value of a variable, say 99, as meaning that the value is missing, rather than just leaving the cell blank (e.g., you could use 99 to mean "missing" for the academic interest scores, because none of the real values can be that high), you can then attach a value label to that value, such as "never answered the questionnaire." Note that clicking on one of the icons at the top of the Data View allows you to switch back and forth between displaying the code numbers in the spreadsheet or the value labels you created.

The column that is called **Label** allows you to create a "label" associated with your variable name, and this label may contain spaces and symbols that are not allowed in your variable names. It is a good idea to keep your variable names meaningful but short and use the label to create a longer description of your variable, if needed.

To create new variables that are based on ones already in your spreadsheet, which some of our computer exercises will ask you to do, click on the **Transform** menu, then **Compute**. In the Compute Variable box that opens up, *Target Variable* is a name that you make up (and type into that box) for the new variable; when you have filled in a *Numeric Expression* and then click **OK**, the new variable will suddenly appear in the rightmost column of your Data View spreadsheet. We will leave it to your instructor, or an SPSS guidebook, to teach you various ways to create numeric expressions that transform, or combine, your existing variables into new ones.

# Chapter 2
# Frequency Distributions and Graphs

## PREVIEW

### The Purpose of Descriptive Statistics

What is the primary purpose of descriptive statistics?

What are the most useful types of descriptive statistics?

### Regular Frequency Distributions

What is a regular frequency distribution, how is it constructed, and why is it useful?

### Cumulative Frequency Distributions

How does a cumulative frequency distribution differ from a regular frequency distribution, and how is it constructed? How does a cumulative percentage frequency distribution differ from a cumulative frequency distribution, and how is it constructed?

What are percentile ranks and percentiles?

What is the primary purpose of percentile ranks and percentiles?

Why must we take careful note of the reference group to interpret a percentile rank correctly?

What is a raw score?

Why is it often necessary to compare a raw score to the specific group of scores in which it appears?

What is a transformed score?

### Grouped Frequency Distributions

How does a grouped frequency distribution differ from a regular frequency distribution, and how is it constructed?

What is gained by using a grouped frequency distribution?

What is lost by using a grouped frequency distribution, and why are such distributions usually not used when computing means and other statistics?

### Real and Apparent Limits

How do real limits differ from apparent limits, and how does that difference influence how data are assigned to groups?

### Graphic Representations

What are bar charts, histograms, frequency polygons, stem-and-leaf displays, and boxplots? What are the advantages of each graphic display? When is a bar chart preferable to a histogram?

### Skewing of Frequency Distributions

What is meant when we say that a distribution is symmetric? Skewed? Unimodal? Bimodal? Normal? Rectangular? A J-curve?

### Summary

### Exercises

*Introductory Statistics for the Behavioral Sciences*, Eighth Edition. R. Brooke Lea and Barry H. Cohen.

Companion website: www.wiley.com/go/ISBS8e

**PREVIEW** *(continued)*

**Thought Questions**

**Computer Exercises**

**Bridge to SPSS**

## The Purpose of Descriptive Statistics

The primary goal of descriptive statistics is to bring order out of chaos. For example, consider the plight of a professor who has given an examination to a class of 85 students and has recorded the score for each student's exam. In order to decide what represents relatively good and less good performance on the examination, the professor must find a way to organize and interpret 85 numbers. Or consider a researcher who runs 60 participants in a memory experiment for 20 faces and records a rating of familiarity for each face for each participant during the recall phase. This scientist must contend with the problem of interpreting a total of 1,200 (i.e., $60 \times 20$) numbers!

In addition to causing problems for the professor or researcher, the large quantity of numbers also creates difficulties for the audience they intend to inform. The students in the first example are likely to request information about the distribution of test scores so they can interpret their own performance; otherwise, they will have trouble trying to grasp the meaning of 85 unorganized numbers. The readers of the scientific paper on face recognition will have an even harder time trying to interpret a table with more than 1,000 numbers in it.

Descriptive statistics help resolve problems such as these by allowing us to *summarize and describe large quantities of data*. Among the various techniques that you will find particularly useful are the following, which are described in this chapter and Chapters 3 and 4:

***Frequency distributions and graphs***—Procedures for describing all (or nearly all) of the data in a convenient way.

***Measures of "central tendency"***—Single numbers that describe the location of a distribution of scores: where the "center of gravity" of the scores generally falls within the infinite range of possible values.

***Measures of variability***—Single numbers that describe how "spread out" a set of scores is: whether the numbers are similar to each other and vary very little, as opposed to whether they tend to be very different from one another and vary a lot.

***Transformed scores***—New scores that replace each original number and show at a glance where any particular score is in comparison to the other scores in the group.

***Areas under the normal curve***—Proportions that tell you the probability of randomly selecting a score smaller or larger than yours from a particular normal distribution.

Each of these procedures serves a different (and important) function. Our discussion of descriptive statistics begins with frequency distributions and graphs; the other topics are treated in subsequent chapters.

## Regular Frequency Distributions

One way of making a set of data more comprehensible is to write down every possible score value in numerical order and record next to each score value the number of times that the score occurs. For example, imagine that you are interested in the psychology of social

**Table 2.1** **Number of Friend Requests Within 24 Hours of Opening a Facebook Account by 80 Sixth Graders**

| | | | | | | | |
|---|---|---|---|---|---|---|---|
| 5 | 8 | 3 | 6 | 5 | 6 | 3 | 7 |
| 6 | 8 | 7 | 4 | 6 | 5 | 8 | 7 |
| 5 | 3 | 6 | 1 | 8 | 6 | 8 | 5 |
| 8 | 9 | 6 | 9 | 6 | 5 | 9 | 9 |
| 8 | 5 | 8 | 4 | 8 | 6 | 7 | 4 |
| 8 | 7 | 6 | 8 | 5 | 7 | 6 | 9 |
| 4 | 8 | 6 | 8 | 6 | 5 | 4 | 7 |
| 7 | 4 | 5 | 5 | 9 | 6 | 6 | 7 |
| 10 | 3 | 5 | 7 | 9 | 10 | 6 | 7 |
| 6 | 9 | 8 | 7 | 8 | 5 | 7 | 4 |

networking sites such as Facebook and LinkedIn. In particular, you have a theory about the rate at which people are friended on opening a Facebook account. You randomly sample 80 sixth graders who are about to open a Facebook account and measure how many friend requests they receive in the first 24 hours. The scores (i.e., number of friend requests) of the 80 students are shown in Table 2.1.

As you can see, the table of 80 numbers is difficult to interpret. A *regular frequency distribution*, however, will present a clearer picture. The first step in constructing such a distribution is to list every *possible score value* (typically denoted by the symbol $X$) in the first column of a table, with the highest score at the top. The *frequency* (denoted by the symbol $f$) of each score, or the number of times a particular score was obtained, is listed to the right of the score in the second column of the table. To arrive at the figures in the "Frequency" column, you could go through the data and count all the 10s (assuming 10 is the highest score), go through the data again and count all the 9s, and so forth until all frequencies were tabulated. A more efficient plan is to go through the data just once and make a tally mark next to the appropriate score in the score column for each score, and add up the tally marks at the end.

The complete regular frequency distribution is shown in Table 2.2 (ignore the "Cumulative Frequency" column for the moment; it will be discussed in the next section). The table reveals at a glance how often each score was obtained. For example, eight of the students were friended 9 times each, and only four were friended 3 times each. This method of organizing the data makes it easier to interpret the friending behavior because you can conveniently ascertain (among other things) that 6 was the most frequently obtained score, scores distant from 6 tended to occur less frequently than scores close to 6, and that the majority of people received at least five friend requests.

**Table 2.2** **Regular and Cumulative Frequency Distributions for Data in Table 2.1**

| Score ($X$) | Frequency ($f$) | Cumulative frequency ($cf$) |
|---|---|---|
| 10 | 2 | 80 |
| 9 | 8 | 78 |
| 8 | 15 | 70 |
| 7 | 13 | 55 |
| 6 | 17 | 42 |
| 5 | 13 | 25 |
| 4 | 7 | 12 |
| 3 | 4 | 5 |
| 2 | 0 | 1 |
| 1 | 1 | 1 |
| 0 | 0 | 0 |

## Cumulative Frequency Distributions

We wouldn't want to encourage this kind of competitiveness, but if someone in this group wanted to know how many students had fewer friend requests than they did, the easy way would be to create a *cumulative frequency distribution* from the regular frequency distribution already produced. Just start with the *lowest* score in the regular frequency distribution and form a new column of *cumulative frequencies* (as we did in Table 2.2) by adding up the frequencies as you go along. For example, the cumulative frequency (denoted by the symbol *cf*) for the score of 4 is equal to 12. This value was obtained by adding the frequency (*f*) of that score, namely 7, to the *cf* of the next lower score (i.e., 3), which is 5. The other *cf*s were obtained in a similar fashion starting from the bottom. Quite simply, the *cf* of 12 for a score of 4 means that 12 people obtained a score of 4 *or less*. This can be readily verified by looking at the frequencies (*f*): seven people scored 4, four people scored 3, and one person scored 1, for a total of 12 people with scores at or below 4.

If you are really competitive, and you want to know how many of your classmates scored *lower* than you, you would look at the *cf* for the score immediately below yours. For example, if your score was 7 friends, and you wanted to know how many of your classmates received fewer friend requests, you would look at the *cf* for a score of 6, which is 42 in Table 2.2. Thus, while 13 students are tied with you (the *f* for a score of 7 is 13), 42 students scored lower than you.

Everyone obtained a score of *10 or less*. So, the *cf* for a score of 10 equals 80, the total number of people counted in this distribution. (The *cf* for the *highest* score should always equal *N*.) Although no unique information is presented by the *cf* distribution, it allows you to arrive at certain needed information more quickly and conveniently than is possible using the regular *f* distribution. However, while it can be interesting to know how many people in a distribution had scores lower than yours, it is usually more informative to know the *percentage* of people (or, cities, for instance) that have lower scores (e.g., crime rates for cities). This can be found easily by adding one more column to the data in Table 2.2: the *cumulative percentage* (*cp*) column, as we did in Table 2.3.

### Cumulative Percentage Distributions and Percentile Ranks

To get the *cp* corresponding to each *cf*, you need only divide by the number of data points in the distribution ($N = 80$, in this example) and then multiply by 100. As a formula, $cp = 100 \times (cf/N)$. For example, you can see in Table 2.3 that the *cp* for a score of 7 friends equals $100 \times (55/80) = 68.75\%$. Just a little more than two-thirds of the students received 7 or fewer friend requests. The *cp* for a score is called its percentile rank (PR). Therefore, the PR for a score of 7 equals 68.75. The usefulness of the PR is that it tells you immediately the percentage of scores that are either tied with your score or less than your score. A PR of 68.75 is pretty good,

**Table 2.3 Cumulative Frequency and Cumulative Percentage Distributions for Data in Table 2.1**

| Score (*X*) | Cumulative frequency (*cf*) | Cumulative percentage (*cp*) |
|---|---|---|
| 10 | 80 | 100 |
| 9 | 78 | 97.5 |
| 8 | 70 | 87.5 |
| 7 | 55 | 68.75 |
| 6 | 42 | 52.5 |
| 5 | 25 | 31.25 |
| 4 | 12 | 15 |
| 3 | 5 | 6.25 |
| 2 | 1 | 1.25 |
| 1 | 1 | 1.25 |
| 0 | 0 | 0 |

but, of course, a PR of 87.5 is better. If you want to know the percentage of scores that are *lower* than yours, you need to look at the *cp* for the score just below yours. If you had 7 friend requests, you beat 52.5% (just over half) of the group.

### Percentiles

A *percentile* is the flipside of a PR. That is, a percentile is a score that has a particular PR. For example, looking at Table 2.3, you can see that a score of 4 has a PR of 15. Therefore, a score of 4 is said to be at the 15th percentile. If you want to know what score corresponds to the 87.5 percentile, you can see that 8 is the 87.5 percentile. Percentiles play an important role in some types of statistics. The most useful percentiles are the deciles (10th percentile, 20th, etc.) and the quartiles (25th, 50th, and 75th). Unfortunately, these particular percentiles are not likely to appear in your *cp* column. For instance, if you want to find the 50th percentile for the present distribution, the closest you can get in terms of a Table 2.3 entry is a score of 6, which has a PR of 52, rather than 50. You could estimate that the exact 50th percentile would be about 5.9, but although some textbooks may show you how to make this estimation, the calculation is best left to statistical software. The 50th percentile plays an important role in some types of statistics as a measure of central tendency because half the scores in the distribution are above the median and half are below. The median will be discussed along with other measures of central tendency in Chapter 3.

### Interpreting a Raw Score

The need for cumulative frequency and percentage distributions arises because most raw scores are difficult to interpret in isolation. For example, consider the raw scores of 28, 950, and 8. The first score represents the number of words you were able to remember from a list of 50 words, the second is your reaction time in milliseconds to decide that the letter string TUBTER is not an English word, and the third is the number of people who friended you in the first 24 hours of your Facebook account. Are these "good" scores? Without knowing anything about other scores, it is difficult to know. "Good compared to what?" you might ask. Twenty-eight out of fifty words does not sound too impressive. But what if you found out that 28 was the highest score and that most participants recalled fewer than 14 words? A reaction time of 950 ms sounds pretty fast—less than a second! If most people responded within 500 ms, however, then your response starts to look leisurely. You can probably see where we are going with this: scores often lack meaning—they are difficult to interpret—out of context. Without additional information, it is hard to know whether 8 friend requests in 24 hours are a lot of requests, relatively few, or about average.

Alternatively, you might know what part of the distribution you are interested in and want to know which score falls there. For example, which raw score falls in the exact middle of the distribution? Which score cuts off the bottom 25%? (Or the top 75%?) One way to obtain this additional information is to transform the original score (called the *raw score*) into a new score, such as one of the percentiles you just learned about, which will show at a glance where the score fits in with others. There are other kinds of transformed scores, such as the standardized scores described in Chapter 4.

Although PRs and percentiles show directly how an individual score compares to the scores of a specific group, in order to interpret a PR correctly, you must take careful note of the reference group in question. A college senior who obtains a test score with a PR of 90 would seem to have done well because his score places him just within the top 10% of some reference group. But if the reference group consists of *high school* seniors, then perhaps the student should not feel so proud of his performance! A score at the 12th percentile is usually considered poor because only 12% of the reference group did as badly or worse. But if the score was obtained by a high school ninth grader and the reference group consists of college graduates, the score may actually represent good performance relative to other ninth graders.

Indeed, there are many practical situations where misleading conclusions can easily be drawn. If you score at the 60th percentile on a midterm examination in statistics and a friend in a different class scores at the 90th percentile on his statistics midterm, she is not necessarily better at stats. The students in her class might have a relatively weak background in basic mathematics, which could make it easier for her to obtain a high standing in comparison to that reference group. Remembering that a percentile *compares* a score to a *specific group of scores* will help you avoid pitfalls such as these.

## Grouped Frequency Distributions

When the number of *different scores* to be listed in the score ($X$) column is not too large, regular frequency distributions are an excellent way to summarize a set of data. If, however, there are more than 15 or 20 values of $X$ to be written, interpreting a regular frequency distribution is likely to prove tedious. One way of avoiding a summary that does not summarize enough is to use a *grouped frequency distribution*. Instead of listing single scores in the score column (e.g., 0, 1, 2, 3, 4, 5, 6, 7, ...), several score values are grouped together into a *class interval* (e.g., 0–4, 5–9, 10–14, 15–19, ...), and frequencies are tallied for each interval.

The data in Table 2.4 represent the scores of 85 students on a questionnaire that ranges from 0–50. Inspection of the data, however, shows that the actual scores range from a low of 9 to a high of 49. Even so, if a regular frequency distribution were to be used, some 41 separate scores and corresponding frequencies would have to be listed. To avoid such a cumbersome table, a grouped frequency distribution has been formed in Table 2.5. An *interval size* of 3 has been chosen, meaning that there are three score values in each class interval; successive intervals of size 3 are formed until the entire range of scores has been covered. Next, frequencies are tabulated, with all scores falling in the same interval being treated equally. For example, a score of 39, 40, or 41 would be entered by registering a tally mark

**Table 2.4** **Scores of 85 Students on a 50-Point Academic Interest Questionnaire**

| | | | | | | | | | |
|---|---|---|---|---|---|---|---|---|---|
| 39 | 42 | 30 | 11 | 35 | 25 | 18 | 26 | 37 | 15 |
| 29 | 22 | 33 | 32 | 21 | 43 | 11 | 11 | 32 | 29 |
| 44 | 26 | 30 | 49 | 13 | 38 | 26 | 30 | 45 | 21 |
| 31 | 28 | 14 | 35 | 10 | 41 | 15 | 39 | 33 | 34 |
| 46 | 21 | 38 | 26 | 26 | 37 | 37 | 14 | 26 | 24 |
| 32 | 15 | 22 | 28 | 33 | 47 | 9 | 22 | 31 | 20 |
| 37 | 40 | 20 | 39 | 30 | 18 | 29 | 35 | 41 | 21 |
| 26 | 25 | 29 | 33 | 23 | 30 | 43 | 28 | 32 | 32 |
| 34 | 28 | 38 | 32 | 31 | | | | | |

**Table 2.5** **Grouped and Cumulative Frequency Distributions for Data in Table 2.4**

| Class interval | Frequency ($f$) | Cumulative frequency ($cf$) | Cumulative percentage ($cp$) |
|---|---|---|---|
| 48–50 | 1 | 85 | 100 |
| 45–47 | 3 | 84 | 98.8 |
| 42–44 | 4 | 81 | 95.3 |
| 39–41 | 6 | 77 | 90.6 |
| 36–38 | 7 | 71 | 83.5 |
| 33–35 | 9 | 64 | 75.3 |
| 30–32 | 14 | 55 | 64.7 |
| 27–29 | 8 | 41 | 48.2 |
| 24–26 | 10 | 33 | 38.8 |
| 21–23 | 8 | 23 | 27.1 |
| 18–20 | 4 | 15 | 17.6 |
| 15–17 | 3 | 11 | 12.9 |
| 12–14 | 3 | 8 | 9.4 |
| 9–11 | 5 | 5 | 5.9 |

next to the class interval 39–41. When the tabulation is completed, the frequency opposite a given class interval indicates the number of cases with scores in that interval.

Note that grouped frequency distributions lose information, because they do not provide the exact value of each score. They are very convenient for purposes of summarizing a set of data but generally should not be used when computing means and other statistics.

In the illustrative problem, the interval size of 3 was specified. In your own work, it will be up to you to construct the proper intervals. The conventional procedure is to select the intervals according to the following three guidelines:

1. Have a total of approximately 8–15 class intervals.
2. Use an interval size of 2, 3, 5, or a multiple of 5, selecting the smallest size that will satisfy the first rule. (All intervals should be the same size.)
3. Make the lowest score in each interval a multiple of the interval size.

Notice how Table 2.5 follows these guidelines: 14 intervals; an interval size of 3; and the lowest scores of the intervals (9, 12, 15, etc.) are all multiples of the interval size, 3. However, we could also have satisfied these guidelines by using the following set of intervals: 5–9; 10–14; 15–19; 20–24; 25–29; 30–34; 35–39; 40–44; 45–49. The best way to decide between these two possible sets of intervals is usually to create the table in both formats and then see which seems to better convey the information in the data. (For further information on the construction of grouped frequency distributions, see B. H. Cohen, 2013.)

Grouped distributions make it even more difficult to estimate percentiles and PRs, especially for scores within an interval. So, this task is best left to statistical software. However, in anticipation of their use in Chapter 3, we will estimate the quartiles for these data at the end of the next section.

## Real and Apparent Limits

So far, the scores we have been dealing with, such as number of friends or points on an exam, come only as whole numbers (teachers can give half points, but that has not been the case yet for our examples). However, when measuring physical quantities, like height or weight, we can round off to whole numbers, or we can keep more accurate measures that include decimal points. For example, imagine that the scores in Table 2.4 represent the amounts of time taken by 85 participants, all of whom solved the same puzzle. In that case, it could be that the data have been rounded off to the nearest minute, though the original scores included one digit past the decimal point. In fact, it is important to understand that Table 2.5 could have been constructed in the same way even if the scores in Table 2.4 were all presented to the nearest 10th of a minute (e.g., 37.2, 17.8).

At first glance, it may appear that some fractional scores would fall between the cracks in Table 2.5. For instance, in which interval would you place a score of 38.7? It seems that this score is too large to be in the 36–38 interval but too small to be in the 39–41 interval. But that is because the intervals in Table 2.5 are described in terms of their *apparent limits*. When dealing with continuous data, the apparent limits provide a convenient way to express the intervals, but they are not the *real limits* of the intervals. For continuous data, the real limits of the 36–38 interval are 35.5–38.5, and the real limits of the next higher interval (39–41) are 38.5–41.5. Note that there are no gaps between the real limits of the intervals. So, there is no ambiguity about where to place a score of 38.7. This score falls clearly in the interval with the apparent limits of 39–41 because it is between 38.5 and 41.5. Any score between 38.5 and 39.0 belongs in the 39–41 interval, and any score between 38.0 and 38.5 belongs in the 36–38 interval. Of course, there is potential ambiguity for a score of exactly 38.5. For such borderline cases, it would be acceptable to decide between the two possible intervals at random, separately for each case, or to adopt a consistent rule, such as always rounding up (e.g., 38.5 would always be placed in the 39–41 interval).

We will now estimate the quartiles for these data, taking into account the real limits of the intervals. The 75th percentile is easy because a *cp* of 75.3 is in Table 2.5 and it corresponds to the 33–35 interval. Note that the *cp* of 75.3 corresponds to the upper real limit of that interval, which is equal to 35.5. We can estimate that the 75th percentile is just a bit less than that. So, we will estimate that the third quartile is 35.4. The first quartile falls in the 21–23 interval, which ranges from a *cp* of 17.6–27.1. A *cp* of 25 is somewhere between these two values, but obviously much closer to 27.1, which corresponds to a score of 23.5 (the upper real limit of the interval). Because 25 is about three-fourths of the way from 17.6 to 27.1, we will estimate that the 25th percentile is about three-fourths of the way from 20.5 to 23.5, which is about 22.8. Finally, the second quartile, the 50th percentile (also the median) is in the 30–32 interval, but much closer to the lower real limit of 29.5. In fact, a reasonable estimate for the 50th percentile is 30.0.

## Graphic Representations

It is often effective to express frequency distributions pictorially as well as in tables. Five procedures for accomplishing this are discussed next.

### Bar Charts

Suppose that Jackson tabulated the number of previous outstanding teachers that each of the 25 Geology students in their class had prior to enrolling in statistics, and the results were as follows:

| Number of outstanding teachers ($X$) | $f$ |
|---|---|
| 7 or more | 0 |
| 6 | 1 |
| 5 | 0 |
| 4 | 3 |
| 3 | 4 |
| 2 | 8 |
| 1 | 5 |
| 0 | 4 |

A **bar chart** (or *bar graph*) of these data is shown in Figure 2.1. To construct a bar chart, the $Y$-axis (vertical axis) is marked off in terms of *frequencies*, and the $X$-axis (horizontal axis) is marked off in terms of *score values*. The frequency of any value is expressed by

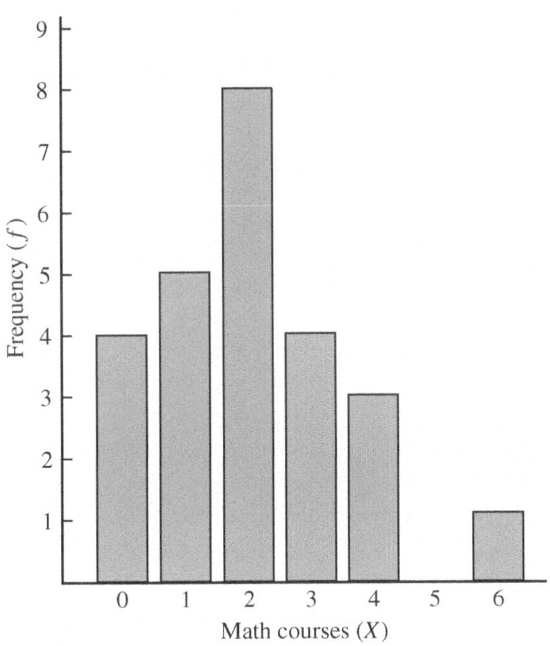


Figure 2.1

**Bar Chart Expressing Number of Prior Outstanding Teachers by 25 Geology Students**

the height of the bar above that value. For example, to show that a score of 3 (teachers) occurred 4 times, a bar 4 units in height is drawn above this score.

Bar charts can be used for any kind of data, but they are particularly appropriate for *discrete* data, where results between the score values shown *cannot* occur. In the present example, it is impossible for a student to have had 2.4 outstanding teachers, and this fact is well expressed in the bar chart by the separate and distinct bars above each score value. Categorical data (e.g., each student's undergraduate major) are always discrete and are therefore good candidates for being displayed as bar charts.

### Histograms

A **histogram** is similar to a bar chart except that adjacent bars are allowed to touch because the data being graphed are considered *continuous*. When dealing with continuous data, results between whole score values can occur, or could if it were possible to measure with sufficient refinement. Interval and ratio data are often treated as continuous, regardless of whether fractions can occur. Ordinal data (somewhat agree, strongly agree, etc.) are sometimes graphed as a histogram but are usually more appropriately displayed in a bar chart.

Treating the scores in Table 2.4 as numbers of minutes to solve a problem, and therefore as measures that are continuous, the frequencies in Table 2.4 are presented as a histogram in Figure 2.2. As in the bar chart, the *Y*-axis is marked off in terms of frequencies, and the frequency corresponding to a score (or class interval, as found in Table 2.5) is expressed by the height of the bar above the *X*-axis. However, because the scale being marked off on the *X*-axis is treated as continuous, it is marked off in terms of the *real limits* of each class interval, rather than its *apparent limits*. For instance, whereas the apparent limits of the two lowest class intervals in Table 2.5, 9–11 and 12–14, do not appear to touch, if the scale is considered continuous, the real limits of those intervals are 8.5–11.5 and 11.5–14.5; these intervals *do* touch, and therefore do not allow any fractional scores to fall between the intervals.

### Frequency Polygons

In some cases, line graphs are easier to interpret than a series of bars and they can be used to represent various kinds of frequency distributions. A line graph that can be used as an alternative to the histogram is the **frequency polygon** (like the histogram, it is appropriate only for continuous data). The frequency data in Table 2.4, which were graphed in Figure 2.2, are presented again as a *regular frequency polygon* in Figure 2.3. The strategy is the same as in the case of the histogram: the frequency of a score (or score interval) is expressed by the height of the graph vertically above the *X*-axis (as measured along the *Y*-axis), and the scores

Figure 2.2

**Histogram Representing the Grouped Data in Table 2.4**

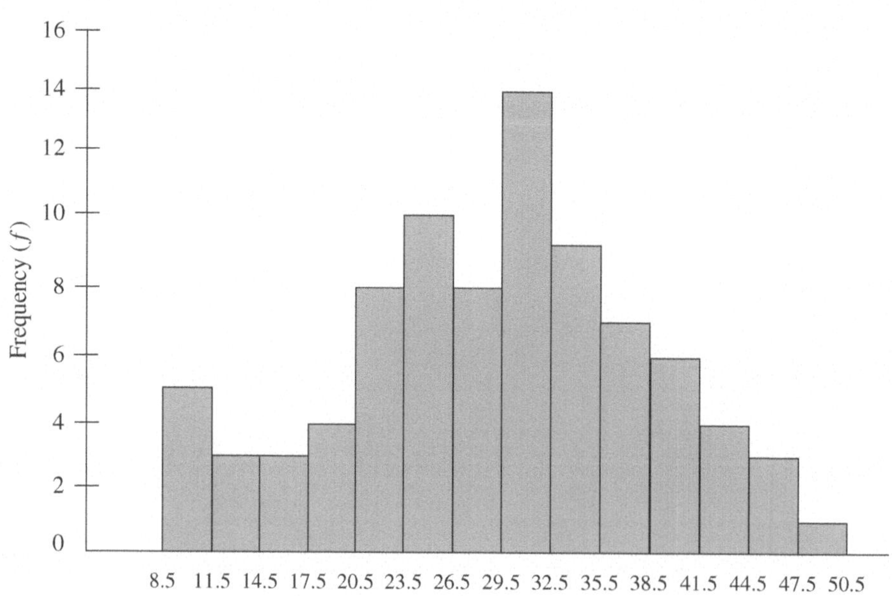

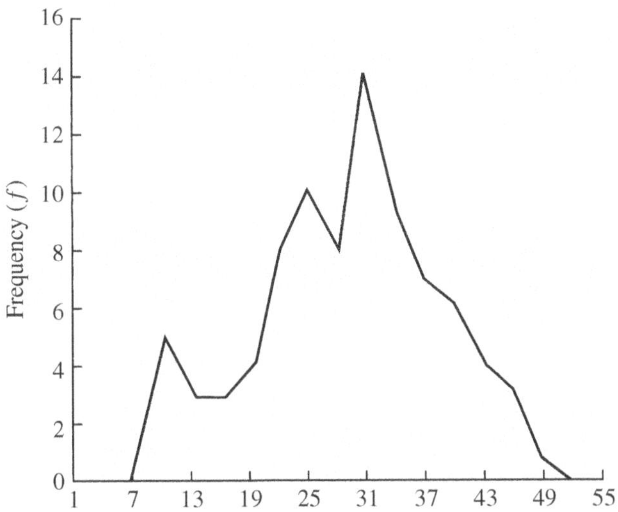


# Figure 2.3

**Regular Frequency Polygon for Data in Table 2.4**

are measured horizontally along the *X*-axis. (Note that to avoid clutter, the *X*-axis displays the values for every other interval midpoint.)

The difference is that points, rather than bars, are used for each entry. Thus, the frequency of 5 for the class interval of 9–11 is shown by a dot 5 units up on the *Y*-axis above the score of 10 (the midpoint of the 9–11 interval); with a regular frequency distribution, as in Table 2.2, each score would be entered on the *X*-axis. All dots are connected with straight lines, and the leftmost and rightmost dots are connected to the *X*-axis at the midpoints of the next intervals to the left and right, respectively (see Figure 2.3). Regular frequency polygons are particularly appropriate when many different scores (or intervals) are being represented on the *X*-axis because a histogram with many very narrow bars is particularly tedious to look at. Also, it is relatively easy to draw two frequency polygons on the same graph if, for example, you wanted to compare the quiz distribution for this semester with the corresponding distribution from the previous semester. Imagine how hard it would be to look at and compare two histograms drawn on the same set of axes.

## Stem-and-Leaf Displays

A simple and useful technique for summarizing a set of data, the **stem-and-leaf display** is a hybrid that combines features of the frequency distribution and the histogram (see Tukey, 1977). The "stems" consist of class intervals, while the "leaves" are strings of specific values within each interval.

To illustrate, consider once again the scores of the 85 students shown in Table 2.4. Let us choose an interval size of 5 for the stem-and-leaf display, and order the stems (intervals) from small to large going down the page (see Table 2.6). Each leaf is made up of the scores within a given interval, with each score represented solely by its unit's digit, and the scores in each leaf are ordered from low to high. For example, the interval 15–19 has leaves of 5, 5, 5, 8, and 8, which means that it contains the following five scores: 15, 15, 15, 18, and 18. Because each observation takes up one space, the stem-and-leaf display provides the same graphic representation as the histogram (after the stem-and-leaf display has been rotated 90° counterclockwise), while at the same time explicitly displaying the value of each score. Thus, you can see both the overall shape of the distribution and such particulars as the largest and smallest observations. The stem-and-leaf display may be supplemented by the frequency distribution and/or the cumulative frequency distribution, as needed.

Since the stem-and-leaf display is used primarily to help visualize and understand a set of data, you may use or invent whatever variations best accomplish this purpose. For data that cover a wide range of values, you can use the last two digits to represent each score in the leaf (e.g., 460–479 | 63, 68, 74, 74, 78). Or, if there are a great many observations, you might choose to let each leaf entry represent not one but two or more cases.

## Table 2.6 Stem-and-Leaf Display for Data in Table 2.4

| Stems (intervals) | Leaves (observations) | Frequency ($f$) | Cumulative frequency ($cf$) |
|---|---|---|---|
| 5–9 | 9 | 1 | 1 |
| 10–14 | 0111344 | 7 | 8 |
| 15–19 | 55588 | 5 | 13 |
| 20–24 | 00111122234 | 11 | 24 |
| 25–29 | 55666666688889999 | 17 | 41 |
| 30–34 | 00000111222222333344 | 20 | 61 |
| 35–39 | 5557777888999 | 13 | 74 |
| 40–44 | 0112334 | 7 | 81 |
| 45–49 | 5679 | 4 | 85 |

## Skewing of Frequency Distributions

Before we introduce the fifth graphic procedure, we will describe for displaying data, we need to discuss an important aspect of the general shape of a frequency distribution: *skewness*, because this last display is particularly good at illustrating it in a very compact way.

A distribution is *symmetric* if and only if it can be divided into two halves, each the mirror image of the other. For example, Distributions A, B, C, and G in Figure 2.4 are symmetric. Distributions D, E, F, and H are *not* symmetric, however, since they cannot be divided into two mirror images.

A markedly asymmetric distribution with a pronounced "tail" is described as *skewed* in the direction of the tail. Distribution E is *skewed to the left* (or negatively skewed) because the long tail is to the left of the distribution. Such a distribution indicates that most people obtained high scores but some (indicated by the tail) received quite low scores, as might happen if an instructor gave an exam that proved easy for all but a few students. Distribution F, however, is *skewed to the right* (or positively skewed), as indicated by the position of the tail. This distribution indicates many low scores and some quite high scores, as could occur if an instructor gave a difficult examination on which only relatively few students excelled.

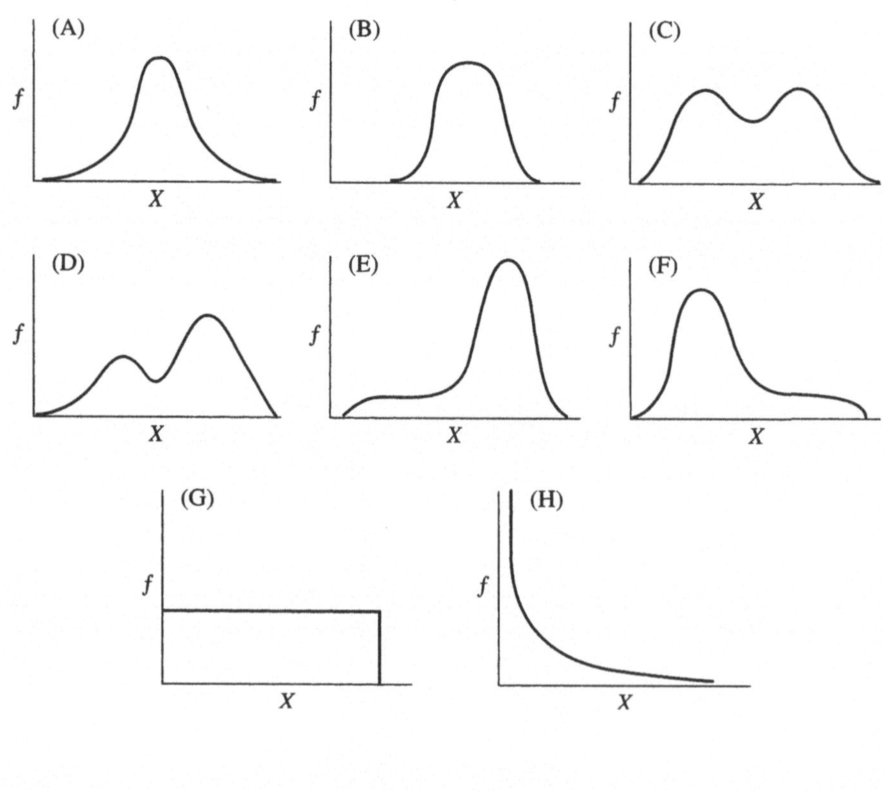


Figure 2.4

**Shapes of Frequency Distributions.**
**(A) Normal curve (symmetric, unimodal).**
**(B) Symmetric, unimodal.**
**(C) Symmetric, bimodal.**
**(D) Asymmetric, bimodal.**
**(E) Unimodal, skewed to the left.**
**(F) Unimodal, skewed to the right.**
**(G) Rectangular.**
**(H) J-curve**

If you were to inspect the distribution of baseline anxiety scores for all of Jackson's students, you would see a positive skew. Most of the students have fairly low anxiety scores when not facing a threat (like a statistics quiz!), but a few of the students have elevated anxiety levels at baseline.

### Box-and-Whisker Plots

One more way to picture a distribution of data is the graphic method known as the box-and-whisker plot, or boxplot, for short. The boxplot uses the 50th percentile (i.e., the median) as its center, and the top and bottom of the "box," which are called the hinges, are essentially, but not exactly the 25th and 75th percentiles (the exact calculation of the hinges is best left to statistical software). The sides of the box do not have any function other than creating the appearance of a box. The median will be somewhere between the hinges, of course, but if the distribution is not symmetric, it won't be exactly midway. In fact, the position of the median with respect to the two hinges is related to the direction and the amount of skewing. The lengths of the "whiskers" depend, to some extent, on the height of the box; the height of the box is called the H-spread, because it is the separation of the two hinges. The boxplot lets you see the spread of a distribution, and the symmetry of that spread, at a glance.

The first step in constructing a boxplot is finding the median. Next, the two hinges are found—approximately, the 25th and 75th percentiles. In a typical boxplot, as shown in Figure 2.5, the two hinges are the top and bottom of the box, though the boxplot is sometimes constructed horizontally. The H-spread is found by subtracting the 25th percentile from the 75th percentile. Inner fences are then set above and below the box at a distance of 1.5 times the H-spread away from each hinge. The inner fences are there to constrain the length of the "whiskers" that extend vertically away from each hinge of the box. Data points that are outside of the inner fences are considered to be so far from the bulk of the distribution that they can be labeled and treated as "outliers." We will have more to say about how to deal with outliers in later in the book. At this point, it should be abundantly clear why you will want to let your statistical software create boxplots for you (the boxplot in Figure 2.5 was created by SPSS, version 29).

In Figure 2.5, you can see that the median is 30, and it is nearly midway between the hinges. There is a slight negative skew (the median is a bit closer to the upper hinge), which can also be seen easily in Figure 2.3. The inner fences are at 9 and 49, which are the highest and lowest scores in the distribution, indicating that there are no outliers. The whiskers extend all the way to the inner fences in this example, which is not always the case.

### Modality

In addition to skewness, another important feature of a distribution's shape is its *modality*. This term refers to the number of clearly distinguishable high points, or peaks, in a distribution. In Figure 2.4, Distributions A, B, E, and F are described as **unimodal** because there

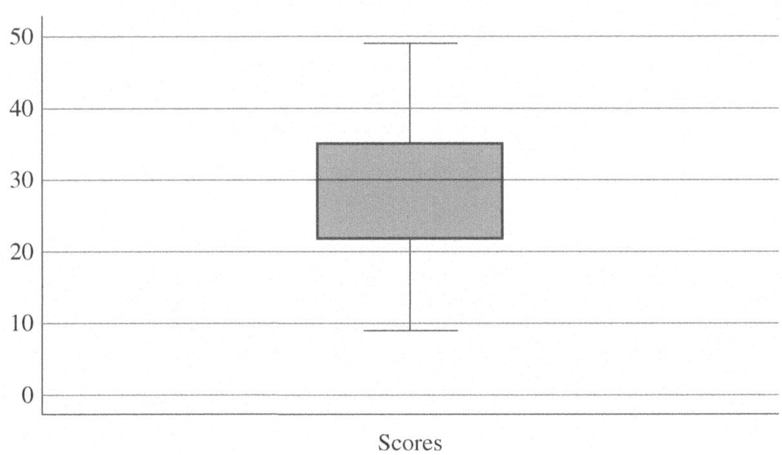


Figure 2.5

**Boxplot for Data in Table 2.4**

is one peak; Distributions C and D are **bimodal** because there are two clearly pronounced peaks; and Distribution G has no modes at all. Be careful, however, not to be influenced by minor fluctuations when deciding whether to consider a distribution unimodal, bimodal, or even multimodal.

### Special Distributions

Certain frequency distributions have names of their own. A particular bell-shaped, symmetric, and unimodal distribution shown in Example A of Figure 2.4 is called the **normal distribution**, about which much more will be said in subsequent chapters. In this distribution, frequencies are greatest at the center, and more extreme scores (both lower and higher) are less frequent. Many variables of interest to the behavioral scientist, such as spatial ability, are approximately normally distributed. The distribution of math quiz scores, as shown in Figure 2.4, may look more like Example B than Example A in Figure 2.4, but it is similar enough to the normal distribution for most statistical purposes. This will be made clear in Chapter 4

Example G illustrates a *rectangular* distribution, in which each score occurs with the same frequency. A distribution approximately like this would occur if one fair die were rolled a large number of times and the number of 1s, 2s, and so on were recorded.

Example H is an illustration of a *J-curve,* wherein the lowest score is most frequent and the frequencies decrease as the scores become larger. An example of such a distribution would be the number of industrial accidents per worker over a two-month period: The majority of workers would have no accidents at all, some would have one accident, a few would have two accidents, and very few unfortunate souls would have a larger number of accidents. The mirror image of this distribution, which is shaped almost exactly like a J, is also called a *J-curve*.

## Summary

In *regular frequency distributions,* all score values are listed in the first column and the frequency corresponding to each score is listed to the right of the score in the second column. *Grouped frequency distributions* sacrifice some information for convenience by combining several score values in a single *class interval* so that fewer intervals and corresponding frequencies need to be listed. *Cumulative frequency distributions* are obtained by starting with the frequency corresponding to the lowest score and adding up the frequencies as you go along. Cumulative frequencies can be easily converted into cumulative percentages that help us find the PR of any one score—that is, the percentage of the scores that are at or below that particular score. *Percentiles* are scores in the distribution with particular PRs, the most useful of which are *deciles* and *quartiles.* For instance, the fifth decile, which is also the second quartile and is best known as the median, is the score whose PR is 50. *Bar charts* and *histograms* are two similar methods for graphing frequency distributions, but bar charts are particularly appropriate for discrete data and histograms are preferred for continuous data. *Frequency polygons* are a useful alternative to histograms when you want to compare the shapes of two distributions. The *stem-and-leaf display* combines features of the frequency distribution and histogram and facilitates the comprehension of a data set by allowing the reader to see it all at once and with as much detail as is needed. Box-and-whisker plots (*boxplots*) allow you to visualize the amount and direction of skewness at a glance and identify *outliers*.

### Regular Frequency Distributions

List every score value in the first column, with the highest score at the top.
List the *frequency* (symbolized by *f*) of each score to the right of the score in the second column.

### Cumulative Frequency and Percentage Distributions

To the right of the frequency column, form a column of cumulative frequencies (symbolized by *cf*) by starting with the frequency for the lowest score and adding up the frequencies as you go along. The cumulative percentages (*cp*) are then found by dividing each *cf* by N (the total number of scores in the distribution, which is also the highest *cf*) and multiplying by 100.

The *PR* corresponding to a given score refers to the percent of cases in a given reference group *at or below* that score. A specified raw score may be converted to the corresponding PR to express its standing relative to the reference group, or the raw score corresponding to a specified percentile may be determined, preferably by statistical software.

### Grouped Frequency Distributions

List the *class intervals* in the first column and the frequencies in the second column. It is usually desirable to do the following:

- Have a total of about 8–15 class intervals.
- Use an interval size of 2, 3, 5, or a multiple of 5, selecting the smallest size that will satisfy the first rule. (All intervals should be the same size.)
- Make the lowest score in each interval (or, perhaps, the highest) a multiple of the interval size.

Do not use grouped frequency distributions if all scores can be quickly and conveniently reported because grouped frequency distributions lose information.

### Graphic Representations

*Bar charts*, in which the frequency of any score is expressed by the height of the bar above that score, and the bars do not touch, are particularly appropriate for discrete data (where results between the score values shown cannot occur). *Histograms* are similar to bar charts except that adjacent bars are allowed to touch. So, these types of graphs are particularly appropriate for continuous data (in which results between the score values shown can occur, or could if it were possible to measure with sufficient refinement). *Frequency polygons* are a good alternative to histograms, especially when the frequencies of many scores (or class intervals) are being represented on the *X*-axis and you want to see the shape of the distribution at a glance, or you want to compare two distributions on the same graph. *Stem-and-leaf displays* are formed by listing class intervals ("stems") in a column at the left, with specific values within each interval ("leaves") on a horizontal line next to that interval (often represented solely by the unit digit).

## Exercises

1. Create regular and cumulative frequency and percentage distributions for the data from the Turck, Kirk, and Dupre dormitories (see first exercise of Chapter 1).
2. Create grouped and cumulative grouped frequency and percentage distributions for Turck and Kirk Halls.
3. Plot a histogram corresponding to the grouped frequency distribution that you created for Turck Hall in the previous exercise. What kind of distribution shape do you see?
4. Using the grouped frequency distributions you created for Turck and Kirk Halls in Exercise 2, plot the two corresponding frequency polygons on the same set of axes (i.e., in one graph).
5. Create a stem-and-leaf plot for the data from Turck Hall, using an interval size of 3.
6. For each of the frequency distributions shown in the following table, state whether it is
   (a) (approximately) normal
   (b) unimodal, skewed to the right

(c) unimodal, skewed to the left
(d) bimodal, approximately symmetric
(e) bimodal, skewed to the right
(f) bimodal, skewed to the left
(g) (approximately) rectangular
(h) J-curve

| (1) | | (2) | | (3) | | (4) | | (5) | | (6) | |
|---|---|---|---|---|---|---|---|---|---|---|---|
| *X* | *f* | *X* | *f* | *X* | *f* | *X* | *f* | *X* | *f* | *X* | *f* |
| 10 | 0 | 45–49 | 2 | 27–29 | 2 | 55–59 | 0 | 10 | 3 | 80–89 | 5 |
| 9 | 0 | 40–44 | 3 | 24–26 | 8 | 50–54 | 1 | 9 | 4 | 70–79 | 4 |
| 8 | 1 | 35–39 | 1 | 21–23 | 17 | 45–49 | 3 | 8 | 10 | 60–69 | 4 |
| 7 | 0 | 30–34 | 4 | 18–20 | 24 | 40–44 | 3 | 7 | 6 | 50–59 | 3 |
| 6 | 1 | 25–29 | 2 | 15–17 | 16 | 35–39 | 8 | 6 | 2 | 40–49 | 5 |
| 5 | 3 | 20–24 | 5 | 12–14 | 6 | 30–34 | 13 | 5 | 5 | 30–39 | 6 |
| 4 | 0 | 15–19 | 12 | 9–11 | 8 | 25–29 | 19 | 4 | 11 | 20–29 | 5 |
| 3 | 2 | 10–14 | 9 | 6–8 | 2 | 20–24 | 12 | 3 | 6 | 10–19 | 4 |
| 2 | 6 | 5–9 | 3 | 3–5 | 3 | 15–19 | 10 | 2 | 2 | 0–9 | 4 |
| 1 | 14 | 0–4 | 0 | 0–2 | 1 | 10–14 | 4 | 1 | 1 | | |
| 0 | 21 | | | | | 5–9 | 2 | 0 | 0 | | |
| | | | | | | 0–4 | 0 | | | | |

7. Use the regular (ungrouped) frequency distribution that you created for Turck Hall in Exercise 1 to solve this exercise. (a) Estimate the percentile rank corresponding to a score of 8. (b) Estimate the percentile rank corresponding to a score of 12.
8. Use the grouped frequency distribution that you created for Kirk Hall in Exercise 2 to solve this exercise.

   (a) Estimate the percentile rank corresponding to a score of 16.
   (b) Estimate the percentile rank corresponding to a score of 7.
9. Use the same distribution for this exercise that you used for Exercise 7.

   (a) Approximately what score corresponds to the 25th percentile?
   (b) Approximately what score corresponds to the 75th percentile?
10. Use the same grouped distribution for this exercise that you used for Exercise 8.

    (a) Estimate the score that corresponds to the second decile.
    (b) Estimate the score that corresponds to the 50th percentile.
    (c) Estimate the score that corresponds to the 68th percentile.

## Thought Questions

1. Why would it be a bad idea to create a *grouped* frequency distribution for the Doty Hall data from Chapter 1?
2. For which of the following should you use a bar chart, and for which should you use a histogram? Why? (a) Scores on a midterm examination in psychology taken by 50 students. (b) The number of children in each of the families of the 50 students taking the psychology course.
3. What is a symmetric distribution?
4. In a *positively* skewed distribution, are the unusual scores extremely large or extremely small? At what end of the graph is the "tail?"

5. In a *negatively* skewed distribution, are the unusual scores extremely large or extremely small? At what end of the graph is the "tail?"
6. What is a bimodal distribution? If a distribution of grades in a college course is bimodal, what does this imply about the students who took the course?
7. Why might you want to convert a raw score into a percentile rank?
8. What is the difference between a percentile rank and a percentile?
9. For each of the following, state which one represents better performance or if they are the same: (a) The 3rd quartile and the 70th percentile. (b) The 1st quartile and the 3rd decile. (c) The median and the 50th percentile. (d) The median and the 2nd quartile. (e) The 6th decile and the 80th percentile.
10. Rachel's percentile rank on a psychology test is 93. Darcy's percentile rank on a physics test is 80. Why is it potentially incorrect to conclude that Rachel did better than Darcy?

## Computer Exercises

1. Request a frequency distribution and a bar chart for the variable Undergraduate Major across all 100 of Jackson's students.
2. Repeat Exercise 1 for the variables Outstanding Teachers and Procrastination. Would it make sense to request a histogram instead of a bar chart for Procrastination? Discuss.
3. Request a frequency distribution and a histogram for the variable Academic Interest across all students. Describe the shape of this distribution.
4. Request a frequency distribution and a histogram for the variables Baseline Anxiety and Baseline Heart Rate for all students. Comment on your statistical software's choice of class intervals for each histogram.
5. Repeat Exercise 4 separately for students who regularly exercise and those who do not. Describe the shapes of all four of the resulting distributions.
6. Request a stem-and-leaf plot for the variables Baseline Anxiety and Baseline Heart Rate for all students. Did your statistical package use the stem-and-leaf arrangement that you would have used for maximum clarity? Explain.
7. Request the deciles for the variable Academic Interest across all students.
8. Request the quartiles for the variables Baseline Anxiety and Prequiz Anxiety for all students. Describe the shift in anxiety scores between these two measurement points.
9. Repeat Exercise 8 separately for students who regularly exercise and those who do not.
10. Request the following percentiles for the variables Baseline Heart Rate and Prequiz Heart Rate for all students: 15, 30, 42.5, 81, and 96.

## Bridge to SPSS

Frequency distributions can be obtained from SPSS by clicking on **Analyze/Descriptive Statistics**, and **Frequencies** .... The *Display frequency tables* box is checked by default. So, if you keep that box checked, then for whichever dependent variable(s) (DVs) you move into the *Variable(s)* box, you get a table with five columns, the first of which contains every different score on that DV. That is, SPSS gives you a regular frequency distribution and does not create a grouped frequency distribution, no matter how many different scores you have. The second column is Frequency (the number of times each different score occurs). Note that at the bottom of this table there is a row for Total, and if there are any missing values, there is another row below it for the number of Missing values, and then another Total (which includes the number of missing values). If your DV has missing values for any of your participants, as is the case for the Academic Interest variable in Jackson's data set, the two totals will differ, and the values in the three bottom rows become critical for understanding the remaining columns.

The third column from the left is called Percent; each of its entries is the Frequency divided by the Total number of cases (i.e., rows) in your spreadsheet—the bottom-row total that includes any missing values—multiplied by 100. The fourth column is Valid Percent, which differs from the third column in that each frequency is divided by the total number of actual scores in your data (i.e., the total that does not include the missing values—we will call this the Valid Total). This is the percentage you are more likely to be interested in. Finally, the last column contains the cumulative values of the Valid Percents. If you want the cumulative frequency distribution, you will just have to multiply each of these entries by the Valid Total, divide by 100, and round off to the nearest whole number.

Going down the right side of the **Frequencies** dialog box are five buttons. We will discuss the results that can be obtained by clicking on the **Statistics ...** button in Chapter 3. For now, let us consider your choices if you click on the **Charts ...** button. The two choices that are relevant to this chapter are *Bar Charts* and *Histograms*. If you select *Bar Charts*, SPSS will create a graph based on a regular frequency distribution of your variable; class intervals will not be created, no matter how many different score values your data contain. Moreover, a bar chart will treat your data not only as discrete (inserting slim spaces between adjacent bars), but as though it were measured on a nominal or ordinal scale. For instance, no place is held for a value within your variable's range that happens to have zero frequency. Selecting *Bar Charts* gives you two choices with respect to the scaling of the vertical axis: frequencies (the default choice) and percentages. The relative heights of the bars will look the same, but if you choose percentages, the $Y$-axis will be marked off to correspond with the fact that the frequencies are being divided by the valid $N$ and multiplied by 100.

If your variable has been measured on a scale that can be considered quantitative (especially interval or ratio rather than just ordinal), you will most likely want to choose *Histograms*, instead of *Bar Charts*, in the **Frequencies** dialog box on the next page. If you choose *Histograms* for your selected variables, each variable will be treated as though measured on an interval/ratio scale: Adjacent bars will touch, and if there are many different values, they will be grouped into convenient class intervals (a full bar width will be left for each empty class interval within your range of scores). However, the bars are labeled in terms of the midpoints of the intervals; the real limits of the intervals are not shown (we put them in Figure 2.2 more for educational than descriptive purposes). You also have the option of checking a box to have a normal curve superimposed on your histogram to see how "normal" your distribution is. If, instead, you would like to see stem-and-leaf or box plots for any of your DVs, you cannot request them from the **Frequencies** dialog box. Instead, after clicking **Analyze** and **Descriptive Statistics,** click on **Explore** .... If you click on **Plots** ... in the **Explore** dialog box, you will have the choice of selecting *Boxplots* and *Stem-and-leaf* displays, as well as *Histograms*.

Percentiles can be obtained from SPSS by opening the **Frequencies** ... dialog box just described and moving the DVs for which you want to find percentiles into the *Variable(s)* box. Then click on the **Statistics** ... button. The upper-left quadrant of the *Frequencies: Statistics box* that opens presents three choices for obtaining percentiles (see Figure 2.6). The topmost choice, **Quartiles**, will give you, of course, the 25th, 50th, and 75th percentiles. The second choice, **Cut points** for ..., will give you the deciles, if you use the default number of 10. If you change that number to 5, for instance, you will obtain the 20th, 40th, 60th, and 80th percentiles. The third choice, **Percentile(s)**, allows you to specify any number of particular percentiles that you would like to see—just click **Add** after typing in each one. Click **Continue** to return to the main **Frequencies** dialog box and then click **OK**. You will get a table of all of the percentiles you requested in numerical order (e.g., if you requested quartiles as well as the particular percentiles 35 and 60, the table will list the scores corresponding

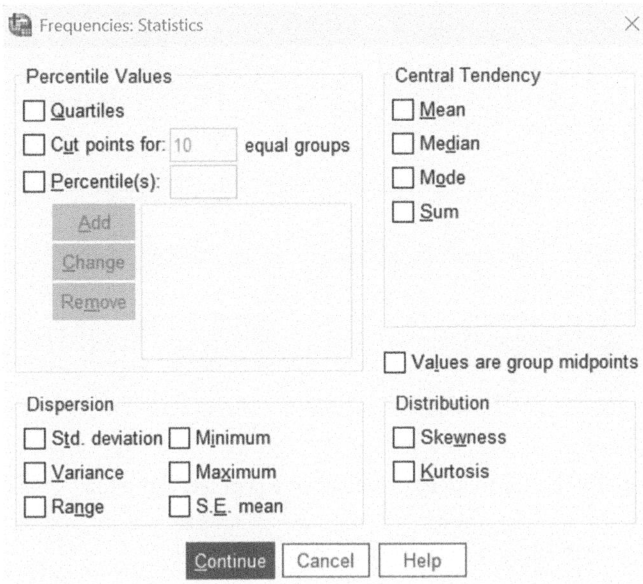


# Figure 2.6

**The Frequencies: Statistics Dialog Box**

*Source:* Courtesy of International Business Machines Corporation, © SPSS, Inc. / IBM company

to the 25th, 35th, 50th, 60th, and 75th percentiles, in that order). Following the table of requested percentiles, SPSS will print the regular frequency distribution that we described earlier in this section, unless you uncheck the little box labeled *Display frequency tables* under the list of all your variables on the left side of the Frequencies dialog box. Most of the other choices in the *Frequencies: Statistics* box will be explained in Chapter 3.

# Chapter 3

# Measures of Central Tendency and Variability

## PREVIEW

### Introduction

What is a measure of central tendency?

What do we gain by using a measure of central tendency instead of a regular frequency distribution?

What important information is not conveyed by a measure of central tendency?

### The Mode

What is the mode?

Why is the mode usually not used as the measure of central tendency?

When is the mode your only choice as a measure of central tendency?

### The Median

What is the median and how is it computed?

How do very large or very small scores affect the median?

In what kind of distribution is the median larger than the mean?

In what kind of distribution is the mean larger than the median?

When should the median be used as the measure of central tendency?

### The Mean

How do we compute the mean of a set of scores?

Is the population mean computed any differently from the sample mean?

How is the mean computed from a regular frequency distribution?

How is a weighted mean calculated?

What must the sum of the deviations of all scores from the mean be equal to?

What does this imply about the way in which very large or very small scores affect the mean?

When should the mean be used as the measure of central tendency?

### The Concept of Variability

What is meant by the variability of a set of scores?

Why is it usually essential to know about the variability of a set of scores rather than just the central tendency? For example, how might your letter grade on a midterm exam be strongly affected by the variability of the examination scores?

### The Range

What is the range, and how is it computed?

Why is the range usually not used as the measure of variability?

What are the interquartile and semi-interquartile ranges?

How are these measures computed, and when should they be used?

*Introductory Statistics for the Behavioral Sciences*, Eighth Edition. R. Brooke Lea and Barry H. Cohen.

Companion website: www.wiley.com/go/ISBS8e

**PREVIEW** (*continued*)

### The Standard Deviation and Variance

Why cannot we use the average of the deviation scores as a measure of variability? How does squaring all of the deviations solve this problem?

What is the sum of squares of a set of scores?

What is the variance of a set of scores?

What is the standard deviation of a set of scores?

What advantage does the standard deviation have over the variance as a measure of variability?

When computing the variance or standard deviation, when should the sum of the squared deviations be divided by $N$? When should you divide instead by $N - 1$?

What was the original purpose behind the creation of computing formulas, and why are they less needed now?

What are some of the useful mathematical properties of the standard deviation?

### Summary

### Exercises

### Thought Questions

### Computer Exercises

### Bridge to SPSS

## Introduction

The techniques presented in Chapter 2 are useful when you wish to provide a detailed summary of all of the data in a convenient format. Often, however, your primary objective will be to highlight certain important characteristics of a group of data. For example, suppose that a prospective student at a University wants to know how wheelchair-accessible your campus is. You might go to all of the buildings and compile a regular frequency distribution such as the one in Table 3.1.

This would show that four of your campus buildings have no wheelchair-accessible entrances, nine have one, and so forth. But all this detail is not essential in order to answer the question, and it would undoubtedly prove tiresome both for you and for your audience. In addition, presenting the data in this form would make it awkward for the student to compare your university to other universities, if they wished. A better plan would be to select one or two important attributes of this set of data and summarize them so that they could be reported quickly and conveniently.

One item of information that you would want to convey is the general *location* of the distribution of grades. You could simply state that your campus buildings typically have

**Table 3.1 Wheelchair-Accessibility Distribution for 20 Buildings On One University Campus**

| # of Entrances | *f* |
|---|---|
| 0 | 4 |
| 1 | 9 |
| 2 | 6 |
| 3 | 1 |
| 4+ | 0 |

one accessible entrance. If you wished to be precise, you would report a numerical average—a single number that describes the general location of this set of scores. In either case, you would summarize your campus by referring to a central point of the distribution. It would be misleading to describe your campus as overall having 0 or 3 accessible entrances per building, even though some of the buildings have that many.

This is one of many situations that benefit from the use of a measure of **central tendency**—*a single number that describes the general location of a set of scores.* Other examples include the average income of families in the United States, the number of dollars gained or lost by an average share of stock on the New York Stock Exchange in a single day, and the number of minutes taken by the average six-year-old child to complete a particular jigsaw puzzle.

It should be stressed, however, that the *overall magnitude of a set of data* (i.e., the size of the scores) is *not* its only important attribute. Suppose that the average score on a statistics quiz that you have just taken is 5.0. This average provides information as to the general location of these scores along the possible range of quiz scores, but it does not tell you how many high and low scores were obtained. Consequently, you cannot determine what score will be needed to ensure an A (or to just pass with a D!). In a distribution such as the following one, a score of 7 would rank very highly:

7 7 6 5 4 4 4 3

However, a score of 7 is not so impressive in a distribution like this:

10 10 9 7 5 4 3 2 0 0

In both examples, however, the average, measured as the arithmetic mean (i.e., the sum of the scores divided by the number of scores), is equal to 5.0. This indicates that there are important aspects of a set of data that are *not* conveyed by a measure of central tendency; a second vital characteristic, called *variability*, is considered later in this chapter.

## The Mode

The simplest measure of central tendency is the **mode**. The mode is the score that occurs most often. For example, the mode of the data in Table 3.1 is B because this score was obtained more often (nine times) than any of the others. In a bar chart or histogram (e.g., Figure 2.1 or 2.2), the mode is the score on the *X*-axis that has the highest bar over it. A distribution can have two or more modes because two or more scores can have exactly the same frequency (e.g., Figure 2.4C). The mode can also be viewed in a less exact, relative sense, as when a distribution is referred to as bimodal even though its two peaks are not equally high (e.g., Figure 2.4D on page 32). If one were to examine the relationship between traffic accident occurrence in a metropolitan area and time of day, a bimodal frequency distribution is likely to result. Note that accidents most commonly occur during the morning and the evening rush hours, with relatively few accidents occurring at other times. It would be misleading to pick a single time of day during which most accidents occur; the distribution is bimodal. This manner of describing the shape of a distribution is one advantage that the mode has over the median and mean (which represent distributions with a single number). Distributions can have more than two modes, in which case they can be described as multimodal. Of course, when a distribution has more than five or six modes, it is usually best to simply say that it has no mode.

The mode is a crude descriptive measure of location that ignores a substantial part of the data. Therefore, it is not often used in research in the behavioral sciences. However, when the variable you are dealing with has only categorical levels, the mode is the only measure of central tendency available. For instance, a marketing researcher may allow the children in a study to take home only one toy each but may offer that toy in a choice of five bright colors. If a manufacturer wants to produce the toy in only one color, it would make sense to use the "modal" (i.e., most frequently chosen) color from the research study, especially if there were many children participating in the study.

## The Median

The **median** (Mdn) is defined as the score corresponding to the 50th percentile (see Chapter 2). Half of the scores in a distribution are larger than the median, and half are smaller. The median is the *middle* score in the distribution when scores are put in order of size. For example, there are an odd number of scores in Table 2.3, so when they are placed in numerical order, there is a single middle score (the 43rd score from the lowest) with 42 scores below it and 42 scores above it. For those data, the median is 30. If there is an even number of scores (so that there is no single middle score), the median is computed by averaging the two middle scores.

### Properties of the Median

When the distribution of scores is symmetric, the mean and the median will be equal. In a *positively skewed* distribution, where there are extreme values at the higher end, the mean will be pulled upward by the extremely high scores and will therefore be larger than the median:

3 4 4 4 5 6 7 39

The *mean* of these data is 9.0. The *median* is only 4.5. However, it is *not* affected by the size of any extremely large (or extremely small) values. In fact, the median will remain at 4.5 even if the value of 39 is changed to a huge number, such as 39,000. (The median will also remain 4.5 if the value of 3 is changed to an extremely small number, such as −1,000,000. But the median *will* change somewhat if the value of 39 is reduced to a small number like 3, for there will no longer be 4 cases above it and 4 cases below it.)

In a *negatively skewed* distribution where the extreme values are at the lower end, the mean will be pulled downward by the extremely low scores and will therefore be smaller than the median:

3 6 25 26 27 27 27 29

Here the mean is equal to 21.25, while the median is equal to 26.5.

In some cases, the actual size of the extreme scores may be unknown. For example, suppose that a few participants in a learning experiment do not learn the task even after a great many trials. In that case, you might have to terminate the experiment after, say, 75 trials. Such participants would therefore have learning scores of "at least 75 trials." They would be the slowest learners (and should not be discarded from the experiment, as the results would then be biased); but their exact scores would be unknown, so you could not compute a sample mean. Undetermined scores also occur in surveys when the highest point on a scale includes a phrase such as "or above." The scale for annual household income, for example, might include the category "$500,000 or above"; the psychologist may not be interested in income distinctions among people making more than half a million dollars per year. But if she wanted to compute the mean income reported, what number should she use for those who chose the $500 k + category? This is not a problem for the median because it does *not* take the exact value of extreme scores into account. As long as undetermined scores can be located with respect to other scores in the data set (e.g., "at least 75 trials" is definitely more trials than a score of "74 trials"), then the median can be computed.

### Usage

For descriptive purposes, use the median when either (1) the data are highly skewed or (2) there are inexact data at the extremes of the distribution. This will enable you to profit from the fact that the median is not affected by the size of extreme values. In almost all other cases, use the mean so as to benefit from its numerous advantages. Except for a few infrequent situations that require the median (e.g., summarizing data based on ranks), the mean is the measure of location that is usually used in inferential statistics, with the median reserved for situations where the objectives are purely descriptive.

## The Mean

When people use the word *average* in everyday language, they are usually referring to what statisticians call the **arithmetic mean**. There are other types of means in statistics (e.g., the *harmonic mean*, which will be introduced in Chapter 7), but in this text, the term *mean* by itself will always refer to the arithmetic mean. In the case of populations, the Greek letter mu ($\mu$) is used to represent the *population mean*. Nevertheless, the computational procedure is the same—sum all scores and divide by $N$.

### Computation

The mean of a set of scores is computed by adding up all the scores and dividing the result by the number of scores added. In symbols,

$$\mu = \frac{\sum X}{N}, \qquad \text{Formula 3.1}$$

where

$\mu$ = population mean (it is pronounced "myoo")
$\sum X$ = sum of the $X$ scores (see Chapter 1)
$N$ = total number of scores

In the case of the first set of quiz scores given in the introduction to this chapter, the mean is equal to 5.0:

$$\frac{7 + 7 + 6 + 5 + 4 + 4 + 4 + 3}{8} = \frac{40}{8} = 5.0$$

Note that simply computing $\sum X$ is not sufficient to identify the location of these scores. What is further required is to divide by $N$ (the number of scores). This step ensures that the means of two different samples will be comparable even if they are based on different numbers of scores. The mean of the second set of quiz scores presented in the introductory section is equal to 50/10 or 5.0; $\sum X$ is different, but the mean correctly shows that the overall location is the same.

### The Weighted Mean

The mean can be viewed as a **weighted mean** (also called "weighted average") of the different possible scores in a distribution, each weighted by its frequency. For example, a student's *grade point average* (GPA) is usually found by first assigning a numerical value or score to each possible letter grade (commonly A = 4, B = 3, C = 2, and so on, and sometimes with intermediate grades, such as B+ = 3.3), and then weighting each possible grade by the number of courses that received that grade. (We are assuming that all courses carry the same number of credits. Otherwise, you would have to use the number of credits for your weights.) The weighted mean will become important in Chapter 7 when we deal with two unequal sample sizes and need to decide how to arrive at the best overall average (solution: use the weighted mean!).

### Properties of the Mean

One important property of the mean is that if you add a *constant* (i.e., a fixed number) to every score in a distribution, the mean of the new distribution will be the original $\mu$ plus that constant, which we will call $k$ (as we did in Chapter 1). It is easy to prove that this is true by expressing each new score as $(X + k)$ and then following these steps:

$$\mu_{\text{new}} = \frac{\sum(X + k)}{N}$$

$$= \frac{\sum X + Nk}{N} \qquad \text{(Rule 6, Chapter 1)}$$

$$= \frac{\sum X}{N} + \frac{Nk}{N}$$

$$= \mu_{\text{old}} + k$$

Similarly, we could use Summation Rule 7 from Chapter 1 to show that if we subtract $k$ from every score, $\mu_{\text{new}} = \mu_{\text{old}} - k$. A related property of the mean follows from Summation Rule 8. If every score is multiplied by $k$, then the mean of the new scores (i.e., the mean of $kX$) can be shown to be $k$ times the old mean by the following steps:

$$\mu_{\text{new}} = \frac{\sum kX}{N}$$

$$= k\frac{\sum X}{N} \qquad \text{(Rule 8, Chapter 1)}$$

$$= k\mu_{\text{old}}$$

Because dividing by $k$ is the same as multiplying by $1/k$, it is just as easy to show that when all of the scores are divided by $k$, $\mu_{\text{new}} = \mu_{\text{old}}/k$. We will make use of these properties in Chapter 4.

Another important property of the mean is that the sum of the distances (or *deviations*) of all scores from the mean is zero. That is,

$$\sum(X - \mu) = 0$$

It can readily be proved that this must always be true.[1] As an illustration, consider once again the small set of quiz scores discussed previously.

| **Score** | **Deviation from mean $(X - \mu)$** |
|---|---|
| 7 | +2 |
| 7 | +2 |
| 6 | +1 |
| 5 | 0 |
| 4 | −1 |
| 4 | −1 |
| 4 | −1 |
| 3 | −2 |
| $\mu = 5$ | |

[1] $\sum(X - \mu) = \sum X - N\mu$ (Rule 7, Chapter 1) but $\sum X = N\mu$ because $\mu = (\sum X)/N$. Therefore, $\sum(X - \mu) = N\mu - N\mu = 0$.

The mean balances or equates the sums of the positive and negative deviations. It is in this sense that it gives the location of the distribution. This implies that the mean will be sensitive to extreme values on one side that are not balanced by extreme values on the other side, as the following example shows:

| **Score** | $X - \mu$ |
|---|---|
| 39 | +30 |
| 7 | −2 |
| 6 | −3 |
| 5 | −4 |
| 4 | −5 |
| 4 | −5 |
| 4 | −5 |
| 3 | −6 |
| $\mu = 9$ | |

As a result of the change in one score from 7 to 39, the mean shows a substantial 4-point increase. Given that change, all of the other seven scores now fall below the mean and, of course, balance the effects of the large positive deviation introduced by the 39. One might well question the use of the mean to describe the location of a set of data in a situation where it is so influenced by one extreme score!

As another example of the sensitivity of the mean to unbalanced extreme values, consider the case of a manufacturing company in which the president earns $60 million per year and the 99 assembly-line workers each earn only $56,500 per year. The president might attempt to refute criticism of the company's miserly tactics by arguing that the mean annual income of all 100 people in the company (the 99 workers plus himself) is $655,935. The workers would undoubtedly object to the appropriateness of this figure, which is more than 11 times their actual salaries! Here again, the use of the mean as the index of location is questionable. It almost always is with income data, since almost all income distributions are positively skewed. (The workers would prefer, of course, to use the median [$56,500] to describe the central tendency of the salaries in this case.)

### Usage

The mean has many advantageous properties: (a) It takes all of the scores into account. So, it makes the most of the information provided by the data; (b) the mean is also the most stable of the measures of central tendency for most distributions encountered in practice: It is the most consistent across different samples drawn from the same population. For these and other reasons, many of the procedures of inferential statistics make use of the mean. Thus, a third major advantage of the mean is that it is usable as a datum in further statistical analyses, while other measures of central tendency usually are not. For these reasons, the mean is very often the most suitable measure of central tendency.

At times, however, the first of the advantages just mentioned becomes a liability instead of an asset. As we have seen, extreme scores at one end of a distribution exert a strong influence on the mean and cause it to give a misleading picture of the location of the distribution. Therefore, when a distribution is highly skewed and when you do not intend to use the measure of central tendency in subsequent statistical analyses, you should consider an alternative to the mean that will not be affected by unbalanced extreme scores, such as the median.

As discussed in the section on the median, the actual size of extreme scores is sometimes undetermined. Scores such as "at least 75 trials" or "$500,000 or above" are difficult to use with the mean because they do not represent discrete values. How could you plug these numbers into $\sum X$? In such cases, the median is a more reasonable measure of central tendency because it is based on the number of scores (ordered by value), not the specific values of the scores.

In addition to general location, there is a second important attribute of a distribution of scores—its *variability*. Measures of variability are used extensively in the behavioral sciences. So, it is essential to understand the meaning of this concept as well as the relevant computational procedures.

## The Concept of Variability

Variability refers to how *spread out or scattered* the scores in a distribution are (or how like or unlike each other they are). If an average is thought of as representing the typical score in a distribution, a good measure of variability might be thought of as representing the typical distance the scores fall from the average. As an illustration, six different distributions, each involving a small number of scores, are shown in Table 3.2.

The minimum possible variability is zero. This will occur only if all of the scores are exactly the same, as in Distribution 1, and there is no variation at all.

In Distribution 2, there is a small amount of variability. The scores are somewhat spread out, but only to a slight extent.

Distribution 3 is equal in variability to Distribution 2. The locations of these two distributions differ, but variability is not dependent on location. The distance between each score and any other, and hence the amount of spread, is identical.

Each of the remaining three distributions is more variable than the ones that precede it. At the opposite extreme of Distribution 1 would be a distribution with scores spread out over the entire range from −1,000,000 to +1,000,000 (or more). Such extreme variabilities, however, are rarely encountered in practice.

Variability is important in many areas, although frequently it is not reported (or is described vaguely in words) because it is less familiar to nontechnical audiences than is central tendency. We list a few examples next.

### Testing and Variability

Suppose that you score 75 on a statistics midterm examination and that the mean of the class is 65; the maximum possible score is 100. Although your score must be pretty good because it is above average, its relative value in comparison to the rest of the class will be strongly influenced by the variability of the distribution of examination scores.[2]

**Table 3.2 Six Distributions to Illustrate Variability**

| Distribution | | | | | |
|---|---|---|---|---|---|
| 1 | 2 | 3 | 4 | 5 | 6 |
| 7.0 | 7.2 | 40.2 | 7.0 | 10.0 | 97.8 |
| 7.0 | 7.1 | 40.1 | 7.0 | 10.0 | 88.5 |
| 7.0 | 7.1 | 40.1 | 6.0 | 9.0 | 83.4 |
| 7.0 | 7.1 | 40.1 | 5.0 | 7.0 | 76.2 |
| 7.0 | 7.1 | 40.1 | 4.0 | 5.0 | 69.9 |
| 7.0 | 7.0 | 40.0 | 4.0 | 4.0 | 67.3 |
| | | | 4.0 | 3.0 | 58.4 |
| | | | 3.0 | 2.0 | 44.7 |
| | | | | 0.0 | |
| | | | | 0.0 | |

[2] For this example and others, we will assume that test grades are assigned based on relative, not absolute, performance. If the best score on a 0–100 point exam is, for example, 80, then that exam should get an A, not a B− as it would if grades were assigned on absolute scale (e.g., 90–100 = A; 80–90 = B, etc.).

# Figure 3.1

**Frequency Polygons of Two Distributions With the Same Mean but Different Variability**

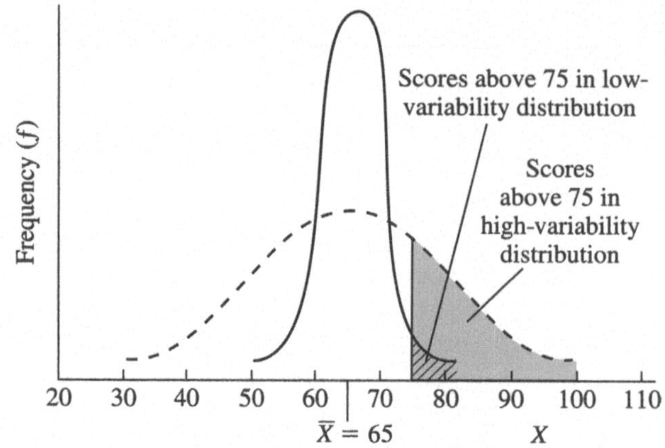


If most scores are clustered tightly around the mean of 65, your score of 10 points above average will stand out as one of the highest (and may well merit a grade of A). But if the scores are widely scattered, and values in the 80s and 90s (and 40s and 50s) are frequently obtained, being 10 points above average will not be exceptional because many scores fall above it. In this case, your score may result in a B− or C+. A mean of 65 together with low variability would indicate that you did very well, whereas a mean of 65 together with high variability would imply that your performance was less than outstanding compared to the group that took the test. These two possibilities are illustrated graphically in Figure 3.1. (Note that the spread of a frequency polygon indicates the variability of the distribution.)

Consider Distributions 4 and 5 in Table 3.2, which happen to be the two sets of scores discussed at the end of the introduction to this chapter (to illustrate how impressive—or not—a score of 7 can be). The mean of both distributions is 5.0. Yet a score of 7 ranks higher in Distribution 4, where the scores are less variable than in Distribution 5, where the scores are more variable. Looking at the lower end of these distributions, a score of 3 is lower in comparison to the group in Distribution 4. Being two points above or below average stands out more in a distribution in which there is less spread.

### Sports and Variability

Suppose that two professional basketball players average 20.3 points per game. Although their mean performance is the same, they may be different in other respects. One player may be consistent (low variability) and always score close to 20 points in every game, rarely scoring as high as 25 but also rarely falling to 15 or less. The second player may be "streaky" (high variability); he scores more than 30 points in some games but drops to below 10 in others. The second player is likely to be a greater source of frustration to his coach and to the fans. Here again, a measure of variability would provide useful additional information to that given by the mean.

### Psychology and Variability

Modern psychology is based on the idea that behavior varies. The psychologist's goal is to understand and explain this variability with theories or models. Some psychological research focuses on person-to-person variability. On any trait of interest—musical ability, throwing a baseball, introversion, height, verbal ability, resilience—people are distributed over the entire range from low to high. Most psychological science, however, is not so much interested in why one particular individual differs from another; instead it is primarily concerned with how various manipulations affect behavior. For example, does imagery improve memory for words? Does a new antidepressant drug decrease negative thoughts? Does conformity vary depending on the social status of the experimenter? Are children able to conserve volume (e.g., realize that the amount of liquid does not change when poured from a short, wide beaker into a tall, skinny one) better depending on how the question is posed to them? In each of these cases, the variability of interest is between groups of individuals who were randomly assigned to participate in different conditions of the study. Were the people asked to visualize words able to remember more, on average, than the group that was asked to think of rhymes for the words? Was the antidepressant group less depressed, on average, than the group that received a placebo instead of the drug? Did people in the lab-coat-experimenter group conform more than those in the Hawaiian-shirt-experimenter group? Were the children in the new-wording group more likely to conserve volume than those in the old-wording group? In all of these examples, the researcher is interested in how a group of people randomly assigned to be treated one way is systematically different (or not) from a group of people treated a different way. Now, of course, individual differences do exist for all of these variables—people do differ with respect to how good their memory is—but those individual differences are usually not the chief object of study in this sort of research. In fact, person-to-person variability is sometimes thought of as error or "unexplained" variance. The goal of much research is to maximize "explained" variability. The more variance a model or theory can explain, the better it can predict behavior. As you may be learning in your psychology classes, we are far from explaining all variability in behavior. As you will learn later in this book, you do not need to understand all or even most variability in order to contribute importantly to science or to help people and society. For now, we simply want you to appreciate how central the concept of variability—and our ability to measure it—is to the behavioral sciences.

### Statistical Inference

Another reason for the importance of variability will become evident when statistical inference is discussed beginning in Chapter 5. Other things being equal, the more variable a phenomenon is, the less precise the estimate you can get of the population's location (e.g., mean) from sample information.

Rather than describe variability in ambiguous terms such as *small* or *large*, it is preferable to summarize the variability of a distribution of scores in a single number. Techniques for accomplishing this are discussed next. As with measures of central tendency, we will start with the simplest measures.

## The Range

One possible way of summarizing the variability of a distribution is to look at the distance between the smallest and largest scores. The *range* of a distribution is defined as the largest score minus the smallest score. For example, the range of Distribution 4 in Table 3.2 is (7.0 − 3.0) or 4.0.

Although this procedure makes intuitive sense, it is likely to give misleading results on many occasions because frequently the extreme values are atypical of the rest of the distribution. Consider the following two distributions:

| | | | | | | | | | | | | | |
|---|---|---|---|---|---|---|---|---|---|---|---|---|---|
| DISTRIBUTION A: | 10 | 10 | 10 | 9 | 7 | 6 | 5 | 4 | 4 | 3 | 2 | 0 | 0 |
| DISTRIBUTION B: | 10 | 6 | 6 | 5 | 5 | 5 | 5 | 5 | 5 | 5 | 4 | 4 | 0 |

In both of these examples, the range is equal to 10. Yet Distribution A, with scores spread out over the entire 10-point range, is more variable than Distribution B, where all but two scores are concentrated near the middle of the distribution. The range is a poor measure of the variability of Distribution B because the extreme values are not typical of the total variation in the distribution; if the two extreme scores are excluded, the range drops to 2. Because this type of distortion occurs fairly often, the range (like the mode) is best regarded as a crude measure that generally should not be used in behavioral science research. However, it is always a good idea to take note of the minimum and maximum values obtained for each of your variables, especially to check for errors and anomalous findings.

### The Semi-Interquartile Range

You may recall that a major advantage of the median as a measure of central tendency is that it is not affected by extreme scores. In fact, the median is still valid if there are some scores on either end of the distribution that are undetermined. For example, on a survey question about how many different jobs the respondent has held, several participants may have checked "10 or more." Obviously, the ordinary range is much affected by extreme scores and cannot be used with open-ended responses.

There is a type of range measure, called the *interquartile range* (IQR), which has advantages similar to those of the median. It is found by subtracting the score at the 25th percentile (the 1st quartile) from the score at the 75th percentile (the 3rd quartile). The median, of course, is the score in the 2nd quartile. This range measure tells you the total spread for the middle 50% of the scores, and it is clearly not affected by extreme or indeterminate scores (unless the latter comprise as much as 25% of the scores on either end of the distribution). For example, in Chapter 2, at the end of the Real and Apparent Limits section, we estimated the 25th percentile for the academic interest questionnaire to be a score of 22.8. The 75th percentile was estimated to be 35.4. Thus, the IQR for those quiz scores is 35.4 − 22.8, which equals 12.6.

Dividing the value for the IQR by 2 results in a useful variability measure called the *semi-interquartile range* (SIQR). The main advantage of the SIQR is that it gives you an immediate idea of the distance of the typical score from the median. (About half the scores in the distribution are farther from the median than the SIQR and about half are closer than the SIQR.) For the academic interest scores, the SIQR is 12.6/2 = 6.3. This means that if you need to come up with a single number that characterizes how much the scores tend to vary from the median of that distribution, 6.3 points would serve the purpose. Although the SIQR is a good descriptive measure of variability, it does not play a role in more advanced statistical methods. So, it will not be discussed further in this text. A more sophisticated way of determining how much the "typical" score differs from the center of the distribution is described in detail next.

## The Standard Deviation and Variance

As we have mentioned, the mean, an average that takes all of the scores into account, is usually the best measure of central tendency for inferential purposes. Similarly, an "average" variability that is based on all the scores will usually provide the most useful information for purposes to be described in later chapters. Before an *average* variability can be computed, however, this concept must be defined in terms of an individual score.

Variability actually refers to the difference between each score and every other score, but it would be quite tedious to compute this in practice (especially if $N$ is large). If there are 100 scores, you would have to compute the difference between the first score and each of the 99 other scores, compute the difference between the second score and each of the 98 remaining scores, and so on—4,950 differences in all.

A more reasonable plan, which will serve the purpose equally well, is to define the "differentness" or *deviation* of a single score in terms of how far it is from the center of the distribution. In a distribution of scores that are closely packed together, most scores are close to each other and hence close to their center. Conversely, in a highly variable distribution, some scores are quite a distance from each other and hence far from their center. Since the mean is the most frequently used measure of central tendency, a reasonable procedure is to define the deviation of a single score as its *distance from the mean*:[3]

$$\text{Deviation Score} = X - \mu$$

[3] A reference point other than the mean could be used, but the choice of the mean has certain statistical advantages as well as making good intuitive sense. For example, it can be proved that the mean is the value of $c$ about which $\sum(X - c)^2$, the sum of squared deviations, is a minimum (the importance of which will become apparent in the following discussion). This is called the Least Squares Property of the mean.

Extremely deviant scores (ones very far from the mean) will have numerically large deviation scores, while scores close to the mean will have numerically small deviation scores.

The next step is to derive a measure of variability that will take into account the deviations of all of the scores. There are several possible ways to do this. If we were to average the deviation scores by the usual procedure of summing and dividing by $N$, we would get

$$\frac{\sum(X - \mu)}{N}$$

It will prove frustrating to try to use this as the measure of variability because $\sum(X - \mu)$ is *always* equal to zero. The problem is that the deviation scores cancel each other out (this is one of the properties of the mean discussed a few pages ago). As a result, this "measure" cannot provide any information as to the variability of any distribution.

This problem could be overcome if we were to focus on the *size* of the deviations and ignore whether they are positive or negative. We might therefore first take the *absolute value* of each deviation, its numerical value ignoring the sign, and then compute the average variability. That is, we might compute

$$\frac{\sum |X - \mu|}{N}$$

where $|X - \mu|$ is the absolute value of the deviation from the mean. This is not an unreasonable procedure, and in fact yields a good descriptive measure called the *mean* or *average deviation* (often symbolized as MD, or MAD, for the mean absolute deviation). But, unfortunately, measures based on absolute values are unsuitable for use in further statistical analyses (one awkward property of the mean deviation (MD) is that it is smaller when deviations are calculated from the *median*, rather than from the mean). The measure that is most frequently used circumvents this difficulty by *squaring* each of the deviations prior to taking the average. The sum of the squared deviations from the mean, $\sum(X - \mu)^2$, is symbolized by $SS$ and is called the **sum of squares**. The measure of variability produced by taking the average, instead of the sum, of the "squares" is called the **variance** and is symbolized by $\sigma^2$:

$$\sigma^2 = \frac{\sum(X - \mu)^2}{N} = \frac{SS}{N} \qquad \text{Formula 3.2}$$

This is a basic measure of the variability of any set of data. However, when the data of a sample are used to estimate the variance of the population from which the sample was drawn, the *population variance estimate*, or **sample variance** (symbolized by $s^2$), is computed instead:

$$s^2 = \frac{\sum(X - \overline{X})^2}{N - 1} = \frac{SS}{N - 1} \qquad \text{Formula 3.3}$$

Note that in Formula 3.3, the mean is symbolized as $\overline{X}$, instead of $\mu$. Even though $\overline{X}$ is always calculated in exactly the same way as $\mu$, we will use the symbol $\overline{X}$ whenever the mean has been calculated for a group of scores that are being viewed as a sample of a population, rather than as an entire population in themselves. This is an important distinction because sometimes both symbols for the mean will appear in the same formula. In general, you will see that $\overline{X}$, calculated for a sample, is usually used as an estimate of $\mu$ for the entire population from which that sample was drawn. Because $s^2$ is the symbol we will be using whenever the variance is calculated for a sample and not a population, the mean that you will see represented in formulas for $s^2$ will be the sample mean ($\overline{X}$), rather than the population mean ($\mu$).

In each of the preceding variance formulas, the order of operations is as follows: (1) subtract the mean from each score; (2) square each result; (3) sum; (4) then divide. The sample estimate of the population variance, $s^2$, is computed just a little differently from $\sigma^2$; the sum of squared deviations is divided by $N - 1$ instead of $N$. This is to enable $s^2$ to be an *unbiased* estimate of the population variance—that is, an estimate that on the average will be too large no more often than it is too small. Hence, it is common to refer to $s^2$ simply as the *unbiased variance* and to $\sigma^2$ as the *biased variance*.

There is one remaining difficulty. Having squared the deviations to eliminate the negative numbers that otherwise would have led to a sum of zero with annoying regularity, the variance is expressed in terms of the original units *squared*. For example, if you are measuring height, the variance indexes variability in terms of square inches, rather than inches. Especially for descriptive purposes, it is preferable to have a measure of variability that is in the same units as the original measure, and this can be accomplished by taking the positive square root of the variance. This yields a very commonly used measure of variability called the **standard deviation**, symbolized by $\sigma$ or $s$, depending on whether it is based on the biased or unbiased version of the variance:

$$\sigma = +\sqrt{\sigma^2} = \sqrt{\frac{\sum(X-\mu)^2}{N}} = \sqrt{\frac{SS}{N}} \qquad \text{Formula 3.4}$$

$$s = +\sqrt{s^2} = \sqrt{\frac{\sum(X-\overline{X})^2}{N-1}} = \sqrt{\frac{SS}{N-1}} \qquad \text{Formula 3.5}$$

Following the shorthand-naming convention applied to $\sigma^2$ and $s^2$, $\sigma$ is often called the *biased standard deviation*, and $s$ is called the *unbiased standard deviation*. Of course, $\overline{X}$ appears in the formula for $s$, which is always based on a sample, whereas $\mu$ appears in the formula for $\sigma$, the standard deviation of a population.

Whereas the mean represents the "average score," the standard deviation represents a kind of "average variability"—similar to the average of the deviations of each score from the mean $(X - \mu)$—with two minor complications: (1) squaring the deviations before averaging, and then taking the positive square root, to return to the original unit of measurement; and (2) dividing by $N - 1$ instead of $N$ when estimates of the population are involved. Although the standard deviation is not literally the average deviation, it can be thought of, less precisely, as a typical deviation. The formulas we have given are called the *definitional* formulas for $\sigma$ and $s$ because their primary function is to define the meaning of these terms; they are not necessarily the formulas by which $\sigma$ and $s$ are most easily computed.

Illustrative examples of the computation of $\sigma^2$ and $\sigma$ and of $s^2$ and $s$ using the definitional formulas are shown on the left-hand side of Table 3.3. If the eight scores in Table 3.3 represent the entire population, $\sigma$ is computed; however, if these eight scores are a sample from a larger population, $s$ is computed. Note that a partial check on the calculations is possible in that $\sum(X - \mu)$ should always equal zero. As expected from the previous discussion concerning these distributions, Distribution 5 (from Table 3.2) has a larger standard deviation (is more variable) than Distribution 4. The typical deviation from the mean is 3.66 points in Distribution 5, but only 1.5 points in Distribution 4.

### Computing Formulas

Using the definitional formulas to calculate $\sigma$ and $s$ can be awkward for several reasons. First, if the mean is not a whole number, subtracting it from each score will yield a deviation score with decimal places, which when squared will produce still more decimal places. This can make the computations quite tedious. Second, the mean must be calculated before the formula can be used, which necessitates two steps in the computation of the standard deviation (although normally the mean will be desired anyway). Therefore, before the advent of handheld statistical calculators and personal computers made these calculations so easy to perform, *computing formulas* for $\sigma$ and $s$ were derived by manipulating the definitional

# Table 3.3 Computation of $\sigma^2$ and $\sigma$ and of $s^2$ and $s$, for Two Small Samples Using the Definition and Computing Formulas

**Example 1. Distribution 4 (where $\mu = 5.0$)**

| $X$ | $X - \mu$ | $(X - \mu)^2$ | $X$ | $X^2$ |
|---|---|---|---|---|
| 7 | $7 - 5 = 2$ | 4 | 7 | 49 |
| 7 | $7 - 5 = 2$ | 4 | 7 | 49 |
| 6 | $6 - 5 = 1$ | 1 | 6 | 36 |
| 5 | $5 - 5 = 0$ | 0 | 5 | 25 |
| 4 | $4 - 5 = -1$ | 1 | 4 | 16 |
| 4 | $4 - 5 = -1$ | 1 | 4 | 16 |
| 4 | $4 - 5 = -1$ | 1 | 4 | 16 |
| 3 | $3 - 5 = -2$ | 4 | 3 | 9 |
| | | $\sum(X - \mu)^2 = 16$ | $\sum X = 40$ | $\sum X^2 = 216$ |

*A. Definition formulas*

$$\sigma^2 = \frac{\Sigma(X - \mu)^2}{N} = \frac{16}{8} = 2.00$$

$$\sigma = \sqrt{2.00} = 1.41$$

$$\text{MD} = \frac{\Sigma\,|\,X - \mu\,|}{N}$$

$$= \frac{2 + 2 + 1 + 0 + 1 + 1 + 1 + 2}{8}$$

$$= \frac{10}{8}$$

$$= 1.25$$

$$s^2 = \frac{\Sigma(X - \overline{X})^2}{N - 1}$$

$$= \frac{16}{7}$$

$$= 2.29$$

$$s = \sqrt{2.29} = 1.51$$

*B. Computing formulas*

$$\sigma^2 = \frac{1}{N}\left[\Sigma X^2 - \frac{(\Sigma X)^2}{N}\right]$$

$$= \frac{1}{8}\left[216 - \frac{(40)^2}{8}\right]$$

$$= \frac{1}{8}(216 - 200)$$

$$= \frac{1}{8}(16) = 2.00$$

$$\sigma = \sqrt{2.00} = 1.41$$

$$s^2 = \frac{1}{N - 1}\left[\Sigma X^2 - \frac{(\Sigma X)^2}{N}\right]$$

$$= \frac{1}{7}\left[216 - \frac{(40)^2}{8}\right]$$

$$= \frac{1}{7}(216 - 200) = \frac{1}{7}(16) = 2.29$$

$$s = \sqrt{2.29} = 1.51$$

**Example 2. Distribution 5 (where $\mu = 5.0$)**

| $X$ | $X - \mu$ | $(X - \mu)^2$ | $X$ | $X^2$ |
|---|---|---|---|---|
| 10 | $10 - 5 = 5$ | 25 | 10 | 100 |
| 10 | $10 - 5 = 5$ | 25 | 10 | 100 |
| 9 | $9 - 5 = 4$ | 16 | 9 | 81 |
| 7 | $7 - 5 = 2$ | 4 | 7 | 49 |
| 5 | $5 - 5 = 0$ | 0 | 5 | 25 |
| 4 | $4 - 5 = -1$ | 1 | 4 | 16 |
| 3 | $3 - 5 = -2$ | 4 | 3 | 9 |
| 2 | $2 - 5 = -3$ | 9 | 2 | 4 |
| 0 | $0 - 5 = -5$ | 25 | 0 | 0 |
| 0 | $0 - 5 = -5$ | 25 | 0 | 0 |
| | | $\sum(X - \mu)^2 = 134$ | $\sum X = 50$ | $\sum X^2 = 384$ |

*(continued)*

## Table 3.3 *(continued)*

| *A. Definition formulas* | *B. Computing formulas* |
|---|---|
| $\sigma^2 = \dfrac{\Sigma(X-\mu)^2}{N} = \dfrac{134}{10} = 13.40$ | $\sigma^2 = \dfrac{1}{N}\left[\Sigma X^2 - \dfrac{(\Sigma X)^2}{N}\right]$ |
| $\sigma = \sqrt{13.40} = 3.66$ | $= \dfrac{1}{10}\left[384 - \dfrac{(50)^2}{10}\right]$ |
| $\text{MD} = \dfrac{\Sigma\lvert X-\mu\rvert}{N}$ | $= \dfrac{1}{10}(384 - 250)$ |
| $= \dfrac{5+5+4+2+0+1+2+3+5+5}{10}$ | $= \dfrac{1}{10}(134) = 13.40$ |
| $= \dfrac{32}{10} = 3.2$ | $\sigma = \sqrt{13.40} = 3.66$ |
| $s^2 = \dfrac{\Sigma(X-\overline{X})^2}{N-1}$ | $s^2 = \dfrac{1}{N-1}\left[\Sigma X^2 - \dfrac{(\Sigma X)^2}{N}\right]$ |
| $= \dfrac{134}{9}$ | $= \dfrac{1}{9}\left[384 - \dfrac{(50)^2}{10}\right]$ |
| $= 14.89$ | $= \dfrac{1}{9}(384 - 250) = 14.89$ |
| $s = \sqrt{14.89} = 3.86$ | $s = \sqrt{14.89} = 3.86$ |

formulas algebraically. Because the computing and definition formulas are algebraically equivalent to each other, they always yield identical results. Although there is no longer much need for them, for historical purposes, as well as to facilitate comparison with the many texts that still present them, the computing formulas for both the biased and unbiased versions of the standard deviation and variance are shown next.

$$\sigma^2 = \frac{1}{N}\left[\sum X^2 - \frac{(\sum X)^2}{N}\right], \quad \sigma = \sqrt{\frac{1}{N}\left[\sum X^2 - \frac{(\sum X)^2}{N}\right]} \qquad \text{Formula 3.6A \& B}$$

$$s^2 = \frac{1}{N-1}\left[\sum X^2 - \frac{(\sum X)^2}{N}\right], \quad s = \sqrt{\frac{1}{N-1}\left[\sum X^2 - \frac{(\sum X)^2}{N}\right]} \qquad \text{Formula 3.7A \& B}$$

Examples of the use of the computing formulas for $\sigma^2$ and $\sigma$ and for $s^2$ and $s$ are shown on the right-hand side of Table 3.3. For comparison purposes, we have included the calculation of the MD, which is often similar to, but can never be larger than, the standard deviation.

### Properties of the Standard Deviation

The properties we discuss here for $\sigma$ also hold for its unbiased cousin, $s$. Although $\sigma$ shares several important properties with the mean, one property it does *not* share is its response to adding a constant to all of the scores in the distribution. (The properties that are associated with addition also hold for subtraction, which can be viewed as the addition of a negative constant.) We demonstrated earlier in this chapter that if the same constant ($k$) is added to all of the scores, $\mu_{new} = \mu_{old} + k$. However, it is easy to show that adding (or subtracting) a constant to all of the scores has no effect on $\sigma$. Consider any one deviation score: $X - \mu$. If you add $k$ to all of the scores, each new deviation score will look like this: $(X + k) - (\mu + k)$. (Remember that the mean is also increased by $k$.) So, the new deviation score equals $X + k - \mu - k = X - \mu$, which is exactly the same as the old deviation score.

Because all of the deviation scores remain the same, $\sigma$, which is based on these deviation scores, also remains unchanged. It is as though everyone in a large crowd were to move exactly 12 inches in exactly the same direction. After the move, the people are standing in exactly the same way in relation to each other; only the overall location of the crowd (like the mean of the distribution) has changed. Yet, multiplying by a constant *does* have the same effect on $\sigma$ as it does on the mean: $\sigma_{new} = k\sigma_{old}$. (The same properties hold for division.) Again, we look at what happens to any one deviation score when all of the scores are multiplied by $k$. The new deviation equals $kX - k\mu = k(X - \mu) = k$ times the old deviation. Because the size of each deviation score is multiplied by $k$, $\sigma$ is multiplied by $k$ as well.

Like the mean, the standard deviation has the statistically desirable property of being based on *all* of the scores in the distribution. This implies, however, that $\sigma$ also has the often *un*desirable property that it can be strongly influenced by one very extreme score. It can therefore produce a value for variability that gives a misleading picture of the entire distribution. Unlike the case of the mean, equally extreme scores on both sides of the distribution will not cancel each other out (because their deviations are squared before being added), so $\sigma$ can be misleading even with symmetrical distributions. Extreme scores have less of an impact on the MD than they do on $\sigma$, because large deviations are not squared in the process of calculating the MD as they are in finding $\sigma$. Therefore, the MD could be a useful measure of variability for purely descriptive purposes. However, because it is not suitable for further statistical analyses, it is rarely used at all. In cases where you want to use the median instead of the mean to describe the central tendency of your distribution, it would make sense to use the SIQR as your measure of variability rather than the standard deviation.

As we mentioned earlier, $\sigma$ has an undesirable property (i.e., bias) that can be corrected by calculating $s$ instead when we wish to estimate the variability in the population from which a sample was drawn. It is the latter measure, the unbiased standard deviation (and its square, the unbiased variance), that is used most frequently as the basis for the more advanced statistical procedures that we begin to describe in Chapter 5.

## Summary

### Central Tendency

One important attribute of a set of scores is its *location*—that is, where in the possible range between minus infinity and plus infinity the scores tend to fall. This can be described in a single number by using either the *mean*, the best measure in most instances, or the *median* (the score corresponding to the 50th percentile), which is preferable when data are highly skewed or there are extreme data whose exact values are unknown, and when the objectives are purely descriptive. When a variable is not associated with scores at all, but rather distinct categories, a cruder measure, the *mode*, is the only measure that can be used to summarize the frequencies of that variable's various values.

#### *1. The Mode*

This is the most frequently obtained score. The mode is at best a rough measure and is usually inappropriate when dealing with quantitative data. However, it is the only measure of central tendency that can be used with qualitative levels of a variable (e.g., different colors, different diagnoses for mental illness).

***2. The Median***
For either grouped or ungrouped data, the median is the score that corresponds to the 50th percentile.

***3. The Population Mean***

$$\mu = \frac{\sum X}{N} \qquad \text{Formula 3.1}$$

**Variability**
A second important attribute of a set of scores is its *variability*, or how much the scores *differ from one another*. In the behavioral sciences, this is customarily summarized in a single number by computing the *variance* or its positive square root, the *standard deviation*. The larger the variance or standard deviation, the more different the numbers are from one another. The concept of variability has many practical applications and is particularly important in statistical work.

***1. The Range***
The simplest (and crudest) measure of variability encountered in the behavioral sciences is the range, which is equal to the highest score minus the lowest score. Although it is sometimes useful to know the maximum separation of your scores, as a measure of variability, the range is rather unreliable and is therefore generally less useful than the other available variability measures.

***2. The Interquartile and Semi-Interquartile (SIQ) Ranges***
The IQR is the distance from the 1st to the 3rd quartile of the distribution, calculated as the 75th percentile minus the 25th percentile. The SIQR is half of that distance (i.e., IQR/2). These measures are useful for descriptive purposes, especially when your distribution contains extreme or indeterminate scores, but are not used for inferential statistics.

***3. The Mean Deviation***
This is the average of the absolute values of all the deviations from the mean. It is less susceptible to extreme scores (called *outliers*) than the standard deviation, and it makes intuitive sense, but it is not used for drawing inferences about populations. The formula for the MD is:

$$\text{MD} = \frac{\sum |X - \mu|}{N}$$

where $|X - \mu|$ = the absolute value of the deviation from the mean

***4. The Biased Variance and Standard Deviation***
Variance (definitional formula):

$$\sigma^2 = \frac{\sum (X - \mu)^2}{N} \qquad \text{Formula 3.2}$$

Variance (computing formula):

$$\sigma^2 = \frac{1}{N}\left[\sum X^2 - \frac{(\sum X)^2}{N}\right] \qquad \text{Formula 3.6A}$$

Standard deviation formula (in terms of variance):

$$\sigma = +\sqrt{\sigma^2}$$

When calculated for samples, $\sigma^2$ and $\sigma$ tend to underestimate their corresponding values in the population. Therefore, these measures are acceptable only for descriptive purposes; they are not used to make estimates of, or draw inferences about, population parameters.

***5. The (Unbiased) Population Variance Estimate and the Unbiased Standard Deviation*** When the data of a sample are to be used to estimate the variance of the population from which the sample was drawn, compute the *population variance estimate.*

Unbiased variance (definitional formula):

$$s^2 = \frac{\sum (X - \overline{X})^2}{N - 1} \qquad \text{Formula 3.3}$$

Unbiased variance (computing formula):

$$s^2 = \frac{1}{N - 1}\left[\sum X^2 - \frac{(\sum X)^2}{N}\right] \qquad \text{Formula 3.7A}$$

Unbiased standard deviation formula:

$$s = +\sqrt{s^2}$$

## Exercises

1. Calculate the friendliness means of Turck, Kirk, Dupre, and Doty Halls (see first exercise of Chapter 1).
2. Compute the median of Turck Hall from the regular frequency distribution you created for that dorm when solving the first exercise of Chapter 2.
3. Find the mode of Kirk Hall by inspecting the regular frequency distribution you created for that dorm when solving the first exercise of Chapter 2.
4. Compute the median of Turck Hall from the *grouped* frequency distribution you created for that dorm when solving the second exercise of Chapter 2.
5. Using the raw data for Dupre Hall, demonstrate that the sum of the deviations from the mean equals zero.
6. Compute the median and mean for Distribution 6 in Table 3.2. Why cannot you find the mode of this distribution?
7. Insert the sums you found for Exercise 3 of Chapter 1 into the *computing* formulas given in this chapter to find $\sigma^2$, $\sigma$, $s^2$, and $s$ for the four dorms. (We highly recommend that you also learn to obtain these results directly with a handheld calculator.)
8. Recalculate $\sigma^2$, $\sigma$, $s^2$, and $s$ for Dupre Hall using the *definitional* formulas, and compare them to the corresponding results from the previous exercise. Which type of formula (computing or definitional) usually leads to greater error due to rounding off before the final result?
9. The mean of an exam is 57.3 and the standard deviation is 9.6. What will happen to the standard deviation of these scores if the instructor (a) adds 5 points to each score? (b) subtracts 4 points from each score? (c) multiplies each score by 2? (d) divides each score by 3?
10. Compute the range for each of the four dorms.
11. Using the results you found for Exercise 9 in Chapter 2, compute the IQR and SIQR for Turck Hall. Using the same methods, calculate these measures for Kirk Hall, as well.
12. Using the definitional formula, calculate the (biased) "variance" of Doty Hall in terms of deviations from the median, rather than the mean, of that distribution

(just put the five scores in numerical order, and use the middle one as the median). (a) How does this "variance" compare to the $\sigma^2$ that you calculated in Exercise 7? (b) What property of the variance from the mean is illustrated by this exercise?

13. Calculate the mean deviation for Doty Hall, and compare it to $\sigma$ (as calculated in Exercise 7). What do you think would happen to the difference between these two measures if you were to add an extreme score to the data of Doty?
14. Compute the biased and unbiased standard deviations for Distribution 6 in Table 3.2.

## Thought Questions

1. For each of the following problems, which measure of central tendency should be computed? (a) An experimental psychologist wants a measure of central tendency for the number of trials taken to learn a task by 45 eight-year-old children. The maximum number of trials allowed is 30. Most of the children took between 14 trials and 30 trials to learn the task, but five children did not learn the task at all even after 30 trials. (b) A student who has taken 25 college courses, each of which is worth 3 credits, wants to compute her GPA. (c) A student who has taken 25 college courses, some of which are worth 3 credits and some 4 credits, wants to compute his GPA. (d) A researcher wants a measure of central tendency for the income of a sample of 100 people that includes Bill Gates and Tiger Woods. (e) The editor of a college student yearbook wants to decide the color of the cover and asks the 30 students working on the yearbook to vote for orange, blue, green, or black.
2. (a) In a *positively* skewed distribution, which is larger, the mean or the median? (b) In a *negatively* skewed distribution, which is larger, the mean or the median? (c) In a *symmetric* distribution (such as the normal distribution), which is larger, the mean or the median?
3. Consider the following set of data, which represents 15 scores on a 10-point quiz: 0, 1, 3, 3, 4, 4, 4, 5, 5, 5, 7, 8, 9, 9, 10. (a) If the score of 10 is changed to 225,000,000, but the other numbers remain the same, what is the general effect on the mean and the median? (b) If the score of 10 is changed to 0, what is the general effect on the mean and the median? Why does the median change in this example, but not in the preceding example?
4. Why is the range much less informative (and much less useful) as a measure of variability than the standard deviation? Give an example of when the range could really be useful.
5. The mean of a statistics examination is 77.0. (a) If your score is 85, would you prefer the standard deviation to be large or small for purposes of getting a better grade? Why? (b) If your score is 69, would you prefer the standard deviation to be large or small? Why?
6. (a) When computing the standard deviation of a set of scores, when do we divide by $N$, and when do we divide by $N - 1$? (b) What is the difference between $\sigma$ and $s$?
7. If the standard deviation of a set of scores is zero, what does this imply about the scores? Can a standard deviation be negative? Why or why not?
8. (a) When is it desirable to use the SIQR as the measure of variability? (b) When is it desirable to use the mean deviation (MD) as the measure of variability?

## Computer Exercises

1. Use your statistical package to find the mode, median, and mean for each of the quantitative variables in Jackson's data set.
2. Use your statistical package to find the mode for each of the categorical (i.e., qualitative variables) in Jackson's data set.

3. Use your statistical package to find the range, SIQR, unbiased variance, and unbiased standard deviation for each of the quantitative variables in Jackson's data set. (Does your statistical package offer you the option of obtaining the biased versions of the variance and standard deviation?)
4. Use your statistical package to create a new variable that is equal to 2 times the academic interest score, so that the new variable is measured on a 0–100 scale. How do the measures of variability found in the previous exercise compare between the new variable and the original academic interest score? What general principle is being illustrated with respect to the standard deviation?
5. Use your statistical package to create a new variable that adds 50 points to the academic interest score. How have the measures of variability changed for this new variable as compared to the same measures calculated for the original academic interest score? What general principle is being illustrated with respect to the standard deviation?

## Bridge to SPSS

The three measures of central tendency discussed in this chapter can be obtained from SPSS by opening the *Frequencies: Statistics* box described in the previous chapter (click on **Analyze**, **Descriptive Statistics**, and **Frequencies** ..., and then click on the **Statistics** ... button). The upper-right quadrant of this box (see Figure 3.2) is labeled *Central Tendency* and allows you to select any number of the following choices: Mean, Median, Mode, and Sum (we do not know why SPSS includes the sum of the scores under the heading of Central Tendency). The mean of a distribution can actually be obtained from quite a few of the choices under **Analyze**; we point out a number of these options in later chapters.

There are three choices on the **Analyze/Descriptive Statistics** menu that will give you measures of variability (along with measures of central tendency). The one that is labeled **Descriptives** ... is intended for use with continuous, quantitative data only. For measures of variability (SPSS refers to variability as *dispersion*), its **Options** ... box offers only the standard deviation, variance, and range. Note that with respect to the first two measures just mentioned, only the "unbiased" versions are given. In response to how infrequently the biased versions are used in the social sciences, it appears that SPSS does not offer the biased variance or biased standard deviation from any of its menus. (It could be computed, however, by multiplying the unbiased value by $(N - 1)/N$).

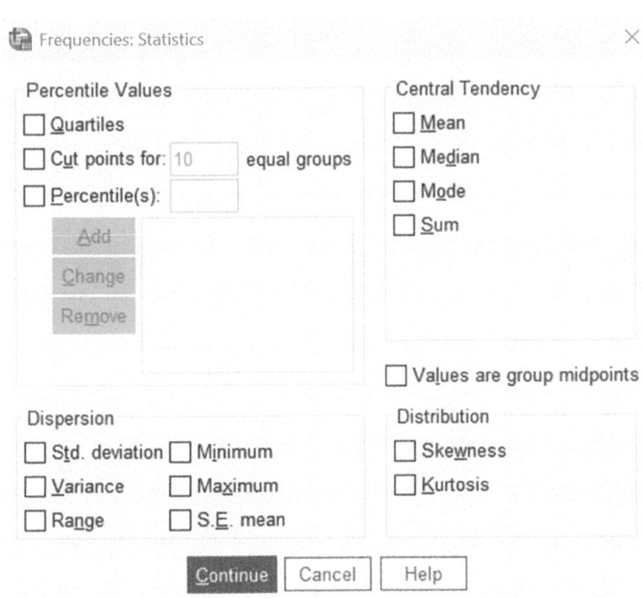


Figure 3.2

**The Frequencies: Statistics Dialog Box**

*Source:* Courtesy of International Business Machines Corporation, © SPSS, Inc. / IBM company

## Figure 3.3

**The Explore Dialog Box**

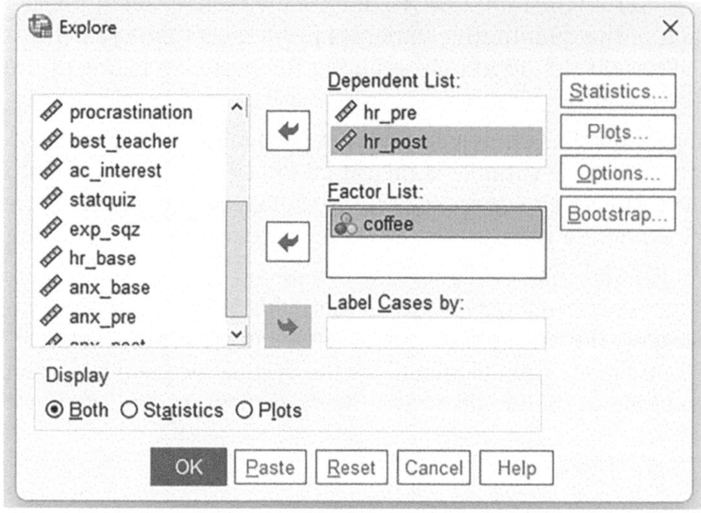


*Source:* Courtesy of International Business Machines Corporation, © SPSS, Inc. / IBM company

The **Frequencies** dialog box is intended to accommodate a broader range of data types, including ordinal data and open-ended distributions. After clicking on the **Statistics ...** button, you can select *Quartiles* (see upper-left quadrant of Figure 3.2), and then calculate the interquartile and SIQ ranges from those results.

The third relevant choice on the **Analyze/Descriptive Statistics** menu is **Explore** ...; this selection is intended for the thorough exploration of distributions that may be problematic. If you select the **Statistics ...** button in the Display section of the **Explore** dialog box (see Figure 3.3) and then click on **Descriptives** in the *Explore: Statistics* box, you will get the standard deviation, variance, range, and IQR (plus a few measures that are too advanced to be mentioned in this text). If you want the SIQ range, just divide the IQR by 2.

# Chapter 4

# Standardized Scores and the Normal Distribution

## PREVIEW

### Interpreting a Raw Score Revisited

What is the advantage of using a transformed score, such as a percentile rank?

When comparing two (or more) scores from different distributions, why is it useful to refer each score to its mean and standard deviation?

How are we likely to be misled if we look only at the raw scores?

### Rules for Changing $\mu$ and $\sigma$

What will happen to the mean and standard deviation of a set of scores if we do the following: Add or subtract a constant to every score? Multiply or divide every score by a constant?

How can these rules be used to obtain transformed scores with any desired mean and standard deviation?

### Standard Scores ($z$-Scores)

What are the mean and standard deviation of a set of $z$-scores? Why are these particular values desirable?

How are $z$-scores computed?

When raw scores are transformed into $z$-scores, what happens to the shape of the distribution?

### *T* Scores, SAT Scores, and IQ Scores

Why might we prefer to use $T$ scores or SAT scores instead of $z$-scores?

How are $T$ scores and SAT scores computed?

What is the mean and standard deviation of a set of $T$ scores or a set of SAT scores?

What is the mean and standard deviation of a set of IQ scores, and how are these scores computed?

### The Normal Distribution

What is meant by saying that the normal distribution is a theoretical distribution?

How can the normal curve model be useful when dealing with real data?

What is the shape of the normal curve? Is it unimodal? Is it symmetric?

What is the relationship between the proportion of area within a section of a population distribution, and the probability of randomly selecting a score from that section?

What are the two parameters of the normal distribution?

### Table of the Standard Normal Distribution

How is the normal curve table used?

How do we find the percent of the area under the normal curve that falls between the mean and any particular $z$-score?

*Introductory Statistics for the Behavioral Sciences*, Eighth Edition. R. Brooke Lea and Barry H. Cohen.

Companion website: www.wiley.com/go/ISBS8e

**PREVIEW (*continued*)**

Where in the normal curve are values more likely to occur?

What percent of the area under the normal curve falls between the mean and one standard deviation unit? Between the mean and two standard deviation units?

### Illustrative Examples

How do we calculate the proportion of the population that can be expected to have scores between two specified values, or above or below one specified value?

How do we calculate the score that is required to be within a specified proportion of the population (e.g., the top 10% or the middle 60%)?

### Summary

### Exercises

### Thought Questions

### Computer Exercises

### Bridge to SPSS

## Interpreting a Raw Score Revisited

In Chapter 2, we saw that it can be helpful to transform a raw score into a percentile rank (PR). The PR shows at a glance how the score stands in comparison to a specific reference group.

In Chapter 3, we identified two important characteristics of a population of scores: its location, usually summarized by the mean ($\mu$), and its variability, for which the standard deviation ($\sigma$ or SD) is most often used. It is possible to deduce how well a given score compares to the group of which it is a part by using the mean and SD of that group. As was the case with percentiles, we can build this information into the score itself. Specifically, we can derive a transformed (standardized) score that shows at a glance where the original raw score lies with respect to the mean, using the SD of the reference group as the unit of measurement.

To illustrate, let us suppose that a college student takes three midterm examinations in three different courses and obtains the following raw scores:

| | English | Mathematics | Psychology |
|---|---|---|---|
| $X$ | 80 | 65 | 75 |
| ___ | ___ | ___ | ___ |
| ___ | ___ | ___ | ___ |

On the surface, it might seem as though the student's best score is in English and his lowest score is in mathematics. But if one defines "best" in the context of other scores in the class, it would be unwise to jump to such a conclusion, however, because there are several reasons why the raw scores may not be directly comparable. For example, the English examination may have been relatively easy and resulted in many high scores, while the mathematics examination may have been extremely difficult. Or the English examination may have been based on a total of 100 points and the mathematics examination on a total of only 70 points. The raw scores provide information about the absolute number of points earned, but they give no clear indication as to how good the performance is and certainly no indication of how good the performance is compared to others.

Suppose that we find out that the mean and standard deviation of each test are as follows:

| | English | Mathematics | Psychology |
|---|---|---|---|
| $X$ | 80 | 65 | 75 |
| $\mu$ | 85 | 55 | 60 |
| $\Sigma$ | 10 | 5 | 15 |

This additional information changes the picture considerably. Looking at the means, we can see that the class scores on the English examination were high, so much so that the score of 80 is below average. However, our student's midterm scores in both mathematics and psychology are above average. Therefore, the student's score in English was actually their lowest midterm score in the context of other students' scores.

The unwary observer might now conclude that the student's best score is in psychology because that score is 15 points above average while the mathematics score is only 10 points above average. But as we pointed out in Chapter 3, the variability of a distribution of scores also influences the relative standing of a given score. The standard deviation (SD) indicates that a "typical" deviation on the psychology test was 15 points from the mean; so the student's psychology score of 75, which is 15 points (or 1 SD) above average, was certainly exceeded by a number of higher scores[1] (though it surpassed quite a few, as well). The average variability on the mathematics examination, however, was only 5 points from the mean. Hence, the student's mathematics score of 65 is 10 points (or 2 SDs) above average. That makes the math score unusually far above the mean, and it is therefore likely to be one of the best scores in that class.[2]

The picture presented by the raw scores was quite misleading in this instance. It turns out that the student's best score is in mathematics, the next best score is in psychology, and the lowest score is in English, relative, that is, to the other students in each course.

The raw scores of 80, 65, and 75 cannot be compared directly to each other because they come from distributions with different means and different SDs (and maybe even different shapes). That is, the units in which the raw scores are measured are not comparable from test to test. This difficulty can be overcome by transforming the scores on each test to a common scale with a specified mean and SD. This new scale will then serve as a "common denominator," enabling us to compare directly the transformed scores of different tests.

Two questions remain: How do we change the scores so that they will have the desired common mean and SD? And what values of the mean and SD are useful choices for the "common denominator"?

## Rules for Changing $\mu$ and $\sigma$

Suppose that the mathematics instructor in the previous example suffers a pang of conscience about the low mean and decides to add 5 points to everyone's score. She has added a *constant* amount to each score (one that is exactly the same for all the scores), so she does not have to recompute a new mean via the usual formula; we already showed in Chapter 3 that adding 5 points to everyone's score increases the mean by 5 points. Thus, the mean of the transformed scores (symbolized by $\mu_{new}$) will be 60.

We also know from the properties of the mean described in Chapter 3 that if the English instructor subtracts 7.5 points from every score, the new mean will be 77.5. If he multiplies every score by 4, the new mean will be 340. And if she divides every score by 2, the new

[1] The exact number depends on the shape of the distribution of scores. For example, if the scores are normally distributed, approximately 16% of the scores are more than 1 standard deviation above the mean (as is shown in Chapter 8).

[2] If the distribution is normal, less than 2½% of the scores are more than 2 SDs above the mean.

mean will be 42.5. Recall that these rules for conveniently determining the new mean work only if every original score is altered by exactly the same amount.

Insofar as the variability of the distribution is concerned, we saw in Chapter 3 that adding a constant to every score or subtracting a constant from every score does *not* change the SD or variance. However, we also showed the following:

If every score is *multiplied* by a positive constant $k$,

$$\sigma_{new} = k\sigma_{old}$$

and if every score is *divided* by a positive constant $k$,

$$\sigma_{new} = \sigma_{old}/k$$

Therefore, if the English instructor adds 10 points or subtracts 6 points from every score, the SD remains 10 (and the variance remains $10^2$ or 100), but if she multiplies every score by 4, the new SD equals $4 \times 10$ or 40 (and the new variance equals $4^2 \times 100$ or 1,600). If she divides every score by 2, the new SD equals 10/2 or 5 (and the new variance equals $100/2^2$ or 25).

These rules make it possible to obtain transformed scores with any desired mean and SD. For example, scores on the English test can be transformed into scores comparable to those on the mathematics test in two steps:

| Procedure | New $\mu$ | New $\sigma$ |
|---|---|---|
| 1. Divide every score on the English examination by 2.0. | $85/2 = 42.5$ | $10/2 = 5$ |
| 2. Add 12.5 to each of the scores obtained in Step 1. | $42.5 + 12.5 = 55$ | 5 (no change) |

The first step is to change the SD to the desired value by multiplying or dividing each score by the appropriate constant, which affects the value of both $\mu$ and $\sigma$. Then the desired mean is obtained by adding or subtracting the appropriate constant, which does not cause any further change in $\sigma$. When the student's English score of 80 is subjected to these transformations, it becomes $(80/2) + 12.5$ or 52.5, and it is evident that this score is not nearly as good as the student's mathematics score. Note that both the original and the transformed English scores are half an SD below the means of their respective distributions.

## Standard Scores (*z*-Scores)

The techniques discussed in the preceding section make it possible to switch to any new mean and SD, which is extremely useful when comparing scores from different distributions. Therefore, the next logical step is to choose values of $\mu_{new}$ and $\sigma_{new}$ that facilitate comparisons among the scores. One useful procedure is to convert the original scores to new scores with a mean of 0 and an SD of 1; these new scores are called ***z*-scores** or *standard scores*.

Standard scores have two major advantages. Since the mean is zero, you can tell at a glance whether a given score is above or below average; an above-average score is positive and a below-average score is negative. Also, because the SD is 1, the numerical size of a standard score indicates *how many SDs* above or below average the score is. We saw at the beginning of this chapter that this information offers a valuable clue as to how good the score is; a score 1 SD above average (i.e., a standard score of 1) would demarcate approximately the top 16% in a normal distribution (ND), while a score 2 SDs above average (a standard score of 2) would demarcate approximately the top $2\frac{1}{2}$% in an ND.

To convert a set of scores to standard scores, the first step is to subtract the original mean from every score. According to the rules given in the previous section, the new mean is equal to:

$$\mu_{new} = \mu_{old} - \mu_{old} = 0$$

while the SD is unchanged. Next, each score obtained from the first step is divided by the original SD. As a result:

$$\mu_{new} = \frac{0}{\sigma_{old}} = 0$$

$$\sigma_{new} = \frac{\sigma_{old}}{\sigma_{old}} = 1$$

That is, the mean remains zero, while the SD becomes 1. Summarizing these steps in a single formula yields:

$$z = \frac{X - \mu}{\sigma} \quad \text{Formula 4.1}$$

where $z$ is the symbol for a standard score.

Converting each of the original examination scores given at the beginning of this chapter to $z$-scores yields the following:

| | English | Mathematics | Psychology |
|---|---|---|---|
| $X$ | 80 | 65 | 75 |
| $\mu$ | 85 | 55 | 60 |
| $\sigma$ | 10 | 5 | 15 |
| $z$ | $\frac{80-85}{10} = -0.50$ | $\frac{65-55}{5} = +2.00$ | $\frac{75-60}{1.5} = +1.00$ |

The standard scores show at a glance that the student was half an SD below the mean in English, two SDs above the mean in mathematics, and one SD above the mean in psychology.

It is important to remember that when raw scores are transformed into $z$-scores, *the shape of the distribution remains exactly the same*. We are just measuring all the scores with reference to a new point (the mean instead of zero) and with a new unit size (the SD instead of the original units). To illustrate this point, the baseline heart rates (HRs) for the 20 low-procrastinators that do not read for enjoyment in their leisure from Jackson's statistics class are presented in Table 4.1 along with the corresponding $z$-scores; we will be treating these 20 scores as a population. The scores have been arranged in decreasing order for clarity.

The mean of the baseline HR scores is 67.80 beats per minute (bpm) and the SD is 1.78 bpm; the $z$-scores were obtained using Formula 4.1. For example, the $z$-score corresponding to 72 bpm is equal to

$$z = \frac{72 - 67.80}{1.78} = +2.36$$

In accordance with the previous discussion, the mean of the $z$-scores is 0.00 and the SD is 1.00. The raw score and $z$-score distributions are plotted in Figure 4.1; note that a single graph suffices because the shape of the distribution is the same for both sets of scores. Thus, the relationship of the scores to one another is *not* changed by transforming them to standard scores. The only things changing are the location (i.e., central tendency) and the scaling (i.e., spread) of the distribution.

Although the HR scores are measured in bpm, identical $z$-scores would be obtained (within rounding-off error) for each individual if HR were measured in beats per *second*, as is shown in the rightmost column of Table 4.1. For example,

| **HR in beats per minute** | ***z* score** | **HR in beats per second** |
|---|---|---|
| $\frac{70 - 67.80}{1.78} =$ | $+1.24$ | $= \frac{1.167 - 1.13}{0.0296}$ |

**Table 4.1** **Distribution of 20 Baseline Heart Rate Scores Expressed in Beats per Minute (bpm), *z*-Scores, and Beats per Second (bps)**

| Low-Procrastinating Non-Reader | Heart rate (bpm) | Heart rate *z*-score | Heart rate (bps) |
|---|---|---|---|
| 1 | 72 | +2.36 | 1.200 |
| 2 | 70 | +1.24 | 1.167 |
| 3 | 70 | +1.24 | 1.167 |
| 4 | 70 | +1.24 | 1.167 |
| 5 | 69 | +0.67 | 1.150 |
| 6 | 69 | +0.67 | 1.150 |
| 7 | 68 | +0.11 | 1.133 |
| 8 | 68 | +0.11 | 1.133 |
| 9 | 68 | +0.11 | 1.133 |
| 10 | 68 | +0.11 | 1.133 |
| 11 | 67 | −0.45 | 1.117 |
| 12 | 67 | −0.45 | 1.117 |
| 13 | 67 | −0.45 | 1.117 |
| 14 | 67 | −0.45 | 1.117 |
| 15 | 67 | −0.45 | 1.117 |
| 16 | 67 | −0.45 | 1.117 |
| 17 | 66 | −1.01 | 1.100 |
| 18 | 66 | −1.01 | 1.100 |
| 19 | 66 | −1.01 | 1.100 |
| 20 | 64 | −2.14 | 1.067 |
| μ | 67.80 | 0.00 | 1.13 |
| Σ | 1.78 | 1.00 | 0.0296 |

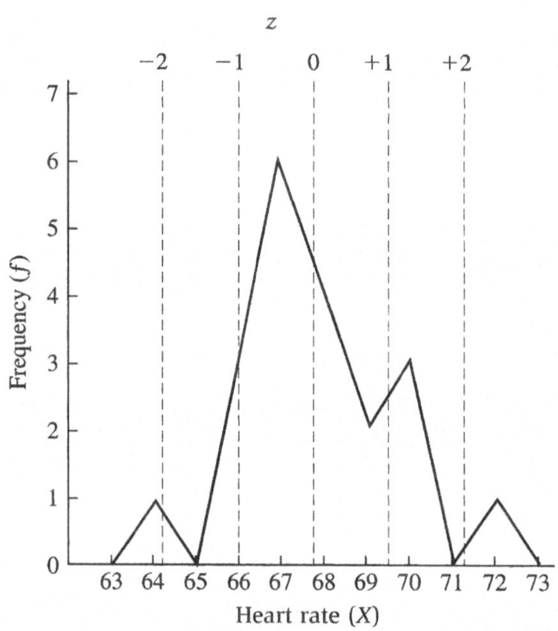


**Figure 4.1**

**Frequency Distribution of Raw HR Scores and Their *z*-Scores from Table 4.1**

This indicates that *z*-scores give an accurate picture of the standing of each score relative to the reference group even after the original scores have been divided by a constant (60, in this case). In fact, the properties of the mean and SD are such that the original scores can be subjected to any combination of addition, subtraction, multiplication, or division by constants without affecting the corresponding *z*-scores.

Standard scores are used extensively in the behavioral sciences. They also play an important role in statistical inference, as we will see in subsequent chapters.

## *T* Scores, SAT Scores, and IQ Scores

Standard scores have their disadvantages. For example, they can be difficult to explain to someone who is not well-versed in statistics. A college professor once decided to report the results of an examination as *z*-scores. He was quickly besieged by anxious students who did not understand that a *z*-score of 0 represents average performance (and not zero correct!), not to mention the agitated ones who received negative scores and wondered how they could ever make up the negative points.

### *T* Scores

One task of behavioral scientists is to report test scores to people who are not statistically sophisticated. So, several alternatives to *z*-scores have been developed. The mean and SD of each alternative described next have been chosen so that all of the scores will be positive and so that the mean and SD will be easy to remember. One such alternative, called *T scores*, is defined as a set of scores with a mean of 50 and an SD of 10. The *T* scores are obtained from *z*-scores using the following formula:

$$T = 10z + 50 \qquad \text{Formula 4.2}$$

Each raw score is first converted to a *z*-score, then each *z*-score is multiplied by 10, and 50 is added to each resulting score. For example, an HR of 69 in Table 4.1 is converted to a *z*-score of 0.67 by the usual formula. Then *T* is equal to (10)(0.67) + 50 or 56.70. It is easy to prove that Formula 4.2 does produce the desired mean and SD:

| Procedure | New $\mu$ | New $\sigma$ |
|---|---|---|
| 1. Convert raw scores to *z*-scores. | 0 | 1 |
| 2. Multiply each *z*-score by 10. | $10 \times 0 = 0$ | $10 \times 1 = 10$ |
| 3. Add 50 to each score obtained in Step 2. | $0 + 50 = 50$ | 10 (no change) |

Common physical measurements (including HRs) are rarely transformed to *T* scores, because they are already expressed in units that are familiar. However, a *T* score would still provide additional information in locating a particular individual within a distribution. (You may know that a man who is 6 feet tall is taller than average, but a *T* score of 60 for height would tell you immediately that he is 1 SD above the mean.) But *T* scores are even more helpful when dealing with psychological tests that are measured in arbitrary units. For example, scores on the anxiety scale that Jackson used can range from 10 to 50. Whereas it may be clear that 10 represents very little or no anxiety and 50 a great deal of anxiety, it is certainly not obvious just how to interpret a score of, say, 20 or 30. However, once you become familiar with *T* scores, you can tell at a glance that a student whose baseline anxiety *T* score is 45 is below average in anxiety and, more specifically, that they are exactly one-half of an SD below average. (A *z*-score of −0.5 yields the same information but requires dealing with a minus sign and a decimal point.) Although a negative *T* score is mathematically possible, it is very unlikely. It would require that a person be more than 5 SDs below average, and scores more than 3 SDs above or below the mean rarely occur with real data.

### SAT Scores

Scores on some nationally administered examinations, such as the Scholastic Aptitude Test (SAT), the College Entrance Examination Boards, and the Graduate Record Examination, are transformed to a scale with a mean of 500 and an SD of 100. These scores, which we will refer to as SAT scores (based on their best-known application), are obtained as follows:

$$\text{SAT} = 100z + 500 \qquad \text{Formula 4.3}$$

As in the case of *T* scores, the raw scores are first converted to *z*-scores; each *z*-score is then multiplied by 100, and 500 is added to each resulting score. A comparison with the

*T* score formula should make it obvious that a SAT score is just 10 times a *T* score. This explains the apparent mystery of how you can obtain a score of 642 on a test with only 30 or 40 items. And you may well be pleased if you obtain a score of 642 because it is 142 points or 1.42 SDs above the mean (and therefore corresponds to a *z*-score of 1.42 and a *T* score of 64.2).

### IQ Scores

Another common score that is based on transforming *z*-scores to a more convenient form is the *IQ score*. The well-known Stanford–Binet (S–B) test creates its intelligence quotient (i.e., IQ) scores by transforming its *z*-scores according to the following formula:

$$\text{S-B IQ} = 16z + 100 \qquad \text{Formula 4.4A}$$

It is because of this formula, rather than some convenient coincidence, that the average IQ of the population happens to be about 100. You can also see why IQ scores above 148 or below 52 are very unusual: They differ from the mean by 3 SDs. The newer Wechsler Adult Intelligence Scale (WAIS) uses a slightly different formula to create IQ scores. The WAIS is also based on a mean of 100, but unlike the Stanford–Binet test, it uses an SD of 15 instead of 16:

$$\text{WAIS IQ} = 15z + 100 \qquad \text{Formula 4.4B}$$

## The Normal Distribution

So far, we have shown that a standard score can be more informative than a raw score—for example, when trying to interpret an individual's performance on a difficult exam for which the average could be as low as 40 points out of 100. A *z*-score of 1.0 has to be relatively good, no matter what the mean is. However, precisely how good that *z*-score is depends on the *shape* of the distribution you are dealing with.

Suppose that the scores on an exam have an extremely positive skew. The bulk of the scores are a bit below 40, but some scores are over 80, and there are even a few in the 90s. In this case, a *z*-score of 1.0 is not nearly as impressive as it would be in a very negatively skewed distribution (where there are extreme scores in the lower end of the distribution) because there will be some scores whose *z*-scores are quite a bit larger than 1.0. Conversely, suppose that the scores are clustered slightly above 40 with a "tail" that narrows down as you get near a score of zero, but there are no really high scores. Here, a *z*-score of 1.0 is likely to be one of the highest scores in the class and therefore much more impressive than in the preceding case. Two *z*-scores, each from a different distribution, can be more easily compared if we know that the two distributions have similar shapes.

Whereas *z*-scores are almost always more informative than raw scores, they become especially informative when you are dealing with a well-known distribution called the **normal distribution** (less formally referred to as the *normal curve*). Any distribution of real data will never match the ND exactly. The ND is a *theoretical* distribution—a mathematical abstraction like a perfect circle. The normal curve is unimodal and symmetric (as was illustrated in Chapter 2) and is most often described as a bell-shaped curve because it is high in the middle and low at both ends or *tails*. However, the ND is not just any bell-shaped curve; its shape is defined by a particular mathematical equation.[3] As a theoretical distribution, the ND contains an infinite number of scores. Its tails on either side keep falling closer to the *X*-axis but do not actually touch it until you go an infinite distance in either direction. (Mathematicians say that the curve approaches the *X*-axis asymptotically.)

---

[3] The formula for the normal curve is fairly complicated:

$$f(x) = \frac{1}{\sigma\sqrt{2\pi}} e^{-(X-\mu)^2/2\sigma^2}$$

where $\pi = 3.1416$ and $e = 2.7183$, both approximately.

Despite its abstract definition, the ND is a distribution that tends to arise, at least approximately, rather frequently in nature. For example, if you look at the heights of all people in a large population, their distribution will look rather similar to the ND. However, you can tell immediately that your empirical height distribution cannot match the ND exactly. Your distribution will end abruptly on either side when you get to the shortest and tallest people in the population. Nonetheless, for many purposes, the ND will be an acceptable approximation for the variable you are dealing with, and it is much easier to work with mathematically than any finite distribution of real data. SAT scores, for which a large number of data points are available, are used later in this chapter as our example of a population that has a distribution similar to the ND.

The statistical procedures described in this chapter may seem relevant only when measuring dependent variables that have approximately normal distributions in a population. However, we show in the next chapter that the ND has a much broader use as a statistical model when used with groups of scores rather than individuals, even when the variable being measured has a very skewed distribution in the population (e.g., annual income).

As with any frequency distribution, the horizontal axis of the normal distribution in Figure 4.2 represents score values and the vertical axis is related to the relative frequencies of the scores. Also, all values in Figure 4.2 are located *under* the curve. So, the *total population* is expressed in the graph as the *total area under the normal curve*, and we can arbitrarily define this area as 100% without bothering about the exact number of cases in the population. Alternatively, we may move the decimal point two places to the left and define the total area under the normal curve as 1.00. This definition is useful when we wish to answer questions that deal with probability since the sum of all the specific probability values in any distribution is 1.00.

Now suppose that you are asked to find the proportion of the population with values between points $X_1$ and $X_2$ in Figure 4.2. Graphically, the answer is given by the percent *area between* these two points. This is equivalent to the percent frequency of all these values and/or to the probability of occurrence of these values.

If you have taken an elementary calculus course, you know that the numerical value of the area between any two $X$ values may be found by integrating the normal curve equation. However, you also know how tedious it would be to carry out an integration every time you wished to use the normal curve model. Fortunately, all necessary integrations have been performed and summarized in a convenient table, which shows the percent frequency of values within any distance (in terms of SDs) from the population mean. We explain the use of the normal curve table next.

### Parameters of the Normal Distribution

Normal distributions can differ from each other in just two ways: their locations (represented by their central point, $\mu$) and their variability or spreads (as measured by $\sigma$). Just as any rectangle can be specified by giving both its length and width, any ND can be completely

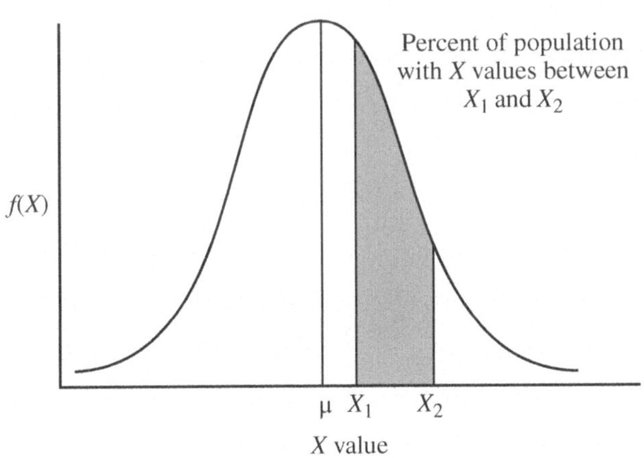


**Figure 4.2**

**The Normal Curve: A Theoretical Distribution**

determined by giving its $\mu$ and $\sigma$. These two quantities are referred to as the *parameters* of the ND. (This usage is consistent with the meaning of a parameter as a population value as defined in Chapter 1.) Note that these two parameters appear in the equation for the ND (see footnote on page 68) and must be specified before that equation can be solved for any value of $X$. Because Parts II and III of this text deal with statistical procedures whose purpose is to estimate or make inferences about the parameters of some ND of interest, procedures in those chapters are referred to collectively as *parametric statistics*.

## Table of the Standard Normal Distribution

Although integral calculus could be used *in theory* to create a table of areas for every possible normal distribution, *in practice*, this would require an infinite number of tables—one table for each possible combination of $\mu$ and $\sigma$. Instead, a convenient table has been created for just one special ND, the one whose parameters are $\mu = 0$ and $\sigma = 1$. This ND is called the **standard normal distribution**, and it arises whenever you transform all the scores in any ND into $z$-scores. It is important to recall that transforming raw scores into $z$-scores does *not* change the shape of a distribution at all. Only if a population of raw scores follows an ND will its $z$-scores also form a normal distribution—specifically, the standard ND.

In order to use the table for the standard ND (presented as Table A in the Appendix) to solve problems about *any* ND, it is necessary to convert the scores you are interested in to $z$-scores. Fortunately, as you have seen in the previous chapter, this is easy to do. To illustrate the way areas can be looked up in terms of $z$-scores in Table A, a portion of that table is reproduced next.

| Percents calculated from area under the normal curve | | | | | | |
|---|---|---|---|---|---|---|
| $z$ | 0.00 | 0.01 | 0.02 | 0.03 | ... | 0.09 |
| 0.0 | | | | | | |
| ⋮ | | | | | | |
| 1.0 | 34.13 | 34.38 | 34.61 | 34.85 | | |
| 1.1 | 36.43 | 36.65 | 36.86 | 37.08 | | |
| 1.2 | 38.49 | 38.69 | 38.88 | 39.07 | | |
| 1.3 | 40.32 | 40.49 | 40.66 | 40.82 | | |
| ⋮ | | | | | | |
| $\infty$ | | | | | | |

The column on the left represents $z$ values expressed to one decimal place, while the top row gives the second decimal place. The values within the table represent the *percent area between the mean and the z value*, expressed to two decimal places. For example, to find the percent of cases between the mean and 1.03 SDs from the mean, select the 1.0 row and go across to the 0.03 column; the answer is read as 34.85%.

The table gives percents on one side of the mean only so that the maximum percentage given in the table is 50.00 (representing half of the area). However, the normal curve is *symmetric*, which implies that the height of the curve at any positive $z$ value (such as 1) is exactly the same as the height at the corresponding negative value ($-1$). Therefore, the percent between the mean and a $z$ of 1 is exactly the same as the percent between the mean and a $z$ of $-1$, and presentation of second half of the table to deal with negative scores is unnecessary. It is up to you to remember whether you are working in the half of the curve above the mean (positive $z$) or the half of the curve below the mean (negative $z$). Also, do not expect the value obtained from the table to be the answer to your statistical problem in every instance. The table only provides values between the mean and a given $z$-score, so some additional calculations will be necessary whenever you are interested in the area between two $z$-scores (neither of which falls at the mean) or the area between one end of the curve and a $z$-score. These calculations are illustrated in the following sections.

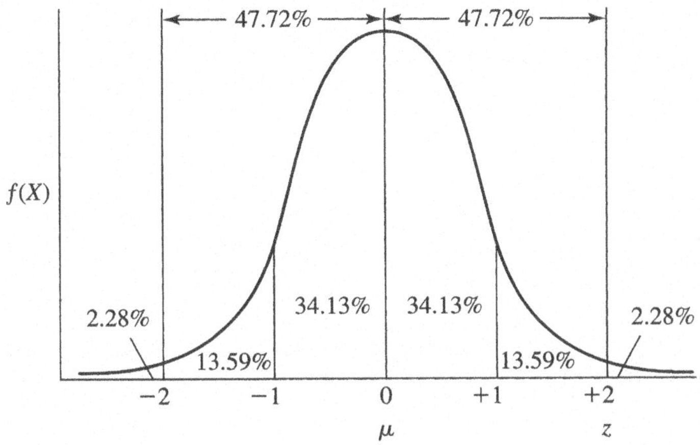


## Figure 4.3

**Normal Curve: Percent Areas from the Mean to Specified *z* Distances**

### Characteristics of the Normal Curve

In Figure 4.3, the normal curve is presented with the *X*-axis marked off to illustrate specified *z* distances from the mean. The percent areas between the mean and those *z* values have been determined from Table A. For example, notice that the percent of the total area between μ (the population mean where $z = 0$) and a *z* of +1.00 is 34.13%, and the area between μ and $z = -1.00$ is also 34.13%. Thus, 68.26% of the total area under the normal curve lies between $z = -1.00$ and $z = +1.00$.

Looking at the shape of the normal distribution, you should expect *z* values between 0 and +1 to occur more frequently than *z* values between +1 and +2. Even though the score distance between the mean and $z = +1$ is the same as the score distance between $z = +1$ and $z = +2$, the *area* between the mean and $z = +1$ is much greater than the area between $z = +1$ and $z = +2$. As you can see from Figure 4.2, the area between a *z* of +1 and a *z* of +2 is only about 13.6%.

Adding up the areas in Figure 4.3, you can see that about 95% of the normal curve lies within approximately two SD units of μ (more exactly, between $z = -1.96$ and $z = +1.96$). For example, if the heights of US adults were assumed to be normally distributed with the mean at 5 foot 8 inches and the SD equal to 3 inches, you would expect close to 95% of these adults to have heights within 5 feet 8 inches ±6 inches, or between 5 feet 2 inches and 6 feet 2 inches. Table A can be used to answer more complex questions about areas of the normal curve, as we see next.

## Illustrative Examples

The exam known familiarly as the SAT is administered each year to many thousands of high school seniors, and the obtained scores are known to have an approximately normal distribution. Therefore, it is reasonable to use the normal curve table (Table A) in the Appendix to answer questions about SAT scores.

However, because scores are given separately for different parts of the SAT, it is simpler for our examples to deal with the score from just one part of the SAT. We chose to express our examples in terms of the math SAT only. Because we know both the μ (500) and σ (100) for the math SAT, we can convert any score on that test to a *z*-score and that can be found into Table A in the Appendix, using Formula 4.1:

$$z = \frac{X - \mu}{\sigma} = \frac{X - 500}{100}$$

In situations where μ and σ are not known, they can be estimated from the sample $\overline{X}$ and *s* provided that the sample size is large; we will demonstrate this in the next chapter. There are many problems that can be solved by the use of this model. The steps required to solve these different problems can be illustrated in the following diagram:

Figure 4.4

**How to go from raw scores to probabilities or the other way**

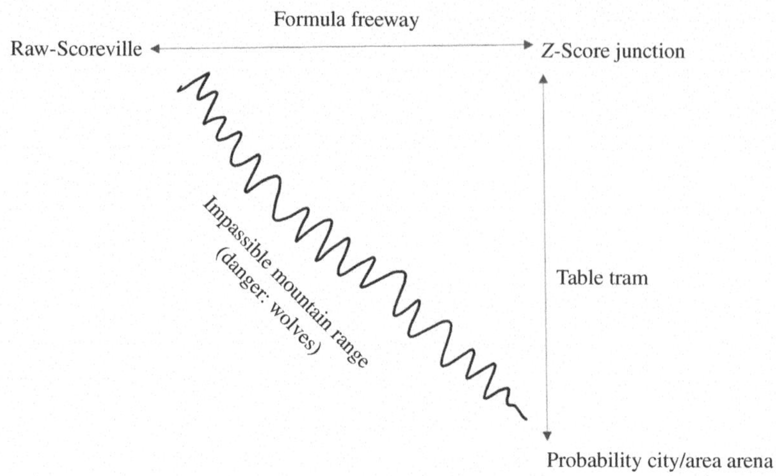


According to Figure 4.4, we have three possible starting locations: Raw-Scoreville, *Z*-Score Junction, and Probability City (sometimes referred to as Area Arena). These locations can also be destinations; you might start at Raw-Scoreville (i.e., with a raw score) and want to get to Probability City (i.e., want to know probabilities associated with that raw score). Note that the direct route between these two locations requires navigating an impassable mountain range, populated with dangerous wolves. The best route, therefore, is to take public transportation—the Formula Freeway to *Z*-Score Junction, where you will pick up the Table Tram that takes you to Probability City. As you may have guessed, the Formula Freeway is the *z*-score formula, and the Table Tram is the normal table (i.e., Table A in the appendix). In the problems that follow, you will start at one location on the map, and the solution will require getting to another location. The *z*-score formula and normal table (together with a little addition or subtraction) will be all you need to get to your destination.

1. What percent of high school seniors in the population can be expected to have math SAT scores between 500 and 670? For this problem and the next five (problems 1–6), you are starting with raw scores and want to find an area (or probability) under the normal curve.

   The first step in any normal curve problem is to draw a rough diagram and find the part of the curve needed to solve the problem; this will prevent careless errors. The desired percent is expressed as the shaded area in Figure 4.5.

   To obtain the percent area between the mean (500) and a raw score of 670, convert the raw score to a *z*-score:

$$z = \frac{670 - 500}{100} = +1.70$$

Figure 4.5

**Percent of Population With Math SAT Scores Between 500 and 670**

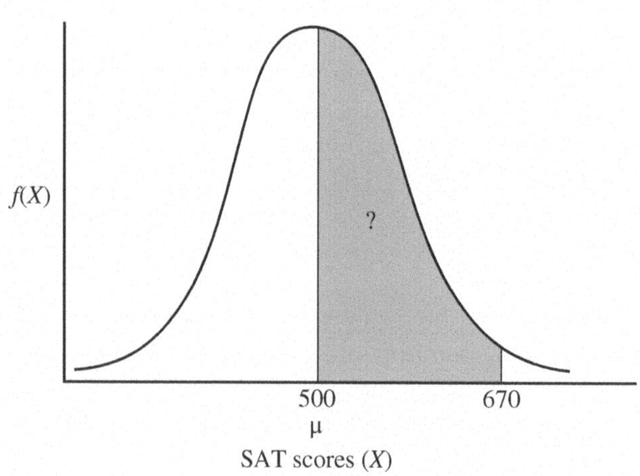

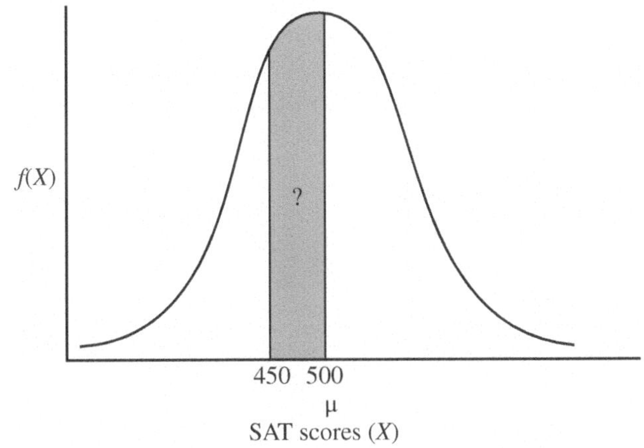


## Figure 4.6

**Percent of Population With Math SAT Scores Between 450 and 500**

So, a SAT score of 670 has a *z*-score of 1.70; you have gotten to *Z*-Score Junction from Raw-Scoreville via the Formula Freeway. Now we get on the Table Tram. Table A reveals that the percent area between the mean and a *z*-score of 1.70 is equal to 45.54%. Therefore, nearly 46% of the population is expected to have scores between 500 and 670. By using the normal curve model as an approximation, you can reach this conclusion even though you cannot measure the entire population.

The area we just found can be used to determine the PR for a score of 670. For any raw score, the PR is just the percentage of the area in the distribution below (i.e., to the left of) that score. For scores above the mean, you would need to add 50% (the area below the mean) to the area between the mean and the score of interest (45.54% in this example). Therefore, the PR for a score of 670 is 50 + 45.54, which equals 95.54—that is, a score of 670 is higher than 95.54% of the entire distribution.

2. What percent of the population can be expected to have scores between 450 and 500?

The diagram for this problem is shown in Figure 4.6. Converting the raw score of 450 to a *z*-score yields

$$z = \frac{450 - 500}{100} = -0.50$$

Table A does not include any negative *z* values, but entering the table at the *z*-score of +0.50 will produce the right answer because of the symmetry of the normal curve. The solution is that 19.15%, or about 19% of the population, is expected to have scores between 450 and 500.To find the PR for a score that is less than the mean, such as 450, we need to find the percentage of the distribution that is below that score, and that amount is not given in Table A. However, we can see from Figure 4.6 that together the percentage below the score and the percentage between the score and the mean must add up to 50% (i.e., the total amount of the normal distribution below its mean), so the percentage below the mean can be found by subtracting the area between the score and the mean from 50%. In this example, we can find the PR of 450 by subtracting 19.15% from 50%: thus, the PR for 450 equals 50 − 19.15, or about 31.

3. What percent of the population can be expected to have scores between 360 and 540?

This problem is likely to produce some confusion unless you use the diagram in Figure 4.7 as a guide. Table A gives values only between the mean and a given *z*-score, so two steps are required. First, obtain the percent between 360 and the mean (500); then obtain the percent between the mean and 540. Then *add* the two percentages together to get the desired area. The *z* values are as follows:

$$z = \frac{360 - 500}{100} = -1.40; z = \frac{540 - 500}{100} = +0.40$$

## Figure 4.7

**Percent of Population With Math SAT Scores Between 360 and 540**

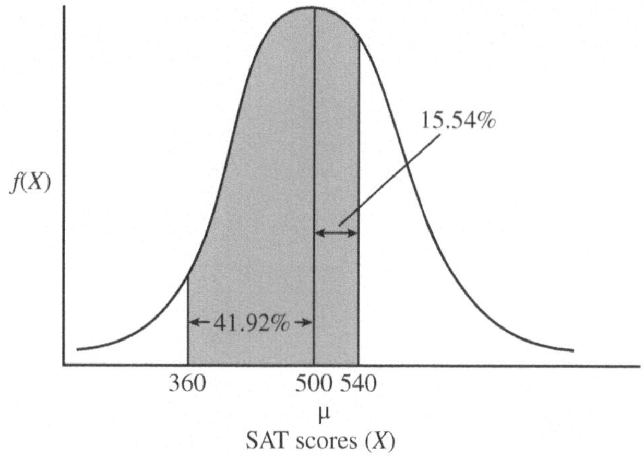


## Figure 4.8

**Percent of Population With Math SAT Scores Between 630 and 700**

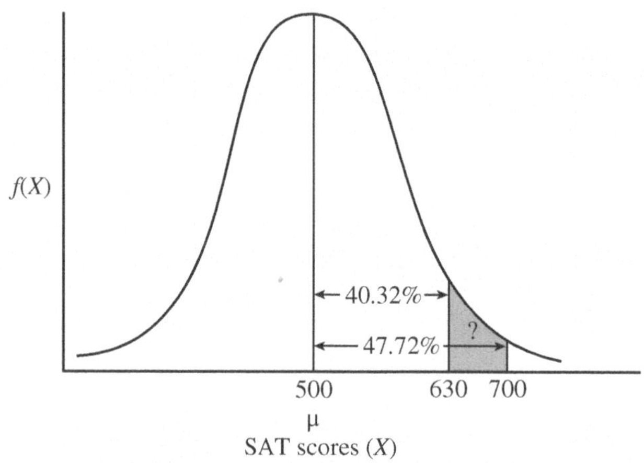


Table A indicates that 41.92% of the total area falls between the mean and a $z$ of $-1.40$, and 15.54% falls between the mean and a $z$ of $+0.40$. So, the percent of the population expected to have scores between 360 and 540 is equal to 41.92% + 15.54%, or 57.46%.

4. What percent of the population can be expected to have scores between 630 and 700? Once again, you are likely to encounter some difficulty unless you draw a diagram such as the one shown in Figure 4.8. The $z$-score corresponding to a raw score of 630 is equal to $(630 - 500)/100$ or $+1.30$, and the $z$-score corresponding to a raw score of 700 is equal to $(700 - 500)/100$ or $+2.00$. According to Table A, 40.32% of the curve falls between the mean and $z = +1.30$, and 47.72% of the curve falls between the mean and $z = +2.00$. So, the answer is equal to 47.72% *minus* 40.32%, or 7.4%. In this example, subtraction (rather than addition) is the route to the correct answer, and drawing the illustrative diagram will help you decide on the proper procedure in each case.
5. What percent of the population can be expected to have scores above 720?The percent to be calculated is shown in Figure 4.9. The $z$-score corresponding to a raw score of 720 is equal to $(720 - 500)/100$ or $+2.20$, and the area between the mean and this $z$-score as obtained from Table A is equal to 48.61%. At this point, avoid the incorrect conclusion that the value from the table is the answer to the problem. As the diagram shows, you need to find the percent *above* 720. Since the half of the curve to the right of the mean is equal to 50.00%, the answer is equal to 50.00% − 48.61%, or 1.39%. A score of 720 is close to the 99th percentile, because just a little more than 1% of the population exceeds it.

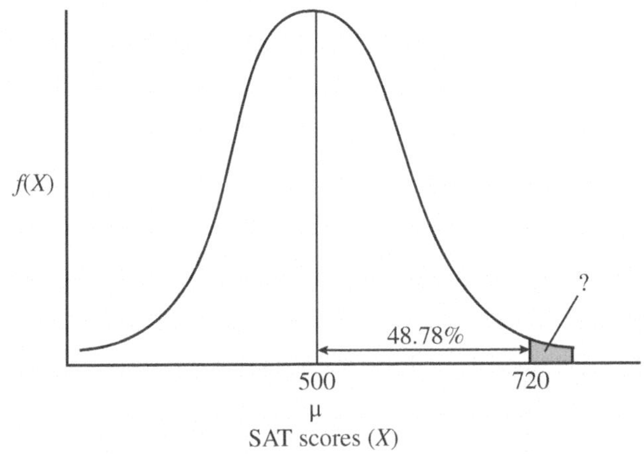


## Figure 4.9

**Percent of Population With Math SAT Scores Above 720**

6. What is the probability that a person drawn at random from the population will have a score of 720 or more?

   This problem is identical to the preceding one, but is expressed in different terminology. As before, convert 720 to a *z*-score, obtain the percent between the mean and 720, and subtract from 50% to obtain the answer of 1.39%. To express this as a probability, move the decimal point two places to the left. The probability that any one individual drawn at random from the population will have a score of 720 or more is .0139, or about .014 (i.e., 1.4%).

7. A math professor wishes to invite only the top 10% of the student population to join a competitive math team that she is forming. What cutoff math SAT score should she use to accept and reject candidates? Now we are starting at Area Arena, and want to get to Raw-Scoreville.

   This problem is the reverse of the ones we have looked at up to this point. Previously, you were given a score and asked to find a percent; this time, the percent is specified and a score value is needed. The problem is illustrated in Figure 4.10.

   The needed value is the raw score value corresponding to the cutoff line in Figure 4.10. The steps are exactly the reverse of the previous procedure:

   (a) If 10% are above the cutoff line, then 40% (50% − 10%) are between the mean and the cutoff line.

   (b) Enter Table A in the *body* of the table (where the percents are given), looking for the value of 40.00 (40%). Read out the *z*-score corresponding to the percent nearest in value to 40.00 ($z = 1.28$, which corresponds to 39.97).

   (c) Determine the *sign* of the *z*-score. The diagram shows that the desired cutoff score is *above* the mean, so the *z*-score is positive and equals +1.28.

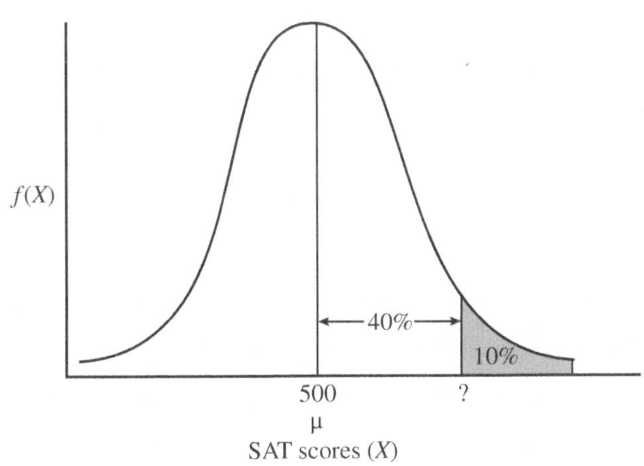


## Figure 4.10

**Math SAT Score That Demarcates the Top 10% of the Population**

(d) Convert the $z$-score to a raw score. Because $z = (X - \mu)/\sigma$, some simple algebraic manipulations yield $X = z\sigma + \mu$. Thus, $X = (+1.28)(100) + 500$, or 628. Any student with a score of 628 or more may try out for the new math team; individuals with scores of 627 or less are not eligible. Note that the score that cuts off the top 10% of the distribution is the 90th percentile (or 9th decile). In general, anytime you need to find a percentile greater than the 50th, you can follow the procedure above, as long as you realize that the $X$th percentile is the score that cuts off the top $(1 - X)$% of the distribution. For percentiles smaller than the 50th, just note that the $X$th percentile is the score that cuts off the bottom $X$% of the distribution, and proceed with the steps listed above, being careful to make your $z$-score negative in Step c, and then use that negative $z$ to find the raw score in Step d. The 50th percentile (i.e., the median) is always equal to the mean in a normal distribution.

8. What two cutoff scores enclose the middle 60% of the population? Again, we are starting with an area/probability and looking for raw scores.

   Because you are given a percentage and asked to find corresponding raw scores, this is another "reverse" problem; you need to find the two raw scores illustrated in Figure 4.11. Finding the middle $X$% of a normal distribution is a fairly common task, as you will see when we describe confidence intervals in Chapter 6. The first step in solving this type of problem is to divide $X$ by 2 and then mark off $(X/2)$% on either side of the mean. In the case of the middle 60%, we mark off 30% on each side, as shown in Figure 4.11.

   Because the area between the mean and the upper cutoff line is 30%, this is one of the values that must be used to enter Table A. As in the previous problem, look in the body of the table to find the value of interest; in this problem, we want the value in the table that is closest to 30.00, which is 29.95. Then read off the $z$-score corresponding to that value (29.95), which is 0.84. Next, convert this value to a raw score: $X = (0.84)(100) + (500)$, or 584. Thus, the upper cutoff score is 584. Thanks to the symmetry of the normal curve, we know that the $z$-score for the lower cutoff in Figure 4.11 is $-0.84$. Converting this value to a raw score, we obtain: $X = (-0.84)(100) + (500) = 500 - 84 = 416$. So, we can now say that the middle 60% of the scores are contained between 416 and 584. (As a check, add the two boundary values just found, $416 + 584 = 1{,}000$; the sum should equal twice the value in the middle. This will always be true when the two boundaries are equally distant from the middle.)

As you can see, the normal curve is useful for solving various types of problems involving scores and percentages of distributions. Of even greater importance is its use with

## Figure 4.11

**Math SAT Scores Enclosing the Middle 60% of the Population**

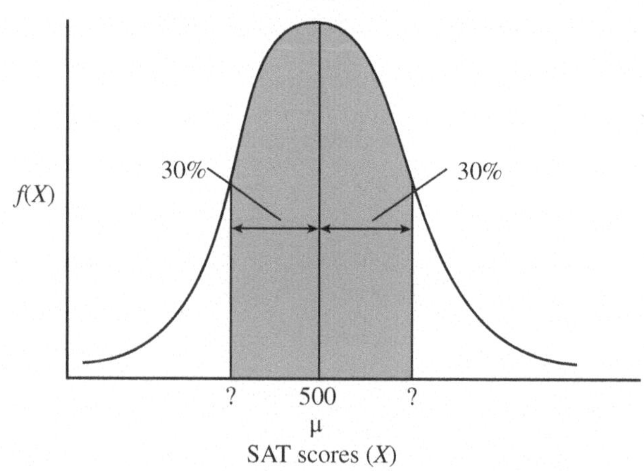

certain sampling distributions, such as the sampling distribution of the mean, which is an important topic in the next chapter.

## Summary

In order to compare scores based on different means and SDs, it is desirable to convert the raw scores to *standardized scores* with a common mean and SD. Some frequently used transformed scores are $z$-scores (sometimes called *standard scores*), with a mean of 0 and an SD of 1; *T scores*, with a mean of 50 and an SD of 10; *SAT scores*, with a mean of 500 and an SD of 100; and IQ scores, with a mean of 100 and an SD of 15 or 16 (depending on the test). A transformed score shows at a glance whether a score is above or below average and how far it deviates from the mean in terms of SD units, with respect to a specified reference group.

### General Transformations

| Operation | Effect on $\mu$ | Effect on $\sigma$ | Effect on $\sigma^2$ |
|---|---|---|---|
| 1. *Add* a constant, $k$, to every score | New mean = old mean + $k$ | No change | No change |
| 2. *Subtract* a constant, $k$, from every score | New mean = old mean − $k$ | No change | No change |
| 3. *Multiply* every score by a constant, $k$ | New mean = old mean × $k$ | New $\sigma$ = old $\sigma \times k$ | New $\sigma^2$ = old $\sigma^2 \times k^2$ |
| 4. *Divide* every score by a constant, $k$ | New mean = old mean/$k$ | New $\sigma$ = old $\sigma/k$ | New $\sigma^2$ = old $\sigma^2/k^2$ |

### *z*-Scores (Standard Scores)

$$z = \frac{X - \mu}{\sigma} \qquad \text{Formula 4.1}$$

### *T* Scores

$$T = 10z + 50 \qquad \text{Formula 4.2}$$

### SAT Scores

$$\text{SAT} = 100z + 500 = 10T \qquad \text{Formula 4.3}$$

### IQ Scores

$$\text{S-B IQ} = 16z + 100 \qquad \text{Formula 4.4A}$$

$$\text{WAIS IQ} = 15z + 100 \qquad \text{Formula 4.4B}$$

### Normal Distributions

One frequently used theoretical distribution is the *normal curve*. One use of this model is as follows: If it is assumed that raw scores in the population are normally distributed and that we know or can estimate the population mean and SD, we can obtain (a) the percent of the population with raw scores above or below or between specified values and/or (b) the raw score that demarcates specified percentages in the population.

***Type 1 Problem: Given a Raw Score, Find the Corresponding Proportion or Percentage of the Normal Curve. In the language of*** Figure 4.4, ***you are starting at Raw-Scoreville and want to get to Probability City/Area Arena.***

(a) Convert the raw score to a *z*-score: $z = (X - \mu)/\sigma$.
(b) Enter the normal curve table in the *z* column, and read out the percent area from the mean to this *z* value.
(c) Compute the answer to the problem. Note that Table A gives only the area from the mean to the *z* value; create an appropriately labeled diagram to make sure your final answer is the one called for by the problem.

***Type 2 Problem: Given a Proportion or Percentage of a Normal Curve, Find the Corresponding Raw Score. According to*** Figure 4.4, ***you want to get from Probability City to Raw-Scoreville.***

(a) Compute the appropriate percentage (if a proportion is given) and enter the normal curve table in the body of the table (where the areas are given). Read out the corresponding *z*-score. If the desired raw score is *below* the mean, assign a negative sign to this *z*-score.
(b) Convert the *z*-score to a raw score: $X = z\sigma + \mu$.

## Exercises

1. The mean of a set of scores is 8 and the standard deviation is 4. What will the new mean, standard deviation, and variance be if you:
   (a) Add 6.8 to every number?
   (b) Subtract 4 from every number?
   (c) Multiply every number by 3.2?
   (d) Divide every number by 4?
   (e) Add 6 to every number, and then divide each new number by 2?
2. For each of the following, compute the *z*-score; then compute the *T* score.
   (a) A Turck Hall student with a score of 17.
   (b) A Turck Hall student with a score of 11.
   (c) A Kirk Hall student with a score of 17.
   (d) A Kirk Hall student with a score of 11.
   (e) A Dupre Hall student with a score of 6.
   (f) A Doty Hall student with a score of 6.
3. Consider the following data: This student believes that he has performed better on the psychology test for two reasons: His score is higher, and he is only 8 points below average (as opposed to 10 points below average on the English test). Convert each of his test scores to a *z*-score. Is he right?

| | **Psychology test** | **English test** |
|---|---|---|
| Student's score | 73 | 67 |
| $\mu$ | 81.0 | 77.0 |
| $\sigma$ | 5.0 | 10.0 |

4. Which of each pair is better, or are they the same? (You should be able to answer by inspection.)
   (a) A *T* score of 47 and a *z*-score of +0.33
   (b) A *T* score of 64 and a *z*-score of +0.88

(c) A *T* score of 42 and a *z*-score of $-1.09$
(d) A *T* score of 60 and a *z*-score of $+1.00$
(e) A *T* score of 50 and a raw score of 26 (mean of raw scores = 26, $\sigma = 9$)
(f) A *z*-score of $+0.04$ and a raw score of 1092 (mean of raw scores = 1113, $\sigma = 137$)
(g) A *z*-score of zero and a *T* score of 50

5. For data set 3 in Chapter 1 (reproduced here), transform each participant's (S) score on *X* to a *z*-score. Then transform each participant's score on *Y* to a *z*-score. ($\mu = 84$ and $\sigma = 10.76$)

| S | X | Y | $z_x$ | $z_y$ |
|---|---|---|---|---|
| 1 | 97 | 89 | ____ | ____ |
| 2 | 68 | 57 | ____ | ____ |
| 3 | 85 | 87 | ____ | ____ |
| 4 | 74 | 76 | ____ | ____ |
| 5 | 92 | 97 | ____ | ____ |
| 6 | 92 | 79 | ____ | ____ |
| 7 | 100 | 91 | ____ | ____ |
| 8 | 63 | 50 | ____ | ____ |
| 9 | 85 | 85 | ____ | ____ |
| 10 | 87 | 84 | ____ | ____ |
| 11 | 81 | 91 | ____ | ____ |
| 12 | 93 | 91 | ____ | ____ |
| 13 | 77 | 75 | ____ | ____ |
| 14 | 82 | 77 | ____ | ____ |

6. For the following data set:

   68.36, 15.31, 77.42, 84.00, 76.59, 68.43, 72.41, 83.05, 91.07, 80.62, 77.83

   Transform each raw score in the preceding data set into a *z*-score, a *T* score, and an SAT score. ($\mu = 72.28$ and $\sigma = 20.07$)

   The remaining exercises are based on SAT scores, with the assumption that they follow the normal distribution with $\mu = 500$ and $\sigma = 100$.

7. What percent of the population obtains scores of 410 or less?
8. What percent of the population obtains scores between 430 and 530?
9. What percent of the population obtains scores between 275 and 375?
10. How do you explain the fact that the answer to Exercise 9 is smaller than the answer to Exercise 8, even though each problem deals with a 100-point range of scores?
11. What is the minimum score needed to rank in the top 5% of the population?
12. A psychologist wishes to test a new learning strategy on the bottom 15% of those who took the math SAT. What cutoff score should she use to select participants for her study?
13. A single student is drawn at random from the population. What is the probability that this student has a score:

    (a) of 410 or less?
    (b) between 430 and 530?
    (c) between 275 and 375?
    (d) in the top 5% of the population?

    (Hint: Base your answers to this exercise on the percentages you found for the previous exercises.)

14. Between which two SAT scores would you find the middle: (a) 70% of the normal distribution? (b) 90% of the normal distribution?

## Thought Questions

1. What are the advantages of having a set of scores that has a mean of zero and a standard deviation of 1 ($z$-scores)?
2. A college student takes two introductory courses for first-year students. On her psychology exam, she scores 73; the mean is 81.0, and the SD is 5.0. On her English exam, she scores 67; the mean is 77, and the SD is 10.0. Both distributions are approximately normal. She concludes that she did better on the psychology exam because her score on that exam is 6 points higher. Is she correct? Why or why not?
3. For the same exams as in the previous question, another student scores 91 on the psychology exam and 93 on the English exam. He concludes that he did better on the English exam because his English score is 16 points above average, while his psychology score is only 10 points above average. Is he correct? Why or why not?
4. Why is Table A given in terms of $z$-scores rather than raw scores?
5. Why is it *not* necessary to know the exact number of cases in the population being studied in order to use the procedures described in this chapter?
6. The distributions of scores on two exams are approximately normal. The mean of the first exam is 80, and the mean of the second exam is 95. The SD of each exam is 15. Draw a figure to show that there *must* be some scores on the first exam that are greater than the mean of the second exam.
7. Suppose that in the previous question, the SD of each exam is changed to 3. How would this affect the likelihood that there are scores on the first exam that are greater than the mean of the second exam? Draw an appropriate figure to help you answer this question.
8. If you know that a distribution of scores is bimodal, or asymmetric, or a J-curve, or any other shape that departs significantly from the normal distribution, would it be correct to use Table A to solve problems, as was done in this chapter? Why or why not?

## Computer Exercises

1. Use your statistical package to create new variables consisting of the $z$-scores for the anxiety and HR measures both at baseline and prequiz in Jackson's data set. Request means and standard deviations of the $z$-score variables to demonstrate that the means and SDs are 0 and 1, respectively, in each case. (Note: If, like SPSS, your statistical software computes only $s$ for the standard deviation and not $\sigma$, you should calculate your $z$-scores based on $s$.)
2. Use your statistical package to create a $z$-score variable corresponding to the academic interest score, and then transform the $z$-score variable to a $T$ score, a SAT score, and an IQ score.
3. Repeat Exercise 2 for the statistics quiz score that was part of Jackson's experiment.

   For the remaining exercises, use your statistical package to find the indicated areas under the normal curve (your answer should include six digits past the decimal point):
4. The area below a $z$-score of 3.1.
5. The area above a $z$-score of 3.3.
6. The area below a $z$-score of $-3.7$.
7. The area between the mean and a $z$-score of 0.542.
8. The area between the mean and a $z$-score of $-1.125$.

## Bridge to SPSS

An easy way to obtain $z$-scores for any variable in your spreadsheet is available from the **Analyze/Descriptive Statistics** menu; just choose **Descriptives** .... Then under the list of variables that appears in the *Descriptives dialog* box, you will see the phrase *Save standardized values as variables* preceded by a small checkbox. Check the box, and move over the variables for which you would like to see the $z$-scores. After you click **OK**, two things will happen. You will get the usual Descriptives output for the variables you moved over, and for each of these variables, SPSS will have added a new variable at the end of your spreadsheet, containing the $z$-scores that correspond to the scores of that variable. SPSS names any new $z$-score variable by putting the letter z at the beginning of the original name. However, we have a warning for those who create $z$-scores automatically in SPSS. Because SPSS never uses or even computes the biased variance or standard deviation, it creates $z$-scores by dividing deviations from the mean by the *unbiased* standard deviation. We described $z$-scores in this chapter as based on $\sigma$ rather than $s$ because that is the more common method. When you are calculating $z$-scores for individuals, it is usually the case that you are viewing the entire set of individuals as a population instead of just a sample. Because SPSS creates $z$-scores based on $s$, calculating $\sigma$ for the set of $z$-scores created will not yield a value of 1.0; it is $s$ that will equal 1.0, in this case. Of course, the mean of the set of $z$-scores will always be zero, even if they were produced by SPSS. To get $z$-scores based on $\sigma$ instead of $s$, you would have to multiply all of the $z$-scores created by SPSS by the square root of the ratio of your total $N$ to $N - 1$.

You can use SPSS to obtain areas under the normal distribution with more accuracy (i.e., more digits past the decimal point) than is available in Table A and for $z$-scores not found in Table A or most other tables available from books. Start by opening a new (i.e., empty) data sheet and entering the $z$-score of interest in the first cell. For convenience, you can assign the simple variable name "z" to this first column. Then open the *Compute Variable* box by selecting **Compute** ... (the first choice) from the **Transform** menu. In the *Target Variable* space, type an appropriate variable name like "area." In the *Numeric Expression* space, type **CDFNORM** and then, in parentheses, the name of the first variable—for example, CDFNORM ($z$). (*Note:* We are using uppercase letters here for clarity, but SPSS does not distinguish between upper- and lowercase letters in variable or function names.) CDFNORM is a function name that stands for the *C*umulative *D*ensity *F*unction for the *NORM*al distribution; therefore, it returns a value equal to all of the area to the left of (i.e., below) the $z$-score you entered. If you multiply this value by 100, you get the PR associated with the $z$-score in question (this works for both positive and negative $z$-scores, as long as you include the minus sign for any negative $z$-score).

If you want the area between the mean and your $z$ value, just subtract .50 from the area obtained by the CDFNORM function (this will work for negative $z$-scores, if you enter them *without* the minus sign). If you want the area above, rather than below, a particular $z$-score, just change the sign in front of the $z$-score (add a minus sign to a positive $z$-score or remove the minus sign from a negative $z$-score); because the normal curve is symmetric around $z = 0$, the area below $-z$ is the same as the area above $+z$. Note that the larger the $z$-score you enter in the first cell of your SPSS datasheet, the more "decimals" you will need to display the answer accurately. The fourth column in *Variable View* lets you set the number of digits that will be displayed to the right of the decimal point—as long as this number is less than the number in the third column (width) for that variable. For instance, if you are looking for the area below a $z$-score of 3 or 4, you will want to set the "decimals" number to at least 6.

# Part II
# Basic Inferential Statistics

# Chapter 5
# Introduction to Statistical Inference

## PREVIEW

### Introduction

The typical population of interest to the behavioral science researcher is extremely large; what serious problem does this create? How does the use of samples help solve this problem?

What new difficulty is created by the use of samples? What type of statistics helps solve this problem?

### The Goals of Inferential Statistics

What are the most common purposes of inferential statistics?

### Sampling Distributions

What is an experimental sampling distribution? Why is this type of statistical model rarely used in inferential statistics?

What is a theoretical sampling distribution? When is it reasonable to use the normal curve as the theoretical distribution (i.e., statistical model)?

### The Standard Error of the Mean

We wish to draw inferences about the mean of one population, based on a statistic whose sampling distribution is normal—namely the sample mean. Why should the standard deviation (SD) of the raw scores not be used as the measure of variability? How does the variability of a distribution of sample means differ from that of raw scores?

What is the standard error of the mean? Why cannot it be measured directly? How is it estimated?

What does the standard error of the mean tell us about the trustworthiness of a single sample mean as an estimate of the population mean?

### The *z*-Score for Sample Means

How is the *z*-score calculated when dealing with sample means instead of individuals?

What is a *p* value?

### Null Hypothesis Testing

What two hypotheses are made prior to conducting the statistical analysis? Which of these hypotheses is assumed to be true?

What is a Type I error?

What is the criterion of significance, and how is it related to a Type I error?

What is the probability of a Type I error?

How does the one-tailed *p* value differ from the two-tailed *p* value?

What are critical values? How are critical values used in null hypothesis testing?

What are the critical values for *z* in a two-tailed significance test using the .05 criterion? What are the critical values using the .01 criterion of significance?

What advantage does the .05 criterion of significance have over the .01 criterion of significance? What is the advantage of the .01 criterion? Which is used more often in behavioral science research?

*Introductory Statistics for the Behavioral Sciences*, Eighth Edition. R. Brooke Lea and Barry H. Cohen.

Companion website: www.wiley.com/go/ISBS8e

**PREVIEW (*continued*)**

What is a Type II error?

Why is the probability of a Type II error harder to determine than the probability of a Type I error?

What is the relationship between the probabilities of Type I and Type II errors?

What is a one-tailed test of significance? What are the advantages and disadvantages of this procedure? Why is the two-tailed test more commonly used?

**Assumptions Required by the Statistical Test for the Mean of a Single Population**

What assumptions underlie the use of the *z*-score for sample means in conjunction with the normal distribution?

What can be done when $\sigma$ is not known?

**Why is Null Hypothesis Testing So Misunderstood?**

**Summary**

**Exercises**

**Thought Questions**

**Computer Exercises**

**Bridge to SPSS**

## Introduction

One daunting fact greatly complicates the lives of behavioral scientists: The populations they wish to study—and from which they seek to collect data—are usually far too large for them to measure every element in the population or even a sizable portion of the population.

Suppose that a psychologist wishes to test the hypothesis that a new interactive video game will improve the performance of students taking high school chemistry. In this study, the population consists of all high school students in the United States who are studying chemistry. A population this size is enough to frighten off even the most dedicated researcher! No behavioral scientist can afford the time, effort, and money needed to obtain and analyze data from so many thousands of people.

Therefore, the psychologist might decide to limit the study to 100 high school chemistry students selected in some way from the population. This procedure will provide them with a group that can be measured in its entirety, but it creates a serious problem. The data are obtained from only a small part of the population. So, there is no assurance that these data will accurately reflect what is happening in that population. If this sample of 100 students is atypical of the population as a whole, the psychologist will suffer some unhappy consequences: The results that they publish, and their decision concerning the merits of the new video game, may be contradicted subsequently by other researchers who try to verify the psychologist's findings with different samples of students drawn from the same population. If that occurs, the overall effort to advance our scientific knowledge in this area will undergo a setback while researchers try to make sense of the conflicting results. Unfortunately, the only sure way to prevent this confusion would be to measure the entire population—which, as we noted above, is simply not feasible. The next best strategy is to try to ensure that the individuals in your sample are selected at random from the population of interest. We will have more to say about the method of drawing random samples shortly.

In sum: Behavioral scientists need to draw conclusions about populations, but they usually can measure only a small part of those populations. This problem affects all areas of the behavioral sciences. For example, an industrial psychologist studying the effects of pay on job satisfaction cannot measure the entire population of all paid employees in the United States. Even if he were to restrict his attention to one type of job, the population would still be far too large to measure in its entirety. An experimental psychologist studying the performance of rats in a maze under varying conditions cannot run all laboratory rats in the world in her experiment, nor can she obtain all possible runs from any one rat. An anthropologist interested in the effects of different cultures on motivation in children cannot study all of the children from each of the cultures in which she is interested. In fact, this predicament is so common that most behavioral science research would not be possible without some effective resolution. As you may have guessed from the title of this chapter, the answer lies in the use of *inferential statistics*—techniques for drawing inferences about an entire population, based on data obtained from a sample drawn from that population.

## The Goals of Inferential Statistics

There are two major kinds of inferences that the behavioral scientist might wish to make about a population. One useful procedure is to estimate the values of population parameters (e.g., the mean and SD). Given a sample whose mean and SD have been computed, we may wish to estimate the mean and SD of the parent population. Any statistic computed from a single sample (such as $\overline{X}$ or $s$) that provides an estimate of the corresponding population parameter (such as $\mu$ or $\sigma$) is called a *point estimate*.

However, for most purposes, a point estimate is not sufficiently informative because it leaves us with no idea of how far off from the parameter it may be. Such information is supplied by an *interval estimate*, a range of values that has a known probability of including the true value of the parameter. For example, suppose that the cell phone records of a sample of teenagers are gathered for a month, and it is found that they send an average of 60 text messages per day. A point estimate would state that the most reasonable value of $\mu$ (per day) is 60. If an estimate in terms of a single number is *not sufficient*, the researcher should use the appropriate techniques of inferential statistics to determine an *interval* that is likely to include the population mean. Suppose she finds this interval to be 54–66, or 60 ± 6. This interval estimate divides all numerically possible values of $\mu$ into two sets: "likely" (54–66) and "unlikely" (less than 54 or more than 66), with "likely" and "unlikely" having precise meanings to be discussed in greater detail in the next chapter.

The other major use of inferential statistics is to assess the probability of obtaining certain kinds of sample results under certain population conditions in order to test a specific hypothesis. The aforementioned psychologist might wish to determine the probability that a sample of 100 chemistry students, who have been using a new educational video game that does not help them learn chemistry at all, would nonetheless have gained an average of at least 5 points on their exam scores. Why anyone would want to know such a probability will be made clear (we hope!) later in this chapter. First, in order to determine probabilities when dealing with samples instead of individuals, you will need to learn about *sampling distributions*. So, it is to this topic that we turn next.

## Sampling Distributions

In the preceding chapter, you learned how to transform a raw score into a new score that shows at a glance how that score stands in comparison to a specific reference group and how to find the proportion of scores above or below that score in a normal distribution (ND). These procedures will be useful in the chapters ahead, but not for the descriptive purposes discussed previously (e.g., describing where a score falls in a distribution). Behavioral researchers are rarely concerned with the score of any one individual in a distribution. Their studies usually involve a group of individuals, all of whom share some common characteristic (e.g., the same psychopathology diagnosis), or all of whom have been subjected to the same experimental treatment. Therefore, the remainder of this text will focus on the performance of a group as a whole—for instance, on some word-recall tasks. This is where descriptive statistics becomes immediately useful, as a first step. To describe the performance of an entire group, rather than an individual, we need a measure of central tendency. As we have seen, the arithmetic mean is usually the most suitable of such measures for advanced statistics. Therefore, throughout this part of the text, groups will be dealt with in terms of their means. Also, our interest will not be in the particular groups we are dealing with but in how these groups serve as samples of (and therefore representations of) the entire populations from which they were drawn. Consequently, we will refer to all of the groups that we deal with from now on as *samples*, and we will focus primarily on the means ($\overline{X}$) of these samples.

Before we can introduce statistical inference, we must first apply the descriptive tools you have just learned to entire collections of groups (i.e., samples), rather than to collections of individuals. We begin with a simple (hypothetical) example in which a researcher (let us call her Elise) is interested in the relationship between diet and physical growth. In particular, Elise has an interest in—and concern about—growth hormones given to cows to increase their production of milk, and the effect those hormones may or may not have on humans who drink that milk (we will call it GH milk, for short). Before going into the lab to conduct an extensive and expensive randomized trial study, Elise wants to see if she can find evidence for such an effect in the general population. So, she obtained permission from the ERC (European Research Council) to examine the medical records of Scandinavian women, who are a particularly useful population for this study because of their unusual height. A simple starting hypothesis is that people who were raised drinking GH milk might be slightly taller than those who were not. Elise begins by finding in the medical records that the height of Scandinavian women is approximately normally distributed with a mean of 68 inches and an SD of 3.2 inches. If very nearly all Scandinavian women were raised on GH milk, Elise can use those population parameters as the parameters for all women in the GH milk population.

To test her hypothesis, Elise will next need to gather a random sample of Scandinavian women who were *not* raised on GH milk (perhaps their parents gave them only organic milk or plant-based "milk"). She would get the most definitive answer to her research question if she could measure the height of every Scandinavian woman not raised on GH milk, but we will assume that even that specific segment of the population contains tens of thousands of people. The practical strategy for a researcher in Elise's position, then, is to select as large a sample as she can manage, given her resources, to measure the mean height of her sample and compare it to the population mean she derived from national medical records. Let us assume for the moment that Elise sampled 25 non-GH milk Scandinavian women, and that the mean of the sample was 66 inches. That result is consistent with her hypothesis: her sample of non-GH women is shorter, on average, than the general population. If Elise had somehow been able to measure the *entire population* of non-GH women in Scandinavia, then her work would be done: 66 inches is less tall than 68 inches; therefore, the populations of non-GH women are shorter than the general population. But when our data come from a sample of the population of interest, then we cannot make a direct comparison between the two means because the sample mean (66 inches) is merely an *estimate* of the population mean. The true non-GH population mean could be 68 inches or 64 inches and Elise's sample of 25 people happened to produce a mean of 66 inches. If Elise had randomly sampled a different group of 25 people from the non-GH population, the sample mean might have turned out to be 67 inches or 65 inches. A skeptic might ask, "If the type of milk a person drinks as a child makes no difference to adult height, then the population mean of the non-GH milk distribution should be the same as the general population mean, 68 inches. If that were true—type of milk makes no difference—then it would not be impossible for Elise to have sampled 25 people and found a mean of 66 inches. The skeptic is right; each time you sample a population, the sample mean will be slightly different. The question becomes, "How rare would it be to sample 25 people and get a sample mean of 66" if the population the sample is drawn from has a mean of 68 inches?

This is a question we can answer, and is why we have been learning about measuring probabilities under a curve. If Elise sampled 25 people many times and graphed all those sample means, then she could see just how unusual a sample mean of 66 inches is. That graph would depict a distribution of sample means. To nail down the probability of sampling 25 people and getting a sample mean of 66 inches—or any other height—Elise would need to know what that distribution of sample means would look like if she collected *an infinite number of samples* of 25 Scandinavian women raised on non-GH milk. If she were to do this for an *infinite* number of sample means, the resulting distribution would be called the **sampling distribution of the mean.**[1] Here's what a hypothetical sampling distribution of the mean might look like for Elise:

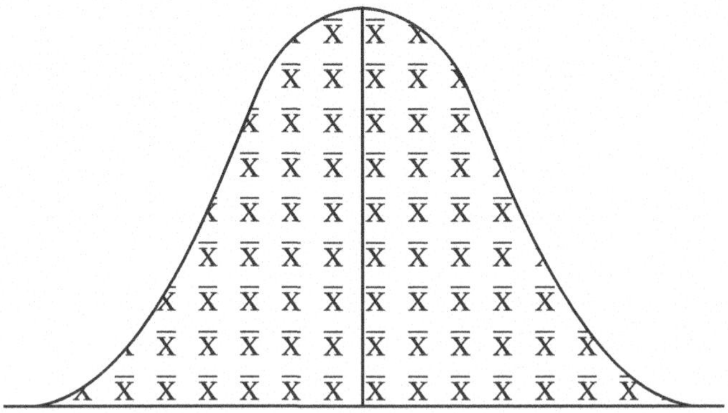

## Figure 5.1

**Mean height of millions of samples each made up of 25 Scandinavian Women raised with non-GH milk**

Given a sampling distribution such as the one in Figure 5.1, one could locate a sample mean of 66 inches and see how unusual it is.

Given that no one is going to draw an infinite number of samples from a population, or even millions of them, we all have to settle for an approximation of the sampling distribution of the mean. There is more than one way to obtain an approximate sampling distribution: (1) directly (i.e., empirically), and (2) indirectly (i.e., theoretically). We'll discuss both, but you will soon see that the indirect method is *much* easier. To obtain an approximate sampling distribution of the mean directly, you would have to draw a large number of samples, each of the same size, from the defined population, and determine on an empirical basis how often the various alternatives occur. It is important that the samples be *randomly selected*—that is, drawn in such a way as to minimize bias and make the sample typical of the entire population insofar as possible. Technically, a *random sample* is defined as a sample drawn in such a way as to (1) give each element in the population an equal chance of being drawn, *and* (2) make all possible samples of that size equally likely to occur. For example, a random sample for the hormone/height study might be obtained by picking names randomly out of a listing of millions of people's names such as one might find in a population census (or a phone book, if you happen to find one); an especially problematic nonrandom sample would be selecting women who were hanging around a basketball or volleyball court.

If the mean of each sample is computed and a frequency distribution of the many means is plotted, then we will have an idea of how often to expect a sample with a particular mean. Such a frequency distribution is called an *experimental sampling distribution* because it is obtained from observed or experimental data.[2] Note that this is not how distributions of sample values are ordinarily determined, but merely how one *could* determine them.

To get concrete about it, suppose that Elise decides that a sample size of 25 Scandinavian women is the largest she can afford to obtain in terms of both time and funding. She could then create an empirical sampling distribution by measuring the height of each woman in her first random sample, computing the mean height of the sample, and then replacing the

[1] If, instead, she found the median of each sample and then plotted all of the sample medians, she would be creating the sampling distribution of the *median*; one can create a sampling distribution for any statistic, though we will focus on the mean, as most behavioral researchers do.

[2] It is also sometimes called a *Monte Carlo distribution*, and it is usually generated by a computer.

**Table 5.1** **Hypothetical Frequency Distribution of Mean Height for 1,000 Samples (Sample $N = 25$)**

| (A) Mean height (in.) | (B) Number of samples ($f$) | (C) Proportion of samples ($p = f/N$) |
|---|---|---|
| 68.55 or more | 0 | .000 |
| 68.50 – 68.54 | 1 | .001 |
| 68.45 – 68.49 | 1 | .001 |
| 68.40 – 68.44 | 4 | .004 |
| 68.35 – 68.39 | 11 | .011 |
| 68.30 – 68.34 | 19 | .019 |
| 68.25 – 68.29 | 31 | .031 |
| 68.20 – 68.24 | 56 | .056 |
| 68.15 – 68.19 | 81 | .081 |
| 68.10 – 68.14 | 112 | .112 |
| 68.05 – 68.09 | 116 | .116 |
| 68.00 – 68.04 | 141 | .141 |
| 67.95 – 67.99 | 131 | .131 |
| 67.90 – 67.94 | 103 | .103 |
| 67.85 – 67.89 | 76 | .076 |
| 67.80 – 67.84 | 57 | .057 |
| 67.75 – 67.79 | 28 | .028 |
| 67.70 – 67.74 | 21 | .021 |
| 67.65 – 67.69 | 7 | .007 |
| 67.60 – 67.64 | 2 | .002 |
| 67.55 – 67.59 | 2 | .002 |
| 67.54 or less | 0 | .000 |
| Total | 1.000 | 1.000 |

entire first sample into the population. Then she would have to draw another random sample of 100 Scandinavian women and compute the mean height of *that* sample. If Elise were to obtain a total of 1,000 samples (each of $N = 100$), the frequency distribution of mean heights of the 1,000 samples might resemble the one in Table 5.1. Figure 5.2 represents the data in Table 5.1 as a frequency polygon.

Of course, anyone who cannot afford to obtain one sample containing more than 100 people would not be able to obtain thousands of samples of 100 each. Fortunately, it is rarely

## Figure 5.2

**Hypothetical Distribution of Mean Height for 1,000 Samples (Sample $N = 100$)**

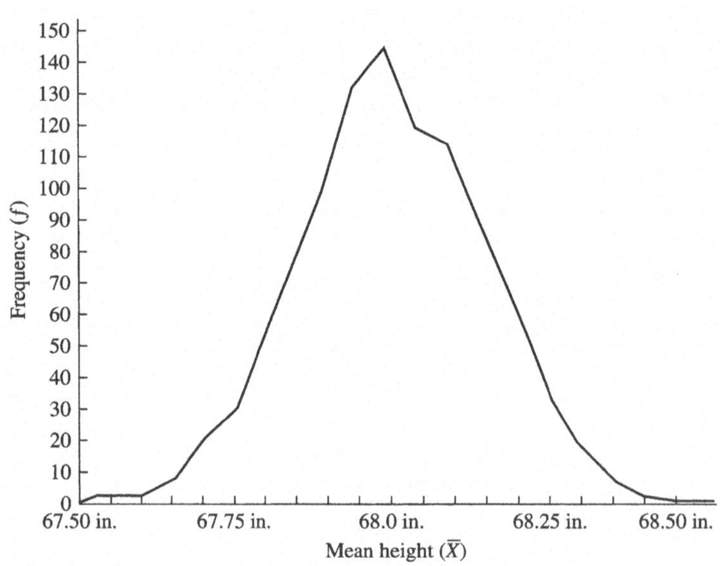

necessary to perform this direct procedure for determining a sampling distribution; we presented it just to make the concept of a sampling distribution more concrete. In practice, we will refer to *theoretical sampling distributions*. But first, let us take a look at a reasonable example of an experimental sampling distribution of height that fits Elise's situation. By inspecting Figure 5.2, you can readily see that most of the sample means cluster around the *grand mean* (i.e., the mean of all of the sample means), which for this example is 68.00; there are fewer and fewer observations as one goes farther away from the grand mean. This distribution closely approximates the theoretical *normal curve*, a statistical model we described in the previous chapter, that can describe the sampling distributions of several statistics of interest, including means. Use of the theoretical normal curve model in the proper situations will enable you to avoid the toil and trouble of determining an experimental sampling distribution in order to ascertain the probabilities in which you are interested.

An important law of statistics tells us that when a population of individuals is normally distributed for a particular variable, say height, the corresponding sampling distribution of the mean will also follow the ND. Although actual populations in behavioral research never follow the ND exactly, variables that have a distribution close to the ND (e.g., height) will lead to sampling distributions of the mean that are approximated very well by the ND. Moreover, a statistical law known as the **Central Limit Theorem** states that as the *size* of the samples increases (remember that a sampling distribution assumes an infinite number of samples *all of which are the same size*), the sampling distribution of the mean more closely resembles the ND, regardless of the shape of the original distribution for individual measurements in the population. However, the more the original population distribution departs from the ND, the larger the samples have to be before the sampling distribution of the mean begins to resemble an ND. In Figure 5.3, we have presented distributions for variables that are non-normal, along with the sampling distribution of the mean for these variables with sample sizes of 2, 4, and 36. The top row depicts how individual scores are distributed in the population.

Consider, for example, the distribution of scores you would get if you rolled a single die an infinite number of times. With a fair die, the probability of rolling a 3 is the same as rolling a 4—or any of the six possibilities; It is ⅙ or 16.67%. A histogram of probabilities would depict a series of six bars, all the same height, like a rectangle. This is an example of a "rectangular distribution," as depicted at the top of the middle column of Figure 5.3. Beginning on the second row, we are graphing the means based on $N$ observations; for our die-throwing example, $N$ = number of throws of the die. In Row B (we need to number these rows), you can see that the sampling distribution of the mean (of just two throws of the die) already looks different from the distribution of individual scores; it has become upside-down V-shaped. Look what happens when you roll the die four times and graph the means (Row C); the distribution has become symmetrically bell-shaped. When you graph an infinite number of means from four rolls of a die, you have a distribution that looks very different from that of the individual scores. By the time you get to sampling distributions of the mean for $N = 36$, you are working with NDs (almost) no matter how non-normal the individual scores are. That is the power of the central limit theorem. It allows psychologists and other social scientists to more easily study variables that are non-normal in the population simply by studying sample means instead of individual scores. There are other advantages to studying groups instead of individuals, as we will see later.

The implications of this theorem for social and behavioral scientists are enormous and impossible to overstate. It means that we can study variables that are not normally distributed in nature so long as we study *groups* of individuals. Even variables that are wildly non-normal among individuals produce sampling distributions of the mean that follow the ND fairly well once the sample size grows above 40 or so. For most variables, a sample size of 20–30 is all that is required to comfortably assume an ND for your sampling distribution.

It should come as no surprise that the mean height of our 1,000 samples of Scandinavian women (summarized in Table 5.1), 68 inches, is the same as the mean of the population of individual Scandinavian women. As the number of samples becomes infinite, the *mean* of this infinite number of sample means (i.e., the *grand mean*) will always equal the mean of

# Figure 5.3

**Three non-normally distributed variables (U-shaped, uniform, and L-shaped) as they appear in the population (individual scores, top row), and their corresponding sampling distributions of the mean for N = 2, N = 4, and N = 30, the second, third, and fourth row, respectively. This illustrates that as the size of samples increases, the sampling distribution of the mean more closely resembles the Normal Distribution, regardless of the shape of the individual scores in the population**

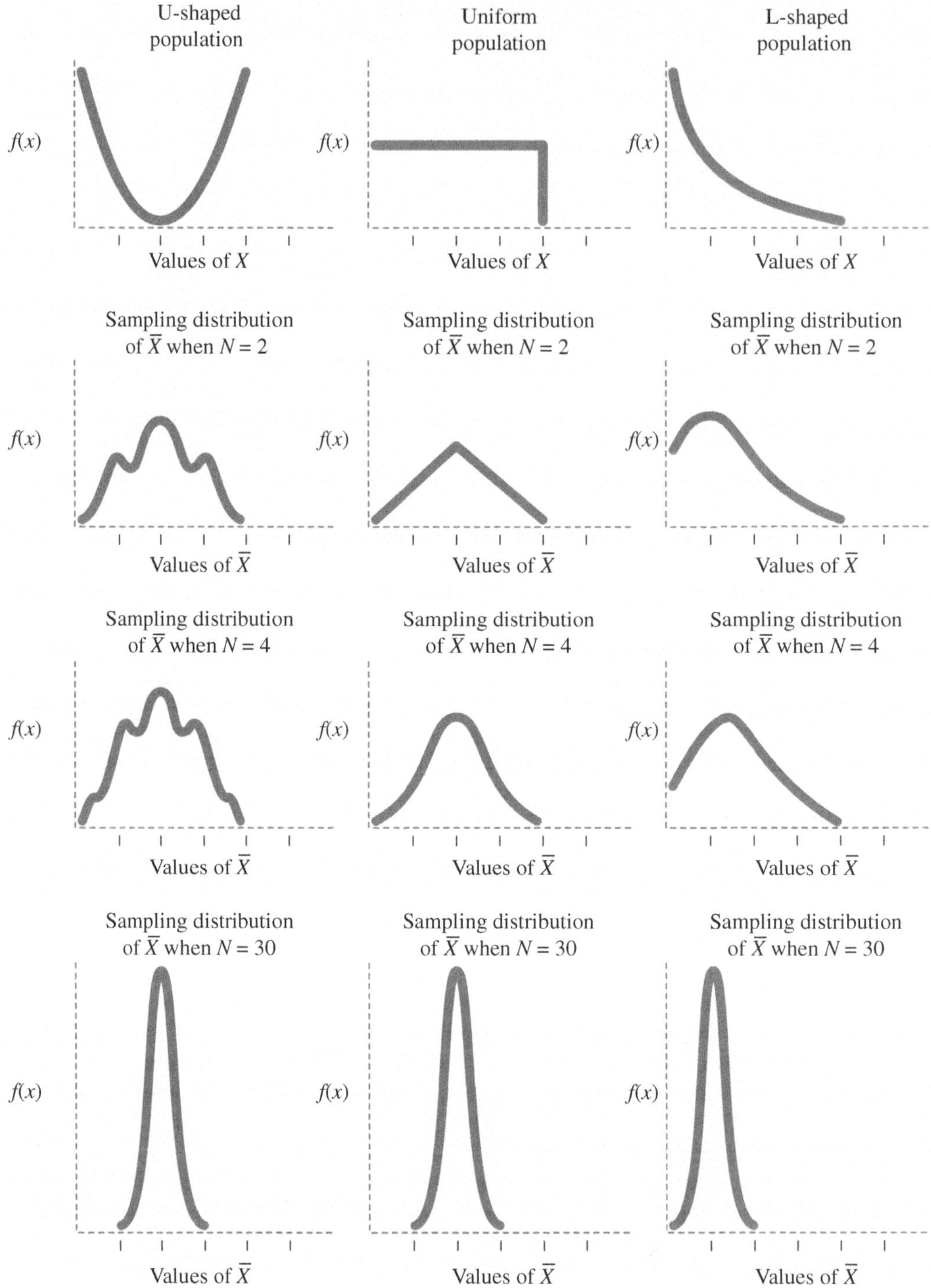


the population of individuals (μ), regardless of the size of each sample ($N$). However, if we want to use $z$-scores to compare the mean of any *one* sample to the whole distribution of sample means, we will also need to know the SD of these sample means (i.e., the SD of the sampling distribution of the mean). It is important to note that the SD of the individual scores, σ (or its estimate $s$), is *not* the correct measure to use when we are working with a distribution of sample means. To see why, consider once again the ND of math SAT scores in the population ($\mu = 500$). Selecting one person at random with a score of 720 is fairly unlikely. But drawing *a whole, totally random, sample of people* with scores so extreme that the sample averages out to a mean of 720 is much less likely because a greater number of low-probability events must occur simultaneously. This implies that the variability of the distribution of sample means is *smaller* than the variability of the distribution of raw scores. In fact, it decreases as the sample $N$ increases.

If a large number of sample means were actually available (as in Table 5.1), you could measure the variability of the means directly. You would simply calculate the SD *of the sample means*, using the usual formula for computing the SD (see Chapter 3) and treating the sample means just like ordinary numbers. However, behavioral scientists rarely have *many* samples of a given size. They usually have only a few. So, instead of measuring the SD of the sampling distribution directly, we will derive it theoretically (much easier!).

## The Standard Error of the Mean

Luckily, there is a simple statistical law that relates the SD of the sampling distribution of means to the sample size $N$ and the SD of the population.

$$\sigma_{\overline{X}} = \frac{\sigma}{\sqrt{N}} \qquad \text{Formula 5.1}$$

where

$\sigma$ = the population SD
$N$ = number of observations in each sample
$\sigma_{\overline{X}}$ = the standard error of the mean

This quantity estimates the variability of means for samples of the given $N$, so it is a *standard deviation* of *means*. But because it does that, it also tells you how trustworthy any single mean that you have is. Note that the symbol for the SD of sample means makes sense; it is $\sigma$ with an $\overline{X}$ subscript, which indicates that it is an SD of means. For historical reasons, the SD for means is called the **standard error of the mean**, but conceptually it provides the same information about the distribution it describes as the SD does for a distribution of individual scores: it tells you how far the scores are, on average, from their mean. Only now the scores are means, and the mean of the distribution is the grand mean.

A *small* value of $\sigma_{\overline{X}}$ indicates that if you were to draw several different random samples of the same size from this population, the various values of $\overline{X}$ would be expected to be relatively close to one another (because the sampling distribution would have a small SD). In this situation, therefore, any single value of $\overline{X}$ that you may have is likely to be a rather accurate estimate of $\mu$. But if $\sigma_{\overline{X}}$ is *large*, and you were to draw a bunch of different random samples of the same size from this population, the various values of $\overline{X}$ would likely differ markedly from one another (because they come from a sampling distribution that has a large SD). Therefore, any one value of $\overline{X}$ would be a less trustworthy indication of the true value of $\mu$.

As you can see from Formula 5.1, the standard error of the mean will always be smaller than the SD of the raw scores. It becomes smaller as the sample size grows larger. The term *standard error* implies that the difference between a population mean and the mean of a sample drawn from that population is an "error," caused by the cases that happened by chance to be included in the sample. If *no* error existed (as would happen if the "sample" consisted of the entire population), all of the sample means would exactly equal each other, and the value of the standard error of the mean would be zero. For example, imagine that your population of interest is made up of the 12 psychology majors who are members of their department's Psych Club (the first rule of Psych Club: Do not talk about Psych Club). Let us imagine that you are interested in the average height of this population, $\mu$. You could draw samples of $N = 12$ students from this population (i.e., you are able to measure every person in the population), and in this case, your sample means would always be exactly the same, and equal to $\mu$. There would be no error due to sampling because each x-bar would be identical, and the standard error of the mean would be zero. A situation such as this would be ideal for purposes of statistical inference because the mean of any one sample would exactly equal the population mean. Of course, this never happens in the real world; we are almost never able to measure the entire population of interest; real data are always subject to sampling error.

## Figure 5.4

**Distribution of Observations of Heights (in Inches) for a Population of 5,000,000 Scandinavian Women**

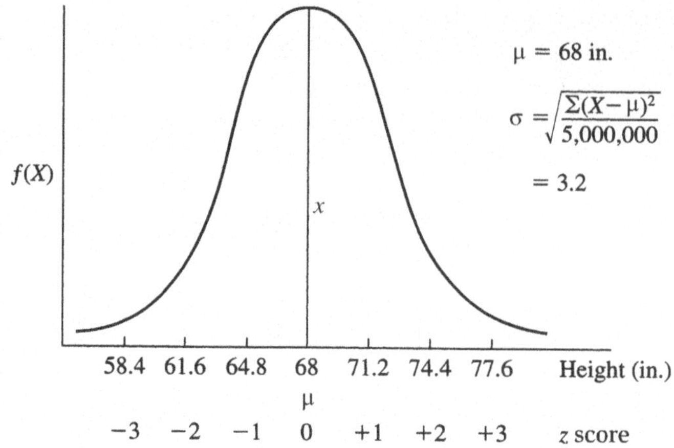


To expand upon the preceding discussion, let us return to our example concerning the heights of samples of Scandinavian women and compare the distribution of raw scores to the distribution of sample means. A hypothetical distribution of five million height observations for Scandinavian women raised on GH milk is shown in Figure 5.4; each $X$ represents one *raw score* (i.e., one individual). Now suppose that one million random samples of $N = 100$ observations each are randomly selected from that population, and the mean of each sample is calculated. The resulting distribution of *sample means* would closely resemble the particular normal curve that is illustrated in Figure 5.5.

Notice that the shape of *both* distributions (the original distribution of raw scores or single observations, and the distribution of sample means) is only approximately normal (all human height distributions have definite endpoints, whereas the ND trails off in each direction indefinitely), and they are both centered around $\mu$ (68 inches). However, the distribution of sample means is *less variable* than the distribution of raw scores. The SD of the raw scores shown in Figure 5.4 could be calculated directly by applying any of the usual formulas for the SD to the 5,000,000 raw scores [e.g., Formula 3.4 (see page 52) for the SD ($\sqrt{SS/N}$)]; let us suppose that $\sigma$ was found to be equal to 3.2 inches. Similarly, the SD of the sample means in Figure 5.5 could be calculated directly by applying Formula 3.4 for the SD, for instance, to the 1,000,000 sample means. Fortunately, we can use the statistical law represented by Formula 5.1 to make our calculation much easier. Because of Formula 5.1, we know that the SD of the sample means (i.e., $\sigma_{\overline{X}}$) will be equal to the $\sigma$ of the individual observations (3.2) divided by the square root of the size of each sample (i.e., sqrt 100 = 10). Therefore, the value of $\sigma_{\overline{X}}$ is $3.2/\sqrt{100} = 3.2/10 = 0.32$. This matches the value shown in Figure 5.5.

## Figure 5.5

**Distribution of Mean Heights (in Inches) for 1,000,000 Randomly Selected Samples of $N = 100$ Height Observations in Each Sample**

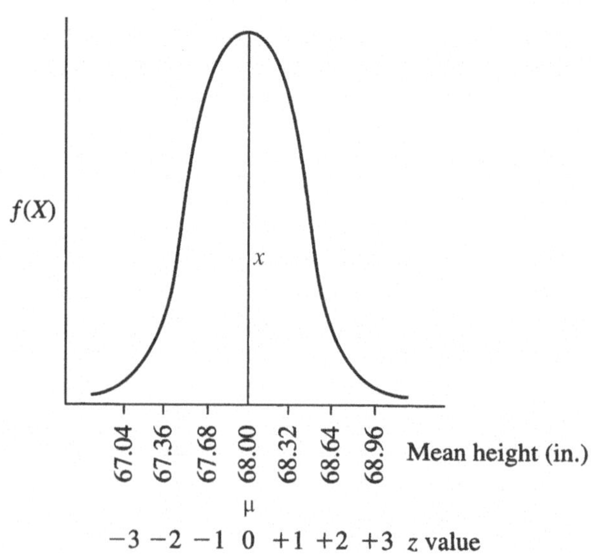

## The *z*-Score for Sample Means

The sampling distribution of the mean for height will closely resemble the ND. Because we know both the mean and SD of this distribution, we can use *z*-scores in conjunction with Table A to answer questions about the proportion of samples that will be shorter or taller than a particular sample. (Why anyone would want to do this will be explained shortly.)

Suppose that Elise has already measured 100 Scandinavian women chosen at random from among those who were *not* raised drinking GH milk, and their average height turns out to be 67.4 inches. This group is shorter than average for Scandinavian women and therefore in line with the researcher's prediction. However, you would expect some samples to be a little taller than 68 inches and others to be a little shorter, by chance. Before the researcher yells "eureka!" she must ask herself a question: How unusual is it to randomly select a group of 100 Scandinavian women from the general population with an average height as short as 0.6 inches less than the population average? Note that the researcher is not interested in the probability that the mean of a randomly drawn sample will be exactly 67.4 inches, like the one she collected; if she just wants to know if her sample is unusually short, she needs to know the proportion of sample means that are as short, or even shorter, than the one she collected. If selecting a group of 100 Scandinavian women at random with a mean of about 67.4 inches (or shorter) is a fairly likely occurrence, then there is little reason to believe that the group of 100 non-GH-milk drinkers is somehow special (e.g., shorter because of the lack of growth hormones).

The researcher's task then becomes one of figuring out exactly how likely it is to randomly select 100 Scandinavian women with an average height of 67.4 inches, or even less. Fortunately, this is precisely the sort of information a sampling distribution of the mean provides (and why we are learning about them). To get a handle on just how unusually short the women in Elise's sample are relative to Scandinavian women in general, we can find out just what proportion of all samples ($N = 100$) of Scandinavian women will have an average height less than (or equal to) 67.4 inches. We saw in the previous chapter that this kind of question can be answered by calculating the corresponding *z*-score. However, in this case, we need to calculate a *z*-score for samples rather than individuals, which means that the formula must be based on the *standard error of the mean* as the measure of variability. That is,

$$z = \frac{\overline{X} - \mu}{\sigma_{\overline{X}}} \qquad \text{Formula 5.2}$$

where

$\overline{X}$ = the observed value of the sample mean
$\mu$ = the hypothesized value of the population mean
$\sigma_{\overline{X}}$ = the standard error of the mean ($= \sigma/\sqrt{N}$)

Given that $\mu = 68$ and $\sigma = 3.2$ for height in the Scandinavian female population,

$$\sigma_{\overline{X}} = \frac{3.2}{\sqrt{100}} = 0.32,$$

and

$$z = \frac{67.4 - 68}{0.32} = \frac{-0.60}{0.32} = -1.875$$

For the sake of simplicity, we will round this z-score to $-1.88$. To find the proportion of the ND that is below $z = -1.88$, draw a diagram of the normal curve, like the ones in the previous chapter, but label it with vertical lines at two *z*-scores: 0 (at the mean) and $-1.88$, nearly 2 SDs to the left of the mean. Then shade in the area to the left of the vertical line at $-1.88$ (a fairly small area). Ignore the negative sign of the *z*-score, and look at the entry for $z = 1.88$ in Table A in the Appendix, which is 46.99. Therefore, the nonshaded area of the left half of your figure (the percent area between the mean and *z*) contains about 46.99% of the distribution. However, we are interested in the shaded area. So, we want to know the percentage of area *below* − 1.88; therefore, we need to subtract the area we just found from 50%. The area below $z = -1.80$ equals 50 − 46.99 = 3.01%. As a proportion, this comes to .0301. Stated as a probability, the chance of randomly selecting 100 Scandinavian women whose height averages out to 67.4 inches or less is only .0301, or just a little over 3%.

Therefore, we know if GH milk consumption were unrelated to height, then the probability of randomly selecting a sample of $N = 100$ with an average height of 67.4 is just 3%. It is not impossible, but it is pretty rare. When compared to groups of Scandinavian women from the general population even though the group average is only a little more than half an inch less than the population mean. As random samples get larger, their means tend to vary less from the population mean. Therefore, a fairly large random sample does not need to have a mean very different from that of the population to be considered unusual. Probabilities such as the value of .0301 calculated here are called ***p* values**, and are a valuable tool for a procedure called *null hypothesis testing (NHT)*, which is discussed next. We introduce this procedure by applying the *z*-score for sample means to another hypothetical study—this one representing a true experiment. (Elise did not randomly assign Scandinavian babies to different milk-drinking conditions, so her study was a quasi-experiment. Even if her non-GH milk group is very unusually short, there may be reasons other than the milk itself that are responsible—for example, babies raised on GH milk may have been raised with other factors that are related to height). Note, however, that NHT is applied the same way to quasi-experiments as it is to true experiments—it is only the causal conclusions that differ.

College Boards and similar standardized tests are usually taken in a group setting with many other students nearby, and this can be distracting for some students. Let us suppose that you want to test the possibility that taking the math section of the SAT in a comfortable private setting will lead to higher exam scores; although it is possible that taking the test in private might lead to *lower* exam scores (and if that is true, you would definitely want to know it), we will begin by expecting that the privacy scores will be higher, not lower, than average. To test this hypothesis, you select 30 names at random from a list of students about to take this test and arrange for these students to take the test individually. After the scores are determined for each of the individually tested students, you calculate the mean math SAT for the group and find that this average is 534.25. This is a welcome result; it would seem unlikely that a random sample of 30 group-tested students would have such a high mean. But we do not have to rely on our subjective impression of how unlikely this sample mean is; we can use the *z*-score for samples to determine this probability precisely. Recall that for SAT scores, the population mean (μ) is set at 500, and the SD (σ) is 100. Therefore,

$$z = \frac{\overline{X} - \mu}{\frac{\sigma}{\sqrt{N}}} = \frac{534.25 - 500}{\frac{100}{\sqrt{30}}} = \frac{34.25}{18.26} = 1.875$$

Because we are interested in the probability of obtaining a sample mean as high or higher than the one we have observed, we want to look up the area above (i.e., to the right of ) a *z*-score of 1.88 (again, we will round the *z*-score to two digits). However, because of the symmetry of the ND, this area will be the same as the area below $z = -1.88$, which we found for the height example to be .0301. Thus, the probability of selecting 30 students at random from the ordinary population (i.e., the population of students taking the exam in the usual setting), which has a mean of 500, and obtaining a sample whose mean is 534.25 or larger is only .0301. Knowing this probability, which, as we just mentioned, is called a "*p* value," will help us make a decision about whether we consider our experimental manipulation to have worked—that is, whether we will conclude that individual testing would produce at least some difference in math SAT scores if given to the entire population.

## Null Hypothesis Testing

The decision process just alluded to, in which we use a *p* value to decide whether our study found a true effect, is known as NHT. You can think of the term "null" as meaning "nothing" or "zero." NHT, therefore, implies that we are testing the hypothesis that the effect being tested is equal to zero. The default hypothesis in NHT is that the treatment you are testing has no effect, and that supposition should be maintained unless a researcher has found convincing evidence that the hypothesis of no effect is false. Therefore, as we will see shortly, researchers usually want to reject the null hypothesis as a means of arguing that their treatment has a significant effect. In terms of the most recent example, the default hypothesis is that taking the SAT in a quiet setting has no effect on SAT scores. More formally, we can say that we will *test the hypothesis* that the mean of the population of individually tested students is equal to (i.e., not different than) 500 on the math SAT.

When the testing of a statistical hypothesis is designed to help make a decision about a population parameter (here, $\mu$), the overall method falls under the category of *inferential parametric statistics*.

### The Null and Alternative Hypotheses

The first step in NHT is to state the **null hypothesis** (symbolized by $H_0$), which specifies the hypothesized population parameter. In the present example,

$$H_0 : \mu = 500$$

This null hypothesis implies that the sample with mean equal to 534.25 is a random sample from the population with $\mu$ equal to 500 (and that the difference between 534.25 and 500 is due to chance factors, and not to the effects of taking the SAT in isolation). The probability of obtaining a sample mean of the observed value (534.25), or even farther above the mean, is calculated under the assumption that $H_0$ is true (i.e., that $\mu = 500$), as we did above ($p = .0301$).

Next, an **alternative hypothesis** (symbolized by $H_A$ or sometimes $H_1$) is formed. It specifies another (usually complementary) value or set of values for the population parameter. In the present study,

$$H_A : \mu \neq 500$$

This alternative hypothesis states simply that the population from which the sample comes does *not* have a $\mu$ equal to 500. That is, the difference of 34.25 between the sample mean and the null-hypothesized population mean is due to the fact that the null hypothesis is false—for example, that individual testing leads to a new distribution of scores with a different mean from the group-testing distribution so that the sample of 30 individually tested students is not actually coming from a population whose mean is 500. (As stated above, the alternative hypothesis is *nondirectional*—no expectation is implied about whether the mean of the tested population will be larger or smaller than the general population. However, in the current example, we actually expect the tested population to have a higher mean. We will deal with directional as compared to nondirectional hypotheses a little later.)

As mentioned above, the null hypothesis gets its name from the idea that nothing is going on. That is, the null hypothesis states that there is *no* effect or *no* difference of the kind that the experiment is seeking to establish. Usually, the theory that the scientist hopes to support is identified with the *alternative* hypothesis. Rejecting $H_0$ will cause the scientist to conclude that her theory has been supported, and failing to reject $H_0$ will cause her to conclude that her theory has not been supported. This is a sensible procedure because scientific caution dictates that researchers should not jump to conclusions. They should claim success only when there is a strong indication to that effect—that is, when the results are sufficiently striking to cause rejection of $H_0$.

### The Criterion of Significance

Having set up the null and alternative hypotheses, we need to define *how unusual* a sample mean must be to cause you to reject $H_0$ and accept $H_A$ (by "unusual" we mean unlikely to occur in the null hypothesis distribution). Intuitively, an observed sample mean of 501 would hardly represent persuasive evidence against the null hypothesis that $\mu = 500$, even though it is not exactly equal to the hypothesized population value. Conversely, an observed sample mean of 700 would do more to suggest that the hypothesized population value was incorrect. Where should the line be drawn? This is where the $p$ value can be helpful.

To simplify our terms, let us refer to an experiment for which the null hypothesis is actually true (e.g., individual testing did *not* affect math SAT scores) as a *null experiment*. The object of NHT can then be described as trying to keep under control the number of null experiments for which the null hypothesis is mistakenly rejected; for such experiments, the results can look good solely because of sampling error (e.g., by chance, our sample included a disproportionate number of students who were very good at math). Rejecting the null hypothesis for an experiment in which the null hypothesis is actually true is called a **Type I error**. Although we can never know whether $H_0$ is true for any particular experiment, and we do not even know how often null experiments are performed by various researchers, we can try to control the rate at which Type I errors are made. We do this by deciding on the proportion of all null experiments we will allow to be declared **statistically significant** (i.e., the null hypothesis has been rejected). This is not something that an individual researcher decides; this is something that a research community decides as a whole.

The proportion of null experiments that will be considered "significant" is called the **criterion of significance** and is symbolized by the first letter of the Greek alphabet, *alpha* ($\alpha$). Due in large part to the pioneering work of Sir Ronald Fisher, the default value commonly used in behavioral research for alpha is .05. Note that setting alpha to .05 fixes only the Type I error *rate* and does not determine the *total number* of Type I errors. Although no attempt is made to control how many null experiments are performed, universal adoption of the .05 level ensures that only (about) 5% of null experiments (i.e., experiments for which the null hypothesis is actually true) that *are* performed will lead to a rejection of $H_0$ and therefore to results that are (mistakenly) regarded as statistically significant (i.e., Type I errors). Routinely setting a lower value for alpha (e.g., .01) is an option that would sometimes result in undesirable consequences that will be discussed shortly, and a larger value for alpha would lead to a rather high rate of Type I errors, so the .05 significance level is considered a reasonable compromise. Now let us see how the $p$ value for a particular experiment can be used to make a decision about the null hypothesis.

### Critical Values

The $p$ value for the math SAT experiment was .0301. This means that if individual testing has no effect at all ($H_0$ is true), about 3% of experiments like ours would nonetheless result in sample means as high as or even higher than ours. It may seem that because .0301 is less than .05, we should reject the null hypothesis using the .05 decision rule. However, .0301 is a *one-tailed* $p$ value, in that it applies only to the upper end (or tail) of the distribution. If our null and alternative hypotheses were set up for a *two-tailed* test, the implication is that we would also have tested our sample mean for statistical significance if it were *below* the population mean. The two-tailed test is the more cautious option, as we will explain shortly, and the one usually required by peer-reviewed journals. To perform a two-tailed instead of a one-tailed test, all you have to do is double your one-tailed $p$ value (which you obtained from Table 5.1) and compare it to your alpha level. For our example, the two-tailed value would be $.0301 + .0301 = .0602$.

Students are often puzzled about why we should double a one-tailed $p$ value to arrive at the comparable two-tailed value. Try thinking about it this way: A two-tailed $p$ is defined as the probability of obtaining a result as *extreme*—in either the positive or negative direction—as the one you got, or more extreme, if the null hypothesis is true. We found a result so extreme (SAT = 534.25) that it left only about 3% of the area of the curve above it. That $p$ value (.0301) indicates that if the $H_0$ is true, you might still come across a sample with a mean *as high or higher* than 534.25 (there is about a 3% chance of that). But what about sample means as or more extreme at the other end of the distribution? Well, if the sample mean for our individual testing group had been 465.75 (i.e., 34.25 points *below* the mean), then it would have a $z$-score of $-1.875$, and .0301 of the curve would fall beyond (lower than) it. Therefore, if you have a two-tailed hypothesis—one for which you would be interested in an extreme result in *either direction* from the mean—then you should add the area beyond one extreme score with the area beyond the opposite extreme score to get the total $p$ value. We have just seen that the $p$ value for sample means 34.25 points *above* the mean is .0301, and the $p$ value for sample means 34.25 *below* the mean is also .0301. Therefore, the two-tailed $p$ value for this experiment is .0301 + .0301 = .0602. Thus, you can see how the symmetry of the normal curve allows you to double your one-sided $p$ value to go from a one-tailed to a two-tailed significance test. Note that .0602 is *not* less than .05, so the $H_0$ cannot be rejected at the .05 level for a two-tailed test. We will elaborate on the distinction between one- and two-tailed tests shortly.

There are other ways to make decisions about the null hypothesis that do not require finding the $p$ value for your experiment. If you are planning to conduct a two-tailed test with, say, $\alpha = .05$, you can find the minimum $z$-score needed for significance ahead of time and then just check the $z$-score for your sample to see if it is larger than the borderline $z$-score. The $z$-scores that fall right at the borderline between the significant and nonsignificant portions of the null hypothesis distribution are called **critical values** (or, less formally, *cutoff scores*). For a two-tailed test, there is a critical value on each side of the distribution, as illustrated in Figure 5.6.

When the normal curve model is appropriate, the two-tailed critical values beyond which $H_0$ should be rejected are easily ascertained using the normal curve table (Table A). Since 2.5% of the area of the curve is in each tail, 50.0% − 2.5%, or 47.5%, falls between the mean and the cutoff score. If we look in the body of Table A for a value of 47.5%, we can see that the critical $z$ values are found to be equal to −1.96 and 1.96. Once the sample mean has been converted to a $z$-score, the .05 criterion of significance states that $H_0$ should be rejected if $z$ is less than or equal to −1.96 or $z$ is greater than or equal to +1.96. Conversely, $H_0$ should be retained if $z$ is between −1.96 and +1.96. Note that the area in each rejection region is equal to $\alpha/2$, or .05/2 or .025, and that the total **area of rejection** (or *critical region*) is equal to the significance criterion ($\alpha$) or .05.

A less common, and more stringent, criterion is the 1% or 0.01 *criterion of significance*. Using this criterion, two-tailed hypotheses about the population mean are rejected if, when $H_0$ is true, an obtained sample mean is so unlikely to occur by chance that no more than 1% of sample means would be so extreme. Here again, $\alpha/2$ or .005 (or 1/2 of 1%) is in each tail. This criterion is illustrated in Figure 5.7.

When the normal curve model is appropriate, the critical values beyond which $H_0$ should be rejected using the .01 criterion are ascertained by looking in the body of Table A for the value 49.5%. After rounding off, these cutoff scores, expressed as $z$-scores, are −2.58 and +2.58. Once the sample mean has been expressed as a $z$ value, the .01 criterion of significance states that $H_0$ should be rejected if $z$ is less than or equal to −2.58 or $z$ is greater than or equal to +2.58. Conversely, $H_0$ should be retained if $z$ is between −2.58 and +2.58. Once again, the area of rejection is equal to $\alpha$ (0.5% + 0.5% = 1% or .01), the criterion of significance.

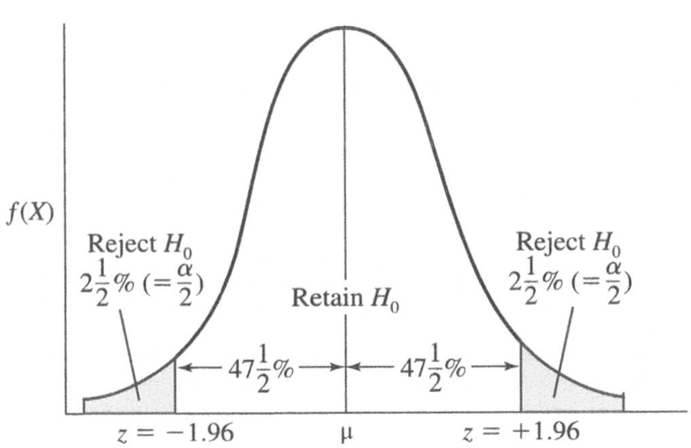


## Figure 5.6

**Areas for Rejecting $H_0$ in Both Tails of the Normal Sampling Distribution Using the 0.05 Criterion of Significance When $H_0$ Actually Is True**

## Figure 5.7

**Areas for Rejecting $H_0$ in Both Tails of the Normal Sampling Distribution Using the 0.01 Criterion of Significance When $H_0$ Is True**

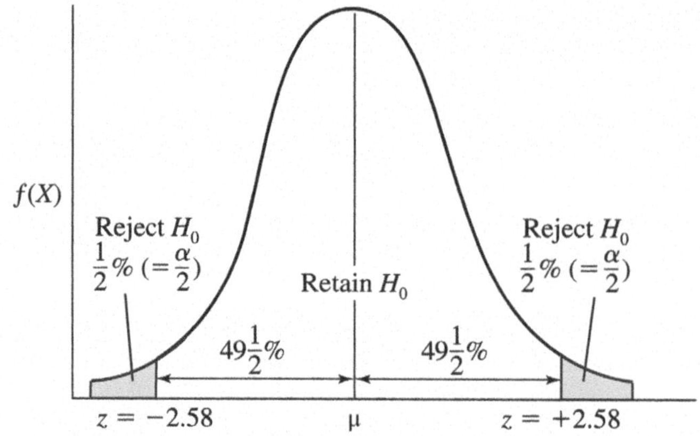


### Consequences of Possible Decisions

Admittedly, an event that occurs 1 in 20 times (i.e., that has a probability of occurrence of .05) is not an extremely rare event. Yet if we declare the result to be statistically significant when we set alpha to .05, we are accepting that probability of committing a Type I error.[3] Given that Type I errors are undesirable, why does the behavioral research community not agree to set .01 as the minimum alpha for ordinary circumstances, rather than 0.05? The short answer is that as you make alpha smaller, the critical value that an experimental result has to exceed to be statistically significant becomes larger. Therefore, more and more experiments for which the null hypothesis is *not* true will nonetheless fail to yield significant results. This is the undesirable consequence of setting a smaller value for alpha that we alluded to earlier.

When the null hypothesis is *not* true but you fail to reject $H_0$ because your experimental results are not extreme enough, you are making an error that can be thought of as the opposite of a Type I error. It is appropriately called a *Type II error* and its likelihood is symbolized by the second letter of the Greek alphabet, *beta* (β). Whereas alpha is fixed by the researcher, the value for beta associated with any particular experiment depends on a number of factors. Usually, it can only be estimated roughly, as we will demonstrate in Chapter 11.

For the math SAT experiment, there are two possibilities with respect to the mean of the theoretical population in which all students are individually tested: Either $\mu = 500$ ($H_0$ is true) or μ is not equal to 500 ($H_0$ is false) and individual testing has at least some effect. There are also just two possible decisions you could make based on the results of this experiment: reject $H_0$, or retain $H_0$. The four possible combinations of the actual mean of the experimental population and the decision that is made about the experiment can be summarized as follows (see Table 5.2):

1. The population mean for individually tested students is actually 500 (i.e., individual testing would not change the mean of the population) and you *incorrectly reject* $H_0$. You have committed a Type I error.
2. The population mean for individually tested students is actually *not* 500 and you *incorrectly retain* $H_0$. Failing to reject a *false* null hypothesis is a Type II error.
3. The population mean for individually tested students is actually *not* 500 and you *correctly reject* $H_0$. The (conditional) probability of reaching this correct decision is called the *power* of the statistical test and is equal to $1 - \beta$.
4. The population mean for individually tested students is actually 500 and you *correctly retain* $H_0$. The (conditional) probability of making this correct decision is equal to $1 - \alpha$.

[3] This is a matter of convention and is not based on any mathematical principle, so common sense is essential in borderline cases. If, for example, $p = .06$, the technically correct procedure is to call the event "not sufficiently unlikely," but .06 is so close to .05 that the best plan is to repeat the experiment and gain additional evidence. The reason that a guide such as the .05 rule is needed is that NHT would quickly become useless if each researcher were free to set alpha to any value they found convenient for each result being reported.

## Table 5.2 Model for Error Risks in Hypothesis Testing

| *Outcome of experiment dictates:* | State of the population: $H_0$ is actually true | State of the population: $H_0$ is actually false |
|---|---|---|
| *Retain* $H_0$ | *Correct decision:* probability of retaining true $H_0$ is $1 - \alpha$ | *Type II error:* probability (risk) of retaining false $H_0$ is $\beta$ |
| *Reject* $H_0$ | *Type I error:* probability (risk) of rejecting true $H_0$ is $\alpha$ | *Correct decision:* probability of rejecting false $H_0$ (power) is $1 - \beta$ |

### One- Versus Two-Tailed Tests of Significance

The statistical tests described thus far in this chapter (see, e.g., Figures 5.6 and 5.7) are called two-tailed, or nondirectional, tests of significance. This means that the null hypothesis regarding the population mean, $\mu$, is rejected if a $z$ value is obtained that is either extremely high (far up in the upper tail of the curve) or extremely low (far down in the lower tail of the curve). Consequently, there is a rejection area equal to $\alpha/2$ in each tail of the distribution. As we have seen, the corresponding null and alternative hypotheses are:

$H_0$: $\mu$ = a specific value
$H_A$: $\mu \neq$ this value (i.e., $\mu$ is greater than the value specified by $H_0$ *or* is less than this value)

Let us suppose, however, that a researcher argues as follows: "My theory predicts that the mean of population $X$ is *less than* 100." (For example, she may predict that people in a distracting environment will score below average on a test of their reaction time.) "It's all the same to me whether $\mu$ is equal to 100 or much greater than 100; my theory is disconfirmed in either case. If I use a two-tailed test, I am forced to devote $2^1/_2$% of my rejection region (using $\alpha$ = .05) to an outcome that is meaningless insofar as my theory is concerned. Instead, I will place the entire 5% rejection region in the lower tail of the curve." (See Figure 5.8.) "In other words, I will test the following null and alternative hypotheses:

$$H_0 : \mu \geq 100$$

$$H_A : \mu < 100$$

$$\alpha = 0.05$$

"Note that I am being properly conservative by assuming at the outset that my results are due to chance (the null hypothesis). Only if I obtain a result very unlikely to be true if $H_0$ is true will I conclude that my theory is supported" (i.e., switch to the alternative hypothesis). "Looking at Table A, I see that a $z$-score of $-1.65$ encloses about 45% of the normal curve between it and the mean, leaving 5% of the distribution in the lower tail" (as shown in Figure 5.8). "Therefore, I will use $-1.65$ as my only critical value." (Actually, $-1.64$ is just as close an approximation to cutting off 5% of the curve, but $-1.65$ is the more cautious choice with respect to controlling Type I errors.) "This means that I will retain $H_0$ if $z$ is anywhere between $-1.65$ and $+\infty$ and reject $H_0$ if $z$ is $-1.65$ or less. Since a two-tailed test of significance would lead to rejection of $H_0$ only if $z$ were $-1.96$ or less (or $+1.96$ or more), the one-tailed value of $-1.65$ quite properly makes it easier for me to reject $H_0$ (and conclude that my theory is supported) when the results are in the direction that I have predicted."

## Figure 5.8

**One-Tailed Test of the Null Hypothesis That $\mu \geq 100$ Against the Alternative That $\mu < 100$**

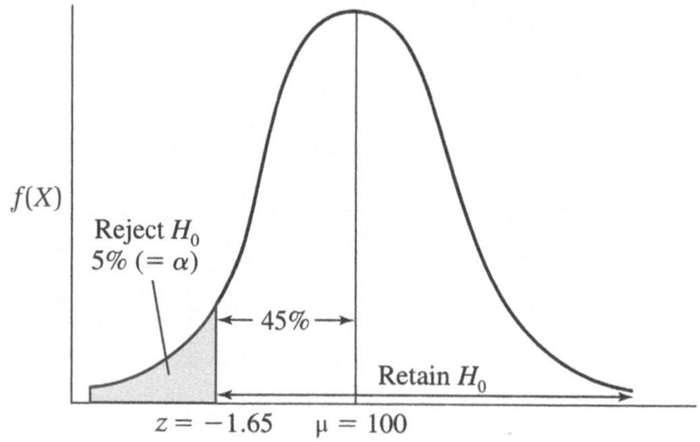


Similar reasoning could be applied if the researcher claimed to be interested *only* in the outcome that the population mean was *greater* than 100. The entire rejection region would be placed at the upper end of the curve, and $H_0$ would be retained if $z$ were between $-\infty$ and $+1.65$ and rejected if $z$ were $+1.65$ or more.

Although behavioral researchers often have a theory that predicts the direction of their results, and could therefore argue that performing a one-tailed test (also called a *directional test*), is sufficiently *conservative* (i.e., in this context, conservative means particularly cautious about Type I errors), they usually test their results in terms of two-tailed tests. Consider once again our math SAT example. The $z$-score calculated in that example, 1.875, is larger than the critical value for a one-tailed test at the .05 level, 1.65. Therefore, the null hypothesis would be rejected using a one-tailed test. We might have had good reason to expect the individually tested sample to have a *higher*, rather than lower, average on the math SAT, because such test conditions are usually more conducive to concentration. So, it would seem fair to use the one-tailed test in this case and declare statistical significance.

However, a one-tailed test implies a promise that we would not have presented our results if our sample mean had been very much *lower* than the mean of the population—even if it were so low as to easily exceed the critical value for a two-tailed test in the other (i.e., lower) direction. If such a promise were routinely made and obeyed, researchers could not ethically report extreme results in the unexpected direction. This could cause serious problems. For example, the individually tested students might feel that they were under close scrutiny and become much more anxious, making them *less* likely to perform well on the test. This could be a finding of interest that deserves to be reported (though it could be argued that the experiment should first be replicated because it yielded unexpected results).

If you were to lower your critical value by putting your entire alpha in one tail and then put even just half of your alpha in the other tail *after* seeing the results, you would be violating the .05 decision rule. That is, you would really be using an alpha of $.05 + .025 = .075$, which would make the probability of a Type I error greater than .05. Students sometimes find this logic difficult to conceptualize. Consider the following example: I have a fair coin that I will flip in a heads or tails bet with you. I flip the coin, catch it in one hand, and immediately slap it against the top of my other arm with my hand covering it. Neither one of us can see which side is up, so the probability of either one of us calling heads or tails correctly is 50%. Now imagine we repeat the flip, but this time I peek under my hand after I slap the coin on my arm and see that it is heads up. The probability of me calling heads or tails correctly has changed from 50% to 100%—I will never call tails after I have seen that the coin is heads up! The same logic applies in hypothesis testing when you "peek" at your data before deciding which tail to test. Let us say you bet on the left tail (extreme scores below the mean) and place all .05 there; the moment you look at your sample mean—cha-ching!—you have "spent" all 5% of your alpha. Why? Any sample mean that ended up in the lower 5% of distribution would lead you to reject the $H_0$. You have exposed 5% of the $H_0$ distribution to rejection. So far so good. But, if after seeing that your sample mean ended up in the other tail, you then decide it would have been wiser to use a two-tailed test, and move half of your alpha to the upper tail, then—cha-ching!—you have just spent another 2.5% and alpha is now .075. Any result that landed in the top 2.5% of the distribution would lead you to reject the $H_0$. Therefore, you have exposed 5% of the $H_0$ when you looked at the data, and another 2.5% when you "moved" half your alpha to the upper tail, so alpha is actually .05 + .025 = .075. One source of confusion about this stems from the illusion that you have "moved" alpha from one tail to the other. Once you look at your data, that original 5% of the $H_0$ is spent—gone. No returns. It is no longer there to move. You have to invest an additional .025 if you want to switch to a two-tailed test, and that extra .025 puts you over your spending limit for Type I errors of 5% of the $H_0$.

Whenever we recognize the significance of a one-tailed test, we are implicitly accepting the promise that the researchers would not have tested and presented their results, no matter how extreme their results might have been in the opposite direction. Because the research community does not want to discourage the reporting of paradoxical findings but also does not want alpha to creep above the .05 level, the two-tailed test is the norm. However, one-tailed tests may be justified if the results would make no sense in the opposite direction and would therefore never be tested in that direction. For example, in a study to test the effectiveness of a new analgesic (pain killer), you would not be interested in the distinction between a result in which (a) the experimental drug did not decrease pain and (b) the drug led to significantly *more* pain. If the only thing you were interested in was testing its effectiveness, then you would test the lower-pain tail of the distribution with a one-tailed test. However, if you were developing psychological theories about pain, then a result in the significantly more pain direction might be of interest. In the world of basic research, one-tailed hypotheses are very rarely tested. There is almost always something meaningful about results that fall in the unpredicted tail. Note that, because the critical *z*-scores for a two-tailed test are larger than for a one-tailed, it is harder to get statistical significance when performing a two-tailed test (a more extreme result is required). Therefore, two-tailed tests are considered more conservative (better at controlling the rate of Type I errors) than one-tailed tests. Using an alpha of .01 rather than .05 is also more conservative, but note that being more conservative leads to making more errors of the second kind, Type II errors. Controlling Type II errors is the main concern of Chapter 11.

## Assumptions Required by the Statistical Test for the Mean of a Single Population

Our use of the *z*-score for sample means, in conjunction with the table of the normal curve, to obtain a *p* value for our hypothetical math SAT experiment was based on two main assumptions. The most important of these is that the individuals in our sample were selected randomly (and independently) from the population. In the case of the math SAT experiment, it is not hard to imagine that individuals could be selected at random from a list of people registered to take the exam, but the validity of the experiment would then depend critically on the compliance of those who were selected for the new experimental condition. If a considerable number of selected individuals refuse to participate (e.g., the ones who are afraid that the new condition will throw them off or, perhaps, jinx them), the results could easily become biased in one direction or the other (e.g., the worried students might be worse at taking standardized tests in general, so by dropping out, their absence would raise the score of the selected group—even if the new condition does not actually improve performance at all). The usual solution is to randomly divide the sample of individuals willing to comply with the experiment into two groups, one of which gets the new condition, while the other serves as a control group (taking the test in the usual way). Such two-group designs are discussed in Chapter 7, but first, we have more groundwork to cover on the one-group case, in the next chapter.

The next most important assumption is that the variable being measured (e.g., math SAT) is normally distributed in the population. The Central Limit Theorem tells us that for fairly large samples, we will get reasonably accurate *p* values even if the original distribution is not very "normal." However, to use the *z*-score formula at all, we had to know $\sigma$, the SD of the individual scores in the population. Unfortunately, we will know $\sigma$ for only a few variables that have been studied extensively in the population. More often we will not know $\sigma$, and we will have to estimate it from the data we do have. Estimating $\sigma$ is an additional step that slightly complicates our test of a single population mean. This complication involves a new theoretical distribution that is similar to the ND and is described in the next chapter.

## Why is Null Hypothesis Testing So Misunderstood?

NHT has been subject to criticism ever since it was introduced by Sir Ronald Fisher in 1928 (see Fisher, 1951, 6th ed.) and soon after modified by the joint work of Neyman and Pearson (1928, 1933). The history of this criticism is covered comprehensively by Nickerson (2000). Many of the criticisms are not about the logic of the tests themselves, but concern how people misuse or fail to draw appropriate conclusions from NHT.

One criticism you can likely relate to is that NHT is confusing. This is due, in part, to the fact that Fisher's approach to NHT was fundamentally different from the Neyman-Pearson approach, and the conflict has never been resolved. Fisher's view was that a $p$ value larger than 0.05 should make us cautious about declaring that there is a real difference between the underlying populations because it would be too easy to get the same result in your sample even if the variable you are studying has no effect whatsoever (e.g., GH milk has no effect on the heights of Scandinavian women). However, Fisher was not in favor of a firm decision when $p$ is larger than .05, and therefore he would not assume that you were making any type of error just by taking a cautious stance with respect to your study. He also understood that a $p$ value less than .05 was not proof that you were dealing with a real effect, but that it was OK to act as though you did, because it would not make sense to be overly cautious about every result. He knew you could be making an error in declaring your result to be "statistically significant," but the line needed to be drawn somewhere, and preventing 5% of "null" studies from being called significant seemed reasonable to Fisher at the time.

Neyman and Pearson took a different view. They felt that there were important consequences to cautiously *not* calling a result significant when there was really something effective going on—you could miss an important intervention or treatment—and that not rejecting the null hypothesis when it should be rejected was another type of error, which they designated as a Type II error. By contrast, they named the more fundamental mistake of falsely rejecting the null hypothesis a Type I error. Estimating the probability of committing a Type II error is tricky because it depends on not just whether the null is true or not, but on how far from true it may be, something you can only guess at. The probability of *not* making a Type II error is called the *power* of a statistical test; power is the topic of Chapter 11.

It is reasonable to "retain" the null hypothesis when you cannot reject it, but making that decision invites thinking that you have good evidence that the null hypothesis is therefore true. However, a single failure to reject the null hypothesis is very weak evidence that the null is true, especially when not using very large samples. This logical mistake is so common that it has its own name: the fallacy of affirming the consequent.[4] This faulty reasoning may have been encouraged by the Neyman-Pearson decision-making approach; it was definitely discouraged by Fisher's approach. He really did not like the Neyman-Pearson modifications to NHT. According to a strict Neyman-Pearson approach, you set your alpha (usually .05) before looking at your results and then you make a decision based on whether $p$ turns out to be less than or greater than .05. In that case, you would report your results as either "$p < .05$" or "$p > .05$." Instead, the approach recommended by the APA style manual

---

[4] In propositional logic, a simple conditional such as if p then $q$ when combined with the premise that $q$ is true, invites the conclusion that $p$ is also true. But that conclusion does not follow logically. There may be other ways $q$ could be true (maybe if $r$ then $q$ is also true), so it's impossible to know whether or not $p$ is true without further information. Consider the statement "When it rains, Jordan uses an umbrella" together with the information that "Jordan is using an umbrella." Does it follow that it must be raining? Not necessarily; maybe it's snowing, or he wants to protect his head from a hot sun. That's why it is a fallacy to infer that $p$ is true simply because $q$ is true. In the context of NHT, the premise "If the null hypothesis is true, then it will be retained when tested," does not imply that a retained null indicates that the null hypothesis must be true. There are many reasons experiments produce nonsignificant results. A true null is only one of them. This fallacy is tempting because simple conditional statements such as if $p$ then $q$ are often interpreted by people as bi-conditionals; if and only if $p$ then $q$, which does allow one to infer $p$ from $q$.

(APA, 2020) is to report the exact $p$ value, which is more in line with Fisher's view, because it gives more information as to how unlikely the null hypothesis is to be true.

Another source of confusion is the use of the word *significant* when describing a result for which the null hypothesis can be rejected. The word "significant" seems to imply that the result is large and/or important, but neither is necessarily true. It helps always qualify that term by writing *statistically significant*, but some people will nonetheless be influenced by the common meaning of the word "significant." In reality, you are only declaring that if the null hypothesis were true, results as good as yours would not occur very often. Moreover, the criterion ($\alpha$) usually used for significance is the .05 level, and obtaining a result that can only occur totally by chance a little fewer than 5 times out of 100 (or once in 20 times) is not really such a terribly rare event. Setting alpha at .05 makes it clear that we are ready to relax our suspicions concerning chance factors fairly easily in the interest of scientific progress. Therefore, we should always be skeptical when first considering a novel statistically significant result. It is always possible that any statistically significant result is due entirely to the chance factors involved in obtaining the sample (e.g., someone accidentally sampled a lot of strong placebo responders) and that there is no effect in the population you are sampling from. In other words, a Type I error has been committed.

Yet another source of confusion concerning $p$ values is that while it is true that they tell you the probability of getting results as extreme as the one you found when the null hypothesis is true, they do not tell you the probability of the null hypothesis being actually true given your results. This confusion about probabilities is another major criticism of NHT, which we will attempt to clarify in Chapter 12, along with some other criticisms, when we introduce some alternative approaches to NHT. However, despite the criticisms, NHT remains the dominant approach to data analysis in many areas of psychological research, so in the following chapters, we will proceed to present some of the most common applications of NHT.

## Summary

Although behavioral researchers are usually interested in what the average response of an entire population would be to some treatment or experimental condition, practical constraints generally lead them to select and study a relatively small, random sample from the population of interest. After the members of the sample have been measured, the following steps can be applied in order to draw some conclusions about the population.

### The Standard Error of the Mean

To test null hypotheses about the *mean of one population*, first compute the *standard error of the mean*. This is a measure of how accurate any one sample mean is likely to be as an estimate of the population mean. Then use the standard error of the mean to determine whether the difference between an observed sample mean and the hypothesized value of the population mean is or is not "sufficiently unlikely" to occur if $H_0$ is true (i.e., if the hypothesized value of the population mean is correct).

The formula for the standard error of the mean is

$$\sigma_{\overline{X}} = \frac{\sigma}{\sqrt{N}} \qquad \text{Formula 5.1}$$

where

$N$ = the number of scores in each sample
$\sigma$ = standard deviation of the population

### The $z$-Score for Sample Means

To draw inferences about the population mean, first calculate

$$z = \frac{\overline{X} - \mu}{\sigma_{\overline{X}}} \qquad \text{Formula 5.2}$$

Then look in Table A to find the percentage of the normal curve that extends beyond your calculated $z$. Converted to a proportion, this is the one-tailed $p$ value corresponding to your sample mean. Double this value when you want to obtain a two-tailed $p$ value.

### Null Hypothesis Testing: General Considerations

1. State the *null hypothesis* (denoted by the symbol $H_0$) and the *alternative hypothesis* (denoted by $H_A$). It is at this point that you should decide whether to perform a one-tailed or two-tailed test. (It is important to understand that you cannot *prove* whether $H_0$ or $H_A$ is true because you cannot measure the entire population.)
2. Begin with the assumption that $H_0$ is true. Then obtain the data and test out this assumption. If this assumption leads to unlikely results, you will abandon it and switch your bets to $H_A$; otherwise, you will retain it.
3. Before you collect any data, it is necessary to define in numerical terms what is meant by "unlikely results." In statistical terminology, this is called selecting a *criterion* (or *level*) *of significance*, represented by the symbol $\alpha$.
   (a) Using the .05 criterion of significance, "unlikely" is defined as having a probability of .05 or less. In symbols, $\alpha = .05$.
   (b) Using the .01 criterion of significance, "unlikely" is defined as having a probability of .01 or less. In symbols, $\alpha = .01$.

   Other criteria of significance can be used, but these are the most common.
4. Having selected a criterion of significance, and either a one- or two-tailed test, obtain your data and compute the appropriate statistical test. If the test shows that $H_0$ is unlikely to be true, reject $H_0$ in favor of $H_A$. Otherwise, retain $H_0$.
5. Though you are siding with the hypothesis indicated by the statistical test ($H_0$ or $H_A$), you could be wrong. You could be making either of the following two kinds of errors depending on your decision:
   (a) Type I error: Rejecting $H_0$ when $H_0$ is in fact true. This error is less likely when the .01 criterion of significance is used. The Type I error rate is a conditional probability; it is the probability of rejecting $H_0$ *given* that $H_0$ is actually true, and it is equal to the chosen criterion of significance, $\alpha$. $p$ is also a conditional probability. $p$ is the probability of obtaining results as extreme or more extreme as the ones you got if $H_0$ is true. It is easy to confuse $p$ with $\alpha$—try to remember that $\alpha$ is fixed before you see your results and it almost always conforms to a convention (e.g., .05), whereas $p$ depends on the results you obtained from your study and it could be any amount between 0 and 1.0.
   (b) Type II error: Retaining $H_0$ when $H_0$ is in fact false. This error is less likely when the .05 rather than the .01 criterion of significance is chosen. The probability of this kind of error, denoted by the symbol $\beta$, is not conveniently determined but is discussed in detail in Chapter 11.

It is the relative importance and consequences of each kind of error that help determine which criterion of significance to use.

### One- Versus Two-Tailed Tests of Significance

The use of one-tailed tests of significance is sometimes justified in behavioral science research. If the results are in the direction predicted by the researcher, it is more likely that statistical significance will be obtained. But if the results are in the opposite direction, the entire experiment should be repeated using a two-tailed design before any conclusions are drawn. Therefore, the two-tailed test is more conservative and much more commonly reported, especially in peer-reviewed psychological journals.

**Required Assumptions**

1. The individuals in the sample were selected randomly (and independently) from the population.
2. The variable being measured is normally distributed in the population. This assumption can be relaxed for very large samples due to the Central Limit Theorem.

## Exercises

1. Assume that the mean age of retirement for people working at a large American company (to be viewed as a population) is 65 years with a standard deviation of 3 years.
   (a) If all of the company's retirees from this year are placed randomly into tables at a celebratory dinner of 36 each, what will be the SD of the table means for age (i.e., the standard error of the mean)?
   (b) Using your answer to part (a), what is the *z*-score for a table whose average age is 64.2 years? What are the one- and two-tailed *p* values for this class?
   (c) If you were testing the null hypothesis (i.e., $\mu = 65$), would you reject $H_0$ for the table in part b, at the .05 level for a one-tailed test? For a two-tailed test?
   (d) Repeat part (b) for a class whose mean age is 67.4 years. Would you reject the null hypothesis for this table with a two-tailed test at the .05 level? At the .01 level?
2. Assume that the mean weight for newborn kittens (i.e., population) is 100 grams with $\sigma = 15$ and that kittens born in a shelter are assigned at random to groups with an *N* of 25.
   (a) What is the *z*-score for a group whose average weight is 104 grams? What is the one-tailed *p* value for this *z*-score? What is the two-tailed *p*?
   (b) Repeat part (a) for a group whose mean weight is 92 grams.
   (c) Perform two-tailed null hypothesis tests for the groups in part (a) and part (b) at both the .05 and .01 levels. In each case explain whether you could be making a Type I or Type II error.
3. Given that for SAT scores, $\mu = 500$ and $\sigma = 100$:
   (a) Test the claim of students at Bigbrain University that they have SAT scores that are statistically significantly higher than the ordinary population because a random sample of 25 of their students averaged 530 on this test. Make your statistical decision by comparing the *z*-score for the Bigbrain sample with the appropriate critical value for a two-tailed test. Would a one-tailed test be significant in this case? Would it be justified?
   (b) Repeat part (a) for a random sample of 64 students, who also have a mean of 530 on the SAT.
4. Referring to the 20-point friendliness measure used in the first exercise of Chapter 1, assume that the mean score for all American college students is known to be 10, with an SD of 5 points.
   (a) Based on the data in Table 1.1, what is the *z*-score for friendliness for the students sampled from Turck Hall? What are the one- and two-tailed *p* values for this sample? (For Turck Hall, $\mu = 13$, $\sigma = 4.36$).
   (b) If you were testing the null hypothesis (i.e., $\mu = 10$), would you reject $H_0$ for the students of Turck Hall at the .05 level for a one-tailed test? For a two-tailed test?
5. Repeat Exercise 4, parts (a) and (b), for the sample from Dupre Hall. What is the same and what is different between your calculations for this exercise, and your calculations for Exercise 4? (For Dupre Hall, $\mu = 7$, $\sigma = 2.58$).

6. Calculate the mean and (unbiased) SD for the following sample: 5, 10, 15, 20, 25, 30, 35, 40, 45, 50, 50, 55, 60, 65, 70, 75, 80, 85, 90, 95. Assume that the SD of the population from which these data were sampled is the same as the SD that you just calculated for the sample. (a) Are these data consistent with the hypothesis that the mean of the population is 60? Explain. (b) Repeat part (a) for a hypothesized population mean of 36.

## Thought Questions

1. What is the difference between descriptive statistics and inferential statistics?
2. (a) Do researchers in the behavioral sciences want to draw conclusions about populations or about samples? Why? (b) Why, then, must researchers use samples? (c) What problems are caused by having to use samples? Include and explain the term *sampling error* in your answer.
3. (a) What is a random sample? (b) Why are random samples desirable in behavioral science research? (c) Why is it often difficult or impossible to obtain a sample that is truly random?
4. Which of the following are random samples from the specified populations and which are not? Why? (a) An experimenter selects every 10th name in the telephone book for her city, starting with one name that is chosen blindly. Population: all those with listed telephone numbers in her city. (b) Teenage American students taking introductory psychology at University X volunteer to participate in an experiment. Population: all American teenagers. (c) The names of all students at one particular college in the United States are written on slips of paper and placed in an extremely large hat. The slips are well shuffled, and 50 names are drawn blindly to serve as participants in an experiment. Population: all students at this college. (d) The procedure is the same as in part (c) but the population is all college students in the United States.
5. (a) What is the difference between the standard deviation of a set of scores and the standard error of the mean? (b) Do behavioral science researchers want the standard error of the mean to be small or large? Why? (c) What happens to the standard error of the mean as the sample size ($N$) becomes smaller? Why? What does this imply about the dangers of using samples in behavioral science research that are very small? (d) What happens to the standard error of the mean as the SD of the original scores in the population becomes larger? Why?
6. You are using the .05 criterion of significance. (a) If the two-tailed $p$ value (i.e., the probability of obtaining the results you got if $H_0$ is true and the hypothesized value of the population mean is correct) is .03, what decision should you make about $H_0$? Why? (b) If the two-tailed $p$ value is .33, what decision should you make about $H_0$? Why? (c) Should you treat a result that has a two-tailed $p$ value of .06 the same as a result that has a two-tailed $p$ value of .45? Why or why not?
7. The probability of making a Type I error when using the .01 criterion of significance is lower than when using the .05 criterion of significance. Why, then, do not researchers always use the .01 criterion, instead of the .05 criterion?
8. (a) What is the difference between a one-tailed test of significance and a two-tailed test of significance? (b) What are the disadvantages of using one-tailed tests of significance? Of using two-tailed tests of significance?

## Computer Exercises

1. Assuming that a normative study has shown that the mean anxiety level for college students ($\mu$) is 18 ($\sigma = 11$) on the scale that Jackson used, what is the $z$-score for the baseline anxiety scores in Jackson's class? What are the one- and two-tailed $p$ values that correspond to this $z$-score?

2. Assuming that the mean resting heart rate for all college-aged noncoffee drinkers ($\mu$) is 70 bpm, with $\sigma = 6$, what is the $z$-score for the baseline HR of the noncoffee drinkers in Jackson's data set? What are the one- and two-tailed $p$ values that correspond to this $z$-score?
3. Repeat Exercise 3 for the coffee drinkers in Jackson's statistics classes.

## Bridge to SPSS

You can use SPSS to obtain accurate $p$ values from the ND using the CDFNORM function described in Chapter 4. For a $p$ value, you will always want the area "beyond" your $z$-score—that is, below a negative $z$-score and above a positive $z$-score. When you enter a negative $z$-score, the area returned by CDFNORM is the one-tailed $p$ value. Just multiply it by 2 if you want a two-tailed $p$ value. If you want the one-tailed $p$ value associated with a positive $z$-score, just enter it with a minus sign in front of it to make it negative. Again, you can obtain the two-tailed $p$ value by doubling the one-tailed $p$.

One-sample tests of a sample mean against a null hypothesis will be the main topic of the Bridge to SPSS section in the next chapter.

# Chapter 6

# One-Sample *t* Test and Interval Estimation



## Introduction

In Chapter 5, we described how you could use a *z*-score for sample means to test a null hypothesis concerning the mean of a population. We explained how to find a *p* value and use it to decide whether it is reasonable to suppose that your sample is just a random selection from the ordinary population (i.e., $H_0$ is true) or whether your sample mean is so

*Introductory Statistics for the Behavioral Sciences*, Eighth Edition. R. Brooke Lea and Barry H. Cohen.

Companion website: www.wiley.com/go/ISBS8e

extreme (and therefore $p$ is so small) that the null hypothesis can be rejected on the basis of being "sufficiently unlikely" to account for your results. The latter decision would support the alternative hypothesis that your sample actually represents a population whose mean differs from the null hypothesized value, either because the treatment applied to the participants in the sample had some effect on your dependent variable (as in an experimental study), or because the characteristic for which the participants were selected (e.g., people who exercise regularly) is related to the variable you are measuring (e.g., subjective level of stress). The second case might be called an observational study or quasi-experiment. The statistical procedure for rejecting the null hypothesis is exactly the same for quasi-experiments as it is for true experiments; it is the interpretation of the results following the rejection of the null hypothesis that differs. You cannot draw the same causal conclusions from a quasi-experiment that you can from a true experiment. We return to this point in the context of other statistical procedures in later chapters of this text.

Unfortunately, the $z$-test requires that you know the standard deviation (SD) of your dependent variable in the population (i.e., $\sigma$), and this is rarely known for the variables studied by behavioral science researchers. The good news is that you can estimate $\sigma$ from the data you do have and then use a formula very similar to the $z$-score for sample means, as we demonstrate next.

When you do not know $\sigma$, the best that you can do is to estimate it by calculating the value of $s$ for your sample. When divided by the square root of $N$, the resulting quantity, expressed symbolically as $s_{\bar{X}}$, can be used as an estimate of $\sigma_{\bar{X}}$ in the $z$-score formula for sample means. The new formula looks like this:

$$\text{est. } \sigma_{\bar{X}} = s_{\bar{X}} = \frac{s}{\sqrt{N}} \qquad \text{Formula 6.1}$$

Notice how similar Formula 6.1 is to Formula 5.1 (page 93). If we use Formula 6.1 as the denominator of Formula 5.2, we get a $z$-score formula for groups that can be used when $\sigma$ is not known.

$$z = \frac{\bar{X} - \mu}{\frac{s}{\sqrt{N}}} \qquad \text{Formula 6.2}$$

Note that the use of the symbol $z$ in Formula 6.2 implies that the values from this formula will form a normal distribution (ND). However, this is not strictly true and is only approximately true when the samples are quite large. That is why Formula 6.2 is sometimes called the *z-test for large samples*. It is also why we need to describe a more general-purpose modification of this formula in the next section.

## The Statistical Test for the Mean of a Single Population When σ Is Not Known: The *t* Distributions

Although Formula 6.2 looks similar to Formula 5.2 (page 95), there is a critical difference. As just mentioned, the values obtained from Formula 6.2 are *not* normally distributed; the normal curve will serve as a good approximation for the $z$-scores obtained *only* if the sample size is quite large (e.g., more than 100). This is because $s_{\bar{X}}$ is only an estimate of $\sigma_{\bar{X}}$. Using the normal curve procedure will give inaccurate answers, especially when the sample size is small because as the samples get smaller, $s$ becomes less reliable as an estimate of $\sigma$. We need a more appropriate statistical model to represent the exact distribution of $(\bar{X} - \mu)/s_{\bar{X}}$, one that is accurate for all sample sizes. This new statistical model is known as the *t distributions*.

In contrast to the normal curve, there is a *different t* distribution for *every sample size*. The sampling distribution of the mean based on a sample size of 10 has one $t$ distribution, while the sampling distribution of the mean based on a sample size of 15 has a different $t$ distribution. Fortunately, you do not have to know the exact shape of each of the various

$t$ distributions. Critical values for the curves corresponding to each sample size have been determined by mathematicians and summarized in statistical tables. All you need do is note the size of your sample and then refer to the appropriate $t$ distribution. A table of critical $t$ values is presented as Table B in the Appendix. (The assumptions that underlie the use of $t$ distributions are discussed in the next chapter.)

Statisticians could have created a table just like Table A (for the normal curve) for each of the $t$ distributions, but that would require a large number of tables. Theoretically, an infinite number of tables would be needed (one for each possible sample size), but the $t$ distribution begins to resemble the ND rather closely for large sample sizes. In fact, when $N$ is about 100 or more, the difference between the corresponding $t$ table and the normal curve table becomes small enough to ignore for most purposes. For example, remember how the critical $z$ value for a two-tailed test at $\alpha = .05$ is 1.96 (see page 71)? Well, the comparable critical $t$ value for a sample size of 121 is 1.98—just a tiny bit larger. By contrast, the critical $t$ for a sample size of 2 is 12.06, considerably larger than 1.96. This difference in critical $t$ values illustrates the convergence between the ND and t distributions as sample sizes get large.

But even if we only needed $t$ tables for samples between 2 and 100, that would still require a cumbersome number of tables. Providing a $t$ table that contains only critical values saves a great deal of space. Unlike Table A, therefore, Table B does not make it possible for you to find the $p$ value that corresponds to a particular value for $t$, but you can look up critical values for several commonly used alphas, for both one- and two-tailed tests. When you use statistical software to compute $t$ values, you will be given the exact $p$ value that corresponds with the observed $t$; in such cases, you do not need critical values or tables at all! Statistical significance is determined by comparing $p$ with alpha (usually set at .05).

### Degrees of Freedom

The $t$ table is given in terms of *degrees of freedom* ($df$) rather than sample size. When you are comparing a sample mean to a hypothesized population mean (and $\sigma$ is not known),

$$df = N - 1$$

That is, the number of $df$ in the present situation is equal to 1 *less* than the sample size. Recall that this is equal to the divisor in the formula for $s^2$ (Formula 3.3; page 51).

*Degrees of freedom* is a concept that arises at several points in statistical inference; you will encounter it again in connection with other statistical tests in later chapters. So, it is worth explaining what it means and why it is important.

Recall that the $t$ distribution is only needed when we use the sample SD, $s$, as an estimate of the population SD, $\sigma$. In general, estimates are only as good as the information they are based on. For example, if you wanted to estimate the average age at which North Americans lose their virginity, you would have more confidence in a sample mean based on the responses of 100 people than if it was based on a sample of just 2 people. By the same logic, we should have more confidence in an $s$ that is based on a large sample compared to a small sample. That is why we need different $t$ distributions; each one is adjusted according to how accurate $s$ is as an estimate of $\sigma$. The more information we have for calculating $s$, the less adjustment is needed, until the $t$ distribution is virtually indistinguishable from the ND. Therefore, the degrees of freedom (which determines the specific $t$ distribution that is appropriate) are based on how many pieces of information (data points) go into $s$. So, why is not $df = N$?

Good question. Recall that (conceptually), the SD is similar to measuring how far each score is from the mean of the scores, and then averaging those deviations[1] (see page 51). Each deviation score can be thought of as providing a piece of information relevant to how dispersed the scores are. You might think that with $N$ data points you would have $N$ pieces of information, but you would be wrong! In Chapter 3 we saw that the sum of the squared

---

[1] The SD is, mathematically, a little more complex than the mean of the deviations scores—which is called the mean deviation (see page 51). But the SD can be thought of as the value of a typical, or average, deviation score.

deviations from the mean must, mathematically, always add up to 0 (page 45). Now, let us say you have a sample of $N = 20$, and you are computing SS by hand. After computing the 19th squared deviation, you notice that the sum of the 19 squared deviations adds up to $-2.5$ (i.e., SS $= -2.5$). What must the squared deviation of the 20th score be? $+2.5$! The last data point is determined after the 19th deviation has been measured—you do not even have to look at the last score—it will always be whatever score creates a squared deviation that makes SS add up to 0. That score, therefore, is fixed, it is not free to vary. So, the score (its squared deviation) provides no new information. That means that whenever you are computing the SD, you only have $N - 1$ scores that are free to vary, and that are therefore informative. The appropriate $t$ distribution to use is the one based on the value of $N - 1$ pieces of information. The number of degrees of freedom tells you how many pieces of information are going into your estimate of $\sigma$.

Note that if you divide the SS of your sample by $N$, instead of $N - 1$, your $s$ will consistently underestimate $\sigma$ (i.e., your sample SD will underestimate the SD of the population from which the sample was drawn). This consistent error is called a *bias*. Dividing by $N - 1$ instead makes $s$ somewhat bigger and thus corrects the bias (for very large samples, dividing by $N - 1$ instead of $N$ makes little difference). You may recall from Chapter 3 that in this textbook, we are adopting the convention of using the symbol $s$ for the SD of a sample that is calculated with $N - 1$ in the denominator and is therefore an *unbiased* estimate of $\sigma$.

### Using the $t$ Distributions to Test Null Hypotheses

Following is an edited section of the $t$ table:

| *Df* | ... | $t_{0.05}$ | $t_{0.01}$ | ... |
|---|---|---|---|---|
| ⋮ | | | | |
| 4 | | 2.78 | 4.60 | |
| ⋮ | | | | |
| 10 | | 2.23 | 3.17 | |
| 11 | | 2.20 | 3.11 | |
| 12 | | 2.18 | 3.06 | |
| 13 | | 2.16 | 3.01 | |
| 14 | | 2.15 | 2.98 | |
| 15 | | 2.13 | 2.95 | |
| ⋮ | | | | |
| 40 | | 2.02 | 2.70 | |
| ⋮ | | | | |
| 120 | | 1.98 | 2.62 | |
| ⋮ | | | | |
| $\infty$ | | 1.96 | 2.58 | |

Each horizontal row represents a separate distribution that corresponds to the *df* listed in the left column. The row labeled with the symbol for infinity ($\infty$) represents the ND, which is what the $t$ distribution becomes when the samples are infinitely large.

To illustrate the use of the $t$ distributions, suppose that you wish to test the null hypothesis $H_0$: $\mu = 17$ against the alternative $H_A$: $\mu \neq 17$ using the .05 criterion of significance. Perhaps a large-scale survey has shown that among people who engage in text messaging, the average number of messages sent per day is 17. You suspect that the average is higher for young people in high school. So, you obtain a sample of $N = 15$ high school students and find that for them, $\overline{X} = 18.80$ and $s = 3.50$. You could try testing the null hypothesis that the average for the entire population of high schoolers is really just 17 by creating a $z$-score with Formula 6.1, but we now know that the mean of a sample size of only 15 is more accurately tested with a $t$ distribution. To create a formula for small sample sizes, we need only to change the $z$ to a $t$ in Formula 6.1, like so:

$$t = \frac{\overline{X} - \mu}{s_{\overline{X}}} \qquad \text{Formula 6.3}$$

where

$\overline{X}$ = the observed value of the sample mean
$\mu$ = the hypothesized value of the population mean
$s_{\overline{X}}$ = the estimated standard error of the mean $\left(= s/\sqrt{N}\right)$

In the present example, the $t$ value would be

$$t = \frac{18.80 - 17}{(3.50/\sqrt{15})} = \frac{1.80}{0.90} = 2.00$$

Now you can compare the $t$ value that you have computed to a critical value obtained from the $t$ table. For 14 $df$ (1 less than your $N$ of 15), and using the two-tailed .05 criterion of significance, the critical value is 2.145. An observed $t$ value larger than this *in absolute value* (i.e., ignoring the sign) is considered unlikely enough (i.e., $p < .05$) to occur if $H_0$ is true, and would therefore allow you to reject $H_0$ in favor of $H_A$. However, the absolute value of the $t$ that you computed was *less than* 2.145. So, you would have to *retain* $H_0$. Because the obtained $t$ value of 2.00 in our example is not larger in absolute value than 2.145, our data do *not* warrant concluding that the high school population mean differs from 17.

If, instead, your sample size were 121, you would have 120 $df$. The corresponding critical value from the table, again using the .05 criterion of significance, would be 1.98 (as we saw in the example above). In this case, calculating a $t$ value of 2.00 *would* allow you to reject $H_0$ in favor of $H_A$. However, the critical value of $t$ using the .01 (two-tailed) criterion of significance with $df = 120$ is 2.617. Therefore, the computed $t$ value of 2.0 is statistically significant at the .05 level but not the .01 level, when $df = 120$.

### The *t* and *z* Distributions Compared

The $t$ distribution for 9 degrees of freedom and the normal curve are compared in Figure 6.1. Note the fatter tails in the case of the $t$ distribution, so that you have to go farther out on the tails to find points that set off the .05 or .01 areas for rejecting $H_0$. While 95% of the normal curve lies between the $z$ values $-1.96$ and 1.96, 95% of the $t$ distribution for 9 $df$ lies between $-2.26$ and 2.26. So, using the two-tailed .05 criterion, 2.26 (and *not* 1.96) must be used as the critical value when $\sigma$ is not known and $df = 9$.

Figure 6.1

**A Comparison Between the Normal Curve and the *t* Distribution for *df* = 9**

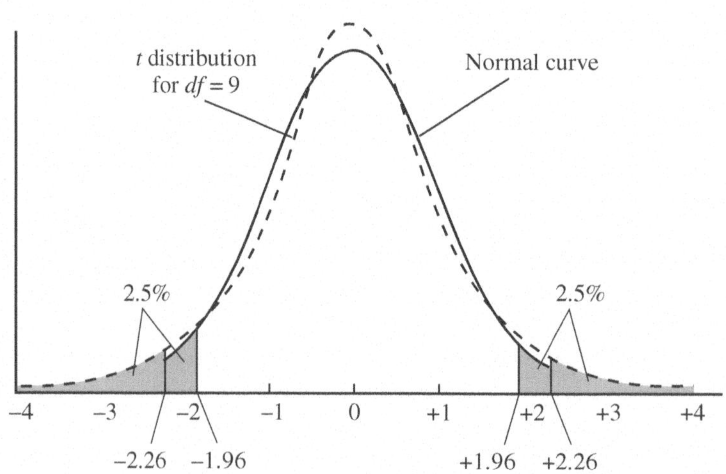

Notice in the $t$ table that the *smaller* the sample size, the *larger* $t$ must be in order to reject $H_0$ at the same level of significance. Also, when the sample size is very large, $z$ and $t$ are virtually equivalent. Thus, for 40 *df*, $t = 2.02$ (at the .05 level). For 120 *df*, $t = 1.98$. For sample sizes in the hundreds, the critical $t$ values become so close to the $z$ value of 1.96 that it makes little practical difference which one you use.

Suppose Jackson suspects that the overall academic interest of statistics students at their school is on the rise and that their class in particular is from a population with a higher academic interest than the population that existed over the previous decade. Jackson finds the means for previous statistics classes on the academic interest questionnaire and calculates an average over the previous 10 years. This 10-year average can be viewed as the population mean against which Jackson wishes to test their current class. If the 10-year average were equal to 27.31, then the null hypothesis that Jackson would be testing is $H_0$: $\mu = 27.31$. Jackson has the means from the previous classes but they do not know $\sigma$; they will have to estimate the standard error of the mean from the current class data. We showed in a previous chapter that for the 85 students who took the academic interest questionnaire in Jackson's class, $\overline{X} = 29.07$ and $s = 9.48$. Therefore:

$$s_{\overline{X}} = \frac{s}{\sqrt{N}} = \frac{9.48}{\sqrt{85}} = \frac{9.48}{9.22} = 1.028$$

and

$$t = \frac{\overline{X} - \mu}{s_{\overline{X}}} = \frac{29.07 - 27.31}{1.028} = 1.71$$

Jackson would have to decide to retain $H_0$ whether they compared 1.71 to the critical $t$ value of 1.99 for $df = 84$ (i.e., $85 - 1$) or to the $z$ value of 1.96. In behavioral science research, it is customary to report the obtained value as a $t$ value even for large samples. The reason is that although $z$ is a good approximation in large samples, it is still an approximation. So, you should use the critical $t$ value obtained from the $t$ table, rather than the $z$ value of 1.96 for the .05 criterion or 2.58 for the .01 criterion unless you happen to know the exact value of $\sigma$. *Retain* $H_0$ if your computed $t$ value (ignoring the sign) is smaller than the critical value, and *reject* $H_0$ otherwise. (If you used a statistical program to calculate your $t$ value, you have the option of reporting the exact $p$ value provided by your software and comparing that $p$ value directly to alpha.)

## Interval Estimation

We have already worked with the simplest form of estimation—point estimation. When we use $s$ to estimate $\sigma$, we are using a sample statistic to estimate a population parameter, and the estimated value is a point on a scale. Estimates are by nature inexact, so it is often helpful to locate a point estimate (e.g., a sample mean) within an interval of scores that are likely to contain the value being estimated (e.g., $\mu$). Interval estimation, therefore, starts with a point estimate, $\overline{X}$—which is your best guess about the value of $\mu$—and then specifies a range of possible values for mu. Your point estimate $\overline{X}$ is more likely to be just a little different from the true value of mu than it is to be very different. These differences can be described by a distribution. Consider Figure 6.2. It depicts $\overline{X}$ in the middle of a distribution of possible values of $\mu$. As noted above, if you had to pick one value for $\mu$, you would pick $\overline{X}$. The distribution adds information (probabilities) about how big or small mu might be. For example, values of mu are much more likely to be fairly close to $\overline{X}$ (lots of area under the curve) than very far from $\overline{X}$ (the tails are thin, so they do not contain a lot of area). The goal of interval estimation is to specify confidence that an interval contains the quantity being estimated.

Null hypothesis testing (NHT) is still popular in the behavioral sciences despite the fact that this method has several serious (and much debated) deficiencies (see Kline,

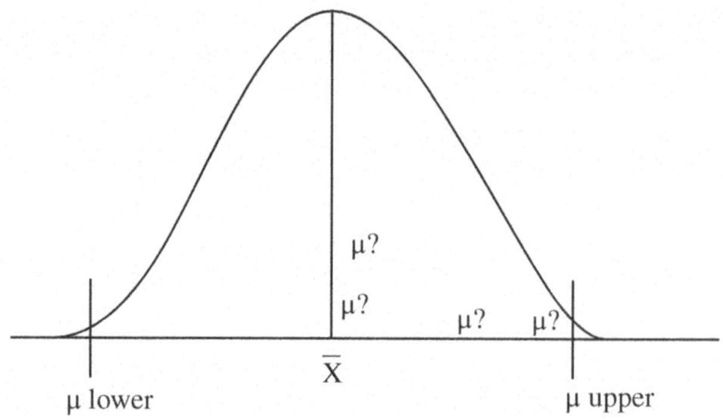


**Figure 6.2**

**The interval surrounding $\overline{X}$ that contains the likely values of μ**

2004). One drawback to NHT is that when the results are statistically significant, the researcher is entitled to reject *only* the exact value specified by the null hypothesis.

For example, suppose that the mean academic interest score for Jackson's class had been 29.81, instead of 29.07. Then their calculated $t$ would have been $2.5/1.028 = 2.43$, and they would have rejected the null hypothesis in favor of the alternative hypothesis and concluded that the mean for the population of current students in this high school is greater than 27.31.

Although in this latest case, Jackson would be able to reject the null hypothesis that $\mu = 27.31$, they could not, without further testing, reject the hypothesis that $\mu = 28$, that $\mu = 27.4$, or even that $\mu = 27.32$. The research study tested one and only one null hypothesis, namely $\mu = 27.31$. So, only this hypothesis can be rejected when the results are statistically significant. To be sure, it is likely that values very close to the null hypothesized value of 27.31 should also be rejected. But an additional statistical procedure would be required to determine what would count as "very close" numerically.

One good way to increase the information beyond what you could get from NHT alone is to estimate a confidence interval (CI). Rather than specifying a single numerical value (as does $H_0$), a CI specifies a *range of values* within which a parameter (here, the population mean) is likely to fall. How likely? If you are using the .05 criterion of significance, you can be 95% confident that the range specified by the CI contains the population mean. The end points of the CI are called *confidence limits*.

### Computation

To illustrate the use of CIs, consider the original example involving the academic interest of Jackson's class. We wish to determine an interval that will include all values of μ that do not pass the threshold of significance (i.e., all values for which $H_0$ should be retained). Using the .05 criterion of significance and $df = 84$, the farthest that any sample mean can lie from μ and still be retained as "not sufficiently unlikely" is $\pm 1.99\, s_{\overline{X}}$ (where 1.99 is the critical value of $t$ for the specified $df$, 84). So, the upper limit of this CI may be found by solving the following equation for μ:

$$+1.99 = \frac{X - \mu}{s_{\overline{X}}}$$

Similarly, the lower limit of the CI can be found from the following equation:

$$-1.99 = \frac{\overline{X} - \mu}{s_{\overline{X}}}$$

Or we can find both of the CI limits by solving for μ in the following equation:

$$\pm 1.99 = \frac{\overline{X} - \mu}{s_{\overline{X}}}$$

which yields this equation:

$$\mu = \overline{X} \pm 1.99 s_{\overline{X}}$$

which can then be expanded like this:

$$\overline{X} - 1.99 s_{\overline{X}} \leq \mu \leq \overline{X} + 1.99 s_{\overline{X}}$$

Thus, for our main example,

$$29.07 - 1.99(1.028) \leq \mu \leq 29.07 + 1.99(1.028)$$

$$29.07 - 2.05 \leq \mu \leq 29.07 + 2.05$$

$$27.02 \leq \mu \leq 31.12$$

You should retain any hypothesized value of $\mu$ between 27.02 and 31.12, inclusive, using the CI corresponding to the .05 criterion of significance (called the *95% CI*). Conversely, reject any hypothesized value of $\mu$ outside this interval.

Note that 27.31 (the hypothesized mean) is in the CI we just calculated, which tells us that it is reasonable to suppose that Jackson's class, with its sample mean of 29.07, comes from a population with a mean of 27.31. This judgment is consistent with *not* rejecting the null hypothesis of 27.31, which was the conclusion reached in the preceding example.

The general formula for computing CIs when $\sigma$ is not known is:

$$\overline{X} - t_{\text{crit}} s_{\overline{X}} \leq \mu \leq \overline{X} + t_{\text{crit}} s_{\overline{X}} \qquad \text{Formula 6.4}$$

where

$t_{\text{crit}}$ = critical value from the $t$ table for $df = N - 1$
$\overline{X}$ = observed value of the sample mean

The CI corresponding to the .01 criterion of significance is called the *99% CI*. In the academic interest example, the critical value of $t$ is based on $df = 84$. So, for the 0.01 criterion, $t_{\text{crit}}$ is 2.62. Thus, the 99% CI is equal to:

$$29.07 - 2.62(1.028) \leq \mu \leq 29.07 + 2.62(1.028)$$

$$29.07 - 2.69 \leq \mu \leq 29.07 + 2.69$$

$$26.38 \leq \mu \leq 31.76$$

Note that the 99% CI is wider than the 95% CI, and contains it completely (i.e., any value in the 95% CI, such as 27.31, will also be in the 99% CI). You can be more sure that the population mean falls in the 99% CI, but you must pay a price for this added confidence: The interval is larger. So, $\mu$ is less precisely estimated. (You could be absolutely certain that the population mean falls within the "confidence interval" $-\infty \leq \mu \leq \infty$, but such a statement is useless—it tells you nothing that you did not know beforehand.) The tradeoff is that by using $\alpha = .05$ to find the critical value in Formula 6.3, you can be only 95% confident that the true population mean falls within the stated interval, but the interval is narrower, and hence more informative, than the 99% CI. By using $\alpha = .01$, instead, you can be 99% confident (more sure) that the true population mean falls within the stated interval, but the interval is now wider and therefore less informative. Note that if $\sigma$ *is known*, you can use the critical $z$ values (1.96 for the 95% CI; 2.58 for the 99% CI) instead of $t_{\text{crit}}$.

### Confidence Intervals and Null Hypothesis Tests

The scientific journals in the behavioral sciences still place a lot of emphasis on testing null hypotheses, but there is a rapidly growing recognition that interval estimation has important advantages.

CIs allow you to do everything that NHT does, and more. If a value of $\mu$ specified by a null hypothesis falls within the CI, that null hypothesis is retained. If a value of $\mu$ specified by a null hypothesis falls outside the CI, that null hypothesis is rejected. But CIs have the advantage that they provide a whole range of population means that should be retained. For example, the 95% CI that Jackson found for the mean of the population from which their statistics class was (theoretically) randomly selected is 27.02 to 31.12. Any value for $\mu$ that falls within this range should be considered a reasonable possibility for the true value of $\mu$. Because the average academic interest score for the previous 10 years (27.31) falls inside the 95% CI, it cannot be rejected as a null hypothesis at the .05 level. This is consistent with the conclusion drawn from Jackson's *t* test. Because their *t* (1.71) was less than the .05 two-tailed, critical value for a *t* distribution with 84 degrees of freedom (1.99), the null hypothesis ($\mu = 27.31$) could not be rejected.

As we have just shown, CIs provide an alternative way of interpreting your data that can be more informative than a simple null hypothesis test. To illustrate this point further, let us consider four very different but possible CIs that Jackson's class data could have led to:

1. $28.8 \le \mu \le 32.9$
2. $27.2 \le \mu \le 31.3$
3. $24.4 \le \mu \le 28.5$
4. $22.0 \le \mu \le 26.1$

The first set of results supports Jackson's suspicion. They can conclude with 95% confidence that the population mean corresponding to their class is well above 27.31; more specifically, it can be expected to fall between 28.8 and 32.9. The second set of results is much less encouraging. Although the population mean seems likely to be greater than 27.31, Jackson cannot reject the hypothesis that $\mu$ is very close to 27.31 or even slightly lower, since the CI extends as low as 27.2. However, because nearly all of the CI is above 27.31, they can be somewhat encouraged that they are on the right track.

The third set of results is downright discouraging. It is unlikely that the population mean is much above 27.31. Therefore, these results provide little support for Jackson's idea that their class comes from a "better" population. Finally, the fourth set of results is completely inconsistent with Jackson's notion about their class. Because 27.31 is well *above* the *upper* confidence limit, these results would lead to the conclusion that, if Jackson's class differs from the previous population at all, it is actually from a population whose mean is *lower* in academic interest than the population of previous students.

CIs are highly recommended by the American Psychological Association (2020) to supplement NHT when submitting a study for publication. Many behavioral researchers feel that CIs should be used more widely than they are at present; some have even suggested that the reporting of CIs should replace NHT entirely (Finch & Cumming, 2009). Because CIs provide considerably more information than the results of testing a specific null hypothesis, we feel that it is safe to say that CIs should be used and reported more widely in behavioral science research.

## Working with Proportions

When the data you are dealing with are quantities of some variable of interest, such as the number of heartbeats per minute, the number of inches of height, or scores on a standard cognitive test, you can calculate means and SDs, and then use the methods of this chapter to test null hypotheses and construct CIs. However, in some cases, the data for each person has only two possible levels—e.g., has had their appendix removed or not; is a vegetarian or not; has a driver's license or does not. For these data, you cannot calculate the mean of your sample. All you can do is count the number of people in your sample that fall into each category, divide by the sample size, and calculate proportions. If your sample consists of city dwellers, you might be interested in knowing the proportion that has driver's licenses and compare them to the proportion in a wider population. A particularly common use of proportions occurs when a random sample of people are surveyed about their attitudes toward political issues. In one of the simplest examples, people are asked which candidate in a two-person race they plan to vote for in an upcoming election.

Suppose that just a few weeks before an election, a worried politician takes a poll by drawing a random sample of 400 registered voters. She finds that 53% intend to vote for her, while 47% prefer her opponent.[2] She is pleased to observe that she has the support of more than half of the sample, but she knows that the sample may not be an accurate indicator of the population because an unrepresentative number of her supporters may have been included by chance. What should she conclude about her prospects in the election?

This problem is similar to the preceding ones in this chapter. There is one population in which the politician is interested (registered voters), and she wishes to draw an inference about the mean of this population based on data obtained from a sample. In particular, she would like to know if the percent supporting her at this point in time is greater than 50%, in which case she is somewhat more likely to win the election. There is one important difference in the present situation, however: Because her data are categorical—people can either vote for this politician or for her opponent—the data consist only of percentages or *proportions*.

While the general strategy is the same as in the case of the sampling distribution of the mean, a modified method must be used to deal with proportions. The needed formula is

$$z = \frac{P - \pi}{\sqrt{\pi(1 - \pi)/N}} \qquad \text{Formula 6.5}$$

where

$P$ = proportion observed in the sample
$\pi$ = hypothesized value of the *population* proportion (not to be confused with $\pi = 3.14$)
$N$ = number of people in the sample

The denominator of this formula, $\sqrt{\pi(1 - \pi)/N}$, is the *standard error of a proportion*, symbolized by $\sigma_P$. It serves a similar purpose to the standard error of the mean, but it is a measure of the variability of *proportions* in samples drawn at random from a population where the proportion in question is $\pi$. (Notice that in this formula $\pi$ is used to represent the hypothesized value of the population proportion. It does *not* represent the usual mathematical value of approximately 3.14159...)

Since the population $\pi$ is specified, and the $\sigma$ of the population in question is considered to be known—it can be proved equal to $\sqrt{\pi(1 - \pi)}$—the results are referred to in Table A in the Appendix. In the case of the anxious politician, the critical population value in which she is interested is .50: She wants to know whether the observed sample proportion of .53 is sufficiently different from .50 to enable her to conclude that a majority of the population of voters will vote for her. Thus,

$\pi$ = hypothesized population proportion = .50
$P$ = proportion observed in sample = .53

The statistical analysis is as follows:

$$H_0: \pi = .50$$
$$H_A: \pi \neq .50$$

[2] For simplicity, we assume that there are no undecided or "won't say" voters in this sample. There will be some such in reality, and in omitting them we make the possibly incorrect assumption that they will break the same way as those who make a choice.

## Figure 6.3

**Sampling Distribution of $P$ when $\pi = .50$ and $N = 400$**

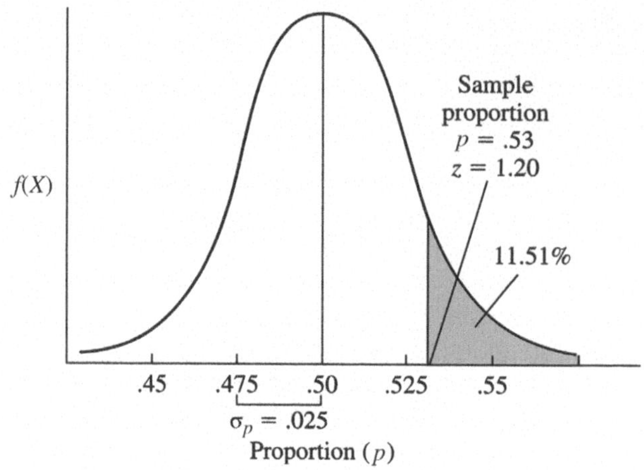


$$\alpha = .50$$

$$z = \frac{.53 - .50}{\sqrt{.50(1 - .50)/400}} = \frac{.03}{.025} = 1.20$$

The value of 1.20 is smaller than the critical $z$ value of 1.96 needed to reject $H_0$. Therefore, the politician *cannot* conclude that she will win the election. It is too likely for 53% of a sample of 400 voters to support her, even if only 50% of the entire population is willing to vote for her. The politician will therefore have a nervous few weeks until the votes are in, and she may well conclude from the sample results that she should increase her campaigning efforts. The results are illustrated in Figure 6.3.

There is one important caution regarding the use of the standard error of the proportion. When the hypothesized value of the population proportion ($\pi$) is quite large or small (i.e., close to 1.0 or 0) and the sample size is small, computing a $z$-score and referring it to the normal curve table will yield incorrect results. As a rough guide, you should *not* use this procedure when either $N\pi$ or $N(1 - \pi)$ is less than 5. For example, if you wish to test the null hypothesis that $\pi = .85$ with a sample size of 20,

$$N\pi = (20)(0.85) = 17 \text{ and } N(1 - \pi) = (20)(0.15) = 3$$

Because $N(1 - \pi)$ is less than 5, the procedures discussed in this section should *not* be used. For a tested value of $\pi$ that is as extreme as .85, the proposed sample size of 20 is too small to yield accurate results. You should either use the appropriate nonparametric statistical technique to deal with this situation (the binomial distribution, which will be discussed in Chapter 16), or increase the size of your sample.

### CIs for $\pi$

Suppose you wish to determine all reasonably likely values of the population proportion ($\pi$) using the .05 criterion of significance. The formula for the appropriate CI is:

$$(P - 1.96_{\sigma P}) \leq \mu \leq (P + 1.96_{\sigma P}) \quad \text{Formula 6.6}$$

There is one difficulty: $\sigma_P$ depends on the value of $\pi$, which is not known. In the previous section, a value of $\pi$ was specified by $H_0$ and used in the calculation of $\sigma_P$. However, there is no single hypothesized value of $\pi$ in interval estimation. One possibility, which yields a fairly good approximation *if the sample size is large*[3] (e.g., *bigger than 100*), is to use $P$ as an estimate of $\pi$. With a large sample, $P$ is unlikely to be so far from $\pi$ as to greatly

[3] Statisticians disagree on how large a sample is needed to use z instead of t; some believe that samples of 30 or 40 are sufficient while others are more conservative and think that samples of 100 or larger are needed to use z.

affect $\sigma_p$ and hence the size of the CI. In the case of the nervous politician in the previous section, $P = .53$; and $\sigma_p = \sqrt{0.53(1 - 0.53)/N} = 0.025$. Then the 95% CI is equal to:

$$0.53 - 1.96(0.025) \le \mu \le 0.53 + 1.96(0.025)$$
$$0.53 - 0.049 \le \mu \le 0.53 + 0.049$$
$$0.48 \le \mu \le 0.58$$

Since the sample size is large ($N = 400$), the politician can state with 95% confidence that between approximately 48% and 58% of the population will vote for her. Note that the value of .50 *is* within the CI, as would be expected from the fact that the null hypothesis that $\pi = .50$ was retained. To determine the 99% CI, the procedure would be the same except that the $z$ value of 2.58 would be used instead of 1.96.

## Summary

The use of the ND to test hypotheses requires that the SD of the population be known. If you have a large sample, its SD can be used in place of $\sigma$ with negligible error. However, if your sample is small, you will need to use one of the $t$ distributions.

### Testing the Mean of a Single Population With the $t$ Distributions

The formula for estimating the standard error of the mean from a single sample is

$$s_{\overline{X}} = \frac{s}{\sqrt{N}} \qquad \text{Formula 6.1}$$

where

$s_{\overline{X}}$ = the estimated standard error of the mean
$s$ = (unbiased) standard deviation of the sample of observations

Then, to draw inferences about a population mean, you can use:

$$t = \frac{\overline{X} - \mu}{s_{\overline{X}}} \qquad \text{Formula 6.3}$$

with $df = N - 1$

Use $z$ (Formula 5.2) instead of $t$ if $\sigma$ is known. For very large samples, you can use Formula 6.2 (same as Formula 6.3, except with $z$ instead of $t$ on the left side of the equation).

### Confidence Intervals

*CIs* establish all reasonably likely values of a population parameter, such as a mean or proportion. CIs provide considerably more information than the results of testing a specific null hypothesis because they show the decision that should be made for every conceivable null-hypothetical value rather than only one.

Use $s_{\overline{X}}$ when determining CIs for the mean of one population, if $\sigma$ is unknown. Obtain a two-tailed critical value of $t$ from the $t$ table ($df = N - 1$; $\alpha = 1.0 -$ confidence), and compute

$$(\overline{X} - t_{\text{crit}} s_{\overline{X}}) \le \mu \le (\overline{X} + t_{\text{crit}} s_{\overline{X}}) \qquad \text{Formula 6.4}$$

where

$t_{\text{crit}}$ = critical value from $t$ table

It is reasonable to use a critical $z$ instead of the critical $t$ for very large samples.

### Test for a Single Proportion

If the data are in the form of proportions, you can begin by computing the appropriate standard error using this formula (the denominator of Formula 6.5):

$$\sigma_P = \frac{\pi(1 - \pi)}{N}$$

Then compute:

$$z = \frac{(P - \pi)}{\sigma_P}$$

where

$\sigma_P$ = the standard error of a proportion
$P$ = proportion observed in the sample
$\pi$ = hypothesized value of the population proportion

Do *not* use this procedure if $N\pi$ or $N(1 - \pi)$ is less than 5. There are more exact tests for such cases (see Chapter 17).

## Exercises

1. Answer these problems by calculating a *t* value and comparing it to the critical value for a two-tailed test at the .05 level. (Note: You can save time by using the means and SDs you calculated for the dormitories called Turck and Kirk Halls for previous exercises.)
   (a) Would you retain or reject the null hypothesis that the population mean for Turck Hall is 16?
   (b) Would you retain or reject the null hypothesis that the population mean for Kirk Hall is 11?
2. For Kirk Hall, compute each of the following:
   (a) The 99% CI.
   (b) The 95% CI.
   (c) The 90% CI.
   (d) Explain why the size of the CIs becomes larger as the level of confidence increases.
3. This exercise is based on the following data set: 1, 3, 6, 0, 1, 1, 2, 1, 4.
   (a) Perform a *t* test in order to decide whether you can reject the null hypothesis of $\mu = 2.5$ at the .05 level (two-tailed) for these data.
   (b) Redo your *t* test in part (a) for a null hypothesis of $\mu = 6.0$.
   (c) Compute the 95% CI for the population mean from which these data were drawn. Explain how this CI could be used to draw conclusions about the null hypotheses in parts (a) and (b).
4. This exercise is based on the following data set: 68.36, 15.31, 77.42, 84.00, 76.59, 68.43, 72.41, 83.05, 91.07, 80.62, 77.83.
   (a) Perform a *t* test in order to decide whether you can reject the null hypothesis of $\mu = 52.3$ at the .01 level (two-tailed) for these data.
   (b) Redo your *t* test in part (a) for a null hypothesis of $\mu = 85.0$.
   (c) Compute the 99% CI for the population mean from which these data were drawn. Explain how this CI could be used to draw conclusions about the null hypotheses in parts (a) and (b).
5. An artist has staked her career on the outcome of the vote for a new mascot at a certain midwestern college. If the highland cow wins out over the current mascot, the Loch Ness Monster, her logo design will be prominently featured throughout the school.

To find out which way the wind is blowing, she obtains a random sample of 100 potential voters a few weeks prior to the vote and finds that 60% of the sample says that they will vote for the new mascot. Assuming that the highland cow will fail if it receives 50% or less of the vote, should she conclude that the voters as a whole will support the highland cow? Perform the appropriate statistical test to answer this question.

6. A certain business concern needs to obtain at least 20% of the market in order to make a profit. A random sample of 200 prospective buyers is asked whether they will purchase the product. What should the company conclude if (find the 95% CI in each case, in addition to performing the null hypothesis test):
   (a) 26 of those asked said they would buy the product?
   (b) 46 of those asked said they would buy the product?
   (c) 58 of those asked said they would buy the product?

## Thought Questions

1. You are drawing inferences about the mean of one population. (a) When should you use the *t* distributions as the theoretical model rather than the normal curve model? (b) Why is there a different critical *t* value for different degrees of freedom?
2. What advantages do confidence intervals have over testing a specific null hypothesis?
3. (a) What happens to the size of a CI as the standard error of the mean becomes larger? Why? (b) What happens to the size of a CI as the sample size becomes larger? Why? (c) What are the advantages of having a smaller CI?
4. (a) Why does the size of the CI become larger when we change from a 95% CI to a 99% CI? (b) We can be more certain that the unknown population mean falls within the 99% CI than within the 95% CI. Why, then, do not we always use 99% CIs?
5. On a test of scholastic ability administered to college students in the United States, the mean is 500 and the SD is 100. Each of the following 95% CIs for the mean of one population was computed from a random sample of students from the university in question: University Y, 601.63–642.75; University Z, 502.42–534.68. Both universities claim that their students are superior to the national average of 500. (a) Is this claim correct? (b) How do these results illustrate the advantages that CIs have over testing a specific null hypothesis, such as whether the population mean is equal to 500?
6. A behavioral science researcher wishes to draw inferences about the proportion of one population. (a) Would she prefer that the standard error of a proportion be small or large? Why? (b) Should she use the normal curve model or the *t* distributions as her theoretical model? Why? (c) When should she *not* use the procedures for drawing inferences about the proportion of one population discussed in this chapter? (d) Why might she prefer to use confidence intervals for the proportion of one population rather than testing a specific null hypothesis?

## Computer Exercises

1. Use your statistical package to perform a one-sample *t* test to determine whether the baseline anxiety of Jackson's students differs significantly ($\alpha = .05$) from the mean ($\mu = 18$) found by a very large study of college students across the country. Find the 95% CI for the population mean represented by Jackson's students.
2. Use your statistical package to perform a one-sample *t* test to determine whether the average baseline heart rate of Jackson's students who regularly exercise differs significantly from the mean HR ($\mu = 70$) for college-aged regular exercisers at the .01 level.

Find the 99% CI for the population mean represented by Jackson's students who regularly exercise.

3. Use your statistical package to perform a one-sample *t* test to determine whether the average *postquiz* heart rate of Jackson's students that *do not* exercise regularly differs significantly ($\alpha = .05$) from the mean resting HR ($\mu = 72$) for college-aged individuals that do not exercise regularly. Find the 95% CI for the population mean represented by Jackson's students who do not exercise regularly.

## Bridge to SPSS

To perform a one-sample *t* test in SPSS, select **Compare Means and Proportions** from the **ANALYZE** menu, and then choose **One-Sample T Test** .... In the dialog box that opens, move the variable you want to test to the *Test Variable(s)* area. Under the *Test Variable(s)* area, you will find a space labeled *Test Value*, which is where you enter the null hypothesis value you want to test the mean of your test variable(s) against. The default value that appears in that space is zero and if you do not change that value and click OK, the results in your Output window will give you *t* and *p* values that test whether the mean of your variable is significantly different from zero. Unless your variable has a mixture of positive and negative values, you will likely get a large *t* and a small *p* value, but these statistics will not tell you anything interesting. There is also a checkbox next to the right of the *Test Value* space, labeled *Estimate effect sizes*, and it is checked by default. Effect sizes will be discussed in the next chapter.

As an example, Figure 6.4 shows the results of testing the mean of baseline anxiety (18.43) for the 100 students in Jackson's data set (the value for the mean appears under the heading of Mean Difference; because the test value is zero, the difference between the mean and the test value is, simply, the mean). Note the very large *t* value, and its associated *df* (degrees of freedom), which is 99, one less than the number of anxiety scores. Because the *t* is so large, the *p* values are so small that they would require quite a few zeros after the decimal point—so following the APA Style manual, the *p* is just given as "< .001". Both one- and two-tailed *p* values are given, but in this case, you cannot see that the two-tailed *p* is just twice as large as the one-tailed *p*. These *p* values are not very interesting, but testing against zero does give you a useful result: the CI for the population mean, centered on your sample mean (see the left side of Figure 6.4). The 95% CI is the default, but you can change that percentage by clicking on the **Options** ... button.

There is another choice you can make after clicking on the **Options** ... button. Suppose you want to find CIs for several of your variables at the same time. You can move many or all of your variables into the *Test Variable(s)* area. If, for instance, you keep the default *Test Value* at zero, that value will apply to all of the variables you are testing. The default choice for handling missing values (**Exclude cases analysis by analysis**) deals with missing values separately for each test variable. However, if you select **Exclude cases listwise**, after clicking on the **Options** ... button, any case (i.e., row in the data sheet) that is missing a value for *any* of the variables that appeared together in the *Test Variables* area will be deleted for *all* of the *t* tests in that run. You would use this option if you want the *t* tests of all your variables to involve exactly the same participants. You will see the **Exclude cases listwise** option presented for a number of other statistical procedures in SPSS, which we describe in subsequent chapters. Finally, there is also a **Bootstrap** ... button, which appears in recent versions of

**Figure 6.4**

**Results of a One-Sample *t* Test of the Mean of Baseline Anxiety Against a Null Hypothesis of Zero**

**One-sample test**

| | Test value = 0 | | | | | | |
|---|---|---|---|---|---|---|---|
| | | | Significance | | | 95% Confidence interval of the difference | |
| | t | df | One-sided p | Two-sided p | Mean difference | Lower | Upper |
| anx_base | 42.559 | 99 | <.001 | <.001 | 18.430 | 17.57 | 19.29 |

**One-sample test**

| | Test value = 20 | | | | | | |
|---|---|---|---|---|---|---|---|
| | | | Significance | | | 95% Confidence interval of the difference | |
| | t | df | One-sided p | Two-sided p | Mean difference | Lower | Upper |
| anx_base | −3.626 | 99 | <.001 | <.001 | −1.570 | −2.43 | −.71 |

Figure 6.5

**Results of a One-Sample *t* Test of the Mean of Baseline anxiety Against a Hypothesized Population Mean**

SPSS in many dialog boxes. Bootstrapping is a more advanced way to conduct statistical tests, which will be explained briefly in Chapter 12.

In this chapter, we described an appropriate situation for a one-sample *t* test in terms of Jackson's desire to test the average academic interest scores of their current statistics class against the mean of previous statistics classes; the mean academic interest scores over the previous 10 years of classes served as the population mean which they wished to test their class against. In the next example, we will take a similar approach using baseline anxiety as our variable of interest. Assuming that Jackson's anxiety measure was a standard one that has been used with students at various colleges many times, the mean anxiety from all of that data can be used as a population mean against which Jackson's students can be compared. Suppose that the anxiety population mean is 20; then, you would type the number 20 into the *Test Value* space, replacing the default value of 0. Figure 6.5 displays the results of testing the anxiety mean of Jackson's students (18.43) against the hypothesized population mean of 20.

The negative *t* value tells you that Jackson's students are less anxious on average than the larger population of college students, and the small *p* values tell you that the difference is easily statistically significant. In this case, the Mean Difference is the difference between the mean of Jackson's sample and the population value that was entered as the *Test Value*. This time, the 95% CI surrounds that mean difference and gives you an estimate of the range in which the mean difference would be expected to fall 95% of the times that you replicated this comparison with a different sample of students at Jackson's college (assuming that Jackson's sample was chosen at random from Jackson's college).

You may be wondering why we did not show you how to perform a one-sample *z*-test at the end of the previous chapter. The reason is that the one-sample *t* test is, for all practical purposes, identical to the *z*-test if the sample is quite large (e.g., > 100). And, it is rare to know the SD of the population to which you are comparing your sample. In any case, the *t* test is the more conservative, and therefore more cautious (about Type I errors) test. However, if your sample size is small and you do know the population SD, you can easily calculate the one-sample *z*-test by hand, using the formulas in the previous chapter.

Finally, you may also be wondering why we are not showing you how SPSS analyzes proportions. This is handled by a fairly new SPSS module and it is too complicated to be worth explaining here.

# Chapter 7

# Testing Hypotheses About the Difference Between the Means of Two Populations

## PREVIEW

### The Standard Error of the Difference

We wish to draw inferences about the difference between the means of two populations. What is the correct standard error term to use in this situation? Why can it not usually be measured directly?

What is an experimental group? A control group?

Why is it desirable to use random samples?

How does sampling error affect the difference between the means of the experimental group and the control group?

What does the standard error of the mean tell us about the trustworthiness of a single difference between two sample means as an estimate of the difference between the population means?

### Estimating the Standard Error of the Difference

What are the procedures for estimating the value of the standard error of the difference (SED)?

How is the SED related to the variance of the population? To the sizes of the samples?

### The *t* Test for Two Independent Sample Means

What is the correct statistical model to use in this situation?

How do we test hypotheses about the difference between the means of two populations based on independent samples?

How are the results of a *t* test usually reported? What are the implications of retaining the null hypothesis? What are the implications of rejecting the null hypothesis?

### Confidence Intervals for $\mu_1 - \mu_2$

How do we estimate an interval in which the difference between the two population means is likely to fall?

What is gained by using confidence intervals?

### Measuring the Size of an Effect

What is a standardized measure of effect size ($d$), and when is it particularly useful?

### Reporting the Results of a *t* Test

According to APA style, how should you report the results of your *t* test, including a CI and effect-size measure?

### The Assumptions Underlying the Proper Use of the *t* Test for Two Sample Means

What assumptions underlie the use of this *t* test?

What is the separate-variances *t* test and under what circumstances would its use be considered?

*Introductory Statistics for the Behavioral Sciences*, Eighth Edition. R. Brooke Lea and Barry H. Cohen.

Companion website: www.wiley.com/go/ISBS8e

**PREVIEW (*continued*)**

### The *t* Test for Matched Samples

How do we test hypotheses about the difference between the means of two matched populations? How does the matched-pairs *t* test compare to the independent-samples *t* test on the same data? What happens to the degrees of freedom (and therefore the critical value) when two sets of scores are matched in pairs?

When is it preferable to measure the same participant twice, and when is it less problematic to match separate participants in pairs? When might you have to give up the idea of conducting a matched-pairs design altogether and use independent samples instead?

### Summary

### Exercises

### Thought Questions

### Computer Exercises

### Bridge to SPSS

In the preceding two chapters, we developed the tools for drawing inferences about the mean of one population. More often, scientists wish to draw inferences about differences between two or more populations. For example, a cognitive psychologist may want to know if the mean reaction time of the population of individuals who are bilingual on a word recognition test of linguistic processing speed is equal to the mean of the population of individuals who are monolingual. Or a behavioral psychologist may wish to find out whether rats perform better on a discrimination learning task to gain a reward or to avoid punishment—that is, whether the mean number of correct responses is greater for the "reward" population or for the "punishment" population. In both cases, the question of interest begins with a comparison between the means of *two* samples.

In this chapter, we discuss techniques for drawing inferences about the *difference between the means of two populations* ($\mu_1 - \mu_2$) based on the data from two samples. Procedures for drawing inferences about the means of more than two populations, or about other aspects of two or more populations, are considered in later chapters.

## The Standard Error of the Difference

Suppose that you wish to conduct a research study to determine whether the use of caffeine increases heart rate (HR). To test this hypothesis, you could obtain two *random* samples of individuals, one to be the *experimental group* and one to be the *control group*. But although random sampling is theoretically the ideal method to avoid obtaining two groups that differ greatly in ability, motivation, or any other variable that might obscure the effects of caffeine, a more common approach is to select one (twice-as-large) convenient sample and then randomly assign the participants to either the experimental or control group. This procedure is called **random assignment**. Unlike random sampling, random assignment is easy to do because it simply involves dividing a sample of participants that you already have into two (or more) groups. For example, you might flip a coin and assign a participant to Group 1 if a "head" comes up and Group 2 if a "tail" comes up (however, we usually constrain the randomness a bit to ensure that the two groups end up as the same size—a variation that can be called "quasi-random" assignment).

Once you have formed two groups, you then give all members assigned to the experimental group a small dose of caffeine in the form of a pill one hour prior to measuring their HR. At the same time, the control group is given a placebo (a pill that, unknown to them, has no biochemical effect whatsoever). This is to control for the possibility that administering *any* pill will affect the students' HR. (For example, they may become more psychologically alert and record higher HRs because they are expecting the pill to have an effect on them.)

Thus, the control group serves as a baseline against which the HRs of the experimental group can be evaluated. Finally, the mean test scores of the two groups are compared. Suppose that the mean of the experimental group equals 81, and the mean of the control group equals 78. Should you conclude that caffeine is effective in increasing HRs?

As was shown in the previous chapter, any sample mean is almost never exactly equal to the population mean because of sampling error; the cases that happened, by chance, to be included in the sample may be unusually high or low on the variable being measured. Therefore, the two populations in question (a hypothetically infinite number of participants given caffeine and a hypothetically infinite number of participants given a placebo) may have equal means even though the sample means are 3 points different. Consequently, you cannot tell what conclusion to reach just by looking at the sample means. It is the same problem we tackled in the last chapter: How big does a difference need to be before you can safely conclude that it is "real" (i.e., statistically significant)? You need additional information: namely, whether the 3-point difference between 81 and 78 is likely to be a trustworthy indication that the population means are different (and that caffeine *has* an effect on HRs), or whether this difference can be easily attributed solely to which cases happened to fall in each sample (i.e., sampling error), in which case you should not conclude that caffeine has an effect.

How can this additional information be obtained? In the previous chapter, we saw that the estimated variability of the sampling distribution of means drawn from one population, $s_{\overline{X}}$, provides information as to how trustworthy any one $\overline{X}$ is as an estimate of $\mu$. The present problem concerns how trustworthy is any one *difference*, $\overline{X}_1 - \overline{X}_2$, as an estimate of the difference between the population means, $\mu_1 - \mu_2$. This problem is also solved by making use of an appropriate sampling distribution: the sampling distribution of the *difference* between *two* sample means.

Suppose that a variable is normally distributed in each of two defined populations and that the populations have equal means and equal variances (i.e., $\mu_1 = \mu_2$ and $\sigma_1^2 = \sigma_2^2$). If many pairs of random samples of equal size were drawn from the two populations, a distribution of differences between the paired means could be established empirically. For each pair, you would subtract the mean of the second sample ($\overline{X}_2$) from the mean of the first sample ($\overline{X}_1$). If we assume that the means of the two populations are equal, any difference between the sample means must be due solely to sampling error.

Figure 7.1 illustrates the procedure needed to obtain an empirical sampling distribution of 1,000 differences in the case where $\mu_1 = \mu_2 = 80$ and the size of each random sample is 30. The resulting frequency distribution is shown in Table 7.1, and the frequency polygon

## Figure 7.1

**Illustration of Procedure for Obtaining the Empirical Sampling Distribution of Differences Between Two Means ($N = 30$)**

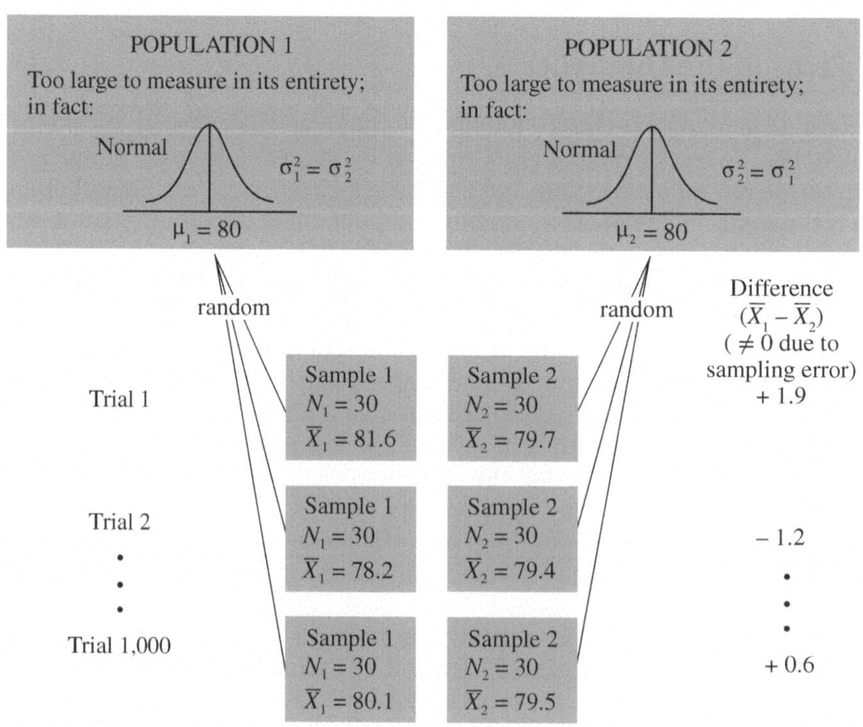

**Table 7.1** **Empirical Sampling Distribution of 1,000 Differences Between Pairs of Sample Means ($N = 30$) Drawn From Two Populations Where $\mu_1 = \mu_1 = 80$ (Hypothetical Data)**

| Difference between sample means $\bar{X}_1 - \bar{X}_2$ | Number of samples ($f$) |
|---|---|
| Greater than +11.49 | 0 |
| +10.50 to +11.49 | 1 |
| +9.50 to +10.49 | 0 |
| +8.50 to +9.49 | 1 |
| +7.50 to +8.49 | 4 |
| +6.50 to +7.49 | 7 |
| +5.50 to +6.49 | 21 |
| +4.50 to +5.49 | 32 |
| +3.50 to +4.49 | 54 |
| +2.50 to +3.49 | 77 |
| +1.50 to +2.49 | 107 |
| +0.50 to +1.49 | 122 |
| −0.50 to +0.49 | 153 |
| −1.50 to −0.51 | 114 |
| −2.50 to −1.51 | 95 |
| −3.50 to −2.51 | 82 |
| −4.50 to −3.51 | 60 |
| −5.50 to −4.51 | 31 |
| −6.50 to −5.51 | 22 |
| −7.50 to −6.51 | 10 |
| −8.50 to −7.51 | 4 |
| −9.50 to −8.51 | 1 |
| −10.50 to −9.51 | 2 |
| Smaller than −10.50 | 0 |
| Total | 1,000 |

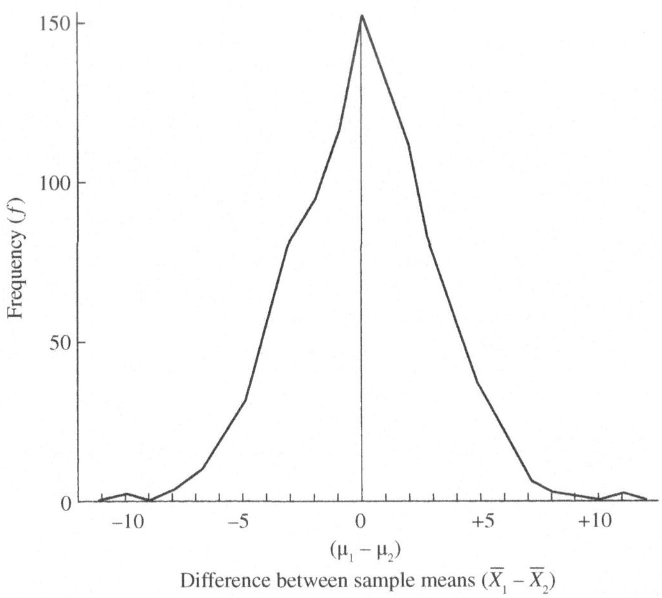


**Figure 7.2**

**Frequency Polygon of the Data in Table 7.1**

plotted from this distribution is shown in Figure 7.2. The sample means will rarely be exactly equal to 80, and the differences between pairs of means will almost never be exactly equal to zero, due to sampling fluctuations, though the means will usually be near 80, and the difference will usually be not far from zero. Notice that the distribution is approximately symmetric, which indicates that chance differences are equally likely to occur in either direction—that is, $(\bar{X}_1 - \bar{X}_2) > (\mu_1 - \mu_2)$ or $(\bar{X}_1 - \bar{X}_2) < (\mu_1 - \mu_2)$. Thus, the mean of the distribution of differences tends to be equal to $(\mu_1 - \mu_2)$, which in this case is zero.

Given the data in Table 7.1, you could calculate the standard deviation (SD) of the distribution of differences for the given sample size by using the usual formula for the SD (see Chapter 3). This SD would tell you how much, on average, a given $\overline{X}_1 - \overline{X}_2$ is likely to differ from the "true" difference of zero (the central point of the distribution of differences). Consequently, it would indicate how trustworthy is the single difference that you have on hand as an estimate of $\mu_1 - \mu_2$. This SD is therefore called the **standard error of the difference (SED)**.

On one hand, a relatively *large* SED indicates that even a moderately sized difference between a pair of sample means must be viewed with grave suspicion. Because a large standard error tells us that large discrepancies between $\overline{X}_1$ *and* $\overline{X}_2$ are likely *even if they come from populations with equal means*, you need quite a large difference before you can safely conclude that $\mu_1$ is not equal to $\mu_2$. If, on the other hand, the SED is relatively *small*, you can place more confidence in any one sample difference as an estimate of the true population difference. In this case, large discrepancies between $\overline{X}_1$ *and* $\overline{X}_2$ are *not* as likely to occur (and mislead you) if in fact $\mu_1 = \mu_2$. So, a fairly large difference between the sample means *is* a trustworthy sign that the population means differ. Of course, we cannot go by our subjective impression of whether a particular difference of sample means is large or not, which is why researchers depend so heavily (perhaps too heavily at times) on null hypothesis testing (NHT) as a fairly objective, quantitative way to make these judgments. We will show you how to apply NHT to the case of two sample means very shortly.

In practice, behavioral scientists never have many paired random samples of a given size with which to compute the SED empirically. Instead, as in Chapter 6, we must make use of formulas developed by statisticians. In the present situation, before you can apply NHT, you will need a formula that enables you to *estimate* the SED based on the *one* pair of samples that you have measured.

## Estimating the Standard Error of the Difference

As was the case with $s_{\overline{X}}$, developing the estimation formula for the SED by means of a formal proof is beyond the scope of this book. Instead, we will attempt to explain the formula in nonmathematical terms.

If it is assumed that the two population variances are equal, the variances of the two samples may be combined into a single estimate of the common value of $\sigma^2$. If the sizes of the two samples are exactly equal, you can obtain this combined or *pooled* estimate by computing the ordinary average of the two sample variances. Often, however, the sample sizes are not equal, in which case greater weight must be given to the larger-size sample because it contains more measurements and therefore more information on which to base an estimate. Therefore, a *weighted* average must be computed. The general formula for the *pooled variance* (symbolized by $s^2_{\text{pooled}}$), which may be used for equal or unequal sample sizes, is

$$s^2_{\text{pooled}} = \frac{(N_1 - 1)s_1^2 + (N_2 - 1)s_2^2}{N_1 + N_2 - 2} \qquad \text{Formula 7.1}$$

where

$s_1^2$ = population variance estimate of sample 1
$s_2^2$ = population variance estimate of sample 2
$N_1$ = number of cases in sample 1
$N_2$ = number of cases in sample 2

Then the estimation formula for the SED is

$$s_{\overline{X}_1 - \overline{X}_2} = \sqrt{\frac{s^2_{\text{pooled}}}{N_1} + \frac{s^2_{\text{pooled}}}{N_2}} = \sqrt{s^2_{\text{pooled}}\left(\frac{1}{N_1} + \frac{1}{N_2}\right)} \qquad \text{Formula 7.2}$$

where

$s_{\overline{X}_1 - \overline{X}_2}$ = estimated SED

This formula should not look totally unfamiliar. We saw in the previous chapter that the estimate for the standard error of the mean ($s_{\overline{X}}$) is equal to $s/\sqrt{N}$. Consequently, $s^2_{\overline{X}}$ is equal to $s^2/N$. Here, because a difference is subject to sampling error from *two* means, the sampling error variance reflects both sources of error. In addition, it uses a more stable estimate of the population variance by pooling the variance information from the two samples.

It should be obvious that the larger the variances of the populations (as estimated by $s^2_{\text{pooled}}$), the larger the estimated SED. Only slightly less obvious is the fact that the *larger* the sample sizes, the *smaller* the SED. Differences based on large samples will have smaller sampling errors (and are thus likely to be more accurate estimates of the population difference) than are differences based on small samples, just as is the case for a single mean.

By inserting Formula 7.1 into Formula 7.2, we can create a convenient formula for the SED:

$$s_{\overline{X}_1 - \overline{X}_2} = \sqrt{\frac{(N_1 - 1)s_1^2 + (N_2 - 1)s_2^2}{N_1 + N_2 - 2}\left(\frac{1}{N_1} + \frac{1}{N_2}\right)} \qquad \text{Formula 7.3}$$

When $N_1 = N_2$, the formula for the SED becomes much simpler:

$$s_{\overline{X}_1 - \overline{X}_2} = \sqrt{\frac{(N - 1)s_1^2 + (N - 1)s_2^2}{2(N - 1)}\left(\frac{1}{N} + \frac{1}{N}\right)} = \sqrt{\frac{s_1^2 + s_2^2}{2}\left(\frac{2}{N}\right)} = \sqrt{\frac{s_1^2 + s_2^2}{N}} \qquad \text{Formula 7.4}$$

## The $t$ Test for Two Independent Sample Means[1]

When you test hypotheses about the difference between two population means, and you do not know the SDs of the populations (which is virtually always the case), the correct statistical model to use is the *t* distributions. Once the estimated SED has been computed, a *t* value (and therefore a probability value) may be found for any given obtained difference, as follows:

$$t = \frac{(\overline{X}_1 - \overline{X}_2) - (\mu_1 - \mu_2)}{s_{\overline{X}_1 - \overline{X}_2}} \qquad \text{Formula 7.5A}$$

Notice that the two sample means yield a single difference score, $\overline{X}_1 - \overline{X}_2$, which is compared to the null-hypothesized mean of the difference scores, $\mu_1 - \mu_2$.

It is possible to test any hypothesized difference between the means of two populations. For example, you could test the hypothesis that the mean of the first population is 20 points greater than the mean of the second population by setting $\mu_1 - \mu_2$ equal to 20 in this equation. Much more often than not, however, you will want to test the hypothesis that the means of the two populations are equal (i.e., $\mu_1 = \mu_2$ or $\mu_1 - \mu_2 = 0$). In this situation, the *t* formula can be simplified to:

$$t = \frac{\overline{X}_1 - \overline{X}_2}{s_{\overline{X}_1 - \overline{X}_2}} \qquad \text{Formula 7.5B}$$

[1] Other names for this test are the "independent-samples *t* test," "Student's two-group *t* test"; the "two-sample *t* test," the "unpaired *t* test," and the "between-subjects *t* test."

By inserting Formula 7.2 into Formula 7.5B, we obtain the following useful formula for computing the two-group $t$ test based on the pooled variances:

$$t = \frac{\overline{X}_1 - \overline{X}_2}{\sqrt{s^2_{\text{pooled}}\left(\frac{1}{N_1} + \frac{1}{N_2}\right)}} \qquad \text{Formula 7.6A}$$

for which $df = N_1 + N_2 - 2$.

Although it is painful to look at, it is convenient to insert Formula 7.3 into Formula 7.5B, to obtain a one-step formula for the $t$ test for two (independent) sample means:

$$t = \frac{\overline{X}_1 - \overline{X}_2}{\sqrt{\frac{(N_1 - 1)s_1^2 + (N_2 - 1)s_2^2}{N_1 + N_2 - 2}\left(\frac{1}{N_1} + \frac{1}{N_2}\right)}} \qquad \text{Formula 7.6B}$$

Note that by inserting Formula 7.4 instead of Formula 7.3 into Formula 7.5B, you would obtain a (much) simplified $t$ test formula for the case when the two sample sizes are equal (see Formula 7.15 near the end of this chapter). In that case, the degrees of freedom ($df$) would equal $2N - 2$.

The number of $df$ on which the two-sample $t$ value is based, which is needed to obtain the critical value of $t$ from the $t$ table, is determined from the fact that the population $\sigma^2$ is estimated using $N_1 - 1$ $df$ from sample 1 and $N_2 - 1$ $df$ from sample 2. Therefore, the estimate is based on the combined $df$:

$$(N_1 - 1) + (N_2 - 1) = N_1 + N_2 - 2$$

Recall the problem of caffeine and HRs posed at the beginning of this chapter. A researcher named Ellen was also interested in this problem and decided to run a randomized control trial with the same procedures. In her experiment, 22 participants were assigned to ingest the caffeine pill and 20 participants were assigned to the placebo group. In order to apply null hypothesis testing (NHT) to the goal of making a statistical decision about the results of such an experiment, it would be convenient to divide the procedure into a series of discrete steps, beginning with a statement of the hypotheses and the selection of a significance criterion.

**Step 1: State the Null and Alternative Hypotheses, and the Significance Criterion, $\alpha$**

$$H_0 : \mu_{\text{caf}} = \mu_{\text{con}}$$
$$H_A : \mu_{\text{caf}} \neq \mu_{\text{con}}$$
$$\alpha = .05$$

$H_0$ states that the caffeine and control samples come from populations with equal means, while $H_A$ states that these two samples come from populations with different means (for the sake of precision, it is customary to state the hypotheses symbolically). The traditional .05 criterion of significance is specified unless there is a special reason to use a different criterion.

Suppose that the HR results for Ellen's two groups were as follows:

| **Caffeine group** | **Control group** |
|---|---|
| $N_1 = 22$ | $N_2 = 20$ |
| $\overline{X}_1 = 81.0$ | $\overline{X}_2 = 68.0$ |
| $s_1 = 12.0$ | $s_2 = 16.8$ |

### Step 2: Calculate the Appropriate Test Statistic

For a comparison of the means of two independent samples, the test statistic most commonly calculated is a $t$ value based on pooling the variances of the two samples, in which case this procedure is usually referred to as a **pooled-variances *t* test**. (The appropriateness of this statistic depends on some assumptions, which are discussed later in this chapter.) Because the $N$s for this experiment were not equal, we need to estimate the SED using Formula 7.3, which pools (i.e., takes a weighted average of) the sample variances, as follows:

$$s_{\overline{X}_1 - \overline{X}_2} = \sqrt{\frac{(21)(12^2) + 19(16.8^2)}{22 + 20 - 2}\left(\frac{1}{22} + \frac{1}{20}\right)}$$

$$= \sqrt{(209.664)\left(\frac{1}{22} + \frac{1}{20}\right)}$$

$$= 4.474$$

The $t$ value can now be found from Formula 7.5B:

$$t = \frac{81.0 - 68.0}{4.474} = 2.91, df = 22 + 20 - 2 = 40$$

### Step 3: Find the Critical Value and Make a Statistical Decision With Respect to Your Null Hypothesis

For this example, $df = N_1 + N_2 - 2 = 22 + 20 - 2 = 40$. Referring to the $t$ table (i.e., Table B in the Appendix), in the row for $df = 40$ and the column for a two-tailed significance level of .05, we find that the critical $t$ value is 2.021—that is, a calculated $t$ greater in absolute value than (approximately) 2.02 is needed to justify rejection of $H_0$ at the .05 two-tailed level of significance. (The two rejection regions for this experiment are illustrated in Figure 7.3.)

Because the obtained value of $t$ (2.91) is greater in absolute value than the critical value found in Table B, we can declare that a difference between sample means as large as the one observed by Jackson in this situation arises from sampling error so rarely (i.e., $p < .05$) that she can reject $H_0$ and therefore conclude that the caffeine and control groups are random samples from populations with *different* means. More specifically, Ellen can declare that caffeine has a statistically significant positive effect on resting HRs.

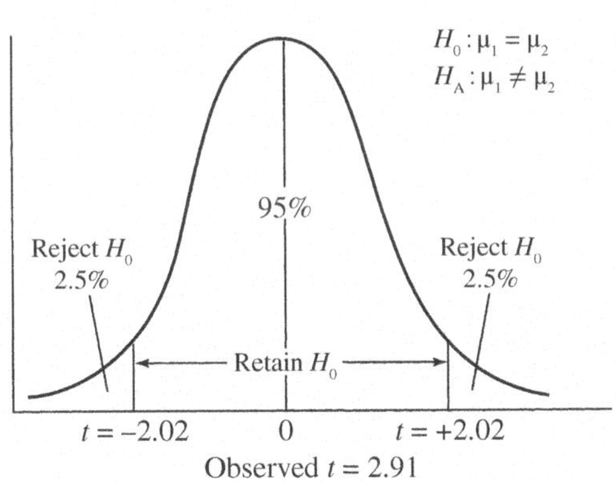


**Figure 7.3**

**Acceptance and Rejection Regions for the Caffeine/Control Comparison in the *t* Distribution for *df* = 40**

### Reporting the Results of a *t* Test

A common way to express the results of a two-sample experiment in a journal article is in the context of a sentence like this one: "The group that was given a caffeine pill an hour before the HR measurements exhibited a higher HR ($M$ = 81, SD = 12.0 bpm) than the group that had been given a placebo pill ($M$ = 68, SD = 16.8 bpm); this difference was statistically significant, $t(40) = 2.91$, $p < .01$.[2] (The number in parentheses after $t$ is, of course, the *df*.) Note that while statisticians prefer to use the symbol $\overline{X}$ to represent the arithmetic mean, the American Psychological Association (2020) dictates that psychologists use an uppercase M to represent the mean when submitting articles for publication. This is a long-standing policy that arose back in the days when printing special characters, like an X with a bar over it, was considerably more costly than printing an ordinary capital letter. Also, note that it is becoming increasingly common to include a measure of effect size, and sometimes a CI, right after the value of $t$, to aid in the interpretation of the results. So, we will return to the format for reporting $t$ test results later in this chapter.

### Interpreting the Results of a *t* Test …

Rejecting or retaining the null hypothesis is not the end but rather just a first step in the process of drawing conclusions from your results. Here, and in the next few sections, we describe the next steps in making sense of your two-group comparison.

#### *When Retaining $H_0$*

Suppose that a different caffeine study yielded an obtained $t$ value of 1.14 for $df = 40$. Since 1.14 is smaller than the critical value of $t$ obtained from Table B, you would decide to retain $H_0$. This does *not* imply that you have shown that the population means are equal. You do not know the probability that your decision to retain $H_0$ is wrong (i.e., how likely you are to have made a Type II error). So, you should limit yourself to a cautious statement such as "We did not obtain sufficient evidence to reject the (null) hypothesis that caffeine has no effect on HR." "Retaining" $H_0$ merely means that we have not refuted it. It does *not* mean that we have shown $H_0$ to be true, which is why we do not even say that we "accept" the null hypothesis.

#### *When Rejecting $H_0$*

When the results of your statistical analysis indicate that you should *reject* $H_0$ (i.e., the results are *statistically significant*), you may then conclude that your independent variable (IV) (e.g., presence versus absence of caffeine) has some effect on your dependent variable (DV), but only if random assignment had been used to create the two groups. But, even so, there is still some chance that you have committed a Type I error and that the population means actually are equal. Therefore, it is usually *not* a good idea to regard any one such statistical finding as conclusive. You should wait to see if repetitions (*replications*) of your study also indicate similar effects before reaching firm conclusions.

It is also important *not* to confuse statistical significance with practical significance. Depending on your sample sizes, effects too small to be of any consequence can end up statistically significant, and effects that look fairly large can fail to reach statistical significance. To give you some tools with which to judge practical significance, we show you in this chapter how to compute both a confidence interval (CI) for two means, and a standardized measure of effect size. We take up the topic of CIs next.

Although, in our example, we were able to conclude that caffeine seems to have *some* effect on HR, we have just learned that the NHT alone does not allow us to estimate the

[2] These days it is common either to report the exact two-tailed $p$ value provided by the computer program that performed the statistical test (e.g., $p = .017$) or to state that $p$ is lower than some potential alpha level that ends in a 1 or a 5, such as $p < .005$ or $p < .001$ (using the lowest alpha that $p$ is less than). The traditionalist sets one alpha level (usually .05) at the outset of a study and then reports $p$ only as less than or greater than that alpha—for example, $p < .05$ or $p > .05$.

amount of that effect. For some well-known variables, however, the difference between two sample means does provide a good idea of the size of effect that we are dealing with. For the present example, $\overline{X}_1 - \overline{X}_2$ equals 13 bpm, which, based on extensive cardiological research, represents a substantial increase in HR associated with caffeine. In any case, $\overline{X}_1 - \overline{X}_2$ is the best point estimate for the difference between the two population means ($\mu_1 - \mu_2$). However, as we pointed out in the previous chapter, a point estimate alone does not tell you how accurate the estimate is. Interval estimation is much more informative. So, we now apply this procedure to the two-sample case.

## Confidence Intervals for $\mu_1 - \mu_2$

The use of CIs to determine all reasonably likely values of the mean of a *single* population was discussed in the previous chapter. In a similar fashion, all reasonably likely values of the *difference between two population means* can be found by using the following formula for the two-group CI:

$$\left[\left(\overline{X}_1 - \overline{X}_2\right) - t_{\text{crit}} s_{\overline{X}_1 - \overline{X}_2}\right] \leq \mu_1 - \mu_2 \leq \left[\left(\overline{X}_1 - \overline{X}_2\right) + t_{\text{crit}} s_{\overline{X}_1 - \overline{X}_2}\right] \qquad \text{Formula 7.7}$$

where

$t_{\text{crit}}$ = the critical value obtained from the $t$ table for the specified criterion of significance and the appropriate $df$

Notice that Formula 7.7 is just Formula 7.5A twisted around algebraically. This CI pertains to the *difference* between two population means. So, it is appropriately based on the SED $\left(s_{\overline{X}_1 - \overline{X}_2}\right)$.

For the present example, the observed sample means were 81.0 for the caffeine group and 68.0 for the control group; the SED was 4.474, and the critical value of $t$ for $df = 40$ and $\alpha = .05$ (two-tailed) was 2.021. Therefore, the 95% CI interval for the difference between the population means is:

$$[(81.0 - 68.0) - (2.021)(4.474)] \leq \mu_1 - \mu_2 \leq [(81.0 - 68.0) + (2.021)(4.474)]$$

$$(13.0 - 9.04) \leq \mu_1 - \mu_2 \leq (13.0 + 9.04)$$

$$3.96 \leq \mu_1 - \mu_2 \leq 22.04$$

You can state with 95% confidence that the difference between the means of the caffeine and control populations is included in an interval that runs from 3.96 points to 22.04 points, with the caffeine mean being the larger. Note that zero is *not* in this interval, indicating that we are permitted to treat $\mu_1 = \mu_2$ as an unlikely possibility.

To determine the 99% CI, the procedure would be the same except that the critical value of $t$ for 40 $df$ for $\alpha = .01$, which is 2.704, would be used instead of 2.021. The 99% CI would therefore be:

$$13.0 - (2.704)(4.474) \leq \mu_1 - \mu_2 \leq 13.0 - (2.704)(4.474)$$

$$(13.0 - 12.1) \leq \mu_1 - \mu_2 \leq (13.0 + 12.1)$$

$$0.90 \leq \mu_1 - \mu_2 \leq 25.1$$

As always, the 99% CI is wider than the 95% CI. You can be more certain that the true difference between the two population means falls in the 99% CI, but you must pay a price for this added confidence: The interval is larger, so $\mu_1 - \mu_2$ is estimated less precisely.

Here again, CIs allow you to do everything that NHT does. If a value of $\mu_1 - \mu_2$ specified by the null hypothesis (such as the frequently used value of zero) falls within the CI, that null hypothesis is retained. If a value of $\mu_1 - \mu_2$ specified by the null hypothesis falls outside the CI, that null hypothesis is rejected. (Note that zero is not included in the 99% CI found above, so the null hypothesis that $\mu_1 = \mu_2$ can be rejected at the .01 level.) By specifying a range of population differences that should be retained, CIs provide us with more information about how different the means might be.

Admittedly, some CIs are more informative than others. Let us look at two rather extreme examples. Imagine that in the caffeine study, the SED was only 2.97 instead of 4.474. In that case, the 95% CI would extend from 7.0 to 19.0, indicating that the effect of caffeine on resting HR is a probable mean increase in resting HR of at least 7 bpm and not more than 13 bpm. However, an SED of 6.2 would lead to a 95% CI of 0.48 to 25.52, which would tell a very different story. In this latter case, caffeine may cause an increase of as much as 25 bpm or as little as half a beat per minute. Since an increase of less than 1 bpm is virtually negligible, the possibility that caffeine has no important practical effect on HR would remain a reasonable one. (Such a wide CI suggests the likelihood of substantial sampling error. The best way to improve accuracy is to repeat the experiment with a larger sample.) Yet in both cases, a null hypothesis test will lead to the same simple conclusion: It will tell us only that the null-hypothesized value of zero between the two population means should be rejected.

However, as useful as CIs can be, they are much less meaningful when your DV is measured on a scale that is not commonly used, as when you measure the ability to memorize verbal material in terms of how many words are recalled from a list that you created for your experiment. An increase in HR of 10 bpm is immediately understandable because bpm is a universally accepted metric for HR, whereas an increase of 10 correctly recalled words is not. A more universally interpretable measure of the size of an experiment's effect will be discussed shortly. In the meantime, we want to make the point that although we agree with recent efforts toward encouraging the reporting of CIs whenever possible, we also think that all students of statistics should know how to calculate a CI from the results usually given for a null hypothesis test, even when a CI is not reported.

For example, suppose you read the following report in a published article: "The caffeine group obtained a higher average score on the motor task ($M = 23.5$) than did the placebo group ($M = 16.1$), and this difference was statistically significant, $t(28) = 3.4, p < .01$." Though it is not obvious, the previous sentence contains all the information you need to use Formula 7.7 to compute a CI for the difference between the caffeine and placebo population means. First, you will need $\overline{X}_1 - \overline{X}_2$, but that is just the difference between the two *M*s, which is $23.5 - 16.1 = 7.4$. Next, you will need the critical *t*, but you can easily look that up in Table B in the Appendix. The *df* you need is in parentheses following the *t* (it is 28), and if you want a 95% CI, you should look in the column for a .05, two-tailed test. The critical *t* equals 2.048. Finally, you need the SED. This is easy to find from Formula 7.5B, which we will rewrite here as: $t = (M_1 - M_2)/\text{SED}$. Remember that you were given the *t* value in the results (it is 3.4), and we just found the difference of the means to be 7.4, so we know that $3.4 = 7.4/\text{SED}$. Therefore, $\text{SED} = 7.4/3.4$, which equals 2.18. Sometimes the SED is reported right after the *t* value, and that would save you a step, but as long as you have the two means along with the *t* value, you can easily find the SED for yourself. Now let us insert these values into Formula 7.7:

$$7.4 - (2.048)(2.18) \leq \mu_1 - \mu_2 \leq 7.4 + (2.048)(2.18)$$

$$(7.4 - 446) \leq \mu_1 - \mu_2 \leq (7.4 + 4.46)$$

$$2.94 \leq \mu_1 - \mu_2 \leq 11.86$$

## Measuring the Size of an Effect for a Difference Between Two Independent Samples

If the 95% CI for a new six-month weight-loss program goes from 10 to 30 pounds lost, you can easily decide for yourself whether you are interested. But if a new memory-improvement program announces, with 95% confidence, that you will recall between 10 and 20 more names from a list of 200 after three weeks of training, should you be impressed? When the scale used to measure the DV is not a familiar one, it can be more informative to standardize the obtained difference in sample means rather than to report that difference in its original units. This standardized measure of effect size in your data is often called *d*, or Cohen's *d*, after Jacob Cohen, one of the original authors of this textbook. Unfortunately, the symbol $d$ has sometimes also been used for the population effect size of which it is an estimate and therefore can lead to confusion. To reduce ambiguity, we will use the symbol $d$ for the effect size of a two-group comparison involving *sample* data, and we will use **d** in boldface, in later chapters to represent the effect size comparing two population means. Because, in this case, we are not adhering to the convention of using a Greek letter for a population value (as is commonly done), we will be printing **d** in boldface to remind you that it is a measure for a population rather than a sample. Because $d$ is a *standardized* measure of effect size, its formula looks like a $z$ score:

$$d = \frac{\overline{X}_1 - \overline{X}_2}{s_p} \qquad \text{Formula 7.8}$$

Note that $s_p$ is the square root of the pooled variance (Formula 7.1) that is calculated as part of a $t$ test.

For the caffeine study, $s_p$ equals $\sqrt{209.664} = 14.48$, so $d = (81 - 68)/14.48 = 13/14.48 = 0.9$.

This tells us that the two sample means are nearly an entire SD apart, which is considered a rather large effect.

The interpretation of effect sizes, and the possibility of creating CIs for effect sizes, will be discussed in Chapter 11. An alternative measure of effect size, based on the principle of correlation, will be described in Chapter 10. Note that if you have already calculated $d$, and your two samples are the same size, you can calculate your $t$ value from $d$ and the sample sizes by using the following formula. (When the two samples are the same size, we can use $n$ to represent the size of just one of those samples—i.e., $n = N_1 = N_2$.)

$$t = d\sqrt{\frac{n}{2}} \qquad \text{Formula 7.9A}$$

More usefully, if you know only the $t$ value and the sample sizes, and the two sample sizes are the same, you can compute the effect size by twisting the preceding formula into this one:

$$d = t\sqrt{\frac{2}{n}} \qquad \text{Formula 7.9B}$$

If $N_1 \neq N_2$, you must compute the **harmonic mean** of the two sample sizes in order to use the preceding formula. The harmonic mean of any two numbers can be calculated as follows:

$$\overline{X}_{\text{harmonic}} = \frac{2N_1N_2}{N_1 + N_2} \qquad \text{Formula 7.10}$$

Combining the formula for the harmonic mean with the formula for $d$ in terms of $t$ and $n$, and applying a bit of algebra, yields the following formula for unequal $N$s:

$$d = t\sqrt{\frac{N_1 + N_2}{N_1N_2}} \qquad \text{Formula 7.11}$$

Applying the above formula to the caffeine study yields the same value for $d$, of course, that we calculated more directly with Formula 7.8:

$$d = 2.91\sqrt{\frac{42}{440}} = 2.91 \times 0.309 = 0.90$$

## Reporting the Results of a *t* Test with a CI and a Measure of Effect Size

As promised, we are returning to the format for reporting the results of a $t$ test in APA style, with more complete information, which can help the reader interpret your results. Adding the results we found for the CI and effect size for the caffeine experiment to our report looks like this: "The group that was given a caffeine pill to ingest an hour before the HR measurements exhibited a higher resting HR ($M = 81$, SD $= 12.0$ bpm) than the group that had been given a placebo pill ($M = 68$, SD $= 16.8$ bpm); this difference of 13 bpm was statistically significant, $t(40) = 2.91$, $p < .01$, 95% CI [3.96, 22.04], $d = 0.90$.

## The Assumptions Underlying the Proper Use of the *t* Test for Two Sample Means

It is easy to perform and report two-sample *t* tests, but some care must be taken in deciding when such a test is appropriate and when an alternative approach should be considered. The following points should help you decide when to use the two-group *t* test.

In theory, the use of the independent-samples *t* test, as well as the two-group CI just described, is justified only if the first two of these assumptions are met:

1. *Independent random sampling*. The observations within each group have been randomly sampled independently of each other, and all of the observations in one group are independent of all of the observations in the other group. (For true experiments, this assumption is usually accommodated by randomly assigning the members of one convenient sample to two different groups.)
2. *Normal distributions*. The DV is normally distributed within each population.

   In practice, however, you should not be automatically deterred from using the *t* test even if the assumption concerning normality does not seem to be met. The two-group *t* test is considered *robust* with regard to this assumption—that is, it gives fairly accurate results even if the assumption is not satisfied—as long as the sample sizes are not very small. When dealing with small sample sizes and questionable distributions, a nonparametric alternative to the *t* test should be considered, such as the one that will be described in detail in the next chapter.

   The pooled-variance version of the two-sample *t* test, as expressed in Formula 7.6, requires an additional assumption:
3. The variances of the two populations are equal $(\sigma_1^2 = \sigma_2^2)$. This is referred to as the *homogeneity of variance* (HOV) assumption.

The HOV assumption is important because when the two population variances are very different, it becomes statistically easier to select two sample means that are far apart *even if the population means (μs) are not*. That is, it is easier to make a Type I error when you cannot assume HOV; the true value of alpha creeps above .05.

Fortunately, the HOV assumption can be easily tested statistically with an HOV test, such as Levene's test, which is available from most statistical software programs. As with the tests we have discussed so far in this book, Levene's test is a null hypothesis test; the major difference is that the null hypothesis of Levene's test concerns the equality of population variances ($\sigma^2$s), not the population means (μs). The $H_0$ for an HOV test is that $\sigma_1^2 = \sigma_2^2$. Unlike most null hypotheses, this is an $H_0$ you would prefer to retain because doing so would mean that HOV can, in fact, be assumed.

For example, if the *p*-value for an HOV test is greater than alpha (typically set at .05), then the test is not significant, you can assume HOV, and you can proceed with the *t* test we have been discussing (which is often called the *pooled-variances t* test). If your HOV test is significant (i.e., produces a $p < .05$), however, then you must reject the $H_0$ that the population variances are equal (the HOV assumption), and proceed with what are generically called **separate-variances *t* tests**. These tests, which are typically conducted with statistical software, make adjustments to the SED, to the degrees of freedom, or to both, to keep the true alpha level to .05. Welch's *t* test is a popular version of the *t* test that does not require the HOV assumption.[3] In R, Welch's test is commonly used instead of a pooled-variances *t* test; SPSS produces values for Levene's test and Welch's test (in the "equal variances not assumed" row of the output) automatically when you run an independent *t* test.

---

[3] Some researchers believe that it is best to run a separate-variances *t* test regardless of evidence for or against HOV (Delacre et al., 2017).

When you are analyzing your own data, we strongly recommend that you use an HOV test along with a separate-variances *t* test, if necessary. If you are reading an article and wondering about whether the HOV was likely met, there are a few heuristics (rules of thumb) you can use to estimate whether or not the assumption was met. If the sample sizes are relatively small (say, neither is larger than 15), *and* they are fairly unequal (say, the larger is more than 1.5 times greater than the smaller), *and* the population variances are markedly unequal (say, one is more than 2.0 times greater than the other), then the HOV assumption has probably not been met. If any one of these three criteria does not obtain, however, then it is likely that you can assume HOV. We would like to emphasize that while these heuristics can be useful when reviewing other people's results, they should not replace a statistical test of HOV, such as Levene's test when that option is available.

### Within-Subjects Designs

Thus far, we have compared two samples using what is called a **between-subjects design** (BSD). With this approach, comparisons are made between two independent samples of subjects (the people in one sample are different from the people in the other sample). BSDs are commonly used within behavioral research, but they are not the only choice to test the significance of an effect. A **within-subjects design** (WSD), when used appropriately, can be statistically more powerful[4] than BSDs, while requiring fewer subjects.

Imagine you are interested in testing a version of the Stroop effect (Stroop, 1935), in which subjects are presented words, one at a time, that are printed in either lowercase or uppercase. The subject's task is to respond as quickly as they can by pressing the "1" key if the word is presented in uppercase, or the "2" key if it is presented in lower case. For example, if the word was "chair," the subject would press the "1" as quickly as possible; if the word was TABLE, then they would press the "2" key. Sounds pretty easy, right? The effect of interest occurs when the stimuli are the words "upper" or "LOWER." These items, often called interference trials, create response competition for subjects who can read the word LOWER as quickly and easily as they can perceive that the letters are presented in upper case (assuming they are proficient readers of English). As with the neutral stimulus, **TABLE**, the correct answer is to press the "upper" key, but it takes people a little longer to press the correct button due to the interference created by the mismatching word; both "lower" and "upper" are automatically activated as correct responses, and the subject must exert executive control to select the correct response, which adds time to the response. A Stroop experiment can be implemented as a BSD by randomly assigning subjects to one of two groups: A control group, in which subjects would see, say, 25 neutral words presented in either upper or lower case (e.g., **ROOF**; **street**; **SCHOOL**; **frame**), or an interference group in which the subjects would see 25 mismatching words (e.g., upper; LOWER; LOWER; upper). (Remember that the task is to report the case the word is presented in, not what the letters spell!) At the end of the experiment, the mean reaction time of the subjects in the control group is compared to the mean reaction time of the subjects in the interference group using an independent-samples *t*-test. The typical result is that reaction times are significantly slower in the interference group. But there is a better way to conduct this experiment!

Stroop experiments can also be implemented as a WSD by asking each subject to respond to both 25 control trials and 25 interference trials. These trials could either be presented in a random order (control and interference trials mixed together randomly), or in randomized blocks (subjects respond to a block of 25 trials in one condition, and then a block of 25 in the other condition, with the order of the blocks alternated so that half the subjects see the control block first and the other half see the interference trials first). At the end of the experiment, each subject would have a reaction time mean computed from their 25 interference trials, and a reaction time mean from their 25 control trials. Below we get into how we would analyze these data statistically, but for now note that in WSDs, each subject is "acting

[4] We will discuss statistical power in more detail in Chapter 11, but for now think of it as the ability to detect an effect and therefore reject a false $H_0$.

as their own control." This feature of WSD allows researchers to control for person-to-person variability. But why would we want to do that?

Individuals vary from each other in many ways that create what are called **individual differences**. In some behavioral research, these differences are interesting and the object of study (e.g., Why do some people have higher self-esteem than others?). For a lot of behavioral research, however, this person-to-person variability is not interesting and can make finding significant results more difficult to detect. For example, in the context of research that uses reaction-time as a DV, such as the Stroop effect, it can be a nuisance that some people tend to have faster overall reaction times than others. The individual differences in response time may not be large in an absolute sense, but they add "noise" to the data that can mask the response-time effect the researcher is trying to detect. A WSD controls for these individual differences by comparing subjects' responses in one condition with their responses in the other condition. In the within-subjects version of our Stroop experiment, for example, if Monica is an especially fast responder, then that extra speed will show up in both her control and her interference trials. Similarly, if Joey just pulled an all-nighter and is exhausted at the time of the experiment, then the effects of that exhaustion will be evident in his responses to both conditions. As we will explain in more detail later, with WSDs, we can subtract each subject's control trials mean from the mean of their experimental (e.g., interference) trials to create a difference score that represents the overall effect of the interference. We will then use those difference scores as our data in a null-hypothesis test designed for data of this kind.

Note that individual differences are eliminated with this approach because they contribute equally to both conditions and therefore disappear when one mean is subtracted from the other. With a BSD, by contrast, if Joey shows up to an experiment hungover, then the effects of his relatively slow mental state will only show up in the overall mean of whichever condition he was randomly assigned to. That extra "hangover variability" will be added to the SD for that group, which will in turn contribute to a larger variance estimate in the denominator of the *t*-test, and ultimately to a smaller observed *t* value. By allowing subjects to act as their own controls, therefore, WSDs can reduce the effects of individual differences and lead to a more powerful hypothesis test.

## The *t* Test for Matched Samples

Suppose that you want to test the hypothesis that in families with two children, the firstborn is more introverted than the second born. Once again, the question of interest concerns a comparison between the means of two populations, and the null and alternative hypotheses are as follows:

$$H_0 : \mu_{\text{firstborn}} = \mu_{\text{secondborn}}$$
$$H_A : \mu_{\text{firstborn}} \neq \mu_{\text{secondborn}}$$

To obtain your samples, you randomly select *matched pairs* of children, with each pair consisting of the firstborn child and the secondborn child from a given family. You then administer an appropriate measure of introversion. The general format of the resulting data would be as follows:

| **Pair** | **Firstborn ($X_1$)** | | **Secondborn ($X_2$)** |
|---|---|---|---|
| 1 | 65 | matched ↔ | 61 |
| 2 | 48 | matched ↔ | 42 |
| 3 | 63 | matched ↔ | 66 |
| ⋮ | ⋮ | | ⋮ |
| $N$ | 66 | matched ↔ | 69 |

The firstborn child from Family 1 has an introversion score of 65, and the secondborn child from Family 1 has an introversion score of 61. Similarly, each pair of children is matched by virtue of coming from the same family.

To illustrate a second (and perhaps more common) use of matched samples, let us suppose that you wish to test the effect of a persuasive message on people's attitudes toward gun control. Each subject's attitude is measured *before* receiving the message ($X_1$) and again *after* receiving the message ($X_2$). For Pair 1, the score of 65 could represent the attitude of the first subject before hearing the persuasive message, and the score of 61 could represent the *same* subject's attitude after hearing the message. Each subject therefore serves as his or her own control, and each pair of scores is matched by virtue of coming from the same subject. (It would be desirable to have a control group that does not receive a message about gun control in this study, but this is beyond the present point.)

The statistical analysis described previously in this chapter involved *independent* random samples. That is, there was no connection between any specific individual in Sample 1 and any specific individual in Sample 2. This approach is not suitable for matched samples, as it does not take advantage of the reduction in the error term that can be produced by focusing on the matched pairs of scores. To analyze matched-pair data (of either of the two types described earlier), the procedure that is used is similar to the one we used with one-sample *t*-tests in Chapter 6. Because each score in Sample 1 has a paired counterpart in Sample 2, you can subtract each $X_1$ from each $X_2$ and obtain a difference score (denoted by $D$). Then you can use the $t$ test for the mean of one population—in this case, the mean of the population of difference scores.

To illustrate this procedure, let us look at the number of correct answers out of 100 total for 10 participants selected at random for a Stroop experiment; the data for the neutral and interference accuracies are shown in Table 7.2. To test whether there is a statistically significant difference in accuracy between the neutral and the interference trial, we will focus on the difference scores shown in the last column of Table 7.2. Two important notes to keep in mind: (1) it does not matter whether you subtract the score from $X_1$ from $X_2$, or vice-versa as long as you are consistent across the full set of scores; and (2) it is critical to retain the *sign* of each difference score.

As with any statistical test, we can organize our procedure into three discrete steps.

### Step 1: State the Null and Alternative Hypotheses, and the Significance Criterion, α

In terms of the difference scores, the null and alternative hypotheses are as follows:

$$H_0 : \mu_D = 0$$
$$H_0 : \mu_D \neq 0$$

**Table 7.2** **Accuracy in # Correct out of 100 in Both Neutral and Interference Trials of a Stroop Task**

| Participant | Neutral Trial Accuracy ($X_1$) | Interference Trial Accuracy ($X_2$) | $D = (\overline{X}_2 - \overline{X}_1)$ |
|---|---|---|---|
| 1 | 70 | 77 | +7 |
| 2 | 66 | 75 | +9 |
| 3 | 73 | 68 | −5 |
| 4 | 62 | 62 | 0 |
| 5 | 74 | 89 | +15 |
| 6 | 63 | 74 | +11 |
| 7 | 64 | 62 | −2 |
| 8 | 65 | 73 | +8 |
| 9 | 71 | 76 | +5 |
| 10 | 72 | 84 | +12 |
| $\sum$ | 680 | 740 | +60 |
| Mean | 68.0 | 74.0 | +6.0 |
| SD | 4.472 | 8.59 | 6.48 |

$H_0$ states that the mean of the population of difference scores is zero, and
$H_A$ states that the mean of the population of difference scores is not zero.

For alpha, we will use the usual .05 level, with a two-tailed test.

**Step 2: Calculate the Appropriate Test Statistic**

We begin by computing the mean difference score obtained from the sample, $\overline{D}$:

$$\overline{D} = \frac{\sum D}{N} = \frac{60}{10} = 6.0$$

Note that $N$ is not necessarily the number of different participants but the number of *paired measurements*, which may be the number of *pairs* of subjects. As is shown in Table 7.2, $\overline{D}$ is equal to $\overline{X}_2 - \overline{X}_1$.[5] Thus, testing the hypothesis that $\mu_D = 0$ is equivalent to testing the hypothesis that $\mu_1 - \mu_2 = 0$, or that $\mu_1 = \mu_2$.

To test the hypothesis about the mean of one population, you compare the sample mean to the hypothesized value of the population mean and divide by the standard error of the mean, as follows:

$$t = \frac{\overline{D} - \mu_D}{s_{\overline{D}}}$$

where

$$s_{\overline{D}} = \frac{s_D}{\sqrt{N}} = \sqrt{\frac{s_D^2}{N}}$$

Note the similarity to Formula 6.3; here we are just changing the symbols in Formula 6.3 to represent difference scores. Thus, $s_D$ represents the ordinary unbiased SD of the difference scores.

Because the hypothesized value of $\mu_D$ is generally zero, the matched $t$ formula may be simplified to:

$$t = \frac{\overline{D}}{\frac{s_D}{\sqrt{N}}} \qquad \text{Formula 7.12}$$

As is normally the case in tests concerning one population mean, the number of degrees of freedom on which $t$ is based is equal to one less than the number of scores, the scores being difference scores, in this case.

$$df = (\text{number of } pairs) - 1$$

where each pair yields one difference score.

If the computed $t$ is smaller in absolute value than the value of the critical $t$, as obtained from the table for the appropriate $df$, retain $H_0$; otherwise reject $H_0$ in favor of $H_A$.

Returning to the Stroop experiment, the analysis of the data can be completed by inserting into Formula 7.12 the summary statistics found in the lower-right corner of Table 7.2, as follows:

$$N = 10\ \overline{D} = 6.0, \quad s_D = 6.48$$

$$t = \frac{6.0}{\frac{6.48}{\sqrt{10}}} = \frac{6.0}{\frac{6.48}{3.16}} = 2.93$$

[5] We are subtracting $X_1$ from $X_2$ in this example to reduce the number of difference scores that are negative. This changes the sign, but not the magnitude, of the $t$ value that we will calculate. Therefore, the order of subtraction does not affect our chances of attaining statistical significance. As with the two-sample $t$-test, the sign merely indicates which sample mean is bigger than the other.

### Step 3: Find the Critical Value and Make a Statistical Decision With Respect to Your Null Hypothesis

The critical value of $t$ obtained from Table B for $(10 - 1)$ or 9 degrees of freedom is 2.262. Because the obtained $t$ value of 2.93 is larger in absolute value than the critical value, you can reject $H_0$: There is sufficient reason to believe that the interference trials of a Stroop experiment would increase the mean reaction time compared to the neutral trials for the population being sampled.

### CI for the Population Mean of Difference Scores

A CI for the mean difference score can be computed using a slight variation of Formula 7.7, as shown next.

$$\overline{D} - t_{\text{crit}} s_{\overline{D}} \leq \mu_D \leq \overline{D} + t_{\text{crit}} s_{\overline{D}} \qquad \text{Formula 7.13}$$

Given that the .05 two-tailed critical value of $t$ is 2.262 and that the standard error of the mean of the differences is equal to 20.5 (i.e., 64.8/3.16), the 95% CI is:

$$6 - (2.262)(2.05) \leq \mu_D \leq 6 + (2.262)(2.05)$$

$$1.36 \leq \mu_D \leq 10.64$$

So, the 95% CI goes from +13.6 to +106.4 milliseconds, suggesting that the increase in reaction time could reasonably be as small as 13.6 milliseconds, or as large as 106.4 milliseconds. As would be expected from the rejection of the null hypothesis at the .05 level, the value of zero does *not* fall within the 95% CI.

### Measuring the Size of an Effect for Difference Scores

The common measure of effect size corresponding to the $t$ test for matched pairs resembles $d$ for independent samples with one small difference:

$$d_z = t\sqrt{1/n} \qquad \text{Formula 7.14}$$

The number 2 in Formula 7.9B has been changed to a 1 because instead of comparing two sample means, we are focusing on a single set of difference scores. The subscript z has been added to indicate that this is an effect size for matched pairs (do not ask us why J. Cohen chose z to be the subscript in this case). For our Stroop example, $d_z = 2.93 * \sqrt{(1/10)} = 2.93 * 0.316 = 0.93$.

### Comparing the *t* Test for Matched Pairs With the *t* Test for Independent Samples

To demonstrate more clearly the usual advantage of the matched-pairs (MP) design, let us perform an independent-samples $t$ test on the Stroop experiment data. For this demonstration, we will ignore the connection between the two columns of Stroop experiment data in Table 7.2. Instead, we will treat the 10 neutral trial measurements as though they come from one group of participants, while the 10 interference trial measurements come from another, completely separate, group of students. Because our two "independent" sets of experimental data are the same size, we can use the following simplified, equal-$N$ formula for the independent-samples $t$ test (using Formula 7.4 as its denominator):

$$t = \frac{\overline{X}_1 - \overline{X}_2}{\sqrt{\dfrac{s_1^2 + s_2^2}{n}}} \qquad \text{Formula 7.15}$$

where $n$ is the size of *each* sample. The numerator of the $t$ test remains the same whether you apply the matched-pairs $t$ test[6] or the independent-samples $t$ test to a given set of data (in this example, the numerator is 6 bpm); it is the denominator that is likely to change. We will now use the SDs we included at the bottom of each column of HR data in Table 7.2. They are squared to yield variances and inserted into Formula 7.15 to obtain the following result:

$$t = \frac{74 - 68}{\sqrt{\frac{73.8 + 20}{10}}} = \frac{6}{3.063} = 1.96$$

We do gain additional degrees of freedom by treating the two sets of measurements as independent. For the MP design, the $df$ was one less than the number of *pairs*, or 9. For the independent-samples design, the $df$ would be $10 + 10 - 2 = 18$. This lowers the critical value of the independent test relative to the MP test from 2.262 to 2.101. However, the $t$ value calculated for the independent-samples $t$ test (1.96) is smaller than even this reduced critical value (2.101). So, the independent-samples $t$ test would *not* be statistically significant. In contrast, matching the reaction times in pairs (based on the two measures coming from the same participant) resulted in a decrease in the denominator of the $t$ test from 3.063 for the independent $t$ test to 2.05 for the MP test, and this increased the calculated $t$ value from 1.96 to 2.93. This increase made the results statistically significant.

Note also that the effect size associated with the $t$ value has increased. For independent samples, $d = 1.96\sqrt{(2/10)} = 1.96 * 0.447 = 0.877$; for the matched pairs, we calculated that $dz = 0.93$. The closer the matching, the more the $t$ value and $d_z$ will be increased relative to the $t$ and $d$ for the corresponding independent samples test. The closeness of the matching can be quantified by calculating the degree of correlation between the matched scores. You will learn to calculate a correlation coefficient in Chapter 9.

The extent to which matching the scores in pairs increases the $t$ value relative to the independent-samples test, and thus makes statistical significance more likely, depends on how highly the two sets of scores are correlated. We will discuss correlation in detail in Chapter 9. For now, we will simply state that when two sets of scores are positively correlated, one score *tends* to increase when the score paired with it increases. (For example, higher neutral trial reaction times *tend* to be associated with higher interference trial reaction times. So, the two measures are positively correlated.) In general, measuring the same experimental participant twice, called the *repeated-measures* (RM) design, leads to higher correlations than matching separate participants into pairs. However, the latter design can be very helpful when the RM design is not feasible.

### The Different Types of Matched-Pairs Designs

The MP (difference-score) $t$ formula we just described can be applied to the results of two very different experimental designs, which are described next.

#### *The Repeated-Measures Design*

The RM design works well if your experiment consists of many similar trials of one type that can be randomly mixed with the same number of trials of some other type. Suppose that participants are told to indicate, as soon as they know it, the missing letter of a word presented on a screen (e.g., POLI_E). Before each word is presented, a priming word is flashed subliminally on the same screen. There are two types of trials: Either the priming stimulus is a word that is relevant to one of the solutions (e.g., "crime" for POLICE, or "manners" for POLITE), or it is an irrelevant word (e.g., "basket"). Because the two types of trials are mixed together in a random order, the experimental participant cannot know which type of trial

[6] The matched-pairs t test is sometimes referred to as the t test for dependent (or correlated) means and if the two matched measurements are taken from the same person, it can be called a within-subject t test or a repeated-measures $t$ test.

will be presented next. The reaction times can then be averaged separately for the two types of trials for each participant. So, each participant ultimately contributes a *pair* of scores that can be used to create a difference score for the MP $t$ test.

The RM design is potentially more problematic when you cannot mix the two conditions together but must present one condition before the other. The simplest type of *successive* RM design is the before-after design. For example, if a clinical psychologist wanted to test the effectiveness of a new anti-depression treatment, he might start with a sample of untreated depressive individuals, measure them on a depression inventory before treatment and then again six weeks after starting the treatment. If he finds that the sample was significantly less depressed six weeks after they began treatment than they were before, can he conclude that the drug was effective? Unfortunately, he cannot. People's level of depression varies naturally over time. So, we cannot tell whether the sample became less depressed due to the passage of time, or if the new treatment actually works (both might be true). He needs a control group that has been given a control treatment that, from the participant's perspective, is indistinguishable from the experimental treatment being tested.[7] In short, to draw conclusions from a before-after design, a separate control group is usually required. We do not discuss this issue further until Chapter 15.

The other type of successive RM design involves two different experimental conditions. Suppose you want to see if more clerical tasks are correctly performed in a given time interval while participants listen to happy, as compared to sad, music. It would not be a fair comparison of the two conditions to present one type of music first for all of the participants; it is possible, for instance, that the participants will perform better in the second condition because of the practice they received during the first condition. The common solution to this problem is **counterbalancing**: Half of the subjects perform the tasks first with happy music and then with sad music, while the other half of the subjects are assigned to the reverse order (i.e., sad music first). Counterbalancing averages out **simple order effects**, such as those due to practice or fatigue, but it does not help if you have **differential carryover effects**. For example, if you want to compare the effects of two different types of strategies for remembering a list of words by comparing one group that is asked to develop a visual image of each word on the list, and a control group that is simply asked to remember the words. When the group that got the visualization strategy first moves on to the control condition, the participants may not be able to stop themselves from employing the visualization strategy. This is the case when the RM design becomes problematic.

To avoid carryover effects while still gaining at least some of the benefit of correlated sets of scores, you can employ the design described in the next section, which involves matching your participants into pairs based on some relevant similarity. Although the same statistical procedure (which we have been calling the MP $t$ test but could just as correctly be called the RM $t$ test) is used for both designs, from now on, we will use the term *RM design* when the same participant is measured twice and reserve the term MP *design* for the case in which two different participants have been matched together based on some relevant similarity.

### *The Matched-Pairs Design*

As just mentioned, for some types of experiments, the RM design is not a reasonable option. Suppose you want to compare two methods for teaching children to do long division by hand. It would make little sense to teach a child long division by one method, measure how long it took for them to master it, and then teach the same child long division with the other method—they already know how to do long division! The potential for misleading carryover effects rules out the possibility of counterbalancing. However, the RM design can be approximated by testing the children on general math aptitude and then matching the children into pairs based on the similarity of their scores, as well as their ages and genders.

[7] With that design, you'd have one between-subjects independent variable (control vs. treatment) and one within-subjects independent variable (before treatment versus after treatment). Experiments with two or more independent variables will be covered in Chapters 15 and 16.

Some random process, like the flip of a coin, should be used to decide which child in each pair is assigned to each long-division method. For example, if Izzy and Solveig have similar math aptitude scores, then they would be considered a matched pair (matched on math ability). Then, each would be randomly assigned to one of the two teaching methods. When the experiment is completed, and all the children have had their long-division abilities measured, the scores of the two children in a pair are treated as though they were two scores from the same child measured twice: The difference is calculated for the two members of each pair, and then these difference scores are subjected to an MP $t$ test as described for the HR example.

There are situations in which the RM design is not feasible but there is no relevant information that can be used to match the participants in pairs. For example, you may wish to test whether an experimenter who acts politely leads participants to be more willing to donate their time for a future experiment than does an experimenter who acts in an unpleasant way. On what basis would you match these subjects? There is no obvious way to match people ahead of time on their willingness to donate their time for future experiments. When there is no apparent basis for matching participants, you will not have the advantage of correlated sets of scores, and therefore you would be better off using a BSD. In that case, you will need to make sure that your samples are large enough to have a good chance of attaining significance with the effect size you expect. We show how you can estimate the sample size that you will need when we discuss the topic of power in Chapter 11.

## Summary

To test null hypotheses about *differences between the means of two populations*, perform one of the following procedures.

### For Independent (i.e., Between-Subjects) Samples

The generic formula for the $t$ test for two (independent) sample means is:

$$t = \frac{\overline{X}_1 - \overline{X}_2}{s_{\overline{X}_1 - \overline{X}_2}} \qquad \text{Formula 7.5B}$$

where

$s_{\overline{X}_1 - \overline{X}_2}$ = the estimated SED

To obtain the denominator of the preceding formula, you can begin by computing the pooled variance:

$$s^2_{\text{pooled}} = \frac{(N_1 - 1)s_1^2 + (N_2 - 1)s_2^2}{N_1 + N_2 - 2} \qquad \text{Formula 7.1}$$

Once you have calculated the pooled-variance from Formula 7.1, you are ready to use the following formula for the two-group $t$ test:

$$t = \frac{\overline{X}_1 - \overline{X}_2}{\sqrt{s^2_{\text{pooled}}\left(\frac{1}{N_1} + \frac{1}{N_2}\right)}} \qquad \text{Formula 7.6A}$$

with

$$df = N_1 + N_2 - 2$$

Reject $H_0$ if the computed value of $t$, ignoring the sign, is greater than the critical value of $t$ obtained from the $t$ table. Note that this $t$ formula may well yield misleading results if $\sigma_1^2$ and $\sigma_2^2$ are markedly unequal *and* $N_1$ and $N_2$ are also markedly unequal. In such a case, the separate-variances $t$ test may be preferable (but see an alternative in the next chapter).

When you reject $H_0$, it is likely that you will want to estimate the true difference between the population means with a CI. The formula for the CI in the two-group case is similar to the CI for a single population:

$$\left[(\overline{X}_1 - \overline{X}_2) - t_{\text{crit}} s_{\overline{X}_1 - \overline{X}_2}\right] \le \mu_1 - \mu_2 \le \left[(\overline{X}_1 - \overline{X}_2) + t_{\text{crit}} s_{\overline{X}_1 - \overline{X}_2}\right]$$

where

$t_{\text{crit}}$ = the critical value obtained from the $t$ table for the specified criterion of significance (e.g., .05, two-tailed for a 95% CI) for the appropriate $df$

Another useful way to describe the size of the effect you found in your data is by calculating a standardized measure of effect size, such as $d$:

$$d = \frac{\overline{X}_1 - \overline{X}_2}{s_p} \qquad \text{Formula 7.8}$$

where $s_p$ is the square root of the pooled variance $\left(s^2_{\text{pooled}}\right)$

**For Matched (or Within-Subjects) Samples**

First compute a difference score (symbolized by $D$) for each pair, where

$$D = \overline{X}_1 - \overline{X}_2$$

Next, compute the mean and SD of the $D$ scores, using the usual formulas for the mean and unbiased SD, or by means of a handheld calculator. Then find the matched $t$ value:

$$t = \frac{\overline{D}}{\frac{s_D}{\sqrt{N}}} \qquad \text{Formula 7.12}$$

with

$$df = (\text{number of pairs}) - 1$$

where

$\overline{D}$ = mean of the $D$ scores
$s_D$ = standard deviation of the $D$ scores
$N$ = number of pairs

As usual, reject $H_0$ if the computed value of $t$ (ignoring the sign) is greater than the critical value of $t$ obtained from the $t$ table.

For matched samples, CIs can be computed in the usual way, using the standard error of the mean of the differences. The formula looks like this:

$$\overline{D} - t_{\text{crit}} s_{\overline{D}} \le \mu_D \le \overline{D} + t_{\text{crit}} s_{\overline{D}}, \qquad \text{Formula 7.13}$$

where

$$s_{\overline{D}} = \frac{s_D}{\sqrt{N}}$$

Note once again that CIs provide considerably more information than do null hypothesis tests.

The effect size for the MP $t$ test can be calculated using the following formula:

$$d_z = t\sqrt{1/n} \qquad \text{Formula 7.14}$$

The formula for the MP $t$ test can be used whether the same individual is being measured twice (RM design) or participants are being matched in pairs (MP design). The RM design usually leads to the larger $t$ value, but it may produce misleading results if there are strong carryover effects. The MP design provides much of the benefit of the RM design without the possibility of carryover effects. As long as there is a reasonably high, positive correlation between the two sets of scores, the MP design can be counted on to yield higher $t$ values than an independent-samples $t$ test on the same data, due to a reduction in the denominator of the $t$ formula.

## Exercises

1. Use the two-sample $t$ test to determine whether the difference in means between Turck and Kirk Halls is significant at the .05 and .01 levels, two-tailed. Given your decision with respect to the null hypothesis, which type of error could you be making: Type I or Type II? Report the results of your $t$ test in a sentence that includes the means of the two dormitories and a measure of the effect size. Compute both the 95% and 99% CIs for the difference of the two population means. Explain how the presence or absence of zero in each of those CIs is consistent with your decision regarding the null hypothesis at the .05 and .01 levels.
2. Repeat the previous exercise, except that this time compare Turck Hall to Dupre Hall.
3. Repeat Exercise 2, but this time compare Kirk Hall to Dupre Hall, and conduct only a one-tailed test at the .05 level, assuming that Kirk will have the larger mean.
4. An industrial psychologist is investigating specific cafe locations of a local coffee chain, Spyhouse, to see if there is a difference in customer satisfaction between the original and newest locations. She obtained customer satisfaction scores from 41 customers who went to the original location and 31 customers who went to the most recently opened location, with the following results: original location, $M = 48.75(SD = 9.0)$; newest location, $M = 46.07(SD = 10.0)$. Test this difference for significance at both the .05 and .01 levels (two-tailed). Determine the effect size for the difference of these two sample means. Does this effect size seem to be large or small?
5. Suppose that the industrial psychologist from the previous exercise is testing the difference in the customer satisfaction of two different independent cafes: Dunn Bros and Roots Roasting. Given the following data, can the psychologist reject the null hypothesis (alpha = .05) that the population means of these two groups are the same for the customer satisfaction scores? If these results are statistically significant, compute the 95% CI for the difference of the population means.

| Group 1 (Dunn Bros) | Group 2 (Roots Roasting) |
|---|---|
| 62 | 46 |
| 54 | 53 |
| 59 | 50 |
| 56 | 52 |
| 59 | 54 |

6. Repeat Exercise 5 for the following two sets of summary statistics. In each case, determine if the results would be significant for a one-tailed as well as a two-tailed test, but compute the CI only if the two-tailed test is significant. Regardless of statistical significance, compute $d$, the standardized measure of effect size in sample data, for each data set.

| (a) | | Group 1 | Group 2 |
|---|---|---|---|
| | $\bar{X}$ | 17.34 | 21.58 |
| | $s$ | 5.83 | 4.42 |
| | $N$ | 32 | 30 |

| (b) | | Group 1 | Group 2 |
|---|---|---|---|
| | $\bar{X}$ | 76.57 | 69.72 |
| | $s$ | 20.15 | 22.87 |
| | $N$ | 17 | 15 |

7. A clinical psychologist has developed a new approach for the treatment of Generalized Anxiety Disorder (GAD) based on findings from recently published research that discovered something new about the underlying mechanisms behind GAD. They wish to know if it is superior to the mainline treatment approach for GAD. They therefore obtain participants with GAD who have had no prior history of treatment and randomly split them into two groups: an experimental group, which receives the psychologists' new approach, and a control group, which receives the mainline treatment approach for GAD. The psychologist is afraid, however, that differences among participants in severity of GAD will lead to large error terms. Therefore, they match their participants on severity of symptoms and form 10 pairs such that each pair is made up of two people roughly equal in GAD symptom severity. After both groups have received treatment, the psychologist measures the amount of symptom reduction by means of a 10-item survey, with a higher score relating to more symptom reduction. The results are as follows:

| Pair | Experimental group (new approach) | Control group (standard approach) |
|---|---|---|
| 1 | 9 | 7 |
| 2 | 6 | 4 |
| 3 | 5 | 6 |
| 4 | 7 | 3 |
| 5 | 3 | 5 |
| 6 | 7 | 3 |
| 7 | 3 | 2 |
| 8 | 4 | 5 |
| 9 | 6 | 7 |
| 10 | 10 | 8 |

Test the results for statistical significance at the .05 level. What should the psychologist decide about their new approach?

8. For the following set of data, assume that the $X$ score represents the participant's performance in the experimental condition and that the $Y$ score represents the same person's performance in the control condition. (Thus, each participant serves as his or her own control.) Compute the *matched t* test for these data. Also compute the 95% CI for the mean of the difference scores and the effect size, $d_z$.

| Data set 3, Chapter 1 | | |
|---|---|---|
| S | $X$ | $Y$ |
| 1 | 97 | 89 |
| 2 | 68 | 57 |
| 3 | 85 | 87 |
| 4 | 74 | 76 |

| | | |
|---|---|---|
| 5 | 92 | 97 |
| 6 | 92 | 79 |
| 7 | 100 | 91 |
| 8 | 63 | 50 |
| 9 | 85 | 85 |
| 10 | 87 | 84 |
| 11 | 81 | 91 |
| 12 | 93 | 91 |
| 13 | 77 | 75 |
| 14 | 82 | 77 |

$\left[\overline{X}_X = 84, s_X = 10.76\right]$, $\left[\overline{X}_Y = 80.64, s_Y = 13.29\right]$

9. Now carry out a matched $t$ test for the same set of data as in Exercise 8, but with the $X$ values rearranged as shown (the means and SDs for X and Y remain the same). Again, include both the 95% CI and $d_z$.

| S | X | Y |
|---|---|---|
| 1 | 92 | 89 |
| 2 | 82 | 57 |
| 3 | 85 | 87 |
| 4 | 81 | 76 |
| 5 | 87 | 97 |
| 6 | 93 | 79 |
| 7 | 68 | 91 |
| 8 | 85 | 50 |
| 9 | 63 | 85 |
| 10 | 100 | 84 |
| 11 | 74 | 91 |
| 12 | 77 | 91 |
| 13 | 92 | 75 |
| 14 | 97 | 77 |

10. Suppose that you have read the following sentence in a psychology journal: "The experimental group obtained a higher score on the recall task ($M$ = 14.7) than did the control group ($M$ = 11.1), and this difference was statistically significant, $t(38) = 2.3$, $p < .05$." Assuming that the two samples are the same size: (a) Find the 95% CI for the difference of the means. (b) Find $d$, the size of the effect in the sample data. (c) Compute the 95% CI.

## Thought Questions

1. (a) What is the difference between *random sampling* and *random assignment*? (b) What similar purpose do these two procedures have? (c) Which one is usually much easier to do? Why?
2. A behavioral science researcher wishes to test a null hypothesis about the difference between the means of two populations. (a) Would the researcher prefer that the SED be relatively small or large? Why? (b) If the size of one or both samples becomes larger, what happens to the SED? Why? (c) If the standard deviation of the individual scores in one or both populations becomes larger, what happens to the SED? Why?
3. Suppose that the difference between two sample means is statistically significant. (a) What decision should the researcher make about the null hypothesis? (b) Assuming that all of the procedures and calculations in this experiment were done correctly, why might this decision still be incorrect? (c) If this decision is incorrect, which type of

error would the researcher be making, Type I or Type II? (d) What are the practical consequences of making this type of error?

4. Suppose that the difference between two sample means is *not* statistically significant. (a) What decision should the researcher make about the null hypothesis? (b) Assuming that all of the procedures and calculations in this experiment were done correctly, why might this decision still be incorrect? (c) Does the criterion of significance give the probability that this decision is incorrect? (d) If this decision is incorrect, which type of error would the researcher be making, Type I or Type II? (e) What are the practical consequences of making this type of error? (f) Should the researcher conclude that the two population means are equal? Why or why not?
5. Students at Rolling Rock University claim that their school is superior to vaunted Bigbrain University with regard to achievement on a national aptitude test, where the mean is 500 and the standard deviation is 100. A random sample of students is obtained from each university, the mean for the Rolling Rock sample is found to be higher than the mean for the Bigbrain sample, and the 95% CI for the difference between two means is computed. What should the Rolling Rock students conclude if this CI is: (a) −7.63 to 13.44? (b) 2.27 to 19.85? (c) 97.22 to 122.44? (d) Compare examples (b) and (c). How does this comparison illustrate the advantages that CIs have over testing a specific null hypothesis, such as whether the difference between the means of Rolling Rock University and Bigbrain University is zero?
6. (a) What is the difference between the $t$ test for two independent sample means and the matched-pairs $t$ test? (b) What is the difference between the matched-pairs design and the repeated-measures design? What are their relative advantages and disadvantages?

## Computer Exercises

1. Use your statistical package to perform a two-sample $t$ test to determine whether there is a statistically significant difference in baseline anxiety between those who do and do not exercise regularly in Jackson's class. Find the 95% CI for this difference. Report your results as they might appear in a journal article.
2. Repeat Exercise 1 for baseline HR.
3. Perform a two-sample $t$ test to determine whether the students in the "impossible-to-solve" condition exhibited significantly higher postquiz HRs than the students in the "easy-to-solve" condition at the .05 level. Is this $t$ test significant at the .01 level? Find the 99% CI for the difference of the two population means and explain its connection to your decision regarding the null hypothesis at the .01 level. (Reminder: the experimental condition only relates to one question with four different difficulty levels that was added to the experimental quiz.)
4. Repeat Exercise 3 for the postquiz anxiety scores.
5. Perform a matched-pairs $t$ test to determine whether there is a significant increase in HR from baseline to the prequiz measurement. Also, test the difference between the pre- and postquiz measurements. (Advanced exercise: Repeat these paired $t$ tests separately for regular and nonregular exercisers.)
6. Repeat Exercise 5 for the anxiety measurements.

## Bridge to SPSS

Before you perform an independent-samples $t$ test in SPSS, you should have one column for your grouping variable (usually, "1"s and "2"s), and one for your DV. The matched-pairs $t$ test requires instead that you have two columns for your DV; each column represents your DV measured under a different condition. There is no separate column to indicate group or condition.

### Independent-Samples *t* Test

To perform a two-sample *t* test in SPSS, select **Compare Means and Proportions** from the **Analyze** menu, and then choose **Independent-Samples T Test** ....In the dialog box that opens, move the variable that distinguishes your two groups (e.g., gender) to the area labeled *Grouping Variable:* Doing that activates the **Define Groups** ... button; click this button and then enter the numeric values of your two groups (the numbers 1 and 2 appears as the default). If your grouping variable has more than two values (e.g., undergrad major), you can compare any two of them by specifying their values as Group 1 and Group 2. (Note: The grouping variable can be a string variable, rather than a numeric one, but then you have to enter the exact letter strings representing the two groups, and we do not recommend this. We recommend that you recode into numeric values any string variable that you want to use as a grouping variable.)

Move the DV you want to test to the *Test Variable(s)* area. The grouping variable is often thought of as the IV, but in many cases, it is not truly an IV, because it is just selected for, and not created by, the experimenter. That is the case in our next example, in which the effect of regular exercise on baseline anxiety is tested. Note that you can test many DVs at the same time. The **Options** ... button allows you to exclude cases listwise (as explained in Chapter 6) and also to select the percentage that will be used to create a CI for the difference between the two population means.

The default output for the two-sample *t* test consists of three tables (see Figure 7.4), the first of which presents descriptive statistics for the two groups. The second table contains both the pooled-variances *t* test (labeled "Equal variances assumed") and the separate-variances *t* test (labeled "Equal variances not assumed"). To the left of the *t* values is a test for HOV ("Levene's Test for Equality of Variances"); if the "Sig." (the term SPSS uses for *p* value) for this test is less than .05, reliance on the separate-variances *t* test is recommended, unless the two samples are the same size. To the right of the "*df*" (which usually involves a fractional value when dealing with the separate-variances test) is the two-tailed *p* value for each version of the *t* test, labeled "Sig. (2-tailed)." The Mean Difference and the

Figure 7.4

**Results of a Two-Sample *t* test Comparing the Effects of Two Levels of Exercise on Baseline Anxiety**

**Group Statistics**

| | exercise | N | Mean | Std. Deviation | Std. Error Mean |
|---|---|---|---|---|---|
| anx_base | 0 | 57 | 19.23 | 4.740 | .628 |
| | 1 | 43 | 17.37 | 3.498 | .534 |

**Independent Samples Test**

| | Levene's Test for Equality of Variances | | t-test for Equality of Means | | | | | | |
|---|---|---|---|---|---|---|---|---|---|
| | | | | | | | | 95% Confidence Interval of the Difference | |
| | F | Sig. | t | df | Sig. (2-tailed) | Mean Difference | Std. Error Difference | Lower | Upper |
| anx_base Equal variances assumed | .787 | .377 | 2.161 | 98 | .033 | 1.8560 | .8589 | .1515 | 3.5605 |
| Equal variances not assumed | | | 2.253 | 97.988 | .027 | 1.8560 | .8239 | .2210 | 3.4909 |

**Independent Samples Effect Sizes**

| | | Standardizer[a] | Point Estimate | 95% Confidence Interval Lower | Upper |
|---|---|---|---|---|---|
| anx_base | Cohen's d | 4.252 | .436 | .035 | .836 |
| | Hedges' correction | 4.285 | .433 | .035 | .830 |
| | Glass's delta | 3.498 | .531 | .116 | .939 |

a. The denominator used in estimating the effect sizes.
Cohen's d uses the pooled standard deviation.
Hedges' correction uses the pooled standard deviation, plus a correction factor.
Glass's delta uses the sample standard deviation of the control group.

Std. Error Difference are the numerator and denominator of the $t$ test, respectively (if you divide them, you will get the value in the $t$ column). The CI is for the difference between the two group means.

The third table contains three measures of the effect size between the two groups in the Point Estimate column. Hedge's correction, sometimes symbolized as $g$, adjusts Cohen's $d$ for bias, but as you can see, it is a minor adjustment, so it is rarely reported. Glass's delta is another alternative that is rarely used. Cohen's $d$ is the Mean Difference from the second table divided by the Standardizer in the Effect Sizes table (1.856/4.252 = 0.436). As an exercise, verify that you can get the same value (within rounding error) for Cohen's $d$ from Formula 7.11. This is close to being a medium-sized effect.

### Matched-Pairs *t* Test

To perform an MP $t$ test in SPSS, select **Compare Means and Proportions** from the **Analyze** menu, and then choose **Paired-Samples *t* Test**.... In the dialog box that opens, to the right of the list of variables, you will see an area labeled *Paired Variables:*, which contains spaces for "Variable 1" and "Variable 2" (see Figure 7.5). In this example, we will be testing whether HR changes significantly from baseline to the prequiz measurement.

1. From the list on the left, choose the first of the two variables that you would like to pair together, and click the arrow to move it to the right. This will also open a blank row for Pair 2.
2. Then choose the second variable that you would like in that pair, and click the center arrow again. (Note: These cells will automatically fill in order: Pair 1, right cell, then left cell, then, if you add a second pair, Pair 2, right cell, then left cell. So, be sure to watch that the pairs you are intending to put together do in fact end up together on the paired variables side.)
3. The possible options are the same as for the independent-samples test.

The default output for the paired-samples $t$ test consists of four tables (see Figure 7.6), the first of which has the same descriptive statistics as the first table for the Independent-Samples test. The second table (not included in Figure 7.6) presents the correlation between

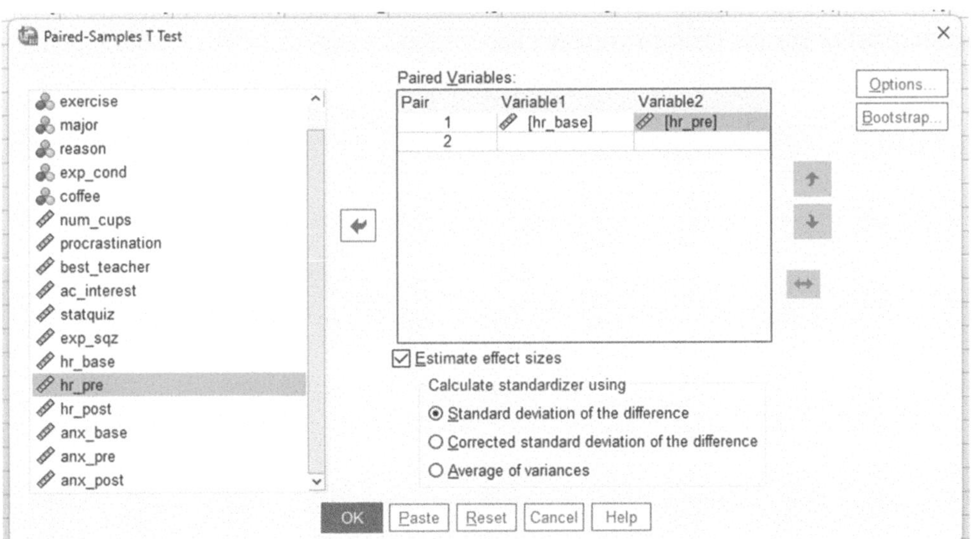


*Source:* Courtesy of International Business Machines Corporation, © SPSS, Inc. / IBM company

## Figure 7.5

**The Dialog Box for the Paired-Samples *t* Test Comparing Baseline HR to Pre-Quiz HR**

## Figure 7.6

**Results of a Matched-Pairs *t* Test Comparing Baseline HR to Pre-Quiz HR**

**Paired Samples Statistics**

| | | Mean | N | Std. Deviation | Std. Error Mean |
|---|---|---|---|---|---|
| Pair 1 | hr_base | 72.27 | 100 | 3.216 | .322 |
| | hr_pre | 73.85 | 100 | 5.131 | .513 |

**Paired Samples Test**

| | Paired Differences | | | | | | | Significance | |
|---|---|---|---|---|---|---|---|---|---|
| | | | | 95% Confidence Interval of the Difference | | | | | |
| | Mean | Std. Deviation | Std. Error Mean | Lower | Upper | t | df | One-Sided p | Two-Sided p |
| Pair 1 hr_base - hr_pre | −1.580 | 4.553 | .455 | −2.483 | −.677 | −3.470 | 99 | <.001 | <.001 |

**Paired Samples Effect Sizes**

| | | Standardizer[a] | Point Estimate | 95% Confidence Interval | |
|---|---|---|---|---|---|
| | | | | Lower | Upper |
| Pair 1 hr_base - hr_pre | Cohen's d | 4.553 | −.347 | −.548 | −.144 |
| | Hedges' correction | 4.588 | −.344 | −.544 | −.143 |

a. The denominator used in estimating the effect sizes.
Cohen's d uses the sample standard deviation of the mean difference.
Hedges' correction uses the sample standard deviation of the mean difference, plus a correction factor.

the two sets of scores and its corresponding *p* value (labeled "Sig.," as usual); we will have more to say about this statistical measure in Chapter 9. The third table contains the *t* value for the MP test, its *df*, and its two-tailed *p* value. From the left, the first three entries are labeled Mean (the mean of the difference scores, which is the numerator of the *t* formula), Std. Deviation (the unbiased SD of the difference scores), and Std. Error Mean (the standard error of the mean for the difference scores, which is the denominator of the *t* formula). These values are followed by the CI for the difference of the two means. The fourth table contains Cohen's *d*, which we are symbolizing as $d_z$. It is found by dividing the mean difference by the Standardizer in the Effect Sizes table $(-1.58/4.553 = 0.347)$. You can get the same result from Formula 7.15 $[3.47 * \text{sqrt}\ (1/100) = 3.47 * 0.1 = 0.347]$.

Note that the *t* and Cohen's *d* are both negative in the results, because we entered hr_base as the first member of the pair and hr_pre as the second, which tells SPSS to subtract hr_pre from hr_base. However, if all of your *t* values turn out to be negative, it is perfectly acceptable to subtract the variables in the opposite direction. Reporting the means will make the direction of your results clear to your reader.

# Chapter 8

# Nonparametric Tests for the Difference Between Two Means

## PREVIEW

### Introduction

What are the main differences between parametric and nonparametric tests?

What are the advantages and disadvantages of using ordinal tests?

What is meant by the power efficiency of a statistical test?

What are some important considerations about significance tests based on ranks?

### The Difference Between the Locations of Two Independent Samples: The Rank-Sum Test

What are the procedures for testing the hypothesis that two independent samples of ranked data come from populations with the same location?

When should the rank-sum test be used in place of the *t* test for two independent sample means?

How should you interpret the results of a rank-sum test? How do we convert a significant result to a measure of strength of relationship?

### The Difference Between the Locations of Two Matched Samples: The Wilcoxon Test

What are the procedures for testing the hypothesis that two matched samples of ranked data come from populations with equal values for location?

What is the power efficiency of the Wilcoxon test?

When should the Wilcoxon test be used in place of the matched *t* test?

How do we convert a significant result to a measure of strength of relationship?

### Summary

### Exercises

### Thought Questions

### Computer Exercises

### Bridge to SPSS

*Introductory Statistics for the Behavioral Sciences*, Eighth Edition. R. Brooke Lea and Barry H. Cohen.

Companion website: www.wiley.com/go/ISBS8e

## Introduction

In this chapter, we introduce the topic of nonparametric statistical tests by way of explaining **ordinal tests**. These are statistical tests that are applied to observations that have been ranked relative to one another—that is, the observations have been assigned ranks. In the last two chapters of this text, we will deal with nonparametric tests of categorical (purely qualitative) data. But first, we will make a few points about nonparametric statistics in general.

### Parametric Versus Nonparametric Tests

A *parametric* statistical test is so labeled because it involves the estimation of the value of at least one population parameter. For example, the pooled within-group sample variance calculated in most $t$ tests is an estimate of the corresponding within-population variance. A *nonparametric* test, on the other hand, does not require such estimation. Also, the mathematical derivations of common parametric tests assume that each population being randomly sampled has a *normal* distribution. A *distribution-free* statistical test, by contrast, requires no assumption about the shape of the distribution in the population. Although the terms *nonparametric* and *distribution-free* have different meanings, most statistical tests that meet one of these conditions also satisfy the other. So, the terms tend to be used interchangeably in practice. For convenience, we will use the term *nonparametric* to refer to any statistical test in which none of the variables has been measured on an interval or ratio scale. Sometimes a nonparametric test is used when one or more variables were *originally* measured precisely, but those measurements have been transformed to a less precise ordinal or ranking scale, as we describe in the next section.

### Ordinal Tests

There are two main types of situations that give rise to ordinal data and that may therefore call for the tests described in this chapter. First, suppose we can determine if one person has a greater amount of a continuous variable than another person, but we cannot measure this variable with any precision. For example, it may be reasonable to conclude that some junior executives at a given company have strong, moderate, or weak leadership qualities, without being able to quantify leadership much more precisely. To use the ordinal statistical procedures discussed in this chapter, we must at least be able to place all of the junior executives in order (i.e., to *rank* them) according to each one's amount of leadership ability. It is acceptable for some of the executives to be tied at the same rank, but a high percentage of ties would make these procedures inappropriate.

Second, the dependent variable may have already been measured on a precise, quantitative scale, but the distribution of the scores strongly violates the usual assumptions of parametric tests. The advantage of ordinal tests in these situations is that they do not require the population being sampled to be normally distributed. When we discussed the assumption of normality in connection with $t$ tests, we stated that these tests were fairly robust, in that moderate departures from normality in the population did not seriously affect their validity. But it is not at all rare in the behavioral sciences to encounter data that greatly depart from normality. Consider, for example, a typical distribution of reaction times involving a cognitive task: The bulk of the reaction times will be quite short (often less than a second), but a fair number will be relatively long, and none will be less than zero. Thus, the distribution is quite skewed and therefore substantially *non*-normal. If our samples are large, the central limit theorem tells us that our sampling distribution will be fairly normal even if the population distributions are not. However, it is not unusual in psychological research to be stuck with rather small samples. So, when we have reason to believe that the population distributions we are sampling from depart greatly from normality, and our sample sizes are not very large, ordinal methods (which require no assumption about population distribution shape) can be desirable. (These methods do generally assume, however, that the population distributions being compared have the *same* shape and variability.)

Of course, it is easy to rank order data that have already been measured precisely, but a good deal of measurement information is lost when scores are transformed into ranks. Whereas the original scores may be at varying distances from each other, their ranks are always one unit apart. For example, suppose that the three lowest resting heart rates within a sample of teenagers are 42, 57, and 59. If these HRs are converted into ranks of 1, 2, and 3, it is no longer possible to tell that the lowest HR was well below the other two, or that the HRs ranked second and third were quite close in value. Thus, a considerable loss of the original interval information generally occurs. The ordinal tests discussed in this chapter compensate for this loss of information by being distribution-free, while suffering only a modest loss of power.

### The Power Efficiency of Statistical Tests

It may have occurred to you to ask, "Why *ever* use parametric tests, which make normality assumptions about unobservable population distributions when ordinal tests with nearly as much power are available? After all, one can never *know* that the shape of a population distribution is normal." Good question! The answer is that it can be quite costly to run participants, and when populations are normally distributed, the sample size required by the parametric test ($N_p$) is smaller than that required by an alternative distribution-free test ($N_d$) in order to obtain the same amount of power. The ratio of these sample sizes expressed as a percent, or $(N_p/N_d) \times 100\%$, is called the *power efficiency* of the distribution-free test. Suppose that for a given difference between population means, a given significance criterion, $\alpha$, and a specified power, a certain ordinal test requires a total sample of $N_d = 80$ cases. Its parametric alternative, however, requires only $N_p = 72$ cases to have the same power. The power efficiency of the ordinal test would be $72/80 = 90\%$, so the parametric test requires only 90% as many cases to have the same probability of rejecting the null hypothesis.

When populations are normally distributed, the power efficiencies of ordinal tests are almost always less than 100%. This means that they require more cases in order to have the same power. Conversely, they have less power for the same number of cases. For this reason, you should *not* rush to use ordinal tests unless substantial non-normality is believed to exist in the population. This is especially true when you are using two-tailed tests and when the sample sizes are *not* very small (say, 20 or more). But if you do have good reason to believe that a population distribution departs substantially from normality, or your data are already in the form of ranks or ordered categories (e.g., very small, small, moderate), an appropriate ordinal test should be performed.

### The Basics of Dealing With Data in the Form of Ranks

As previously mentioned, there may be situations for which you will want to convert your interval/ratio data into ranks, so you can perform an ordinal test instead of a *t* test. Although this is not difficult to do, it is important to keep some basic principles in mind. In converting a set of *N* scores to ranks, it is customary to assign rank 1 to the smallest observation and rank *N* to the largest. If ties occur, the mean of the ranks in question is assigned to each of the tied scores, in a manner soon to be described.

What makes tests based on ranks work so simply is the fact that the data for any problem involving a total of *N* cases (where there are no ties) will be the positive integers 1, 2, 3, …, *N*. Thus, statistical functions of these data are simple functions of *N*. For example, it can easily be proved that the *sum* of the first *N* integers (ranks 1 to *N*) is as follows:

$$\begin{aligned}\sum R &= 1 + 2 + 3 + \cdots + N \\ &= \frac{N(N+1)}{2}\end{aligned}$$

Suppose that there are a total of 22 cases. Whatever their original raw scores may be, the sum of their *ranks* must be (22)(23)/2 = 253. And because the mean is equal to the sum divided by $N$, it therefore follows that the mean of a complete set of ranks (symbolized by $\overline{R}$) is equal to:

$$\overline{R} = \frac{\sum R}{N} = \frac{N(N + 1)}{2N} = \frac{N + 1}{2}$$

Thus, the mean rank for 22 observations is simply 23/2 = 11.5. The mean rank can be thought of as the expected rank of a case drawn at random from the $N$ ranks, in the sense that the mean is the best guess as to the value of this case. Therefore, if we draw two cases at random when $N = 22$, the expected sum of the ranks (symbolized by $T_E$; T for total and $E$ for expected) is 2(11.5) = 23. For three cases, it is 3(11.5) or 34.5, and so on. Generally, if we draw a subset of $N_1$ cases at random from the total $N$ (symbolized as $N_T$), their expected sum of ranks is $N_1$ times the mean rank:

$$T_E = N_1\overline{R} = \frac{N_1(N_T + 1)}{2} \qquad \text{Formula 8.1}$$

It can also be shown that if $N_1$ cases are randomly and repeatedly drawn from the complete set of $N_T$ ranks, and the sum of the ranks ($T$) is obtained for each sample, the standard deviation of the resulting sampling distribution of $T$ values (called a standard error, because in this case, we are not dealing with individual observations) would be:

$$\sigma_T = \frac{N_1 N_2(N_T + 1)}{12} \qquad \text{Formula 8.2}$$

where

$\sigma_T$ = standard error of $T$
$N_2 = N_T - N_1$

The value of 12 in the denominator is a constant and is not affected by $N_T$ or any of the $N_T$ values. We made the effort to derive Formulas 8.1 and 8.2 because we will need them to create a statistical test for ranked data. Regardless of the shape of the population distribution of raw scores, the sampling distribution of a sum of a subset of ranks will be approximately normal in form. This is *not* an assumption but a provable mathematical property.

### An Example of Ranking Data With Ties

As mentioned previously, there are two situations that can produce ordinal data, and both can lead to tied ranks, which must be dealt with. First, there is direct ranking, in which you combine your two subgroups into one group for the purpose of ranking the individuals relative to each other (but you keep track of the group each one came from). For example, participants A, B, C, D, and E have completed a leadership training course, while participants W, X, Y, and Z did not. Now you want to rank them from highest to lowest leadership potential based on presentations each of them gave. Suppose that you place them in the following order: B, D, W, C, A, Y, E, X, and Z, though you feel that, to be fair, D and W are really tied (i.e., your judges cannot distinguish between them with respect to leadership potential), as are E, X, and Z. (One way this could occur is if you really have just six ordered categories in which to place the participants: very high leadership; high; slightly above average; slightly below average; low; and very low; and B falls into the first category; D and W in the second; C, A, and Y in the third, fourth, and fifth categories, respectively; and E, X, and Z in the lowest category.) In order to preserve the sum of ranks, which is 45 for $N = 9$, you must first assign the ranks from 1 to 9, even when some individuals are tied, like this:

| B | D | W | C | A | Y | E | X | Z |
|---|---|---|---|---|---|---|---|---|
| 9 | 8 | 7 | 6 | 5 | 4 | 3 | 2 | 1 |

Then individuals who are tied are assigned the average of their ranks, so both D and W are assigned the rank of 7.5, and E, X, and Z are each assigned the rank of 2:

| B | D | W | C | A | Y | E | X | Z |
|---|---|---|---|---|---|---|---|---|
| 9 | 7.5 | 7.5 | 6 | 5 | 4 | 2 | 2 | 2 |

Note that these ranks have the same sum (45) as when there are no ties; assigning the ranks as we just did ensures that this will be the case. In this example, we made it particularly easy to keep track of the subgroups by using letters at the beginning or end of the alphabet. The sum of ranks for those who took leadership training is $T = 9 + 7.5 + 6 + 5 + 2 = 29.5$; and for those who did not, $T = 7.5 + 4 + 2 + 2 = 15.5$. (The sum of the ranks of both groups is $29.5 + 15.5 = 45$, as it should be for any nine ranks that are properly assigned.)

The other situation that produces ordinal data occurs when you have precise measurements (e.g., reaction time in milliseconds), but the distribution is so far from normality that you choose to assign ranks to the measurements. For very precise measurements, ties will rarely occur, but for a measurement such as the number of children in a family, ties will be common. In any case, ties are dealt with exactly as in the example just described: Ranks from 1 to $N$ are assigned (arbitrarily in the case of tied measurements), and then tied individuals are each given the average of their ranks.

## The Difference Between the Locations of Two Independent Samples: The Rank-Sum Test

Among the most important and useful ordinal tests are those that deal with differences in *location* between populations, based on ranks. We hope you recall from the earlier chapters of this text that the location of a distribution is generally expressed in terms of a measure of its *central tendency*. In parametric statistics, the location of a population or sample is customarily indexed by its arithmetic mean. When we conclude from a $t$ test that scores in Population 1 are larger than those in Population 2, we mean literally that $\mu_1 > \mu_2$. But with ordinal tests, the same statement has a different meaning, which is: If we draw at random a case from Population 1 and another from Population 2 and compare them, and do this repeatedly, more than half the time the case from Population 1 will be larger than the case from Population 2. This can be taken for all intents and purposes as an assertion that $\mathbf{median_1} > \mathbf{median_2}$, or written more compactly, $\mathbf{Mdn_1} > \mathbf{Mdn_2}$ (the boldface is there to remind you that these are medians of populations, not samples). In this chapter, we describe two tests of location that are parallel to the independent-samples and matched $t$ tests, respectively, but are applied to ordinal data and are considered to be more robust (less dependent on making assumptions) than their corresponding $t$ tests. In fact, both of these tests can be categorized as nonparametric and distribution-free, as we explain shortly, and when they can be used as alternatives to $t$ tests, they are nearly as powerful as their parametric counterparts.

The ordinal test most commonly used as a nonparametric alternative to the $t$ test of two independent sample means is called the **rank-sum** test, or sometimes Wilcoxon's rank-sum test (the **Mann–Whitney *U* test** is another version of this test that ultimately yields the same $p$ value). This test requires the type of ranking we just described. We introduce this test in terms of an example for which the use of the ordinary $t$ test would be questionable.

**Table 8.1 The Rank-Sum Test for Two Independent Samples**

| Least procrastinating (Group 1) | | Most procrastinating (Group 2) | |
|---|---|---|---|
| $X$ | $R$ | $X$ | $R$ |
| 55 | 3 | 52 | 1 |
| 62 | 7 | 53 | 2 |
| 69 | 11 | 56 | 4 |
| 78 | 15 | 58 | 5 |
| 79 | 16 | 60 | 6 |
| 84 | 18 | 63 | 8 |
| 92 | 19 | 64 | 9 |
| 96 | 20 | 65 | 10 |
| 98 | 21 | 71 | 12 |
| 99 | 22 | 74 | 13 |
| | | 76 | 14 |
| | | 81 | 17 |
| | $T_1 = 152$ | | $T_1 = 101$ |
| | $N_1 = 10$ | | $N_2 = 12$ |
| | $\overline{R}_1 = \frac{152}{10} = 15.20$ | | $\overline{R}_2 = \frac{101}{12} = 8.2$ |

$$N_T = N_1 + N_2 = 22$$
$$\sum R = N(N + 1)/2 = (22)(23)/2 = 253$$
$$= T_1 + T_2 = 152 + 101 = 253$$

Suppose that at the end of the semester, Jackson decides to use scores on the final exam to compare the performance of their students who procrastinate the most (those who rated their procrastination as more than 6) with that of their students who procrastinate the least (those who gave a rating of 0). The final exam scores for the 10 who procrastinate the least (let us say there were missing data for two of the students who rated their phobia as zero) and the 12 students who procrastinate most are shown in Table 8.1. As you can see, the normality assumption appears to be quite dubious. The samples are also small, so a distribution-free test is appropriate. In particular, we will use the rank-sum test, which allows us to decide whether we can reject the null hypothesis that two independent samples of ranked data come from populations with the *same* location (i.e., central tendency).

From the earlier sections of this chapter, we have all the ingredients we need for hypothesis testing: a sample statistic ($T$) with a known (in this case, approximately normal) sampling distribution, whose standard deviation is known ($\sigma_T$), and a null-hypothetical value to test against ($T_E$).

### Step 1: State the Null and Alternative Hypotheses, and the Significance Criterion, α

The null hypothesis can be stated in terms of the population medians, as follows:

$$H_0 : \mathbf{Mdn}_1 = \mathbf{Mdn}_2$$
$$H_A : \mathbf{Mdn}_1 \neq \mathbf{Mdn}_2$$

The significance (i.e., alpha) level used for nonparametric tests is no different from that which is typically used for parametric tests, so we will set $\alpha$ to .05.

### Step 2: Calculate the Appropriate Test Statistic

First, all 22 observations (ignoring procrastination group) are subjected to a single ranking from the lowest ($X = 52$), which is assigned a rank of 1, to the highest ($X = 99$), which is assigned a rank of 22. The original $X$ scores are ignored from this point on, and the analysis proceeds solely with the ranks. The critical step is to find the sum of the ranks separately for each group, so when ranking observations from both groups together, you do have to keep track of which group each observation came from.

As you can see in Table 8.1, the sum of the 10 ranks in Group 1 (low procrastination), symbolized by $T_1$, is 152. This sum should be compared to the expected sum of any 10 ranks selected at random from the 22 observations because that is the sum you would expect if the null hypothesis were true. This is why we needed to derive Formula 8.1:

$$T_E = \frac{N_1(N_T + 1)}{2} = \frac{(10)(23)}{2} = 115$$

Is the observed $T_1$ of 152 a statistically significant departure from the null hypothesis (expected) $T_E$ value of 115? Because statisticians have shown that $T_1$ is normally distributed around a mean of $T_E$, it is appropriate to use the $z$-score formula for one mean, Formula 5.2, even though our "mean" in this case is actually a sum of ranks. Formula 5.2 can therefore be transformed for our purposes to:

$$z = \frac{T_1 - T_E}{\sigma_T}$$

Like any $z$-score formula, this formula requires knowing not only the mean of the normal distribution of $T$ values but its standard deviation (called a standard error in this case because we are not dealing with individual observations) as well. Fortunately, that can be estimated quite well, if $N$ is not small, by Formula 8.2. Inserting Formula 8.2 into the preceding formula, we get a convenient $z$-score formula for the rank-sum test:

$$z = \frac{T_1 - T_E}{\sqrt{\dfrac{N_1 N_2(N_T + 1)}{12}}} \qquad \text{Formula 8.3}$$

Substituting the appropriate values from Table 8.1 for the symbols in Formula 8.3 yields the following value for $z$:

$$z = \frac{152 - 115}{\sqrt{\dfrac{(10)(12)(23)}{12}}} = \frac{37}{15.17} = 2.44$$

### Step 3: Find the Critical Value and Make a Statistical Decision With Respect to Your Null Hypothesis

Although we would naturally expect the low-procrastination group to obtain the larger (i.e., numerically higher) ranks on the final stats exam, it is more conservative, and more widely accepted, to perform a two-tailed test (e.g., suppose the high procrastinators sought extra help and tutoring and the low procrastinators turned out to be too complacent?). As you may recall, the critical $z$ values for $\alpha = .05$, two-tailed, are $\pm 1.96$. Because the 37 rank-sum units by which Group 1 exceeds $T_E$ place it 2.44 standard deviation units above the mean of a normal distribution of rank-sums for the given sample sizes, the observed $z$-score is $+2.44$, which is larger than $+1.96$. Therefore, you can reject the null hypothesis and conclude that the ranks in the low-procrastination group are significantly higher than chance expectation. The symmetry of the normal distribution implies that the ranks in the high-procrastination group are significantly *lower* than chance expectation and therefore that the ranks for the least procrastinating students are generally higher than those for the most procrastinating.

### Interpreting the Results of a Rank-Sum Test

Rejecting the null hypothesis in a rank-sum test allows us to conclude that the location of the two populations from which these samples were randomly drawn differs on $X$ (in this case, performance on the final stats exam), such that the median of the first (low-procrastination) population is greater than the median of the second (highly procrastinating) population (i.e., their final exam scores are higher).

Note that it makes no difference which of the two sets of ranks you use to compute $T$ (the sum of ranks to be compared in the test) and $T_E$. For Group 2, $T_2 = 101$, and $T_E$ would now be equal to (12)(23)/2, or 138. (Do not forget that $T_E$ depends on the size of the sample associated with the comparison $T$.) The value of $\sigma_T$ is unchanged, so $z$ would equal $(101 - 138)/15.17$, or $-2.44$. And you would reach exactly the same conclusion, with the minus sign indicating that the group being compared (Group 2) is the one with the lower value for location. Although you need to compute only $T_1$ (or only $T_2$) to carry out this statistical test, it is desirable to calculate both values and then verify that $T_1 + T_2 = N_T(N_T + 1)/2$, as a check against errors in ranking and/or calculating.

The rank-sum test for two samples that we just demonstrated depends on the fact that the sampling distribution of $T$ is well approximated by the normal curve. This will be true provided that the samples are not too small, say at least six to eight cases in each. With smaller samples than that, the normal distribution is not an accurate model for the distribution of the $T$ values. Fortunately, when dealing with small samples, it is not difficult to work out all of the possible arrangements of ranks and to calculate the probabilities that correspond to the various possible values of $T$. Even more fortunately, statisticians have created convenient tables of critical values for either: (a) the sum of ranks for one of two small groups (Wilcoxon bases his values on the smaller of the two groups if the two groups are not the same size), or (b) a function derived from the sum of ranks for one of the groups (the test that is credited to both Mann and Whitney is based on a function called the $U$ statistic, so that version of the test is called the "Mann–Whitney $U$ test").

We chose not to include a table of critical values for the rank-sum test in this text, because you are not likely to conduct such a test without statistical software and because, if you do, you should consult a more specialized book devoted to nonparametric statistics, such as the one by Siegel and Castellan (1988). Such books will not only contain the statistical tables you will need but will also cover important refinements to nonparametric procedures with small samples, such as a correction for continuity that is appropriate for two-group comparisons, as well as a procedure for reducing $\sigma_T$ to take the number of tied ranks into account. (You can also find a table for the rank-sum test and some further explanation in B. H. Cohen, 2008.)

As we pointed out earlier, the parametric analog of the rank-sum test is the well-known $t$ test for two independent sample means. Relative to the $t$ test, the rank-sum test has a power efficiency that tends to range from 92% for small samples to 95% for large samples. Applying the pooled-variance formula to the final exam scores ($X$) in Table 8.1 yields a $t$ value of 3.14, which, unlike the $z$ for the corresponding rank-sum test, is significant even at the .01 level—but the $t$ and $z$ values are not very far apart. (We recommend that you calculate the $t$ value for yourself as an exercise.)

For a given set of conditions, when populations are normally distributed, the same power is obtained using the $t$ test with 92% to 95% of the sample size that would be required by the rank-sum test. Compared to other nonparametric tests, those based on ranks have fairly high power efficiencies, so they should be used whenever the normality assumption is in serious doubt. When this is not the case, however, the behavioral scientist usually cannot afford the luxury of what amounts to throwing away 5% to 8% of the cases.

### Measure of Strength of Relationship: The Glass Rank Biserial Correlation

With respect to the two-group *t* test, we have stressed the general insufficiency of computing only a test of statistical significance, which merely helps us to decide whether or not there is some (nonzero) effect in the population and the desirability of calculating an additional measure to index the *strength* of an observed relationship. In the case of the parametric *t* test for two independent sample means, we introduced the statistic known as *d*, as just such a measure in the previous chapter. In Chapter 10, we will describe an alternative measure of effect size for two sample means, which is based on the Pearson correlation coefficient, introduced in the next chapter. The effect-size measure we describe here for the rank-sum test is more closely related to the correlation type of measure but is simple enough to be understood on its own.

A measure of the strength of the relationship between group membership and the *rank* values for two groups is provided by the **Glass rank biserial correlation coefficient**, symbolized by $r_G$. It is computed using the following formula:

$$r_G = \frac{2(\overline{R}_1 - \overline{R}_2)}{N_T} \qquad \text{Formula 8.4}$$

where

$\overline{R}_1$ = mean of ranks in Group 1
$\overline{R}_2$ = mean of ranks in Group 2
$N_T$ = total number of observations

Although the $r_G$ coefficient is *not* a Pearson *r*, it does conveniently have the same limits: It may take on any value from −1 to 1 (with its sign depending on which group is called Group 1), and its magnitude is interpreted much like that of any correlation coefficient. The maximum value of 1 or −1 occurs when there is no overlap in ranks between the two groups, with one group containing all the highest ranks while the other group contains all the lowest ranks. A value of zero suggests a random mixing of the two groups with respect to ranks, with neither group appearing to be superior to the other. It is important to note that a significant rank-sum test can be associated with a tiny value of $r_G$ (e.g., .1) when the sample sizes are very large. In such a case, $r_G$ serves as a useful reminder that statistical significance does not guarantee that there is much separation even in our two samples—only that we are permitted to rule out a *zero* amount of separation in the population.

For the data in Table 8.1, $\overline{R}_1 = 15.20$, $\overline{R}_2 = 8.42$, and $N_T = 22$. Thus,

$$r_G = \frac{2(15.20 - 8.42)}{22} = 0.62$$

This indicates a high degree of relationship between group membership and ranking on *X*, so it should not be surprising that we obtained significant results even though our samples were rather small.

## The Difference Between the Locations of Two Matched Samples: The Wilcoxon Test

The other ordinal test of location we describe in this chapter is the nonparametric alternative to the matched-pairs *t* test. Like the matched *t* test, the ordinal test we are about to describe can be used whether the pairs of scores come from matched participants or the same participant measured twice.

### Rationale and Computational Procedures

As we saw in the previous chapter, when a comparison is to be made between the means of two matched or dependent samples, the procedure that should be used is different from the case in which the samples are independent. For each of the $N$ matched pairs, the difference ($D$) between the $X_1$ and $X_2$ values is found, and the null hypothesis tested is that the population mean of the $D$ values is zero. Rejection of this null hypothesis necessarily implies that $\mu_1$ and $\mu_2$ are not equal.

The same distinction applies in determining the appropriate rank-based test of the difference in location between two samples. The rank-sum test discussed earlier is used with independent samples, while a different technique, the signed-ranks test (often called the **Wilcoxon matched-pairs signed ranks test**) is used with dependent samples. The Wilcoxon test also begins with $D = X_2 - X_1$ values but then operates on only the ranks of the absolute values of these difference scores.

As an example, let us consider an investigation into the question of whether sensitivity training improves extrasensory perception (ESP) scores. A sample of 12 participants is tested for ESP before ($X_B$) and again after ($X_A$) a series of 10 training sessions, and the results are shown in Table 8.2. In this experiment, therefore, the samples are dependent because the two observations in each pair come from the same participant. (It should be noted, however, that this matched design is appropriate whenever there is some connection between the observations making up a pair.)

The first two columns in Table 8.2 give the post-training ($X_A$) and pre-training ($X_B$) ESP scores for the 12 participants. The third column shows each participant's change or difference score, where $D = X_A - X_B$. (You could subtract either score from the other, but it makes sense to subtract the scores in the order that gives you fewer negative scores.) Given the order of subtraction we are using, the $D$ values are positive when the ESP score has increased following sensitivity training and negative when it has decreased. As was the case in the parametric matched $t$ test, all further analyses focus on these $D$ scores.

First, all cases where $D = 0$ are dropped. Case 9 in the table is therefore deleted, and the analysis proceeds on the basis of $N = 11$. (For this test, we will let $N$ stand for the number of *pairs* and not use the symbol $N_T$.) Then, *ignoring the signs* of the $D$ values, they are rank-ordered from the smallest (rank = 1) to the largest (rank = $N$), with ties being resolved as described for the rank-sum test. For example, the differences of +4 and −4 are tied with respect to absolute value and initially occupy the ranks of 5 and 6 (the first four ranks go to +1, +2, and the two −3s); the tie is resolved by assigning the rank of 5.5 to both the +4 and the −4 difference. These ranks appear in the column labeled $R_{|D|}$, and they follow

**Table 8.2** **The Wilcoxon Test for Two Matched Samples**

| Participant | $X_A$ | $X_B$ | $D = X_A - X_B$ | $R_D$ | $R(+)$ | $R(-)$ |
|---|---|---|---|---|---|---|
| 1 | 33 | 38 | −5 | 7 | | 7 |
| 2 | 45 | 43 | +2 | 2 | 2 | |
| 3 | 50 | 42 | +8 | 10 | 10 | |
| 4 | 45 | 44 | +1 | 1 | 1 | |
| 5 | 46 | 49 | −3 | 3.5 | | 3.5 |
| 6 | 45 | 41 | +4 | 5.5 | 5.5 | |
| 7 | 28 | 22 | +6 | 8 | 8 | |
| 8 | 43 | 46 | −3 | 3.5 | | 3.5 |
| 9 | 32 | 32 | 0 | | | |
| 10 | 40 | 31 | +9 | 11 | 11 | |
| 11 | 34 | 27 | +7 | 9 | 9 | |
| 12 | 40 | 44 | −4 | 5.5 | | 5.5 |
| | | | | $\sum R = 66$ $(= T_1 + T_2)$ | $T_1 = 46.5$ | $T_2 = 19.5$ |

$N = 11$ (deleting participant 9 for whom $D = 0$)

$\sum R = N(N + 1)/2 = (11)(12)/2 = 66$

the statistical properties associated with any set of ranks from 1 to $N$; for example, their sum must equal $N(N + 1)/2$, which for $N = 11$ is equal to $(11)(12)/2$, or 66.

If the population locations of $X_B$ and $X_A$ are the same, then (ignoring pairs where $D = 0$) the sum of the ranks of the absolute $D$ values should be approximately equally divided between those that are positive ($R+$) and those that are negative ($R-$). Since the sum of *all* the ranks equals $N(N + 1)/2$, the expectation under the null hypothesis is that the sum of the ranks for the positive $D$s and the sum of the ranks for the negative $D$s will *each* equal half the sum. That is, the expected result under the null hypothesis is:

$$T_E = \frac{1}{2}\frac{N(N + 1)}{2}$$
$$= \frac{N(N + 1)}{4}$$

For the data in Table 8.2,

$$T_E = \frac{(11)(12)}{4}$$
$$= 33$$

which is half of 66, the sum of all the ranks.

In the columns headed $R(+)$ and $R(-)$ in Table 8.2, the $R_{|D|}$ values for positive and negative $D$s have been segregated. The sum of the ranks of the positive $D$ values, $T_1$, is equal to 46.5. The sum of the ranks of the negative $D$ values, $T_2$, is equal to 19.5. Except as a check, only one of these values is needed, so let us choose $T_1 = 46.5$. This value departs 13.5 rank-sum units from the value expected under the null hypothesis, $T_E = 33$. You are then left with the question of whether or not this departure is sufficiently unlikely to reject the null hypothesis, using (say) the $\alpha = .05$ decision rule.

A method for answering this question can be easily derived. In repeated random sampling, $T_1$ (and also $T_2$) is approximately normally distributed when the null hypothesis is true. The standard deviation of the sampling distribution of this matched-pairs signed ranks $T$ is called a standard *error* because it involves samples instead of individual scores, and it is symbolized by $\sigma_{T_M}$. The formula for this standard error is as follows:

$$\sigma_{T_M} = \sqrt{\frac{(2N + 1)T_E}{6}} \qquad \text{Formula 8.5}$$

Thus, we have a sampling distribution that is approximately normal, whose mean is $T_E$ and whose standard deviation is $\sigma_{T_M}$, and we have a null hypothesis to test: that the positive and negative rank sums are equal. Once again, we have all the ingredients for a significance test using the normal curve model. Because **Step 1** of our null hypothesis procedure is exactly the same as for the rank-sum test, we will skip to **Step 2**.

### Step 2: Calculate the Appropriate Test Statistic

The one-group $z$ test, with the symbols appropriately modified for the Wilcoxon signed-ranks test, looks like this:

$$z = \frac{T_1 - T_E}{\sigma_{T_M}}$$

Inserting Formula 8.5 into the preceding formula yields the following useful $z$-score formula for the signed-ranks test:

$$z = \frac{T_1 - T_E}{\sqrt{\frac{(2N + 1)T_E}{6}}} \qquad \text{Formula 8.6}$$

For the data in Table 8.2, we found that the null hypothesis leads to an expected value of $T_1$ (and of $T_2$) as follows: $T_E = 33$. The observed $T_1 = 46.5$. Therefore,

$$z = \frac{46.5 - 33}{\sqrt{\frac{[(2)(11) + 1](33)}{6}}}$$

$$= \frac{13.5}{\sqrt{\frac{(23)(33)}{6}}}$$

$$= 1.20$$

### Step 3: Find the Critical Value and Make a Statistical Decision With Respect to Your Null Hypothesis

Although the investigator would undoubtedly expect the sensitivity training to lead to *higher* ESP scores after training than before, we will again follow the more conservative and widely accepted custom of performing a two-tailed test. Therefore, the critical $z$ values for this problem are $\pm 1.96$. As you can see from the $z$-score just calculated, the observed departure of 13.5 rank-sum units from the null-hypothetical expected value of 33 is only a little more than one standard error and therefore far less than the value of $+1.96$ needed to reject $H_0$, using the two-tailed .05 criterion of significance. The null hypothesis is retained, though certainly not "accepted."

### Interpreting the Results of a Wilcoxon Matched-Pairs Signed-Ranks Test

On one hand, the data do *not* justify the conclusion that 10 sessions of sensitivity training will improve ESP scores in the population from which this sample was randomly drawn. On the other hand, the results do not permit the conclusion that *no* effect exists, of course; in fact, given our small sample, it is likely that the probability of a Type II error is rather high (this issue is discussed in considerable detail in Chapter 11). Only if the population effect were quite large would there be a reasonably good chance of rejecting the null hypothesis with such a small sample.

Relative to the matched $t$ test, the Wilcoxon test has power efficiency ranging from 92% for small samples to 95% for large samples. This means that if the parametric assumptions of the $t$ test for matched samples were valid, our choice of the Wilcoxon test would result in somewhat lower power than the $t$ test would have; our $N = 11$ for the Wilcoxon test is about as powerful as a $t$ test using $N = (.92)(11) = 10$ pairs. A $t$ test applied to the original (not ranked) 12 D values is also not significant; it equals 1.28, very close to the Wilcoxon $z$ of 1.20. (Again, we recommend calculating the $t$ value yourself, as an exercise.) Note, however, that the Wilcoxon test as presented here should not be used if the sample size is smaller than about 8 because the approximation to normality will then not be sufficiently accurate. As mentioned earlier, statistical tables are available for nonparametric analyses involving small samples (B. H. Cohen, 2008), and it is always a good idea to consult a more specialized text, such as Siegel and Castellan (1988), if you have a large number of zero differences or tied ranks.

### Measure of Strength of Relationship: The Matched-Pairs Rank Biserial Correlation

The matched-pairs rank biserial correlation, symbolized by $r_c$, expresses the strength of the relationship between the condition (such as posttest versus pretest) and the dependent variable. Another way to describe $r_c$ is that it indexes the degree of relationship between the sign of $D$ and its rank. It is computed as follows:

$$r_c = \frac{4(T_1 - T_E)}{N(N + 1)} \qquad \text{Formula 8.7}$$

The $r_c$ index can take on values from $-1$ to $+1$. It is equal to zero when $T_1$ does not differ from the null hypothetical value $T_E$, and it equals $\pm 1$ when all $D$ values have the same sign. The sign of $r_c$ indicates whether the positive or negative sum of ranks is larger. It is not a Pearson correlation coefficient, as described in the next chapter, but rather it is quite similar to $r_G$, for the rank-sum test.

When the Wilcoxon test is not statistically significant, there is usually little interest in computing $r_c$. However, even though the data in Table 8.2 did not yield a significant test, we will demonstrate the calculation of $r_c$ on those data for pedagogical purposes.

$$r_c = \frac{4(46.5 - 33)}{(11)(22)} = .41$$

This value cannot be statistically significant since the Wilcoxon test for the same data did not yield significance. It may be used to describe the relationship within this particular sample, but there is insufficient reason to conclude that the corresponding population correlation index is different from zero. However, it should be noted that an $r_c$ of .41 indicates a fairly consistent change from before to after, and would be associated with a statistically significant result were the sample size not so small.

### The Sign Test

A considerably cruder alternative to both the Wilcoxon and matched-pairs t tests is the sign test, presented in Chapter 16 of this text, following a description of the binomial distribution, on which it is based. Because the sign test considers only whether a difference is positive or negative, and therefore ignores virtually all of the quantitative information in the data, it has a power efficiency that is usually below 70% with respect to the matched-pairs t test. The sign test should only be used when the difference of each pair cannot be quantified precisely enough for the differences to be ranked but the directions for most of the differences can be determined. Nonetheless, when the effect size is quite large, the sign test may be all you need to demonstrate your point.

## Summary

When there is reason to believe that the shape of a population distribution is substantially non-normal and the samples are not very large, nonparametric statistical tests, such as the two ordinal tests described in this chapter, may be desirable.

### Basic Considerations

The main advantage of nonparametric and distribution-free statistical tests is that they do *not* require the population(s) being sampled to be normally distributed, so they are applicable when gross non-normality is displayed in your data and therefore suspected in the population. The primary disadvantage of these methods is that when normality does exist, they are less powerful than the corresponding parametric tests (more likely to lead to a Type II error). A numerical measure of the *power efficiency* of a nonparametric test is given by:

$$\frac{N_p}{N_d} \times 100\%$$

where

$N_p$ = sample size required by a parametric test to obtain a specified level of power for a specified criterion of significance and a specified difference between population means

$N_d$ = sample size required by the corresponding nonparametric test to obtain the same power under the same conditions

The power efficiency of a nonparametric test is almost always less than 100%, but the power of the ordinal tests described in this chapter is quite good. Nonetheless, it is wasteful to use these methods when parametric tests are applicable.

**The Difference Between the Locations of Two Independent Samples: The Rank-Sum Test**

1. Parametric analog: *t* test for the difference between two independent means (Chapter 7).

   Power efficiency: Approximately 92–95%

2. Computational procedures: Rank *all* scores from 1 (smallest) to *N* (largest), regardless of which group they are in. In case of ties, follow the usual procedure of assigning the mean of the ranks in question to each of the tied scores. Then compute a *z*-score using the following formula:

$$z = \frac{T_1 - T_E}{\sqrt{\dfrac{N_1 N_2 (N_T + 1)}{12}}} \qquad \text{Formula 8.3}$$

where

$T_1$ = sum of ranks in group 1
$T_E = N_1(N_T + 1)/2$
$N_1$ = number of observations in Group 1
$N_2$ = number of observations in Group 2
$N_T$ = total number of observations

Do *not* use the preceding formula if there are fewer than six cases in either group. Use an exact table of critical values, such as Table A.15 in B. H. Cohen (2008).

3. Measure of the strength of relationship: If the value of *z* computed in the rank-sum test is statistically significant, or otherwise interesting, a measure of the strength of the relationship between group membership and the rank values may be found by computing the *Glass rank biserial correlation* ($r_G$):

$$r_G = \frac{2(\overline{R}_1 - \overline{R}_2)}{N_T} \qquad \text{Formula 8.4}$$

where

$R_1$ = mean of ranks in Group 1
$R_2$ = mean of ranks in Group 2
$N_T$ = total number of observations

The $r_G$ coefficient falls between the limits of $-1$ and $+1$.

**The Difference Between the Locations of Two Matched Samples: The Wilcoxon Test**

1. Parametric analog: matched *t* test (Chapter 7). Power efficiency: Approximately 92–95%
2. Computational procedures: First, obtain the *D* score for each participant by subtracting $X_2$ from $X_1$. Next, discard any case where $D = 0$ (and reduce *N* accordingly). Then, *ignoring the signs* of the Ds, rank them from the smallest (rank = 1) to the largest (rank = *N*). Finally, compute a *z*-score using the following formula:

$$z = \frac{T_1 - T_E}{\sqrt{\dfrac{(2N + 1)T_E}{6}}} \qquad \text{Formula 8.6}$$

where

$T_1$ = sum of ranks for those $D$ values that are positive (note that $T_2$, the sum of the ranks for those $D$ values that are negative, may be used instead)
$T_E = N(N + 1)/4$
$N$ = number of pairs (excluding cases where $D = 0$)

Do *not* use this procedure if $N$ is less than about 8. Instead, use an exact table of critical values, such as Table A.16 in B. H. Cohen (2008).

3. Measure of the strength of relationship: If the value of $z$ computed in the Wilcoxon test is statistically significant or otherwise interesting, a measure of the strength of the relationship between the independent variable (e.g., condition) and the dependent variable may be found by computing the *matched-pairs rank biserial correlation* ($r_c$):

$$r_c = \frac{4(T_1 - T_E)}{N(N + 1)} \qquad \text{Formula 8.7}$$

where $T_1$, $T_E$, and $N$ have the same meanings as in the Wilcoxon test. The $r_c$ coefficient is very similar to the $r_G$ coefficient, and it also falls between the limits of $-1$ and $+1$.

## Exercises

1. An operator of a certain machine must turn it off quickly if a danger signal occurs. To test the relative effectiveness of two types of signals, a small group of operators is randomly divided into two groups. Those in Group 1 operate machines with a newly designed signal, while those in Group 2 use machines with the standard signal. The signal is flashed unexpectedly, and the reaction time of each operator (time taken to turn off the machine after the signal occurs) is measured in seconds. The actual results (not ranks) are shown here. (a) Use the rank-sum test to determine whether there is a significant difference between the two groups. (b) Compute the appropriate measure of strength of relationship.

| Group 1 | Group 2 |
|---|---|
| 5 | 3 |
| 3 | 17 |
| 4 | 13 |
| 10 | 2 |
| 1 | 8 |
| 3 | 16 |
| 5 | 6 |
| 2 | 9 |
| 7 | 11 |
| | 8 |

2. Redo Exercise 5 from Chapter 7 using the rank-sum test instead of a $t$ test. Does your statistical conclusion differ from the one you made in the previous chapter? Why would the rank-sum test not be very accurate for the data in that exercise?
3. In Experiment 1 (see the following table), Jackson is comparing the attendance at the psychology department's Friday afternoon social events of students in their class

between the days when cookies were offered (Group 1) and the days when mango cake was offered (Group 2). Experiment 2 also compares attendance of the afternoon social events, but in this case, Group 1 consists of days that Jackson had frequently announced the social event during class time throughout the week and Group 2 contains days where Jackson made no mention of the social event throughout the week. (a) Perform the rank-sum test for each experiment. Can you reject the null hypothesis in each case? (b) Calculate the appropriate measure for the strength of relationship between the two variables, for each experiment. How strong does the relationship appear to be in each case?

| Experiment 1 | | Experiment 2 | |
|---|---|---|---|
| **Group 1 (cookies)** | **Group 2 (mango cake)** | **Group 1 (frequent mention)** | **Group 2 (no mention)** |
| 43 | 80 | 96 | 43 |
| 53 | 45 | 78 | 54 |
| 57 | 62 | 41 | 59 |
| 41 | 83 | 60 | 70 |
| 62 | 46 | 53 | 45 |
| 54 | 87 | 54 | 83 |
| 63 | 56 | 63 | 57 |
| 46 | 70 | 46 | 80 |
| 59 | 75 | 80 | 50 |
| 50 | 78 | 56 | 87 |
| 54 | 89 | 75 | 57 |
| 57 | 96 | 46 | 62 |
| 57 | | 57 | |
| 80 | | 62 | |
| 60 | | 89 | |

4. A psychology professor uses three different methods of instruction in three small classes, with the assignment of students to classes being random, and gives each class the same final examination. The results are in the table that follows. (a) Perform the appropriate test to compare each possible pair of groups and, in each case, state which group had the higher scores and whether the results were statistically significant at the .01 as well as the .05 level. (b) Calculate the appropriate measure of the strength of relationship between method of instruction and examination scores for each pair of groups, and comment briefly on the apparent size of the effect.

| Group 1 | Group 2 | Group 3 |
|---|---|---|
| 94 | 60 | 86 |
| 97 | 97 | 42 |
| 100 | 96 | 61 |
| 72 | 57 | 73 |
| 99 | 93 | 40 |
| 96 | 90 | 63 |
| 98 | 92 | 87 |
| 97 | | 65 |
| | | 67 |

5. For each of the two (separate) experiments that follow, perform the Wilcoxon test. If an experiment yields statistically significant results, also compute $r_c$.

| (a) Experiment 3 | | | (b) Experiment 4 | | |
|---|---|---|---|---|---|
| $S$ | $X_1$ | $X_2$ | $S$ | $X_1$ | $X_2$ |
| 1 | 12 | 9 | 1 | 15 | 15 |
| 2 | 13 | 8 | 2 | 25 | 8 |
| 3 | 23 | 16 | 3 | 9 | 16 |
| 4 | 17 | 21 | 4 | 13 | 9 |
| 5 | 19 | 14 | 5 | 11 | 23 |
| 6 | 20 | 9 | 6 | 19 | 14 |
| 7 | 15 | 17 | 7 | 10 | 21 |
| 8 | 14 | 9 | 8 | 12 | 18 |
| 9 | 22 | 14 | 9 | 15 | 17 |
| 10 | 18 | 13 | 10 | 14 | 17 |
| 11 | 25 | 10 | 11 | 9 | 14 |
| 12 | 11 | 10 | 12 | 11 | 15 |
| 13 | 20 | 15 | 13 | 20 | 13 |
| 14 | 15 | 15 | 14 | 10 | 10 |
| 15 | 10 | 11 | 15 | 20 | 22 |

6. There are 10 participants in a problem-solving experiment. Two judges, Steve and Annie, are asked to rate the solutions on an ordinal scale with regard to their creativity (1 = highly creative, 10 = barely creative). The experimenter wishes to know if Steve and Annie were equally lenient in their ratings. Given the ratings in the table, what should the experimenter decide?

| Participant | Steve | Annie |
|---|---|---|
| 1 | 8 | 4 |
| 2 | 7 | 9 |
| 3 | 2 | 1 |
| 4 | 9 | 8 |
| 5 | 2 | 7 |
| 6 | 5 | 1 |
| 7 | 10 | 9 |
| 8 | 5 | 7 |
| 9 | 6 | 3 |
| 10 | 8 | 6 |

7. A psychology instructor develops a training method that is designed to improve the examination scores of students. The performance of a sample of 12 students on the post-test ($X_1$) following training, and the pretest ($X_2$) prior to training, is shown. (a) Test the null hypothesis that the training method has no effect, using the Wilcoxon test. (b) Compute the appropriate measure of the strength of relationship.

| $S$ | $X_1$ | $X_2$ |
|---|---|---|
| 1 | 77 | 68 |
| 2 | 64 | 64 |
| 3 | 56 | 52 |
| 4 | 56 | 57 |
| 5 | 71 | 69 |
| 6 | 54 | 58 |
| 7 | 76 | 70 |
| 8 | 57 | 60 |
| 9 | 63 | 61 |
| 10 | 68 | 61 |
| 11 | 71 | 73 |
| 12 | 66 | 66 |

## Thought Questions

1. Parametric statistical tests are based on certain assumptions about unobservable population distributions. If these assumptions are incorrect, the parametric tests are likely to yield misleading results (and should therefore not be used). Yet we never know for certain what the population distributions are, so we cannot know for certain if these assumptions are justified. Nonparametric tests do *not* require such assumptions. Why, then, do psychological researchers often use parametric tests?
2. Give an example of when each of the procedures discussed in this chapter should be used: (a) rank-sum test; (b) Wilcoxon test.

## Computer Exercises

1. Separately for each of the five majors in Jackson's data set, use the rank-sum test to determine whether students who exercise regularly differ significantly from students who do not with respect to (a) statistics quiz scores, (b) prequiz anxiety scores, (c) procrastination ratings.
2. (a) Separately for students who do and do not exercise regularly, perform the Wilcoxon test to determine whether there is a significant increase in heart rate from baseline to the prequiz measurement. Also, test the difference between the pre- and postquiz measurements. Compare your results to those you obtained in Computer Exercise 5 in the previous chapter, if you did the advanced part of the exercise; otherwise, perform the Wilcoxon test on the entire group (all students included), and then compare to the previous chapter. (b) Repeat part (a) for the anxiety scores.

## Bridge to SPSS

To perform ordinal tests on your variables, select **Nonparametric Tests** from the **Analyze** menu, and then, from the new menu that opens up, click on the last choice, **Legacy Dialogs**. This action will open up a list of choices, two of which will be discussed in this section. Some of the other choices will be described in later chapters. The reason we are describing the use of SPSS modules that come under the heading of **Legacy Dialogs** is that they are easy to explain and will give you the results needed for the computer exercises in this chapter while avoiding the unnecessary (for our purposes) complexity of the newer modules.

### The Rank-Sum Test

From the **Analyze/Nonparametric Tests/Legacy Dialogs** menu, select **2 Independent Samples**. The first (and default) choice in the dialog box under *Test Type* is the one that produces the rank-sum test, and it is labeled **Mann–Whitney U** (see Figure 8.1). Just as in the Independent-Samples *t* test dialog box, you must specify the two levels of the *Grouping Variable* that you wish to compare (even if there *are* only two levels), as well as the list of *Test Variables* (DVs) on which you would like the two groups to be compared.

This test produces two tables of output by default (see Figure 8.2). The first is descriptive and contains the *N*s, sums, and means of the ranks for each group. The second table contains three Test Statistics: the Mann–Whitney $U$ (mentioned only briefly in this chapter), the Wilcoxon $W$ (the lower of the two sums of ranks), and $Z$, which is the $z$-score we showed you how to calculate for the rank-sum test. The $p$ value associated with $Z$ (SPSS uses an uppercase $Z$ to represent the same statistic for which we have been using the lowercase $z$) comes from the normal distribution and is labeled **Asymp. Sig (2-tailed)**. Note that the asymptotic $p$ value given in the results table is identical to the $p$ from the rank-sum test. The $U$ and $W$ statistics are holdovers from the old days when you needed to look up critical values in a statistical table. The tables for the $U$ and $W$ statistics are no longer needed because the **Exact Sig.** is also provided whenever your sample sizes are small enough to

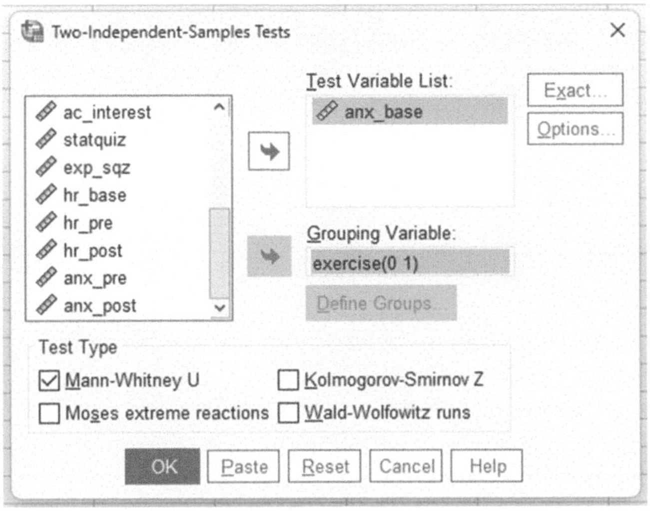


Source: Courtesy of International Business Machines Corporation, © SPSS, Inc./IBM company

## Figure 8.1

**The Dialog Box for Ordinal Tests for the Difference of Two Independent Samples**

**Ranks**

| | exercise | N | Mean Rank | Sum of Ranks |
|---|---|---|---|---|
| anx_base | 0 | 57 | 55.94 | 3188.50 |
| | 1 | 43 | 43.29 | 1861.50 |
| | Total | 100 | | |

**Test Statistics[a]**

| | anx_base |
|---|---|
| Mann-Whitney U | 915.500 |
| Wilcoxon W | 1861.500 |
| Z | −2.168 |
| Asymp. Sig. (2-tailed) | .030 |

a. Grouping variable: exercise

## Figure 8.2

**The Results of the Mann–Whitney U test for the Difference of Two Independent Samples**

need those more exact tables (i.e., both sample sizes less than 25). (Note that the "exact sig." [i.e., *p* value] is always the same for the *U* and the *W* statistics, as one is just an algebraic transformation of the other.)

If you compare the *Z* and *p* values for these nonparametric results, you will see that they are very similar to the *t* and *p* values for the ordinary *t* test shown in the previous chapter for the same variables. This will often be the case, unless you are dealing with a very skewed distribution or some extreme (outlier) scores.

### The Wilcoxon Test

Any time that your data have been entered in such a way that a paired-samples *t* test can be run, you can perform the *Wilcoxon Signed Ranks test* (as SPSS refers to it) instead. Just select **2 Related Samples** from the **Analyze/Nonparametric Tests/Legacy Dialogs** menu, and make sure that the first choice under *Test Type*, **Wilcoxon**, has been checked in the dialog box that opens (see Figure 8.3). Choose *two* variables from the list on the left, and move the pair to the *Test Pair(s)* List, as you would for a paired-samples *t* test. Repeat this process for each pair of variables that you would like to test. In this example, we are comparing baseline heart rate to the pre-quiz heart rate.

## Figure 8.3

**The Dialog Box for Ordinal Tests for the Difference of Two Paired (Matched) Samples**

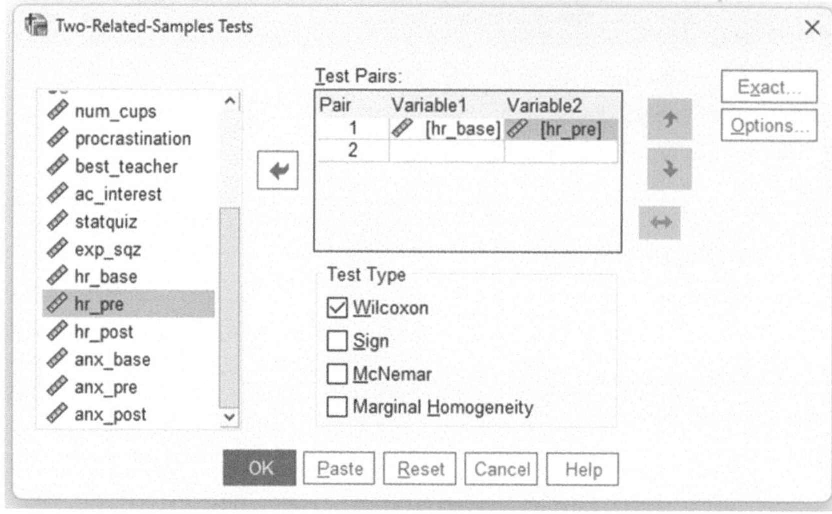


*Source:* Courtesy of International Business Machines Corporation, © SPSS, Inc./IBM company

## Figure 8.4

**The Results of the Wilcoxon Test for the Difference of Two Matched Samples**

**Ranks**

| | | N | Mean Rank | Sum of Ranks |
|---|---|---|---|---|
| hr_pre - hr_base | Negative Ranks | 28[a] | 44.43 | 1244.00 |
| | Positive Ranks | 63[b] | 46.70 | 2942.00 |
| | Ties | 9[c] | | |
| | Total | 100 | | |

a. hr_pre < hr_base

b. hr_pre > hr_base

c. hr_pre = hr_base

**Test Statistics[a]**

| | hr_pre - hr_base |
|---|---|
| Z | –3.373[b] |
| Asymp. Sig. (2-tailed) | <.001 |

a. Wilcoxon Signed Ranks Test

b. Based on negative ranks.

The Wilcoxon test creates difference scores for each pair of variables to be tested and then ranks these differences by magnitude, temporarily ignoring the signs. SPSS does not show you these difference scores or their actual ranks, but the first output table for the Wilcoxon procedure gives the $N$, sum, and mean rank separately for the positive and negative differences (see Figure 8.4). The number of ties (i.e., differences that are equal to zero—not identical difference scores) is also noted. The second output table (see Figure 8.4) contains only the $z$-score (labeled **Z**) obtained from the Wilcoxon approximation formula we included earlier in this chapter, along with its two-tailed $p$ value. A more exact $p$ value can be obtained by clicking on the **Exact** … button, and then checking *Exact*, but this is a very time-consuming task, and won't give you much different results if the sample is fairly large.

Note the similarity of this $z$-score to the matched-pairs $t$ value found in the previous chapter for the same variables. This will usually be the case unless some of the difference scores are relatively extreme compared to the others.

# Chapter 9
# Linear Correlation

## PREVIEW

### Introduction

What is meant by the correlation between two variables?

What is a positive correlation? A negative correlation?

What is a scatter plot?

How is the ability to make predictions related to correlation?

What is a linear relationship?

### Describing the Linear Relationship Between Two Variables

What does the $z$-score product formula for $r$ tell us about the meaning of the correlation coefficient?

Why is it easier to determine the numerical value of $r$ by using the computing (or raw score) formula?

### Interpreting the Magnitude of a Pearson $r$

If two variables are significantly correlated, does this mean that one of the variables causes the other?

How can restriction of range lead to misleading conclusions about the correlation between two variables?

How is it possible for two variables to have a Pearson $r$ near zero and yet be closely related? How can the presence of bivariate outliers in a data set lead to misleading conclusions about the correlation between two variables?

### When Is It Important That Pearson's $r$ Be Large?

What are the different types of reliability for which the Pearson $r$ is frequently used?

What is the validity of a measure, and why is it important that it be large?

### Testing the Significance of the Correlation Coefficient

What are the procedures for testing hypotheses about the population correlation coefficient (usually, that it is equal to zero)?

How much confidence can we have in our conclusion, if we must retain the null hypothesis?

Under what condition can we obtain a statistically significant Pearson $r$ that is too small to be of any practical importance?

### The Relationship Between Two Ranked Variables: The Spearman Rank-Order Correlation Coefficient

What is the procedure for computing the correlation between two variables when both consist of ranked data? When would you want to compute the Spearman correlation instead of the ordinary Pearson $r$?

What is the procedure for testing the significance of the rank-order correlation coefficient? How does this procedure differ from the significance test for the Pearson $r$?

*Introductory Statistics for the Behavioral Sciences*, Eighth Edition. R. Brooke Lea and Barry H. Cohen.

Companion website: www.wiley.com/go/ISBS8e

## Introduction

The independent-samples *t* test described in Chapter 7 can be viewed as a test of association between two variables: a categorical (i.e., discrete) grouping variable (e.g., smoking status) and a continuous dependent variable (e.g., lung function). This two-group example is just a special case of a more general principle: Knowing an individual's value on one variable can help you predict their value on another, related variable (e.g., knowing if someone smokes or not is helpful if you need to guess that individual's lung function). In this chapter, we show how you can quantify the degree of the relationship between two continuous variables, and in the next chapter, we show how you can use that relationship to make predictions about the values of one variable given values on the other variable.

For example, you probably learned early in your scholastic career that the amount of time spent studying is related to grades. True, there are exceptions. Some students may study for many hours and obtain relatively low grades, while others may achieve higher grades despite short study periods. Nevertheless, the general trend holds. In the majority of cases, you can accurately predict that little or no studying will lead to relatively low grades, while more studying will result in better grades. So, if you are dissatisfied with grades that you feel are too low, one likely way to improve this aspect of your environment is by studying more. In statistical terminology, the two variables of "hours studying" and "grades" are said to be co-related or *correlated*.

Many pairs of variables are correlated, while many are unrelated or *uncorrelated*. In the example mentioned above, when college students are surveyed, those who report spending more hours studying each week, also tend to have higher GPSs. These variables are said to be **positively correlated**. This relationship is depicted graphically in Figure 9.1, which is called a **scatter plot**. Note that each point on the graph represents two values for one family, hours spent studying ($X$ variable) and GPA of the student ($Y$ variable). Also, in the case of a positive correlation, the straight line summarizing the points slopes *up* from left to right; as the $X$ values increase in value, so do the $Y$ values. As we discuss later in the chapter, correlation does not imply causation. Therefore, the data presented in Figure 9.1 do not imply that hours spent studying directly affect the GPA of students.

Golf enthusiasts will readily tell you that the number of years of play is negatively related to their golf score: The more years playing golf, the fewer strokes needed to complete a round of 18 holes. These two variables are said to be **negatively correlated**. (See Figure 9.2.) Notice that in the case of a negative correlation, the straight line summarizing the points slopes *down* from left to right. This "negative slope" indicates that as values on one variable are increasing, values on the other variable are decreasing. Of course, golf scores are getting *better* with more years of play, because *better* golf scores are *lower* golf scores.

This example can also be used to illustrate the importance of choosing your variables carefully. Note that "number of years of play" is a relatively imprecise measure that

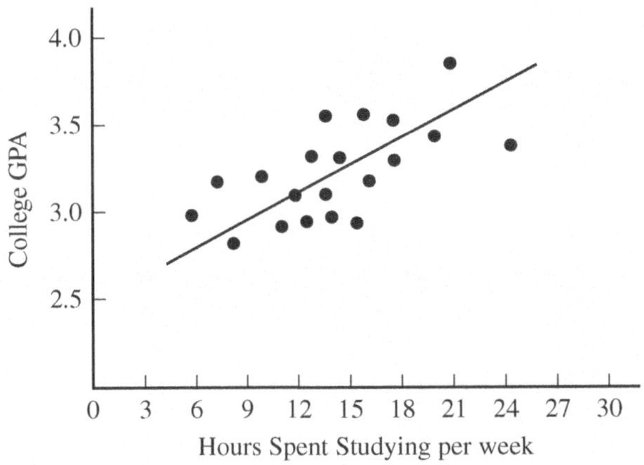


# Figure 9.1

**Possible Relationship Between Hours Spent Studying and the Student' GPA**

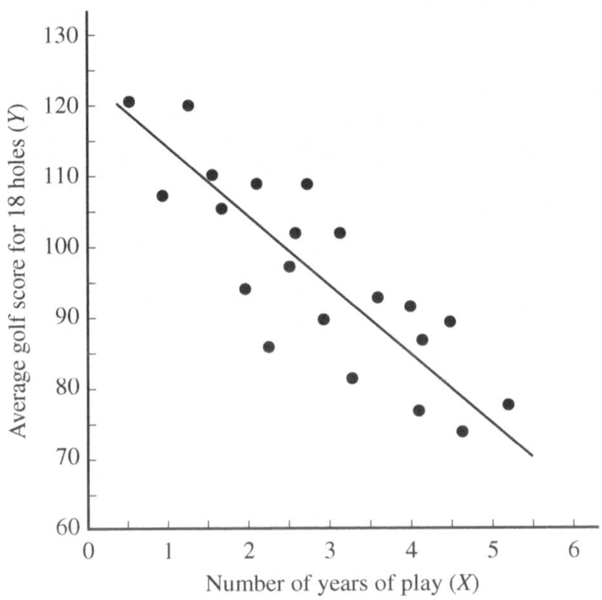


# Figure 9.2

**Possible Relationship Between Years of Play and Average Golf Score**

represents a more specific behavior, perhaps something like experience with or practice playing golf. Its imprecision can be explored by looking at the two players in our sample who have played for about 2.25 years (Figure 9.2). One, we'll call him Dave, is already shooting in the mid-80s—an impressive achievement for such an inexperienced golfer. The other, Sean, is still shooting around 110, much less impressive. You might be tempted, based on this information, to conclude that Dave is a better athlete than Sean, or perhaps just that he learns sports faster. But if you then discover that Dave plays 50 rounds of golf each year and Sean plays only 5, then "number of years of play" suddenly seems like a poor proxy for "experience and practice." Dave has played 10 times as much golf as Sean! A more precise version of the predictor variable, such as number of rounds played, would likely tighten the scatter plot around the line and reduce error in prediction. The lesson here is that one should think as precisely as possible about the variables under consideration and choose measures that are as indicative of those variables as possible. Of course, precision can come with a cost; number of shots made (including practice) might be the most precise variable to use in this example. But obtaining the number of shots made would be much more tedious than counting the number of rounds played!

A different picture occurs when two variables are uncorrelated. For example, length of big toes among adults is uncorrelated with their IQ scores, and the corresponding scatter plot is shown in Figure 9.3.

Figure 9.3

**Possible Relationship Between the Length of Big Toe and IQ Scores**

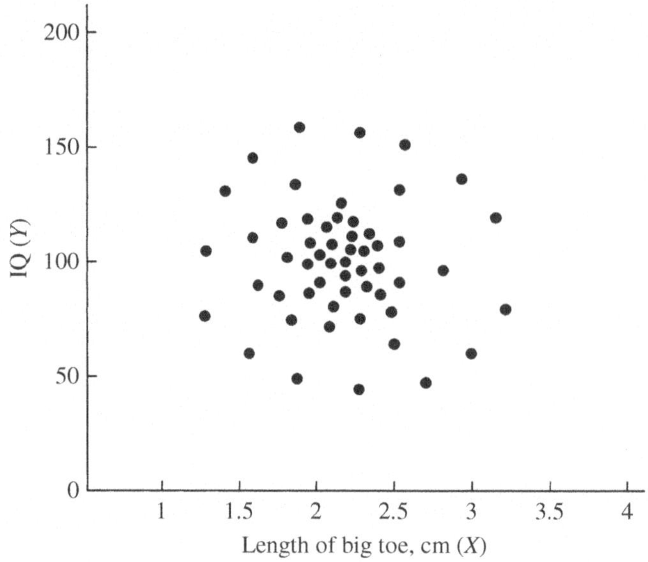


When you can demonstrate that two variables are correlated, you can use the score of an individual on one variable to *predict* or *estimate* their score on the other variable. In Figure 9.2, a fairly accurate prediction of an individual's golf score can be made by knowing the number of years they have played golf. The more closely the two variables are related, the better the prediction is likely to be; if two variables are uncorrelated (as in the case of toe length and IQ), you cannot accurately predict an individual's score on one of them from their score on the other. Thus, the concepts of correlation (the topic of this chapter) and the concepts of prediction (the topic of the next chapter) are closely related.

The main topic of this chapter can be divided into two separable aspects: (1) the *description* of the relationship between two variables for a set of observations; (2) making *inferences* about the strength of the relationship between two variables in a population, given the data of a sample. We will deal with one very common kind of relationship between two variables, namely a *linear* or straight-line relationship. That is, if values of one variable are plotted against values of the other variable on a graph, the trend of the plotted points is well represented by a straight line when the relationship is strong. Notice that the data plotted in Figures 9.1 and 9.2 tend to fall near or on the straight line drawn through the scatter plot; this indicates that the two variables in question are highly linearly correlated. The points in Figure 9.3, by contrast, are scattered randomly throughout the graph and cannot be represented well by a straight line. Therefore, these two variables are linearly uncorrelated.

Many pairs of variables that are important to behavioral scientists tend to be linearly related. Although there are many other ways of describing relationships, the linear model is generally the one most frequently used in such fields as psychology, education, and sociology.

## Describing the Linear Relationship Between Two Variables

Suppose you want to measure the degree of linear relationship between the scholastic aptitude test (SAT) and college grade point average (GPA). Since the SAT was designed to predict college GPA, it would be reasonable to expect these two variables to be positively correlated. This means that students with higher SAT scores, on average, obtain relatively high GPAs and students with lower SAT scores, on average, obtain low GPAs. The phrase "on average" alerts us to the fact that there will be exceptions: Some students with low SAT scores will do very well in college and obtain high GPAs, while some with high SAT scores will do relatively poorly and receive low GPAs. That is, the relationship between SAT scores and GPA is not perfect. Thus, the question arises: Just how strong *is* the relationship? How can it be summarized in a single number?

### The $z$-Score Product Formula for $r$

We have seen in Chapter 4 that in order to compare scores on different variables (such as a geology test and a psychology test), it is useful to transform the raw scores into standardized scores. These transformed scores allow you to compare paired values directly. For the same reason, to obtain a coefficient of relationship that describes the similarity between paired measures in a single number, it is convenient to transform the raw score into $z$ units.

$$z_X = \frac{X - \mu}{\sigma_X} \qquad z_Y = \frac{Y - \mu}{\sigma_Y} \qquad \text{Expressions 9.1A and 9.1B}$$

In Table 9.1, a distribution of SAT scores ($X$) and GPAs ($Y$) is presented along with the corresponding descriptive statistics for 25 students at a college in the upper midwestern United States. (Note that we are using *population* notation with a collection of scores, which could be viewed as either a population or a sample because our emphasis is on description rather than inference at this point; we would like to describe these students, not use them to make inferences about a larger population of students.) The SAT scores were obtained when the students were high school seniors, and the GPAs are those received by the students after one year of college. In addition to the raw scores, the $z$ equivalents are also shown in the table. Notice that students with high SAT scores do tend to have high GPAs and consequently both large $z_X$ and $z_Y$ values. Conversely, students with low SAT scores tend to have low GPAs. So the paired $z$ values are similar for most students, and the two variables are

**Table 9.1 Raw Scores and $z$-Scores on SAT ($X$) and GPA ($Y$) for 25 Students in an Upper Midwestern US College**

| Student | $X$ | $Y$ | $z_X$ | $z_Y$ |
|---|---|---|---|---|
| 1 | 650 | 3.8 | 1.29 | 1.67 |
| 2 | 625 | 3.6 | 1.08 | 1.31 |
| 3 | 480 | 2.8 | −0.16 | −0.09 |
| 4 | 440 | 2.6 | −0.50 | −0.44 |
| 5 | 600 | 3.7 | 0.86 | 1.49 |
| 6 | 220 | 1.2 | −2.37 | −2.89 |
| 7 | 640 | 2.2 | 1.21 | −1.14 |
| 8 | 725 | 3.0 | 1.93 | 0.26 |
| 9 | 520 | 3.1 | 0.18 | 0.44 |
| 10 | 480 | 3.0 | −0.16 | 0.26 |
| 11 | 370 | 2.8 | −1.09 | −0.09 |
| 12 | 320 | 2.7 | −1.52 | −0.26 |
| 13 | 425 | 2.6 | −0.62 | −0.44 |
| 14 | 475 | 2.6 | −0.20 | −0.44 |
| 15 | 490 | 3.1 | −0.07 | 0.44 |
| 16 | 620 | 3.8 | 1.04 | 1.67 |
| 17 | 340 | 2.4 | −1.35 | −0.79 |
| 18 | 420 | 2.9 | −0.67 | 0.09 |
| 19 | 480 | 2.8 | −0.16 | −0.09 |
| 20 | 530 | 3.2 | 0.27 | 0.61 |
| 21 | 680 | 3.2 | 1.55 | 0.61 |
| 22 | 420 | 2.4 | −0.67 | −0.79 |
| 23 | 490 | 2.8 | −0.07 | −0.09 |
| 24 | 500 | 1.9 | 0.01 | −1.67 |
| 25 | 520 | 3.0 | 0.18 | 0.26 |
| | $\mu = 498.4$ | $\mu = 2.85$ | $\sigma_x = 117.3$ | $\sigma_y = 0.57$ |

## Figure 9.4

**Perfect Linear Relationship Between Two Variables: (A) Perfect Positive Linear Relationship; (B) Perfect Negative Linear Relationship**

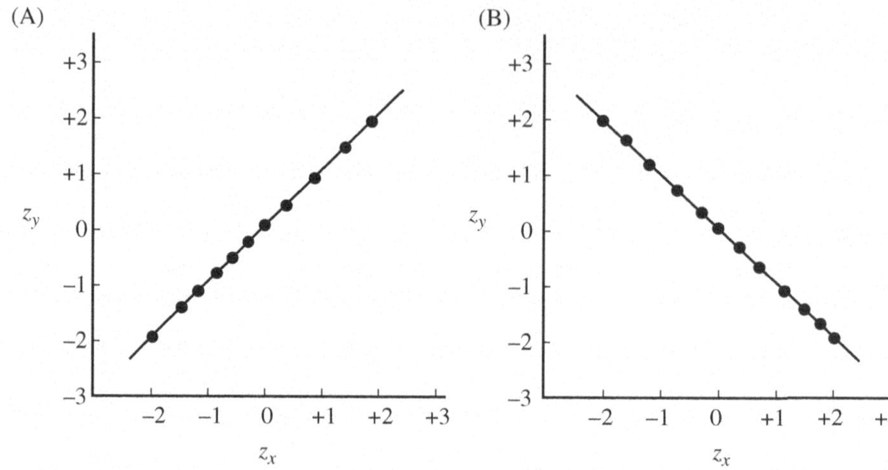


therefore positively correlated. The *raw scores*, however, need not be numerically similar for any pair because they are measured in different units for each variable.

If the association between the selected variables was perfect and in a positive direction, each person would have exactly the same *z*-score on both variables (see Figure 9.4A). If the relationship was perfect but in a negative direction, each $z_X$ value would be paired with an identical $z_Y$ value but they would be *opposite in sign* (see Figure 9.4B). These perfect relationships are offered for illustration, but almost never occur in practice.

So, if the pairs of *z*-scores are identical, then we have a perfect correlation. But how can we quantify the strength of relationship when correlation is less than perfect? A handy mathematical law states that when you multiply two sets of numbers (say, the *z*-scores for *X* and *Y* in Table 9.1), *the sum of the products increases as the similarity of the items in each pair increases* (i.e., large scores on one variable paired with large scores on the other variable, and low scores paired with low scores). The highest sum of products occurs when the numbers in each pair are identical. We can use this law, in light of the fact that higher correlation involves similar *z*-scores, to create a measure of correlation, beginning like this:

$$\sum z_X z_Y \qquad \text{Expression 9.2}$$

The larger this sum of $z_x * z_y$ products, relative to the number of them, the stronger the correlation. We can divide Expression 9.2 by the number of pairs of scores (i.e., the number of cross products) to get the average of the products of the *z*-scores. That average turns out to be an excellent measure of correlation. It was first devised by Karl Pearson more than 100 years ago, so it is called **Pearson's correlation coefficient, *r***:

$$r = \frac{\sum z_X z_Y}{N} \qquad \text{Formula 9.1}$$

In sum, correlation is perfect when two measurements, such height and weight, fall in exactly the same place in their respective distributions. But if height and weight were perfectly correlated we would not expect someone who is 72 inches tall to weigh 72 pounds! That is because height and weight are measured on different scales. Transforming the raw scores in inches and pounds into *z*-scores solves that problem; someone who is one standard deviation (SD) taller than the average person ($z_x = 1$) would be expected to be one SD heavier than the average person ($z_y = 1$) if the two variables were perfectly correlated. Now we can use the mathematical fun-fact that when you multiply two sets of numbers, the sum of the products increases as the similarity of the items in each pair increases, and that the highest sum occurs when the two numbers are identical. Thus, the average of the products of *z*-scores (i.e., Formula 9.1) is a measure of how similar the *z*-scores are, and therefore how strongly they are correlated.

As we have seen, when correlation is positive and perfect, $z_x$ and $z_y$ are always identical, so the cross product in every case becomes $z$-squared:

$$\frac{\sum z_X^2}{N} \qquad \text{Expression 9.3}$$

Does this expression look familiar at all? You may not have noticed it, but Expression 9.3 is equivalent to the equation for the *variance of a set of z-scores*! Recall the formula for the variance:

$$\sigma^2 = \frac{\sum (X_i - \mu)^2}{N} \qquad \text{Formula 3.2}$$

If we express this formula in terms of $z$-scores, then we replace the $X$ scores with $z$-scores and the mean of the $X$ scores with the mean of the $z$-scores. You may recall from Chapter 4 that the mean of a set of $z$-scores is always zero, so that term is dropped and we end up with Expression 9.4:

$$\sigma_z^2 = \frac{\sum (z_i - \mu_z)^2}{N} = \frac{\sum z^2}{N} = r \qquad \text{Expression 9.4}$$

As we saw in Chapter 4, the variance of a set of $z$-scores is always 1.0, so we have shown that when correlation is perfectly positive, Pearson $r$ must also equal +1.0. When correlation is not perfect, some cross products will be negative and others positive; if the negatives balance out the positives perfectly, then $r$ will equal zero. Otherwise, you will get some value between +1.0 and −1.0.

### Computing Formulas for *r*

The preceding procedure for obtaining $r$, although helpful for understanding the meaning of $r$, is too tedious computationally. (Proofs of the equivalence for various formulas for $r$ are given in the appendix at the end of this chapter.) We can avoid the necessity for converting raw scores to $z$ values by rewriting Formula 9.1 as:

$$r = \frac{\sum(X - \mu_X)(Y - \mu_Y)}{N\sigma_X\sigma_Y} \qquad \text{Expression 9.5}$$

Expression 9.5 can be rearranged algebraically to produce the following *raw-score formula* for the Pearson correlation coefficient:

$$r = \frac{\frac{\sum XY}{N} - \mu_X\mu_Y}{\sigma_X\sigma_Y}, \qquad \text{Formula 9.2}$$

where $N$ is the number of *pairs* of observations (i.e., the number of participants, or cases).

Formula 9.2 is useful because it serves as a relatively straightforward extension of the $z$-score formula for Pearson $r$, which is made up of population parameters (i.e., mus and sigmas). If you are computing Pearson r by hand, however, you are more likely to be working with sample statistics (i.e., x-bars and *s*s). Recall that when we wish to use sample data to draw inferences about a population (e.g., studying the correlation between height and weight among a sample of 30 to infer what you might get if you measured the entire population), we must use the un-biased version of the SD (i.e., s rather than sigma). That means that we will need to change Formula 9.2 slightly to work with sample statistics. We will do that next.

The numerator of Formula 9.2 (the average of the raw-score cross-products minus the cross product of the two averages) is called the *covariance*. When relatively large scores on one variable are paired with relatively large scores on the other, and small scores with small

**Table 9.2** **Calculation of Pearson Correlation Coefficient Between SAT (*X*) and GPA(*Y*) by the Raw Score and *z* Product Methods**

| Participant | Raw-score method | | | *z* product method | | |
|---|---|---|---|---|---|---|
| | $X$ | $Y$ | $XY$ | $z_x$ | $z_y$ | $z_x z_y$ |
| 1 | 650 | 3.8 | 2,470 | 1.29 | 1.67 | 2.1543 |
| 2 | 625 | 3.6 | 2,250 | 1.08 | 1.31 | 1.4148 |
| 3 | 480 | 2.8 | 1,344 | −0.16 | −0.09 | 0.0144 |
| 4 | 440 | 2.6 | 1,144 | −0.50 | −0.44 | 0.2200 |
| 5 | 600 | 3.7 | 2,220 | 0.86 | 1.49 | 1.2814 |
| 6 | 220 | 1.2 | 264 | −2.37 | −2.89 | 6.8493 |
| 7 | 640 | 2.2 | 1,408 | 1.21 | −1.14 | −1.3794 |
| 8 | 725 | 3.0 | 2,175 | 1.93 | 0.26 | 0.5018 |
| 9 | 520 | 3.1 | 1,612 | 0.18 | 0.44 | 0.0792 |
| 10 | 480 | 3.0 | 1,440 | −0.16 | 0.26 | −0.0416 |
| 11 | 370 | 2.8 | 1,036 | −1.09 | −0.09 | 0.0981 |
| 12 | 320 | 2.7 | 864 | −1.52 | −0.26 | 0.3952 |
| 13 | 425 | 2.6 | 1,105 | −0.62 | −0.44 | 0.2728 |
| 14 | 475 | 2.6 | 1,235 | −0.20 | −0.44 | 0.0880 |
| 15 | 490 | 3.1 | 1,519 | −0.07 | 0.44 | −0.0308 |
| 16 | 620 | 3.8 | 2,356 | 1.04 | 1.67 | 1.7368 |
| 17 | 340 | 2.4 | 816 | −1.35 | −0.79 | 1.0665 |
| 18 | 420 | 2.9 | 1,218 | −0.67 | 0.09 | −0.0603 |
| 19 | 480 | 2.8 | 1,344 | −0.16 | −0.09 | 0.0144 |
| 20 | 530 | 3.2 | 1,696 | 0.27 | 0.61 | 0.1647 |
| 21 | 680 | 3.2 | 2,176 | 1.55 | 0.61 | 0.9455 |
| 22 | 420 | 2.4 | 1,008 | −0.67 | −0.79 | 0.5293 |
| 23 | 490 | 2.8 | 1,372 | −0.07 | −0.09 | 0.0063 |
| 24 | 500 | 1.9 | 950 | 0.01 | −1.67 | −0.0167 |
| 25 | 520 | 3.0 | 1,560 | 0.18 | −0.26 | 0.0468 |

Raw-score method:

$\sum X = 12{,}460 \quad \mu_x = 498.4$

$\sum Y = 71.2 \quad \mu_y = 2.4$

$\sum XY = 36{,}582 \quad N = 25$

$$r = \frac{\dfrac{\sum XY}{N} - \mu_x\mu_y}{\sigma_X\sigma_Y} = \frac{\dfrac{36{,}582}{25} - (498.4)(2.85)}{(117.33)(.57)}$$

$$= \frac{1{,}463 - 1{,}420.44}{66.8781} = \frac{42.84}{66.8781}$$

$$= +.64$$

*z* product method:

$\sum z_X z_Y = 16.35$

$N = 25$

$$r = \frac{\sum z_X z_Y}{N} = \frac{16.35}{25} = +.65$$

scores, the average of the cross products (left side of numerator) tends to be larger than the cross product of the averages (right side of numerator), producing a positive value for the covariance and therefore a positive correlation. When relatively large scores on one variable are paired with relatively small scores on the other, this relationship is reversed, leading to a negative covariance and therefore a negative correlation. When scores are paired randomly, the two parts of the numerator of $r$ tend to be about the same, leading to a covariance and an $r$ that are near zero.

Using the SAT and GPA data in Table 9.1, we can calculate Pearson $r$ using both the raw-score and $z$ product formulas, as shown in Table 9.2. Note that the discrepancy in the results of these two formulas (.64 for raw scores versus .65 for $z$ products) is due to rounding the means and SDs in Table 9.1, which were then used in the calculations in Table 9.2. The raw-score formula is only as accurate as the number of digits you retain for the means and SDs of the two variables.

As noted above, Formula 9.2 is based on using the biased[1] SDs in its denominator. So, if you want to use $s$s rather than $\sigma$s in the denominator of the raw-score formula, the bias of the covariance in the numerator must also be corrected. The resulting formula will always give the same answer for $r$ (except for any differences in rounding):

$$r = \frac{\frac{1}{N-1}\left[\sum XY - N\overline{X}\overline{Y}\right]}{s_X s_Y} \qquad \text{Formula 9.3}$$

Notice that in going from Formula 9.2 to 9.3, we not only changed the $\sigma$s to $s$s, we also changed the $\mu$s to $\overline{X}$ and $\overline{Y}$. Even though $\mu_X$ and $\overline{X}$, for instance, are computed in exactly the same way, we think it is important to use the $\overline{X}$ symbol as a reminder whenever you are treating your data as a sample of a larger population of interest.

The obtained value of .65 indicates that for this group of 25 students, there is a fairly high positive correlation between these two variables. (In fact, this value is higher than most correlations that you would get in the behavioral sciences unless you are measuring essentially the *same* variable in two different ways.) A scatter plot of the data in Table 9.1 is shown in Figure 9.5.

### Stroop Effect Example

In Chapter 7, we mentioned that the larger the correlation between two sets of scores, the more the denominator of the matched-pairs $t$ formula decreases relative to the denominator of the independent-samples test. As another example of calculating $r$, let us see just how large the correlation is between the interference and noninterference trial accuracies of the 10 participants of the Stroop experiment, whose data were used in Table 7.2. Because we have already calculated the $s$s (and not the $\sigma$s) for these data in Chapter 7, we will use the Pearson $r$ formula that is based on the unbiased SDs.

$$r = \frac{\frac{1}{N-1}\left(\sum XY - N\overline{X}\overline{Y}\right)}{s_X s_Y} = \frac{\frac{1}{9}[50{,}553 - 10(68.0)(74.0)]}{(4.472)(8.59)}$$

$$= \frac{\frac{1}{9}(50{,}553 - 50{,}320)}{38.414} = 25.889/38.414$$

$$= 0.674$$

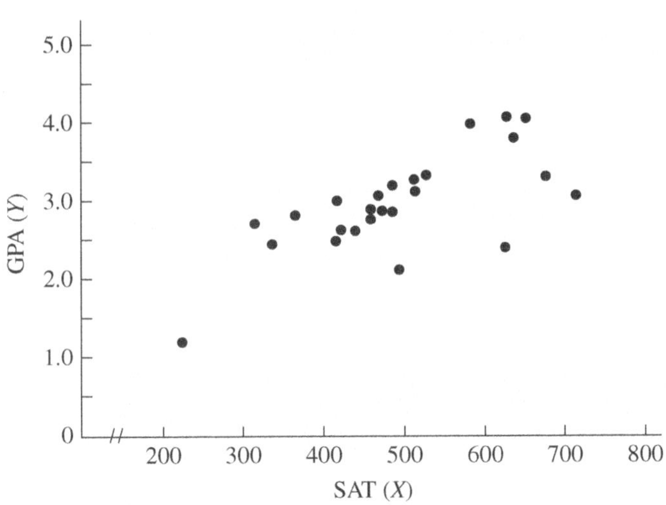


**Figure 9.5**

**Scatter Plot for Data in Table 9.1 ($r$ = +.65)**

[1] Recall from Chapter 3 (page 52) that "biased" in this context refers to the use of a sample SD to make inferences about the population SD. When used only for descriptive purposes, the "biased" form of the SD (i.e., sqroot (SS/N)) is fine to use and not, in fact, biased.

Once you have calculated $r$ for a set of paired scores, you can use an alternative formula for the matched-pairs $t$ test that is based on $r$. This formula is not easier to use than the one we introduced in Chapter 7, but it is instructive to see how it uses the Pearson $r$ to obtain its result. This formula, applied to the accuracy data, yields the following value for $t$:

$$t = \frac{\overline{X}_1 - \overline{X}_2}{\sqrt{\frac{s_1^2 + s_2^2}{n} - \frac{2rs_1s_2}{n}}} = \frac{74 - 68}{\sqrt{\frac{20 + 73.79}{10} - \frac{2(0.674)(4.472)(8.59)}{10}}}$$

$$= \frac{6}{\sqrt{9.379 - 5.178}} = \frac{6}{2.05} = 2.93$$

Note that this alternative formula always yields the same $t$ value as the formula for difference scores (Formula 7.12, page 142). However, an inspection of this alternative formula reveals an important principle: As Pearson's $r$ increases (all else remaining the same), a larger portion is subtracted from the denominator, making the $t$ value larger. When $r$ is zero, the second term in the denominator drops out and you are left with the equal-$N$ formula for the independent-samples $t$ test; in this case, using a matched pairs design has not led to a decrease in person-to-person variability. Also note that a Pearson $r$ does not have to be large or statistically significant to be helpful at reducing the size of the denominator of a matched pairs $t$ test. Moreover, a large, statistically significant $r$ certainly does not guarantee a significant matched-pairs $t$ test. The difference between the two means in the numerator can override either a small or large denominator to determine whether the matched-pairs $t$ test will reach statistical significance.

## Interpreting the Magnitude of a Pearson *r*

A correlation coefficient is a useful way to summarize the relationship between two variables with a single number that falls between −1 and 1. However, any one statistic that is used to summarize a whole set of data is bound to have its limitations, and this is certainly true for Pearson's $r$. A small value for $r$, especially one that is much smaller than expected, should never be interpreted without looking at the corresponding scatter plot. We will point out several problems to look for when $r$ is small, but first, we must caution you about a potential fallacy that is particularly tempting when the magnitude of $r$ is relatively large. Several cautions about the interpretation of Pearson $r$ will be described in the next few sections.

### What does "Correlation does not imply Causation" Really Mean?

You may have heard the expression, "Correlation does not imply causation." We will never disagree with this statement—after all, we made it ourselves just a few pages ago—but we would like to clarify its meaning. People often misunderstand this claim as referring to a characteristic of *statistical tests*, as in "If you analyze your data using a measure of correlation, you cannot make causal claims about the effects of $X$ on $Y$." In fact, it is not the statistical methods you use that determine whether you can make causal conclusions from your results, *it is the research design that sets the limits on causal inference.* In particular, random assignment to groups (or different levels of your independent variable) is the key to a true experimental design that permits causal inferences. If you have randomly assigned your participants to groups, then you can infer causality (assuming that you have done everything else properly); if you have not performed the random assignment, then you cannot.[2]

Most commonly, correlation coefficients are computed between two quantitative dependent variables, both of which are measured on the same randomly (or more often conveniently) selected participants. However, let us consider the case in which you have a truly

[2] There are sophisticated methods (e.g., Structural Equation Modeling) that can help establish the direction of causality among observed variables without random assignment, but random assignment is always preferable when possible, as it allows more definitive conclusions concerning causation.

independent variable with quantitative levels. For example, imagine that you have randomly assigned participants to different quantitative levels of your independent variable (e.g., number of therapy sessions per week, 0, 1, 2, 3, 4, 5) and then measured your participants' improvement on a dependent variable (e.g., level of well-being) at the end of the study. In this case, a significant correlation coefficient computed between your independent variable and your dependent variable *does* allow you to determine that the variance in $X$ *caused* at least some of the variance in $Y$. (For this example, you could conclude that different amounts of therapy *caused* different amounts of improvement in well-being.) Note that in this case, we used a test of correlation to appropriately draw a causal inference.

By comparison, consider what this study might look like if you did not use random assignment. For instance, you might have created two groups based on how often your participants were already going to therapy ("frequent" and "infrequent") and then performed a $t$-test to compare the change in levels of well-being after a given period of time. In this case, a significant $t$ value *does not* permit you to claim that the frequency of therapy visits is causally related to any changes in well-being. Note that the key difference is that in this second version of the study, the participants were not randomly assigned to be in the frequent or infrequent groups; they were already in their "group" based on their own reasons for how often they wanted to go to therapy. The problem, in terms of determining causation, is that maybe it was one of their "own reasons" that was causing the effect. That is, perhaps there is something about people who go to therapy more often (e.g., more disposable income) that is accounting for the difference in improvement in well-being. If you did not assign your participants to particular amounts of therapy per week anyway, you probably would have been better off just computing the correlation between amounts of therapy and amounts of improvement, rather than arbitrarily classifying your participants into two groups. Your correlation coefficient might have a somewhat better chance of reaching statistical significance than the $t$ test, because the data it is based on contains more quantitative information, but neither a significant correlation nor a significant $t$ test would allow you to make a causal inference without assigning your participants to different amounts of therapy.

In sum, it is the design of the study that determines whether causality can be inferred, not the statistical methods you use to analyze the data. As we have just shown, you could use random assignment and infer causation with a significant correlation coefficient, or use previously existing groups with a $t$-test and not be able to make such an inference. Therefore, "correlation" in the expression "Correlation does not imply causation" refers to the correlational (i.e., observational, rather than experimental) design of the study and not to the type of statistics used to analyze the data.

A second point is that, depending on the situation, it can be more or less difficult to resist making causal inferences from large correlations. On one hand, you may have heard well-known examples of bizarre correlations, such as the positive relationship between the volume of ice cream sold and the murder rate in New York City, or a similar relationship between life expectancy and the number of clocks in one's home in medieval Europe. In these cases, it is not particularly tempting to infer that ice cream is setting New Yorkers off on murder sprees or that clocks somehow extended one's time on earth, in addition to measuring it. Instead, it is relatively easy to imagine that in each case a third variable was affecting both of the two variables that exhibited such a surprising relationship (e.g., heat had been affecting both ice cream consumption and the murder rate; wealth affected both longevity and the ownership of clocks).

On the other hand, however, you will encounter correlational results that will be truly difficult to resist. In some cases, the difficulty may be due to the fact that the $X$ variable is indeed causally related to the $Y$ variable, even though the correlational nature of the study does not warrant that conclusion. A classic example of this situation arose with the relationship between cigarette smoking and lung cancer in the 1960s. Mountains of evidence suggested a strong correlational relationship between the two, but because the appropriate experiment could not be conducted (ethically), cigarette companies were able to claim, correctly, that there was no scientific evidence demonstrating a causal connection between cigarette smoking and lung cancer. When researchers finally randomly assigned rats to "smoke" cigarettes, they were able to establish a definitive causal connection between smoking and cancer, but strictly speaking, only for rats. A leap still had to be made to assume that rats are similar enough to humans biologically that the same causal connection between smoking and cancer could be applied to humans.

The relationship between cigarette smoking and lung cancer exemplifies cases in which it is difficult to resist the correlation-causation inference largely because there is a plausible mechanism connecting the two variables (e.g., various chemicals in cigarette smoke are making direct contact with the tissue inside a smoker's lungs). But by far the more insidious case of falling for the correlation-causation fallacy occurs when the correlation between two variables appears to confirm a previously held belief, theory, stereotype, or prejudice. Once people develop a "rooting interest" in an outcome, it can be astonishingly difficult for them to accept that correlational results only indicate that some relationship exists between the two variables, but that the results do not help us understand whether $X$ is causing $Y$, $Y$ is causing $X$, or some third variable is affecting both. Good scientific education—including basic knowledge of statistics—is our best hope for combating these problematic shortcuts in the thinking process. Even seasoned researchers who know about and teach students about the correlation-causation fallacy will occasionally allow causal language to sneak into their work. Articles with titles like "The effect of divorce on later adult relationships" do not need to be examined carefully to know the title's claim cannot be true; it is not ethical to randomly assign people to come from divorced families. But, like strong stereotypes and other cognitive biases, the assumption that correlational studies can imply causation is remarkably difficult to resist, even if one knows better.

### Correlation and Restriction of Range

Another important caution has to do with the effect of the variability of the scores on the correlation coefficient. Suppose you calculate the correlation between achievement test scores and elementary school grades for public school children in New York City. You will probably find a strong linear trend when the data are plotted. But what if you were to calculate the correlation between the same two variables for a group of children in a class for the highly gifted? Here, the range of scores on both variables is considerably narrowed, as is shown in Figure 9.6. Because of this, the correlation between the two variables is markedly reduced. (You can see that the points do not line up nicely within the gray box in Figure 9.6.)

It is much harder to make fine discriminations among cases that are nearly equal on the variables in question than it is to make discriminations among cases that differ widely. More specifically, it is difficult to predict whether a gifted child will be an A or A− student. It is much easier to distinguish among a broad range of A to F students. Thus, *when the variability of scores is restricted* on either or both of two variables, the correlation between them usually decreases in absolute value.

Restriction (or truncation) of range can be a problem when it is difficult to obtain a random sample that represents the variability of scores in the population, as in the

Figure 9.6

**Effect of Restriction of Range on the Correlation Coefficient**

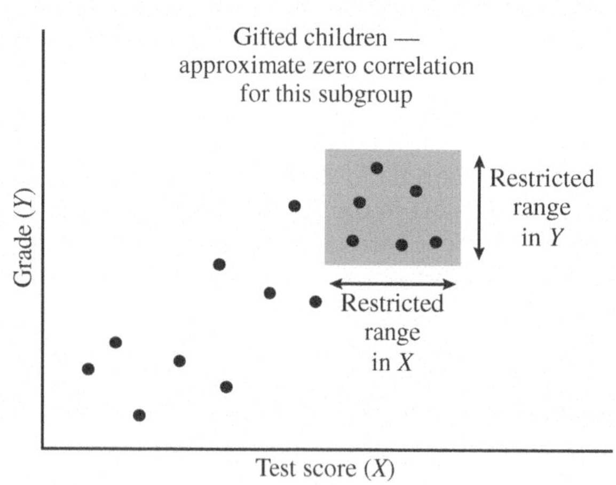

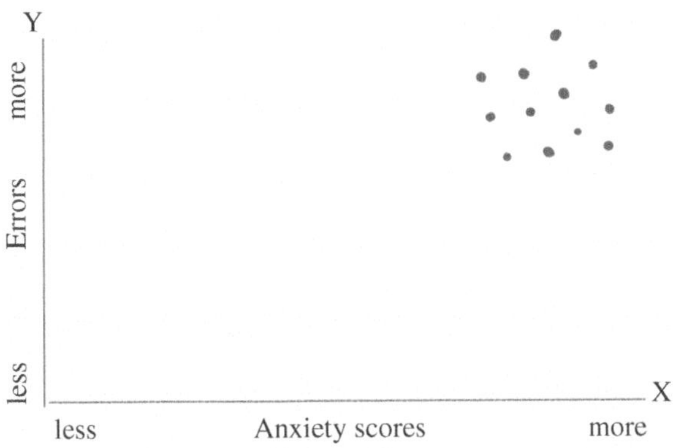


# Figure 9.6A

**Hypothetical data from Jordan's anxiety—academic performance example**

elementary school example above. But occasionally researchers unwittingly box themselves into a restricted range problem due to the focus of their research.

Imagine a researcher, Jordan, who is interested in understanding the relationship between anxiety and academic performance. He begins by finding a group of students who score high on an anxiety scale, and then gives them a difficult academic task and counts the number of errors they make. The data are likely to reveal high scores on the $x$ variable (anxiety) along with high scores on the $y$ variable (errors). Seems like a good study, right? Probably not. The data are likely to turn out like those in Figure 9.6A, in which case Jordan will likely get a value of Pearson $r$ near zero. Why?

By restricting his study to participants high in anxiety, Jordan inadvertently restricted the range of scores on the $x$ variable, and because anxiety is somewhat correlated with academic performance, he ended up with a restricted range in his $y$ variable (errors). "But wait" you might ask. "Doesn't Jordan's study show high scores on anxiety coupled with high scores on errors?" Indeed it does. It shows scores high in an absolute sense (i.e., relative to the full range of possible scores). You continue, "Shouldn't that produce a very high value for r?" Nope. Without any low scores in the data set, Pearson $r$ does not "know" that these raw scores are high on the anxiety or error scales in an absolute sense because it is based on the $z$-scores of the data, not the raw scores! Pearson $r$, therefore, is sensitive only to a score's value *relative to the other scores in the sample*, but it is not sensitive to a score's value relative to all possible scores. Let us take a closer look at why this is so.

Figure 9.6B shows Jordan's data with a normal curve superimposed on it. Let us zoom in on the data point the arrow is pointing to. That individual scored high on the anxiety scale in an absolute sense—the score is about as high as scores can get. It is also high relative to other scores in the group, which means it will have a positive $z$-score on the $x$ variable (anxiety).

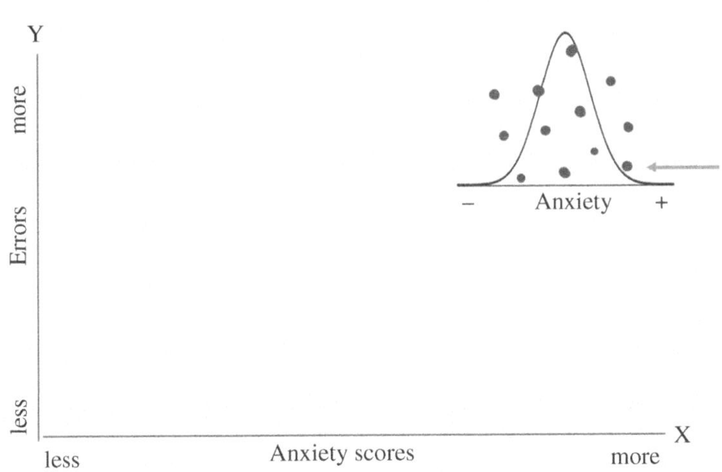


# Figure 9.6B

**Jordan's data with normal curve for anxiety**

## Figure 9.6C

**Jordan's data with normal curve for errors**

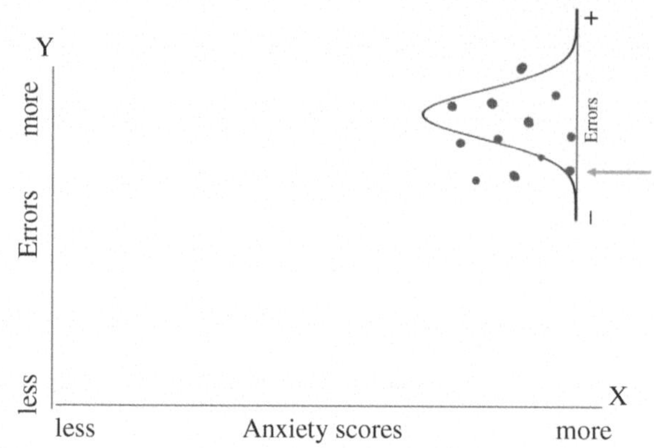


When we superimpose a normal curve on the *y* variable (errors; Figure 9.6C), we can view that data point the way Pearson *r* sees it—as relatively low on errors. That point's *z*-score on errors would be negative—it's among the lowest in the sample—even though the score is relatively high in an absolute sense (compared to all possible error scores). Therefore, this score would have a large positive *z*-score for *x* and a large negative *z*-score for *y* which together would yield a relatively large and negative *z*-score cross product (see Formula 9.1). That cross product would actually count as evidence for a *negative correlation* between anxiety and errors! If we computed all the *z*-score cross products from Jordan's data set there would be some positive, some negative, they would cancel each other out and produce a sum of cross products close to zero. That is why these scores—despite being high on both scales in an absolute sense—would produce a very small value of Pearson *r*.

Note that this truncation of range problem was a self-inflicted error on the part of the researcher. Jordan thought he was making his research more relevant to the population he wanted to understand by selecting only those high on anxiety, when in fact he was creating a sample that was doomed to be insensitive to differences in that group. If he were to repeat the study with a sample that covered a fuller range of anxiety scores, then he would have a chance to find a positive relationship between anxiety and academic performance (or any other variable).

### Correlation and Nonlinear Relationships

Pearson *r* detects only linear relationships. Thus, when *r* turns out to be surprisingly small, the possibility remains that the two variables are related, even closely related, but not in a linear fashion. Indeed, it is possible for two variables to be perfectly related yet to have an *r* equal to zero. In Figure 9.7, for example, you can predict *Y* without error given *X*, but *not* linearly.

The *curvilinear relationship* depicted in Figure 9.7 represents an exception to the usual situation in which restricting the range of either or both of the variables reduces the magnitude of their linear correlation. In this example, restricting the range of *X* to

## Figure 9.7

**Example of Perfect Relationship Between Two Variables Where $r = 0$**

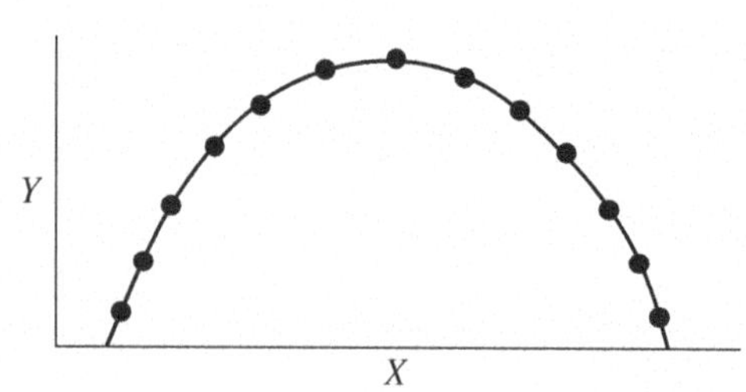

values that are to the left of the center would result in a fairly high positive correlation, whereas restricting $X$ to values to the right of the center would yield an equally strong negative correlation. When $X$ is not restricted, the correlations for each half of the range of $X$ average out to an $r$ near zero. Looking for evidence of a curvilinear relationship is another good reason for inspecting the scatter plot that corresponds to a correlation in which you are interested, especially if the Pearson $r$ is considerably lower than you expected it to be.

### Correlation and Bivariate Outliers

Yet another important caution regarding the interpretation of Pearson's $r$ concerns the disproportionate effect that a single case can have on $r$ if that case represents a *bivariate outlier*. To illustrate this effect, we have reproduced Figure 9.5 as Figure 9.8 with one important difference: We changed the GPA of the student with the highest SAT score (725) from 3.0 to 1.0. A glance at Figure 9.8 makes it easy to identify this case as a bivariate outlier. A case need not be very extreme on either variable to be a bivariate outlier, as long as that case represents an unusual combination of values (e.g., a 6-feet-tall person who weighs 110 pounds). Changing the GPA of this one particular student results in the Pearson $r$ decreasing from .65 to .324. One of the main reasons for creating a scatter plot of your data is to look for such cases.

Bivariate outliers are often, but not always, due to mistakes in recording or entering the data, or to special circumstances involving the participant in question. For example, on further investigation, you might discover that the student with the 725 SAT score got a 1.0 GPA because they fell ill that semester and decided to stay enrolled instead of taking a medical leave of absence. It is acceptable to remove an outlier from your data when you find an objective reason to do so (e.g., you know that a participant was confused about the experiment's instructions), but it is certainly not acceptable to discard a data point merely because it *is* a bivariate outlier, even though it might be tempting to discard it if it would move your statistical results in the direction of your research hypothesis. Note that a single data point can also increase the magnitude of $r$; without the student who obtained a 1.2 GPA and scored 220 on his or her SAT, the correlation calculated in Table 9.2 would have been only .25 for the remaining 24 students. Later in this chapter, we introduce a version of correlation, called Spearman's rank-order correlation, which is much less influenced by extreme scores. If bivariate outliers are a problem in your data set, you should seriously consider Spearman correlation as a substitute for Pearson's $r$.

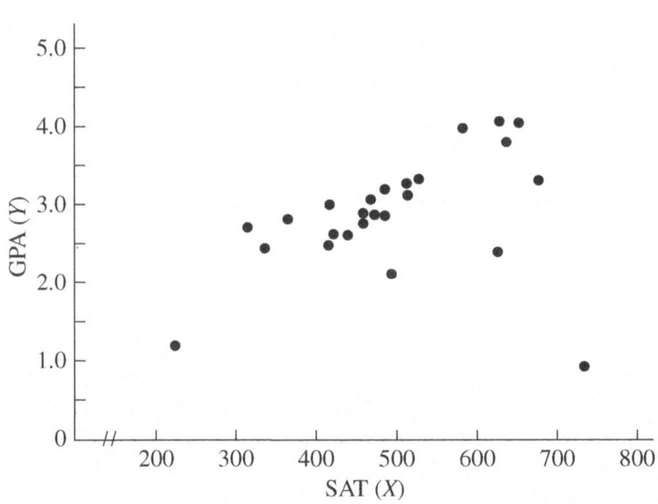


## Figure 9.8

**Scatter Plot for Data in Table 9.1 ($r$ = +.324) With One Data Point Moved to Create a Bivariate Outlier**

## When Is It Important That Pearson's $r$ Be Large?

In the next section, we discuss how to determine whether the $r$ for a sample differs significantly from zero. However, there are a number of important uses for $r$ for which a fairly small $r$, even if statistically significant at a tiny alpha level (e.g., .001), would be quite disappointing. These include measures of reliability and validity, as discussed in the following paragraphs.

### Reliability

The simplest form of reliability involves measuring participants twice with the same scale (often in the form of a self-report questionnaire), always with the (approximately) same time interval between the two measurements. The Pearson $r$ between these two sets of measurements is a measure of **test–retest reliability**. When measuring some trait that is expected to be stable over time (e.g., spatial aptitude, degree of introversion), a scale should have a test–retest reliability that is at least about $r = 0.7$. Otherwise, it would seem that the scale is not measuring the trait in question reliably enough to be trusted. Often, researchers will then modify the scale in an attempt to improve its reliability.

Another commonly used measure of reliability involves the *internal consistency* of the scale. One way to assess this attribute of a scale is to average all of the odd-numbered items together to create an "odd" score and, in an analogous fashion, create an "even" score. The Pearson $r$ between the odd and even scores across all participants is a measure of the scale's **split-half reliability**—an indication of the degree to which the scale is measuring only one distinct trait. Although it is easier to explain split-half reliability, more sophisticated measures of a scale's internal consistency such as *Cronbach's alpha* are often used.

Sometimes measurements are necessarily subjective but are not based on self-reports. For instance, the dependent variable in a study could be the creativity exhibited by a child's painting. To be sure that such a quality can be measured reliably, it is necessary that the measurements of two raters—acting without knowledge of each other—be highly correlated. A low correlation would call into question whether this quality can be measured reliably and would likely spur attempts to use better methods of training to increase the **interrater reliability** of the scale. High interrater reliability can also be important when preparing stimuli for an experiment. A common example is rating the attractiveness of yearbook photos so that the attractiveness of the photo can be controlled on any given trial by the experimenter. Without a high degree of interrater reliability, the experimenter cannot trust the attractiveness ratings of the photo stimuli. (If the raters find it too difficult to assign creativity or attractiveness scores but can place the paintings or photos in rank order, a nonparametric measure of correlation can be used, as described in the last part of this chapter.)

### Validity

Once the reliability of a measurement scale has been established—that is, we know that the scale measures just *one* quality and does so in a stable, reproducible fashion—our attention would turn to establishing whether our scale is measuring what we think it is measuring. For example, items in a questionnaire are often chosen for their obvious **face** (or content) **validity (e.g., an anxiety scale that contains an item that asks how tense you have been feeling)**, but then it is important to confirm these choices by using some objective criteria. One possible goal for a new scale is to measure some human traits in a way that is quicker, less expensive, or more convenient when compared to some well-established measures. In order to replace a traditional scale, a high Pearson $r$ needs to be found between the new and old scales after both are administered to the same set of participants. We do not expect such a measure of **criterion validity** to be larger than the reliability of the new scale, but it should approach that magnitude.

When trying to measure a quality for which no traditional scale exists, a desirable way to establish criterion validity involves correlating the new scale (after establishing its reliability) with some objective measure of actual behavior, if possible. For example, a new scale for measuring generosity could be validated by asking participants (who had already filled out the generosity questionnaire in some different, unrelated context) to contribute their time to some future experiment or to give back part of the money they were given for their participation so that "additional participants could be recruited after the grant runs out." We would not expect our new scale's correlation with an objective criterion to be larger than 0.7, as we would expect for a reliability measure. But we would not be satisfied merely to draw conclusions based on a significance test of whether Pearson $r$ is zero or not. By contrast, when investigating the relationship between two variables that are not obviously connected, except on the basis of the researcher's theory, statistical significance can be very encouraging. We turn to this topic next.

## Testing the Significance of the Correlation Coefficient

The correlation coefficient of .65 for the group of 25 students whose scores are shown in Table 9.1 conveniently describes the linear relationship between SAT scores and GPA *for this group of students*. It would be useful to know, however, whether these two variables are correlated in the population of all students. To answer this question, we must draw an inference about likely values of the **population correlation coefficient** (symbolized by rho, $\rho$). Is it likely that there is actually no correlation in the population and that the correlation in the sample of 25 students was due to sampling error (the cases that happened to wind up in the sample)? Or is the value of .65 large enough for us to conclude that there is some nonzero positive correlation between SAT scores and GPA in the population? The strategy for testing hypotheses about likely values of $\rho$ is similar to that used in previous chapters. The null hypothesis most often tested is that $\rho$ is equal to zero[3]:

$$H_0 : \rho = 0$$
$$H_A : \rho \neq 0$$

A criterion of significance, such as the .05 or .01 criterion, is selected. The correct statistical model to use in this situation is the $t$ distribution, so the critical value for $t$ can be found with degrees of freedom ($df$) equal to $N - 2$. Then $H_0$ is retained if the computed $t$ is less than or equal to the critical value of $t$ from the $t$ table. Otherwise, $H_0$ is rejected in favor of $H_A$, and $r$ is said to be *significantly different from zero* (or, simply, *statistically significant*).

Fortunately, you do not need to compute the $t$ ratio[4] for testing Pearson's $r$. This has (in a sense) already been done for you by statisticians who have constructed tables of significant (i.e., *critical*) values of $r$. Therefore, the procedure for testing a correlation coefficient for statistical significance can be quite simple: You compare your computed value of $r$ to the value of $r$ shown in Table C in the Appendix at the end of the book for $N - 2$ $df$, where $N$ is equal to the number of pairs. If the absolute value of your computed $r$ is smaller than or equal to the tabled value, retain $H_0$; otherwise, reject $H_0$.

As an illustration, suppose that you wish to test the significance of the correlation between SAT scores and GPAs, using $\alpha = .05$. Referring to Table C, you find that for $25 - 2$ or 23 $df$, the smallest absolute value of $r$ that is statistically significant is .396. Since the obtained $r$ of .65 exceeds this value, you reject $H_0$ and conclude that there *is* a positive correlation in the population from which the sample of 25 students was randomly selected. Note that a correlation of $-.65$ would indicate a statistically significant negative relationship; due to the symmetry of the $t$ distribution, the sign of $r$ can be ignored when comparing the computed $r$ to the critical value of $r$. Thus, the test described is the usual two-tailed test—it guards against chance departures of $r$ from zero in either the positive or negative direction.

---

[3] Other null hypotheses are possible, but a different statistical procedure is required in order to test them. See, for example, B. H. Cohen (2008), pp. 281–282.

[4] The formula for the $t$ ratio to test the significance of a Pearson $r$ is

$$t = \frac{r\sqrt{N - 2}}{\sqrt{1 - r^2}}$$

where $N$ = number of *pairs* of scores.

As we observed in Chapters 6 and 7, confidence intervals provide valuable additional information about the probable values of a population parameter. However, the procedures for calculating confidence intervals for correlation coefficients are beyond the scope of this book (see B. H. Cohen, 2008, pp. 280–281).

### Implications of Rejecting $H_0$

If a Pearson $r$ is statistically significant, the significance denotes *some* degree of linear relationship between the two variables in the population. It does *not* denote a "significantly high" or "significantly strong" relationship; it just means that the relationship in the population seems unlikely to be zero.

By inspecting Table C, you can see that the larger the sample size, the smaller the absolute value of the correlation coefficient that is needed for statistical significance. For example, for $N = 12$ ($df = 10$), a correlation larger than .576 is needed for significance using the .05 criterion. For $N = 102$ ($df = 100$), a correlation of only .196 or larger is needed. This implies that the importance of obtaining statistical significance can easily be exaggerated. Suppose that for a very large sample (say, $N = 1000$), a statistically significant Pearson $r$ of .08 is obtained. This statistical test suggests that $\rho$ is greater than zero. But the obtained $r$ of .08 is so close to zero that the relationship in the population, although not zero and not necessarily equal to .08, is likely to be *very* weak (i.e., likely to be quite close to .08)—too weak, in fact, to allow you to make any accurate statements about one variable based on knowledge of the other. Consequently, such a finding would add little or nothing to our practical knowledge, even though it is statistically significant. (It might conceivably be useful in appraising a theory.) Therefore, you need to have *both* statistical significance *and* a reasonably high absolute value of $r$ before you can conclude that $\rho$ is large enough to be useful in applied work.

Correlation coefficients *cannot* be interpreted as percents. For example, you *cannot* conclude that a correlation of .60 is 60% of a perfect relationship, or that it is twice as much as a correlation of .30. As we will see in the next chapter, however, the *squared* correlation coefficient ($r^2$) does permit an interpretation, in percentage terms, of the strength of the relationship between the two variables.

As usual, when $H_0$ is true, the probability of erroneously rejecting it (a Type I error) is equal to the criterion of significance that is selected. Using $\alpha = .05$, when $H_0$ is true, you will, nonetheless, reject it 5% of the time.

### Assumptions Underlying the Use of $r$

A common assumption underlying the use of the Pearson $r$ is that $X$ and $Y$ are linearly related. If $X$ and $Y$ have a curvilinear relationship (as in Figure 9.7), the Pearson $r$ will not detect it. Pearson's $r$ can be calculated, of course, whether or not the best-fitting function is linear. Ordinarily, however, one would not be interested in the best linear fit when it is known that the relationship is not linear.

Otherwise, no assumptions are made at all in using $r$ to *describe* the degree of linear relationship between two variables for a given set of data. When testing a correlation coefficient for statistical significance, it is, strictly speaking, assumed that the underlying distribution is the so-called bivariate normal—that is, scores on $Y$ are normally distributed for each value of $X$, and scores on $X$ are normally distributed for each value of $Y$. However, when the degrees of freedom are greater than 25 or 30, failure to meet this assumption has little consequence for the validity of the test unless your data contain some extreme bivariate outliers—in which case an even larger sample may be required for an accurate significance test.

## The Relationship Between Two Ranked Variables: The Spearman Rank-Order Correlation Coefficient

As we saw with the rank-sum and Wilcoxon matched-pairs tests in the previous chapter, there are times when the assumptions of our parametric tests are not close to being met and we need to consider other options. We now turn our attention to the ordinal alternative to the Pearson $r$, which is called the **Spearman rank-order correlation coefficient**, symbolized as $r_S$.[5] As in the case with other ordinal tests, $r_S$ is commonly used when a dependent variable has been measured precisely, but its distribution makes the use of parametric tests questionable, and the sample size is not large. Even if only one of the two variables that you are correlating (e.g., years of education post-middle school and annual income) seems to violate the distributional assumptions of parametric tests, both variables must be converted to ranks in order to calculate $r_S$. You *cannot* replace one variable with ranks and correlate it with the raw scores of the other variable, and obtain a meaningful correlation.

### Rationale and Computational Procedures

Suppose you wish to determine the relationship between the years of education post-middle school and incomes of 10 entrepreneurs (as in Table 9.3). The income values are extremely skewed, while the years of education post-middle school appear to have a normal distribution. To compute $r_S$ (by hand) you must rank order *both* income *and* years of education. The entrepreneur with the least years of education post middle school gets a rank of 1 for years of education, the next to lowest is assigned the rank of 2, and so on. These ranks then replace the original years of education post middle school, which are no longer used. The same procedure is used to determine rank order by income. Finally, $r_S$ is calculated by applying any formula for the Pearson $r$ to the *ranks* rather than to the original scores.

To make sure that your ranks are properly paired, you can start by listing your participants (in any order), along with their scores on both variables. Then list the participants in years of education order separately, and assign ranks. Next to each year of education score in the original table, write the appropriate rank. Again in a separate place, list the participants in order by income and assign ranks, and then write each participant's income rank next to their income in the original table (see Table 9.3). You can then apply any Pearson $r$ formula to the two sets of ranks or use a shortcut formula for $r_S$ after calculating the difference for each pair of ranks and then summing the squared differences, as shown in Table 9.3.

The shortcut formula for the Spearman correlation coefficient is:

$$r_s = 1 - \frac{6\sum D^2}{N(N^2 - 1)} \qquad \text{Formula 9.4}$$

**Table 9.3** **10 Entrepreneurs Both Measured and Ranked for Year of Education and for Annual Income (1 = Least, 10 = Most)**

| Entrepreneur's first name | Years of education | Ranks for years of education | Annual income (in thousands of dollars) | Ranks for income | $D$ $(X - Y)$ | $D^2$ $(X - Y)^2$ |
|---|---|---|---|---|---|---|
| Carmen | 5 | 3 | 97 | 4 | −1 | 1 |
| Ash | 4 | 2 | 72 | 1 | 1 | 1 |
| Martha | 8 | 5 | 223 | 6 | −1 | 1 |
| William | 14 | 9 | 255 | 7 | 2 | 4 |
| Tyron | 2 | 1 | 88 | 3 | −2 | 4 |
| Sophia | 16 | 10 | 974 | 10 | 0 | 0 |
| Seynab | 12 | 8 | 650 | 9 | −1 | 1 |
| Manuel | 6 | 4 | 85 | 2 | 2 | 4 |
| Malik | 10 | 7 | 148 | 5 | 2 | 4 |
| Casey | 9 | 6 | 403 | 8 | −2 | 4 |
| | | | | | | $\sum D^2 = 24$ |

[5] The Greek letter rho, ρ, has been traditionally used to designate the rank-order correlation coefficient. In order to avoid confusion between this statistic and the population parameter for the Pearson correlation coefficient, which is also symbolized by rho, we will use $r_S$ for the Spearman coefficient.

where

$D$ = difference between a pair of ranks
$N$ = number of pairs

If you applied one of the formulas for the Pearson $r$ (as described earlier in this chapter) to rank-order data by simply treating the ranks as raw scores, and if there were no tied ranks, you would get exactly the same value for $r_S$ as you would get from Formula 9.4. (The short-cut formula for $r_S$ takes advantage of certain relationships that hold when the $N$ scores are just the integers from 1 to $N$.)

Like $r$, then, $r_S$ varies between −1.0 and 1.0. A high positive value of $r_S$ indicates a strong tendency for the paired ranks to be equal. Conversely, a high negative value of $r_S$ indicates a strong tendency for the paired ranks to be opposite, as would occur if entrepreneurs with higher ranks for years of education post-middle school were to have lower ranks for income. A zero value would indicate no relationship between the two sets of ranks.

In our example, the calculation of $\sum D^2$ is shown in Table 9.3; $r_S$ is equal to:

$$r_s = 1 - \frac{6(24)}{10(10^2 - 1)}$$
$$= 1 - \frac{144}{990}$$
$$= 1 - 0.15$$
$$= 0.85$$

### Dealing With Tied Ranks

In the preceding example, if Ash and Carmen had stayed in education for the same amount of time after middle school, both would share second place. This would be expressed numerically by *averaging* the two ranks in question and assigning this average as the rank of each person, just as was done in the ordinal tests described in the previous chapter (see **An Example of Ranking Data With Ties** in Chapter 8). The ranks in question are 2 and 3, so the average of these ranks, or 2.5, is the rank that would be given to both Ash and Carmen, and the next entrepreneur would receive a rank of 4. When there are tied ranks, Formula 9.4 will overestimate the absolute value of $r_S$. But unless there are many ties, and particularly long ties (i.e., three- or more-way ties), this overestimate is likely to be trivial (less than .01 or .02).

### Testing the Significance of the Rank-Order Correlation Coefficient

Although the formula for $r_S$ and the formula for the Pearson $r$ yield the same numerical result in the absence of tied ranks, the significance of the two coefficients *cannot* be tested in the same manner when $N$ is fairly small (especially when it is less than 10). To test the null hypothesis that the ranks are independent in the population from which the sample was drawn, you can use Table I in the Appendix at the end of the book. The minimum values of $r_S$ needed for statistical significance are shown in the table for values of $N$ from 5 to 30. (Note that when using this table, you need only refer to $N$—the number of pairs of ranks—rather than degrees of freedom.)

To use Table I for the problem involving the data in Table 9.3, you would look up the minimum value necessary for $r_S$ to be statistically significant when $N = 10$. Using the two-tailed .05 criterion, this value is .648. Since the absolute value of the obtained $r_S$ (.85) is greater than the tabled value, you can reject $H_0$ and conclude that there is a nonzero relationship between the two sets of ranks in the population.

When $N$ is greater than 30, the critical values for testing a Pearson $r$ for statistical significance will give a very good approximation. That is, you can refer the computed Spearman correlation coefficient to Table C in the Appendix with $N - 2$ degrees of freedom.

This test of significance is not, in general, as *powerful* as the test for the Pearson $r$. That is, it is *not* as likely to detect nonzero relationships. The situation reverses, however, when one of a pair of variables is substantially skewed; transformation into ranks followed by the computation of $r_S$ may well provide a better and more powerful measure of their correlation than the Pearson $r$ computed on the original data. This is the case for the data in Table 9.3. If you calculate the Pearson $r$ for the years of education and income scores, you will see that it equals .78, which is less than $r_S$ (.85).

### When to Use the Spearman Correlation Coefficient

In some instances, the data for both of your variables may be in the form of ranks. Suppose that you wish to determine whether the creativity of abstract paintings can be judged in a reliable way. You ask two art instructors to independently rank the same 10 paintings in order of the creativity they express. The Pearson $r$ for these two sets of ranks *is* a Spearman correlation coefficient (by definition), and therefore its critical value may be found in Table I (unless the number of rank-ordered paintings grows larger than 30). Even if only one of the two variables has been measured ordinally (e.g., you are correlating the creativity ranks from the first art instructor with the price being asked for each painting), it is necessary to compute $r_S$, so you will have to convert the measurements on the second variable to ranks before computing the correlation.

If both variables have been measured on interval/ratio scales and you are worried about the distribution of both variables in the population—or even just one of them, as in the year of education/income example—both sets of measurements should be converted to ranks and $r_S$ should be computed. Even if each variable follows an approximately normal distribution, if the relationship between them follows a monotonic curve (e.g., the curve begins steeply and then levels off but never reverses its direction) rather than a straight line, the Pearson $r$ of the ranks (i.e., $r_S$) will be larger than the Pearson $r$ of the raw scores. This is true because $r_S$ measures the degree to which the relationship of two variables is monotonic (e.g., whenever one variable increases the other variable increases), whereas the Pearson $r$ applied to the raw scores measures the degree to which the relationship conforms to a single, straight line.

Even when you have interval data that are normally distributed for both variables, Spearman correlation can be a good alternative to Pearson's if you have outliers. For example, imagine that instead of making $974 thousand each year, Sophia (in Table 9.3) made $974 *million*. From the perspective of Pearson correlation (think *z*-score formula), Sophia's *z*-score for income is going to change—a lot! Pearson's $r$, therefore, will also be affected considerably by the extreme score. But if you are just ranking scores, Sophia's position does not change when you add three zeros to her income. She is still ranked #1. For this reason, Spearman $r_S$ is not affected by the extreme score. Thus, Spearman correlation, like the median, the interquartile range, and other *robust* (less quantitatively sensitive) statistics, is not affected by outliers because it does not "know" how extreme the score is. If your data set is being spoiled by one or two outliers, consider using Spearman correlation to minimize their effect.

## Summary

*Correlation* refers to the co-relationship between two variables. The Pearson $r$ coefficient is a useful measure of *linear* correlation.

### The Pearson Correlation Coefficient

*z*-score product formula:

$$r = \frac{\sum z_X z_Y}{N} \quad \text{Formula 9.1}$$

Raw score formulas:

$$r = \frac{\frac{\sum XY}{N} - \mu_X \mu_Y}{\sigma_X \sigma_Y} \quad \text{Formula 9.2}$$

$$r = \frac{\frac{1}{N-1}\left[\sum XY - N\overline{X}\overline{Y}\right]}{s_X s_Y} \qquad \text{Formula 9.3}$$

In the foregoing formulas, $N$ = number of *pairs*.

**Points to Remember**

1. The numerical value of $r$ indicates the *strength* of the relationship between the two variables in your sample, while the sign indicates the *direction* of the relationship.
2. $-1 \leq r \leq +1$.
3. The population correlation is symbolized by $\rho$.
4. $r$ *cannot* be interpreted as a percent, but $r^2$ can be (see next chapter).
5. $r$ may be tested for statistical significance by referring the obtained value to Table C with $df = N - 2$.
6. $r$ may also be used as a descriptive statistic.
7. $r$ detects only *linear* relationships.
8. Correlation does not imply causation.
9. A restricted range on one or both of your variables, or a single bivariate outlier, can lead to a correlation considerably lower than you were expecting.
10. Even a tiny, inconsequential correlation can become statistically significant with a large-enough sample.
11. When using correlation for various practical purposes, especially when measuring the various types of reliability and validity, it is important that the Pearson $r$ be not only statistically significant but large in magnitude as well.

**The Spearman Rank-Order Correlation Coefficient**

With ranked data, you can compute the Pearson $r$ for the two sets of ranks, or you can use the shortcut formula for the Spearman rank-order correlation coefficient:

$$r_S = 1 - \frac{6\sum D^2}{N(N^2 - 1)} \qquad \text{Formula 9.4}$$

where

$D$ = difference between a pair of ranks
$N$ = number of pairs

When $N < 31$, $r_S$ should be compared to a critical value from Table I in order to determine statistical significance. However, for $N > 30$, the critical values for Pearson $r$ in Table C provide a good approximation.

## Exercises

1. A college dean would like to know how well he can predict sophomore grade point average for first-semester freshmen so that students who are headed for trouble can be given appropriate counseling. After students have been in school for one semester, the dean obtains their numerical final examination average for the first semester (based on a total of 100 points) and the average number of "cuts" per class during the semester. He then waits for a year and a half, and when the students have finished their second year, he obtains their sophomore grade point average. To keep the computations down to a reasonable level, we will assume that the dean has a sample of

only seven cases. (Note that in a real study of this kind, there would be many more participants, but the same procedures would be used.)

| Student | Test score ($X$) | Cuts ($C$) | Sophomore average ($Y$) |
|---|---|---|---|
| 1 | 70 | 2 | 2.50 |
| 2 | 90 | 1 | 4.00 |
| 3 | 75 | 2 | 3.50 |
| 4 | 85 | 3 | 3.00 |
| 5 | 80 | 5 | 3.00 |
| 6 | 70 | 3 | 2.00 |
| 7 | 90 | 5 | 3.00 |
| Mean | 80 | 3 | 3.00 |
| $\sigma$ | 8.02 | 1.41 | 0.60 |

(a) Convert the test scores ($X$) and sophomore averages ($Y$) to $z$-scores. By inspection of the paired $z$-scores, estimate whether the correlation between these two variables is strong and positive, about zero, or strong and negative. Then verify your estimate by computing $r$ using the *z-score product formula*.

(b) Use (raw-score) Formula 9.2 to compute the Pearson $r$ between the number of cuts ($C$) and the sophomore average ($Y$). Can you reject the null hypothesis at the .05 level with a two-tailed test? (Use Table C.)

(c) Repeat part (b) for the correlation between the number of cuts ($C$) and the test score ($X$). Is this correlation significant at the .01 level with a *one*-tailed test?

2. The data from Exercise 7 in Chapter 7 are reproduced in the following table. Calculate the mean and the *unbiased* SDs for both the experimental and control groups, and then compute the Pearson $r$ with the raw-score formula that uses the *ss* rather than the $\sigma$s in its denominator. Is the correlation coefficient significant? Given your decision with respect to the null hypothesis, what type of error could you be making, Type I or Type II? (Advanced exercise: Recompute the matched-pairs $t$ test with the formula that is based on Pearson's $r$, and compare it to the $t$ value you calculated for these data in the previous chapter.)

| Pair | Experimental group (new GAD approach) | Control group (standard GAD approach) |
|---|---|---|
| 1 | 9 | 7 |
| 2 | 6 | 4 |
| 3 | 5 | 6 |
| 4 | 7 | 3 |
| 5 | 3 | 5 |
| 6 | 7 | 3 |
| 7 | 3 | 2 |
| 8 | 4 | 5 |
| 9 | 6 | 7 |
| 10 | 10 | 8 |

3. The data from Exercises 8 and 9 in Chapter 7 are reproduced in the tables that follow.

(a) Compute the Pearson $r$ for Data Set 3, Chapter 1, and test for significance at the .01 level.

(b) Use the $z$-scores you calculated for these data in the exercises of Chapter 4 to recompute the $r$ with the $z$-product formula. Is it the same?

(c) Compute the Pearson $r$ for Data Set 3, Chapter 1, *with X rearranged*, and test for significance at the .05 level.

**Data set 3, Chapter 1**

| *S* | *X* | *Y* |
|---|---|---|
| 1 | 97 | 89 |
| 2 | 68 | 57 |
| 3 | 85 | 87 |
| 4 | 74 | 76 |
| 5 | 92 | 97 |
| 6 | 92 | 79 |
| 7 | 100 | 91 |
| 8 | 63 | 50 |
| 9 | 85 | 85 |
| 10 | 87 | 84 |
| 11 | 81 | 91 |
| 12 | 93 | 91 |
| 13 | 77 | 75 |
| 14 | 82 | 77 |

**Same data, but with *X* rearranged**

| *S* | *X* | *Y* |
|---|---|---|
| 1 | 92 | 89 |
| 2 | 82 | 57 |
| 3 | 85 | 87 |
| 4 | 81 | 76 |
| 5 | 87 | 97 |
| 6 | 93 | 79 |
| 7 | 68 | 91 |
| 8 | 85 | 50 |
| 9 | 63 | 85 |
| 10 | 100 | 84 |
| 11 | 74 | 91 |
| 12 | 77 | 91 |
| 13 | 92 | 75 |
| 14 | 97 | 77 |

For both tables: [$\overline{X}_1 = 84$, $s = 10.76$], [$\overline{X}_2 = 80.64$, $s = 13.29$]

4. There are 10 participants in a problem-solving experiment. Two judges, Steve and Annie, are asked to rank the solutions with regard to their creativity (1 = most creative, 10 = least creative). The experimenter wishes to know if Steve and Annie are in substantial agreement. Following are the rankings; what should the experimenter decide?

| Participant | Steve | Annie |
|---|---|---|
| 1 | 5 | 4 |
| 2 | 7 | 9 |
| 3 | 2 | 2 |
| 4 | 9 | 8 |
| 5 | 1 | 3 |
| 6 | 4 | 1 |
| 7 | 10 | 10 |
| 8 | 3 | 7 |
| 9 | 6 | 5 |
| 10 | 8 | 6 |

5. (a) Convert the data in part (a) of Exercise 3 to ranks, separately for each variable, in order to compute the Spearman rank-order correlation coefficient. Test $r_S$ for statistical significance by using Table I. Compare $r_S$ to the Pearson $r$ that you calculated for Exercise 3, part (a).
   (b) Redo part (a) for the data in part (c) of Exercise 3 (i.e., *with X rearranged*).

## Thought Questions

1. For each of the following, state whether you would expect the Pearson $r$ correlation between $X$ and $Y$ to be positive, zero, or negative. Assume that each correlation is based on a sample of 50 participants who have scores on both $X$ and $Y$. (a) $X$ = grades in a high school advanced placement psychology course, $Y$ = scores on the advanced placement psychology test. (b) $X$ = intelligence, $Y$ = size of big toe on right foot. (c) $X$ = number of rounds of golf played, $Y$ = average golf score per 18 holes. (d) $X$ = number of hours spent watching television per week, $Y$ = grades in high school. (e) $X$ = self-esteem (low number indicates low self-esteem), $Y$ = depression (higher number indicates more depression). (f) $X$ = number of hours spent studying for a test, $Y$ = number of mistakes made on that test.
2. A student obtains scores of 72 on Test X and 72 on Test Y. The student therefore concludes that there is a high correlation between Test X and Test Y. Why is this conclusion incorrect?
3. In a correlational study, $X$ is the number of hours of violent television programs that participants watch, and $Y$ is the number of violent acts committed by the participants in real life. Suppose that there is a moderately high correlation (say, .48) between $X$ and $Y$ for a sample of 100 American males and that this result is statistically significant. Explain why we *cannot* infer causation from a correlational study by showing that each of the following is possible: (a) $X$ could cause $Y$. (b) $Y$ could cause $X$. (c) The relationship between $X$ and $Y$ could be caused by a third variable. (Hint: Consider physiological causes of violent behavior.)
4. In a research study using a sample of 30 participants, the Pearson $r$ correlation between $X$ and $Y$ is .09 and is *not* statistically significant. Should the researcher conclude that there is little or no relationship between $X$ and $Y$? Why or why not?
5. The SAT is used to predict the success of high school students in college. Suppose that the Pearson $r$ correlation between scores on the SAT and grades at University Z during the freshman year, based on a sample of 500 participants, is +.53. A student complains that a friend of hers did poorly on the SAT but was admitted to University Z anyway and earned high grades during the first year, while another friend did very well on the SAT but flunked out of University Z in the first year. Should the student conclude from these two results that the SAT is inaccurate and worthless? Why or why not?
6. Using a very large sample of participants, a statistically significant Pearson $r$ correlation of .11 is obtained between $X$ and $Y$. Does this mean that there is a strong relationship between $X$ and $Y$? Why or why not?
7. (a) What is the difference between the reliability of a measure and the validity of a measure? (b) What is the difference between test–retest reliability and split-half reliability? (c) What is criterion validity?
8. Describe three situations in which you would consider using Spearman correlation instead of Pearson correlation.

## Computer Exercises

1. (a) Use your statistical package to compute the Pearson $r$ between baseline and prequiz heart rates for all students; also, find the Pearson $r$ between the pre- and postquiz heart rate measurements. (b) Recalculate these two $r$'s separately for those who exercise regularly and those who do not.
2. Compute the Pearson $r$ between baseline heart rate and baseline anxiety for all students; also, find the Pearson $r$ between prequiz heart rate and prequiz anxiety measurements.
3. Create two new variables: prequiz heart rate minus baseline heart rate and prequiz anxiety minus baseline anxiety. Compute Pearson's $r$ between these two difference scores, and interpret the meaning of this correlation.
4. Use your statistical package to create a scatter plot of the relation between the academic interest score and the statistics quiz score for the 85 students who have both scores. Describe the pattern that you see. If possible, use your statistical package to add a linear regression line to the scatter plot.
5. In Exercise 7 of the regular (not computer) exercises in the previous chapter, you were asked to calculate the Wilcoxon matched-pairs test to determine whether a training method improved exam scores for 12 poorly performing students. For this exercise, compute the Spearman correlation between the pre- and post-test scores, and test for significance at the .05 level. How does the magnitude of the Spearman correlation relate to the power of the Wilcoxon test?
6. (a) Separating students that exercise regularly and those that do not, compute the Spearman correlation coefficient between the baseline and the prequiz heart rates. How do these results relate to those you obtained in part (b) of the first exercise? How do the corresponding $p$ values compare? (b) Compute the Spearman correlations for the corresponding anxiety scores, as well.

## Bridge to SPSS

Both the Pearson $r$ and the Spearman $r_s$ can be obtained from the same dialog box in SPSS, as described next.

### Linear Correlation

To compute the Pearson $r$ in SPSS, select **Correlate** from the **Analyze** menu, and then choose **Bivariate** …. In the dialog box that opens (see Figure 9.9), move the variables you wish to see correlated to the area labeled *Variables:* to the right of the variable list, and click **OK**. You can select as few as two variables from the list or as many as all of the numeric variables. Under *Correlation Coefficients*, the *Pearson* box comes already checked as the default. This **Bivariate Correlations** dialog box also allows you to select *Kendall's tau-b* (not discussed in this text and rarely used) and the *Spearman* correlation coefficient instead of, or in addition to, the Pearson $r$. Under *Tests of Significance*, you can select one-tailed rather than the default choice of two-tailed $p$ values (not recommended). Another choice is to suppress (by unchecking) the "flagging" of significant correlations, but especially with a large matrix of correlations the flagging is useful (one asterisk indicates $p$s less than .05 and a pair of asterisks indicates $p$s less than .01).

By default, your output will contain a single matrix (see Figure 9.10) with the variables you chose as the labels of the rows, as well as of the columns; the same Pearson $r$ will appear twice for each possible pair of variables from your selected list (the matrix also contains a diagonal, in which each variable is correlated with itself, always yielding a value of 1). You can suppress the redundant correlations and the diagonal by checking the box labeled *Show only the lower triangle* in the dialog box. The default two-tailed $p$ value (labeled "Sig.," of

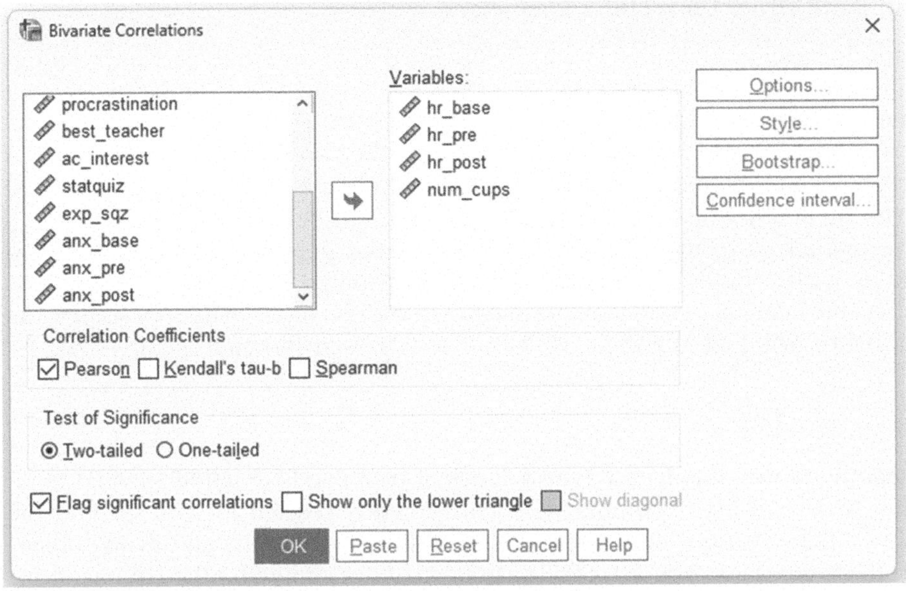


**Figure 9.9**

**The Dialog Box for Bivariate Correlations**

*Source:* Courtesy of International Business Machines Corporation, © SPSS, Inc./IBM company.

course) appears under each Pearson $r$, with the sample size (i.e., the number of pairs, labeled "$N$") under that.

Note that if any of the variables in your list has missing values, the $N$s will not be the same for all of the correlations. For a pair of variables, $N$ cannot be more than the number of participants who have actual (nonmissing) values for both variables. If you want to ensure that all of your correlations are based on the same $N$, you can click on the **Options** ... button in the dialog box and choose to *Exclude cases listwise* instead of pairwise. However, in that case, the $N$ for each correlation will equal the number of cases that do not have a missing value for *any* variable in the selected list. In other words, any case (e.g., participant) with any missing values on any variable on the list will be excluded from every correlation in the matrix. The **Options** ... button also allows you to add a table of descriptive statistics to your output (not shown in Figure 9.10).

**Correlations**

| | | hr_base | hr_pre | hr_post | num_cups |
|---|---|---|---|---|---|
| hr_base | Pearson Correlation | 1 | .483** | .411** | .505** |
| | Sig. (2-tailed) | | <.001 | <.001 | <.001 |
| | N | 100 | 100 | 100 | 100 |
| hr_pre | Pearson Correlation | .483** | 1 | .527** | .461** |
| | Sig. (2-tailed) | <.001 | | <.001 | <.001 |
| | N | 100 | 100 | 100 | 100 |
| hr_post | Pearson Correlation | .411** | .527** | 1 | .404** |
| | Sig. (2-tailed) | <.001 | <.001 | | <.001 |
| | N | 100 | 100 | 100 | 100 |
| num_cups | Pearson Correlation | .505** | .461** | .404** | 1 |
| | Sig. (2-tailed) | <.001 | <.001 | <.001 | |
| | N | 100 | 100 | 100 | 100 |

**. Correlation is significant at the 0.01 level (2-tailed).

**Figure 9.10**

**The Full Matrix of (Redundant) Pearson Correlations**

### The Spearman Rank-Order Correlation Coefficient

Surprisingly, the Spearman correlation cannot be obtained from the **Analyze/Nonparametric Tests** menu, although it could have easily been included in the procedure for two related samples. Instead, as shown in Figure 9.9, **Spearman** is one of the choices when you open the dialog box for **Bivariate Correlations**. Unless the data for the two variables to be correlated have already been entered in terms of ranks, the Spearman correlation coefficient (labeled **Spearman's rho** by SPSS) will almost always differ from the Pearson *r*, as will its "Sig." value. However, either correlation coefficient can be larger, depending on how the data are distributed for each variable and how the data are distributed bivariately (e.g., extreme outliers will affect the Pearson *r* more than the Spearman rho). In terms of its appearance, the default output matrix will look the same as the one for Pearson shown in Figure 9.10.

### Using the Syntax Window

Before there was Windows, one had to type a series of commands to tell SPSS what to do with your data. Now that list of commands is conveniently created for you, behind the scenes, when you make various selections from the pull-down menus and dialog boxes. Depending on the preferences you set, SPSS may or may not show you the command list you created when it displays your output. However, whether you see it or not, SPSS is creating a Syntax file from your command selections, which can be displayed and even modified, in a Syntax Window. This is the third window (along with the Data window and the Output window) that can be saved as a file for future use (the automatic extension given by SPSS is ".sps"). There are two major uses for the Syntax window: saving a long list of complex commands (e.g., computing many new variables) that you may want to repeat in the future (even if you may need to modify them slightly before running them again); and accessing program features that are not included in any of the SPSS menus (obviously, these are features that the SPSS company believes are potentially useful, but not used very often). Because it is relevant to computing and displaying correlations, we will describe one of these not-available-by-menu features next.

For many purposes, it is easier not to type a syntax file from scratch but rather to create one by first making selections in the relevant dialog box and then clicking on **Paste** instead of **OK**. As an example, we will show you how to modify a syntax file for performing correlations so that you do not produce the usual square matrix, when you do not want to see every variable correlated with every other variable in the selected list. You may have in mind a particular target variable and want to see which other variables correlate highly with it. In Figure 9.9, we set up the list of selected variables with the target variable, num_cups, at the end, preceded by the three heart rate measures, in order to see how well each HR measure correlates with the amount of coffee that was drunk before the stats quiz experiment. If we do not want to be bothered seeing any of the pairwise correlations among the HR measures, we would not click OK in that box; instead, we would click on the **Paste** button. A Syntax window will then open automatically, containing the command list needed to perform the analyses you specified in the dialog box.

For instance, if we were to click **Paste** in the dialog box shown in Figure 9.9, the following commands would appear in a Syntax window:

```
CORRELATIONS
/VARIABLES hr_base hr_pre hr_post num_cups
/PRINTTWOTAIL NOSIG
/MISSINGPAIRWISE.
```

Note that this file consists of just one main command, called **Correlations**, followed by several (default) subcommands, each of which must begin with a slash; the command must be closed by including a period at the end of the last subcommand. Running this command would produce the usual square matrix containing 4 × 4 or 16 correlations. However, if you

**Correlations**

| | | num_cups |
|---|---|---|
| hr_base | Pearson Correlation | .505** |
| | Sig. (2-tailed) | <.001 |
| | N | 100 |
| hr_pre | Pearson Correlation | .461** |
| | Sig. (2-tailed) | <.001 |
| | N | 100 |
| hr_post | Pearson Correlation | .404** |
| | Sig. (2-tailed) | <.001 |
| | N | 100 |

**. Correlation is significant at the 0.01 level (2-tailed).

# Figure 9.11

**A Single Column of Pearson Correlations Produced by Running an Appropriately Modified Syntax File**

want SPSS to display only three correlations—num_cups with each of the HR measures—you can add the keyword **with** just before the last variable, so that the Variables subcommand looks like this:

/VARIABLES hr_base hr_pre hr_post **with** num_cups

As you might guess, **with** is a "reserved" word in SPSS and therefore cannot be used as the name of a variable. Running the syntax file after adding the keyword **with** produces an output table consisting of a single column (see Figure 9.11). You would get the same three correlations in your output if you were to put your target criterion variable at the *beginning* of the variables list followed by **with**, except that the output table would be in the form of a single row instead of a single column.

## Appendix: Equivalence of the Various Formulas for *r*

$$\frac{\sum z_X z_Y}{N} = \frac{\sum (X - \mu_X)(Y - \mu_Y)}{N\sigma_X\sigma_Y}$$

*PROOF:* By definition,

$$z_X = \frac{X - \mu_X}{\sigma_X},\ z_Y = \frac{Y - \mu_Y}{\sigma_Y}$$

Therefore,

$$r = \frac{\sum \left(\frac{X - \mu_X}{\sigma_X}\right)\left(\frac{Y - \mu_Y}{\sigma_Y}\right)}{N}$$

$$= \frac{\sum (X - \mu_X)(Y - \mu_Y)}{N\sigma_X\sigma_Y} \qquad \text{((Rule8, Chapter1))}$$

$$\frac{\sum (X - \mu_X)(Y - \mu_Y)}{N\sigma_X\sigma_Y} = \frac{\frac{\sum XY}{N} - \mu_X\mu_Y}{\sigma_X\sigma_Y}$$

*PROOF:* Expanding the numerator yields

$$\sum(X-\mu_X)(Y-\mu_Y) = \sum(XY - X\mu_Y - \mu_X Y + \mu_X\mu_Y)$$
$$= \sum XY - \mu_Y\sum X - \mu_X\sum Y + N\mu_X\mu_Y$$

(Rules 1, 2, 5, and 8, Chapter 1)

Substituting $(\sum X)/N$ for $\mu_X$ and $(\sum Y)/N$ for $\mu_Y$ yields

$$\sum XY - \frac{\left(\sum Y\right)\left(\sum X\right)}{N} - \frac{\left(\sum X\right)\left(\sum Y\right)}{N} + \frac{N\left(\sum X\right)\left(\sum Y\right)}{(N)(N)}$$
$$= \sum XY - 2\frac{\left(\sum X\right)\left(\sum Y\right)}{N} + \frac{\left(\sum X\right)\left(\sum Y\right)}{N}$$
$$= \sum XY - \frac{\sum X\sum Y}{N}$$

Dividing the numerator by $N$ yields

$$\frac{\sum XY}{N} - \frac{\sum X\sum Y}{N\times N} = \frac{\sum XY}{N} - \left(\frac{\sum X}{N}\right)\left(\frac{\sum Y}{N}\right) = \frac{\sum XY}{N} - \mu_X\mu_Y$$

Now that we have divided the numerator by $N$, we must do the same to the denominator:

$$\frac{N\sigma_X\sigma_Y}{N} = \sigma_X\sigma_Y$$

Put the transformed numerator over the transformed denominator to obtain the raw-score formula used in Table 9.2. This formula is sometimes called a raw-score *definition* formula; the terms in the formula are meaningful, but the format may not be optimal for computation.

# Chapter 10
# Prediction and Linear Regression

## PREVIEW

### Introduction

Where does the term *linear regression* come from?

### Using Linear Regression to Make Predictions

What are the procedures for computing the regression line for predicting scores on $Y$ from scores on $X$?

How do we define errors in prediction? Why are some such errors inevitable?

Why is the regression line referred to as the least-squares regression line?

What is the $Y$-intercept of the regression line? The slope?

What are some important properties of linear regression?

How do we predict scores on $X$ from scores on $Y$?

### Measuring Prediction Error: The Standard Error of Estimate

What is the standard error of estimate? Why is it useful?

What formula should be used when you wish to draw inferences about the standard error of estimate in a population?

### The Connection Between Correlation and the *t* Test

What are the procedures for computing the correlation between one continuous and one dichotomous variable (the point-biserial correlation coefficient)?

How is a point-biserial correlation coefficient tested for statistical significance?

When we obtain a statistically significant value of $t$ for the difference between two means, why is it desirable to convert this value to a point-biserial correlation coefficient? How is this done?

What is the relationship between $r_{pb}$ and the effect-size measure known as $g$?

What does squaring $r_{pb}$ tell you about the proportion of the total variance that is accounted for by the dichotomous variable?

### Estimating the Proportion of Variance Accounted for in the Population

How can the formula for $r_{pb}^2$ be adjusted to create an unbiased estimator of the variance accounted for in the population?

What is $\omega^2$, and how is it related to the population effect-size measure known as $d$?

### Summary

### Exercises

### Thought Questions

### Computer Exercises

### Bridge to SPSS

*Introductory Statistics for the Behavioral Sciences*, Eighth Edition. R. Brooke Lea and Barry H. Cohen.

Companion website: www.wiley.com/go/ISBS8e

## Introduction

Behavioral scientists are indebted to Sir Francis Galton for making explicit some elementary concepts of relationships and prediction. Galton wrote his now-classic paper "Regression toward Mediocrity in Hereditary Stature" in 1885. In it, he presented the theory that the physical characteristics of offspring tend to be related to, but are on average less extreme than, those of their parents. That is, tall parents on average produce children *less tall* than themselves, and short parents on average produce children *less short* than themselves. In other words, physical characteristics of offspring tend to move toward or "regress" toward the average of the population. If you were to predict the height of a child from the height of the parents, you should predict a less extreme height—one closer to the average of all children.

Plotting data on the stature of many pairs of parents and offspring, Galton calculated the median height of offspring for each height category of parents. For example, he plotted the median height for all offspring whose fathers were 5 foot 7 inches, the median height for all offspring whose fathers were 5 foot 8 inches, and so forth. By connecting the points representing the medians, Galton found not only that there was a positive relationship between parental height and height of the offspring, but also that this relationship was fairly linear. The line connecting each of the medians (and after Pearson, the means) came to be known as the **regression line**. This term has been adopted by statisticians to indicate the straight line used in predicting or estimating values of one variable from a knowledge of values of another variable with which it is paired. The statistical procedure for making such predictions is called **linear regression**.

## Using Linear Regression to Make Predictions

Predicting behavior is at the core of the behavioral sciences. A truism in psychology is that people's past behavior is the best predictor of their future behavior. Regression generalizes this logic from individuals to populations and formalizes the process by measuring systematic variation in the environment. In general, the procedure begins by measuring people on two variables (e.g., SAT and college GPA), creating a prediction equation from those data, and then using the equation to predict a future event (e.g., college GPA) from a past event (e.g., SAT score). For example, when colleges select applicants, they do not pick names randomly from a hat (though it may seem that way at times). The admissions department wants to identify which applicants are most likely to succeed and be happy at their institution. To help them make accurate predictions, admission staff collect a set of data from each applicant (e.g., SAT scores, high school GPA, letters of recommendation, personal essay), and, based on the relationship these data have with the success of past students, they make informed guesses about who is likely to be a good fit at their college. Because the relationship between variables such as SAT and college GPA is never perfect, their predictions also will not be perfect. However, to the extent that the correlation between the two variables is greater than zero, the admissions committee can make more accurate predictions than simply selecting randomly. Regression not only provides a systematic way to make these predictions but it also provides specific information about how much better your predictions will be compared to random selection.

Predicting behavior is not just of interest to behavioral scientists. Almost everyone would benefit from being able to predict what people will do, even if just a little better than they are able to now. Here's a simple example: Suppose your summer job is to sell ice cream from a portable freezer cart at the zoo. Each morning you carefully decide how much ice cream to take with you to the zoo. This is an important decision: If you take too much, then you will lose money on the unsold (and melted) ice cream at the end of the day. If you take too little, then you will sell out early and turn away hungry customers and their money. A perfect day would consist in selling all the ice cream you brought without turning away any customers. So what can you do to increase the chances that you will have a perfect day? One thing you may have noticed is that ice cream tends to sell faster on warmer days.

Suppose that each day you note the high temperature and how many ice cream cones you sold that day. After a few weeks, you compute Pearson $r$ and discover that there is a $+.47$ correlation between temperature and cone sales—you were right! Now you know to take more ice cream on days that are supposed to be warmer. But how much more ice cream? Using regression, you can plug in the temperature forecast for any particular day, and the prediction equation will produce the number of ice cream cones you are likely to sell that day! For example, the regression equation might tell you that if the temperature is predicted to reach 85° F, then you should take 127 ice cream cones. The predictions will not be perfect, but they will be better than if you did not take into account the relationship between temperature and demand for ice cream.

### Computational Procedures

Now that we have considered a couple of everyday examples, let's think about regression in a more abstract way. Let us assume that we wish to predict scores on $Y$ from scores on $X$. $Y$, the variable being predicted or estimated, is called the *dependent* variable (DV) or **criterion**. $X$, which provides the information on which the predictions are based, is referred to as the *independent* variable (even if it is not manipulated by the experimenter) or **predictor**. In this chapter, we discuss the simplest method of prediction—an equation that produces a *straight line*. One way of writing such an equation is

$$Y' = b_{YX}X + a_{YX} \qquad \text{Formula 10.1}$$

where

$Y'$ = *predicted* score on $Y$
$b_{YX}$ = *slope* of the line, also called the *regression coefficient* for predicting $Y$ from $X$
$a_{YX}$ = *Y-intercept*, or the value of $Y'$ when $X = 0$

The value of $b_{YX}$ summarizes the average rate of change in the $Y$ score per unit increase in the $X$ score, while the value of $a_{YX}$ indicates the $Y$ value at which the regression line crosses the $Y$-axis. For any given set of data, $b_{YX}$ and $a_{YX}$ are constant values in the equation. A straight line has only one slope, and there is only one value of $Y'$ for which $X$ equals zero. We wish to keep our errors in prediction (especially the *squared* errors, as we will see shortly) to a minimum, and the computed values of $b_{YX}$ and $a_{YX}$ are consistent with this goal.

The way in which a regression line is used to obtain a predicted score is illustrated in Figure 10.1. Suppose a person has a score of 20 on $X$, and you want to predict what her score on $Y$ will be. Given the regression line for predicting $Y$ shown in Figure 10.1, you enter the $X$-axis at 20 and proceed up to the regression line. You then move left until you reach the $Y$-axis and read off the predicted $Y$ score; $Y'$ for this subject is equal to 25. In practice, predicted

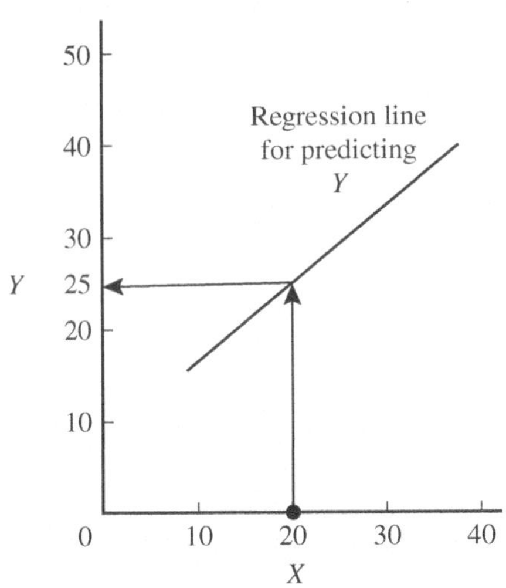


Figure 10.1

**Use of Regression Line to Obtain Predicted Scores on *Y***

scores are computed according to the regression equation because eyeball estimates from a graph are likely to be relatively inaccurate.

The **regression** (or prediction) **line** is the straight line that best represents the trend of the dots in a scatter plot. In any real situation, the dots will *not* all fall exactly on a straight line. Therefore, you will make *errors* when you use a regression line to make predictions. The error in predicting a particular $Y$ value is defined as the difference (keeping the sign) between the *actual* $Y$ value and the *predicted* $Y$ value, $Y'$. That is,

$$\text{error in predicting } Y = Y - Y'$$

A graphic representation of errors in prediction for one set of data is shown in Figure 10.2. For example, the predicted $Y$ value for the individual with an $X$ score of 68 is equal to approximately 159.8. His actual $Y$ value, however, is equal to 148. The difference of approximately −11.8 between the actual and predicted values represents an error in prediction for this individual, and the negative sign indicates that the actual value is smaller than the prediction. These errors are also called **residual scores**, and they always sum to zero (just like deviations from the mean).

In Figure 10.2, vertical lines have been drawn between each observed $Y$ score (actual weight) and the predicted weight score ($Y'$). This has been done for each value of $X$ (actual height) that occurs in the sample. These vertical distances are the errors in prediction (or residuals). All *predicted* scores lie on the (straight) regression line. Since there are two height scores of 70 inches with two different weight scores and two height scores of 69 inches with two different weight scores, the smaller error has been superimposed upon the larger in each of these cases.

The preferred linear regression prediction line is the one that minimizes the sum of *squared* errors in prediction, symbolized as:

$$\sum (Y - Y')^2$$

That is, values are chosen for $b_{YX}$ and $a_{YX}$ that define a particular regression line for predicting $Y$: the line that makes $\sum(Y - Y')^2$ smaller than you would get using any other prediction line for that set of data. This line is therefore called the ***least-squares* regression line** of $Y$ on $X$.

To see why it makes more sense to minimize the sum of *squared* errors rather than the sum of unsquared errors, let us suppose that you wish to predict students' college GPAs from their scores on the SATs. After making your predictions, you wait until the students finish

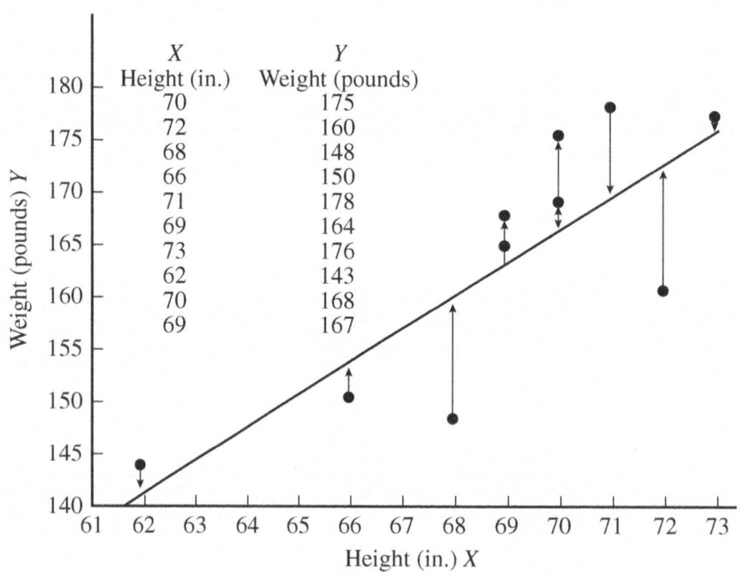


## Figure 10.2

**Regression Line of $Y$ on $X$ Showing Extent of Error (Difference Between Actual Weight Score and Predicted Weight Score)**

college and see what their actual GPAs turn out to be. The results for two students are as follows:

| Student | Actual GPA ($Y$) | Predicted GPA ($Y'$) | Error ($Y - Y'$) |
|---|---|---|---|
| 1 | 3.8 | 3.4 | +0.4 |
| 2 | 2.1 | 2.5 | −0.4 |

The actual GPA of the first student is about half a grade point above the prediction, whereas the GPA of the second student is a similar amount below the prediction for that student. Thus, some error in prediction has made. (This is unavoidable unless $r \pm 1$, which never happens with real data.) Yet if you were to compute the average of the error column, it would always equal zero! The same unhelpful result would be obtained had the predicted $Y'$ values been 2.4 for student 1 and 3.5 for student 2, yielding substantial errors of $+1.4$ and $-1.4$.

As this example shows, it is not sufficient to have a prediction line whose errors balance out (have a mean of zero); *all* lines that go through the point $\overline{X}, \overline{Y}$ have this property. In order to obtain a useful measure of the total amount of prediction error, we must find the line about which the variation of the $Y$ values, and hence the *squared* errors, is as small as possible. (As we see later in this chapter, a good measure of the magnitude of prediction error can be computed by averaging these squared errors and taking the square root.)

As we just mentioned, the values of $b_{YX}$ and $a_{YX}$ in the general formula for a straight line are computed so as to minimize the value of $\sum(Y - Y')^2$. Equations that produce these values have been found (by the laws of statistics) to be:

$$b_{YX} = r_{XY}\frac{\sigma_Y}{\sigma_X} \quad \text{Formula 10.2A}$$

$$b_{YX} = r_{XY}\frac{s_Y}{s_X} \quad \text{Formula 10.2B}$$

$$a_{YX} = \overline{Y} - b_{YX}\overline{X} \quad \text{Formula 10.3}$$

Formula 10.2A for $b_{YX}$ is used when values of $\sigma_X$ and $\sigma_Y$ are available, and the second version of the Formula 10.2B is used when working with the unbiased standard deviations (SDs); the latter is the more common case, but the value for the slope will be the same regardless of which formula is used. For the $Y$-intercept, this distinction does not arise.

To illustrate finding a "best-fit" regression line, let us return to our example of SAT and GPA from the previous chapter (see Table 9.2, page 182). We have already seen that $\mu_x$ (or $\overline{X}$) $= 498.4$, $\mu_y$ (or $\overline{Y}$) $= 2.85$, $\sigma_X = 117.33$, $\sigma_Y = 0.57$, and $r_{XY} = +.65$, so, by Formula 10.2A:

$$\begin{aligned} b_{YX} &= (+0.65)\frac{0.57}{117.33} \\ &= 0.0032 \end{aligned}$$

Because $b_{YX}$ is the slope of the line, this result indicates that each unit of increase in $X$ is associated with 0.0032 units of increase in $Y$. In this case, it means that each extra point on the SAT is associated with an increase in 0.0032 GPA points. This result may not sound meaningful because a difference of 1 point on a scale that ranges from 200 to 800 is tiny, but if you multiply by 100, then you can see that a 100-point increase on the SAT would be associated with a 0.3 increase in GPA, which is more meaningful. Then, according to Formula 10.3:

$$\begin{aligned} a_{YX} &= 2.85 - (0.0032)(498.4) \\ &= 1.26 \end{aligned}$$

This is the value of $Y'$ when $X$ is equal to zero. Such a value need not be a *logical* possibility for a given set of data. This is the case here since SAT scores of zero are outside the possible range of scores. We will have more to say about interpreting $y$-intercepts later in this chapter.

Combining the preceding two results, by using Formula 10.1, yields the linear regression equation for predicting $Y$ (GPA) from $X$ (SAT):

$$\begin{aligned} Y' &= b_{YX}X + a_{YX} \\ &= 0.0032X + 0.26 \end{aligned}$$

This equation can now be used to obtain predicted GPAs ($Y'$), given an SAT score ($X$). For example, the predicted GPA for a student with an SAT score of 600 is:

$$Y' = (0.0032)(600) + 1.26 = 1.92 + 1.26 = 3.18$$

This is the *best* linear prediction you can make, the one that on average would yield the smallest squared error. Using this regression equation you can verify for yourself that an SAT score of 800 leads to a predicted GPA of 3.82.

Note that a sample with scores on *both* the $X$ and $Y$ variables is needed in order to compute a linear regression prediction equation. Thus, to predict GPA from SAT, you must first obtain a sample of college students for whom you have both GPAs and scores on the SAT, and then compute such essential values as $r_{XY}$. Once you have determined the regression line, you can then use it to *make predictions for new cases for which you have data on only the predictor*—that is, to make predictions for graduating high school seniors based on their SAT scores. However, you *must* be careful to ensure that the original sample of college students on whom the regression equation is calculated is representative of the future groups for whom the predictions will be made.

### Properties of Linear Regression

The linear regression procedure has numerous important properties. We have already seen that there is precisely one predictor and one criterion, a straight line is used to make predictions, and the line is such that the sum of squared errors in prediction is minimized. Some additional principles of importance include the following:

1. If there is no good information on which to base a prediction, the same estimate—the mean of the criterion—is made for everyone.

Suppose that a group of graduating high school seniors asks you to predict what each one's college GPA will be, but the only information that they give you is the length of each student's right big toe. Since big toe length is useless for purposes of forecasting someone's GPA (i.e., $r_{XY} = 0$), you should not make any predictions about GPA based on toe size.

When there is no relevant information on which to base a prediction (i.e., $r_{XY} = 0$), linear regression also will not forecast any differences on the criterion. Instead, it "predicts" that each person will have the average score for $Y$. That is, when the correlation between the predictor and the criterion is zero, the linear regression formula becomes:

$$\begin{aligned} b_{YX} &= 0.00\frac{s_Y}{s_X} = 0 \\ a_{YX} &= \overline{Y} - (0.00)\overline{X} = \overline{Y} \\ Y' &= b_{YX}X + a_{YX} \\ &= (0)X + \overline{Y} \\ &= \overline{Y} \end{aligned}$$

Thus, scores on $X$ are ignored since they are irrelevant for purposes of predicting $Y$, and the mean of the criterion variable is predicted for everyone. Of course, no behavioral scientist uses linear regression unless the predictor and criterion are nontrivially correlated, but it is helpful to see this "worst-case scenario" in order to follow the logic of linear regression.

2. When all scores are expressed as $z$-scores, the predicted $z$-score on $Y$ is equal to $r$ multiplied by the $z$-score on $X$.

It can be shown algebraically that[1]

$$z'_Y = r_{XY} z_X \qquad \text{Formula 10.4}$$

where

$z'_Y$ = predicted $z$ value on $Y$
$z_X$ = actual $z$ value on $X$

This equation should help you in understanding linear regression. It shows that the predicted score ($z'_Y$) will be less extreme (i.e., closer to its mean) than the score from which the prediction is made ($z_X$) because $z_X$ is multiplied by a fraction ($r_{XY}$). This illustrates the statement, made previously, about regression toward the mean of $Y$. The equation also shows once again that if $r_{XY}$ is equal to zero, all predicted $z'_Y$ values will equal the mean of $Y$ (they will equal zero, which is the mean of $z$-scores). Finally, the equation indicates that, when dealing with $z$-scores, the regression line passes through the origin of the graph (0,0), because if $z_X = 0$ (the mean of the $X$ scores expressed as $z$-scores), then $z'_Y = 0$ (the mean of the $Y$ scores expressed as $z$-scores). When graphing the original scores, the regression line will pass through the point where the two means meet ($\overline{X}, \overline{Y}$).

Although Formula 10.4 looks refreshingly simple, it is not in general convenient for calculating predicted scores. To use it for this purpose, you would first have to transform $X$ to a $z$ value, then compute the value of $z'_Y$, and then transform $z'_Y$ to a raw score equivalent, $Y'$. Most of the time, it will be easier to use the raw score regression equation.

3. The closer the correlation between $X$ and $Y$ is to zero, the greater is the amount of prediction error.

It should be emphasized that while linear regression *minimizes* the sum of squared errors in prediction, this sum may still be prohibitively large. If the correlation between the predictor and the criterion is numerically small, such as +.07, linear regression will make a substantial amount of prediction error because the relationship between the two variables is so weak. This implies that a correlation near zero is likely to be useless for practical prediction purposes, even if it is statistically significant (as could happen with a very large sample). The *sign* of the correlation coefficient, however, is *not* related to prediction error. A correlation of (say) +.55 and one of −.55 are equally good for prediction purposes because the *strength* of the relationship between the predictor and the criterion is the same.

4. Transforming either $X$ or $Y$ or both by a linear transformation will not affect the linear correlation between them.

Suppose that the correlation between the high temperature for the day in a small town, as measured in degrees Fahrenheit, and the total weight of ice cream eaten that day by the people in that town, as measured in pounds, is equal to .4. If we wished to publish this result in a European journal, and its editors insisted that we convert our temperature measurements to degrees Celsius and our weights to kilograms, the Pearson $r$ would not be affected by these transformations (°C = 0.556 × °F − 17.778; wt. in kg = 0.4 × wt. in lbs.) because they are *linear transformations* (only multiplication, division, addition, and/or subtraction by constants are involved). Linearly transforming an entire set of raw scores will not change the $z$-score for any individual, and the Pearson $r$ depends only on the pairing of the $z$-scores for the two variables and not the raw scores. Of course, a $z$-score is itself just a linear transformation of a raw score, so the correlation between the raw scores on some variable and their corresponding $z$-scores will always be perfect (i.e., $r = +1.0$).

[1] Recall from Chapter 4 that the standard deviation of both $z_X$ and $z_Y$ is equal to 1. Therefore, when dealing with $z$-scores, $b_{YX} = r(1/1) = r$. Because the mean of both $z_X$ and $z_Y$ is equal to zero, $a_{YX} = 0$. Thus, for $z$-scores, $Y' = rX + 0$, so $z_{Y'} = r_{XY} z_X$.

### A Technical Note: Formulas for Predicting *X* From *Y*

Because the assignment of $X$ and $Y$ to the two variables in a linear regression problem is usually up to the researcher, you can usually assign the $Y$ designation to the criterion and $X$ to the predictor and use the formulas given previously. If you should then wish to make predictions in the other direction, you can simply *change the designations* (relabel $X$ as $Y$ and $Y$ as $X$) and use the same formulas. Note that it *is* necessary to recompute a new linear regression equation if you wish to predict in the other direction (unless $r = \pm 1$).

For purposes of reference, we briefly note the linear regression procedure for predicting scores on $X$ from scores on $Y$. The equation is

$$X' = b_{XY}Y + a_{XY},$$

where

$$X' = \text{predicted score on } X$$

$$b_{XY} = r_{XY}\frac{\sigma_X}{\sigma_Y} = r_{XY}\frac{s_X}{s_Y}$$

$$a_{XY} = \overline{X} - b_{XY}\overline{Y}$$

Note that it is the order of the x and y subscripts that indicates which is the predictor variable and which is the predicted variable (e.g., $b_{YX}$ for when you're predicting $Y$ from $X$ and $b_{XY}$ for when you are predicting $X$ from $Y$).

This procedure minimizes $\sum(X - X')^2$, the sum of squared errors, when predicting $X$. While $r_{XY} = r_{YX}$, $b_{XY}$ (the $b$ value or slope for predicting $X$ from $Y$) does *not* equal $b_{YX}$ (the $b$ value or slope for predicting $Y$ from $X$), nor does $a_{XY} = a_{YX}$. In fact, the regression line for predicting $X$ is not the same as the regression line for predicting $Y$ unless the correlation between $X$ and $Y$ is perfect (numerically equal to 1), *and* $\sigma_X = \sigma_Y$, *and* $\overline{X} = \overline{Y}$.

## Measuring Prediction Error: The Standard Error of Estimate

We have seen that when predicting scores on $Y$, the amount of squared error made for a given individual is equal to $(Y - Y')^2$. The *average* squared error for the entire sample can be obtained by summing the squared errors and dividing by $N$. Such a measure of error, the *variance of errors*, is in terms of squared units. As was the case with the SD (Chapter 3), a measure in terms of actual score units can be obtained by taking the positive square root. This gives a useful measure of prediction error that is called the **standard error of estimate**, symbolized by $\sigma_{est}$:

$$\sigma_{est} = \sqrt{\frac{\sum(Y - Y')^2}{N}}$$

In practice, the standard error of estimate (which is also the SD of the residual scores) can more easily be obtained by the following formula, which can be proved to be equivalent to the one above by algebraic manipulation:

$$\sigma_{est} = \sigma_Y\sqrt{1 - r_{XY}^2} \qquad \text{Formula 10.5}$$

Note that if $r = \pm 1$, there is no error ($\sigma_{est} = 0$); if $r = 0$, the standard error of estimate reaches its maximum possible value, which is the ordinary SD of $Y$, or $\sigma_Y$.

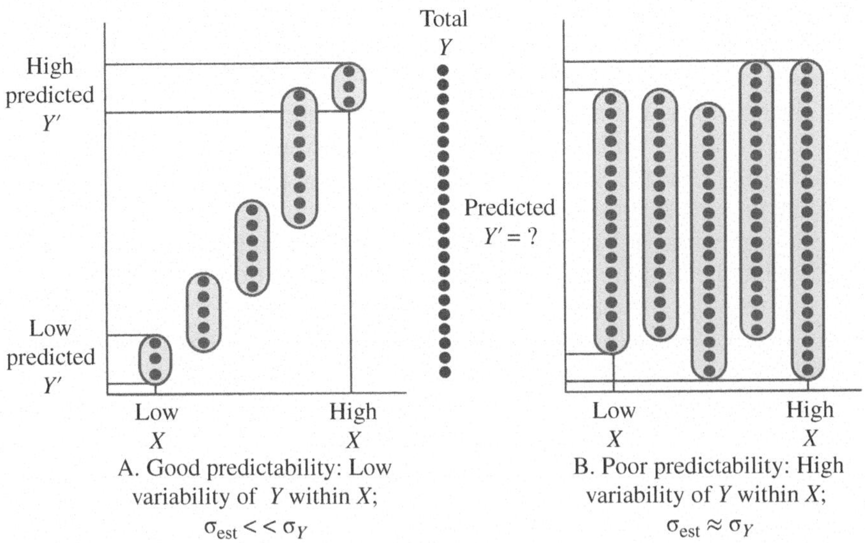


## Figure 10.3

**Relationship Between the Standard Error of Estimate ($\sigma_{est}$), the Standard Deviation of $Y(\sigma_Y)$, and the Accuracy of Prediction**

The standard error of estimate of $Y$ may be thought of as the variability of $Y$ about the regression line for a particular $X$ value, averaged over all values of $X$. This is illustrated in Figure 10.3. In Panel A of Figure 10.3, the variability of the $Y$ scores within each $X$ value is relatively small because the scores within each $X$ value cluster closely together. You can see that $Y$ takes on a relatively narrow range of values for any given $X$. By contrast, the total variability of $Y$ (for all $X$ combined), which is given by $\sigma_Y$, is relatively large. Thus, $\sigma_{est}$ is *small compared to* $\sigma_Y$; the standard error of estimate is always considerably smaller than the SD of $Y$, whenever the correlation is fairly high. On the one hand, in Figure 10.3A, the correlation between $X$ and $Y$ is fairly close to 1.0. This is good for purposes of prediction since an individual with a low score on $X$ is likely to have a low score on $Y$, whereas an individual with a high score on $X$ is likely to have a high score on $Y$.

Figure 10.3B, on the other hand, illustrates a situation in which prediction will be poor. The average variability within each $X$ value is about equal to the total variability of $Y$; that is, the standard error of estimate ($\sigma_{est}$) is *nearly equal to* the SD of $Y(\sigma_Y)$ whenever the correlation is very small. In Figure 10.3B, the correlation is actually close to zero. In this instance, therefore, knowing an individual's $X$ score will *not* permit you to make a good prediction as to what their $Y$ score will be. As a whole, Figure 10.3 shows that prediction will be good to the extent that for any given $X$, $Y$ shows relatively little variability.

### The Proportion of Variance Explained by a Correlation

By solving Formula 10.5 for the square of the correlation coefficient, we can obtain a useful expression that relates the magnitude of the squared correlation to the difference between the variance of errors and the total variance, in terms of a proportion.

$$\sigma_{est} = \sigma_Y\sqrt{1 - r_{XY}^2} \qquad \text{Formula 10.5}$$

Squaring both sides and rearranging terms lead to:

$$r_{XY}^2 = 1 - \frac{\sigma_{est}^2}{\sigma_Y^2}$$

$$= \frac{\sigma_Y^2 - \sigma_{est}^2}{\sigma_Y^2}$$

Thus, the *squared* correlation coefficient shows *by what proportion the variability* (specifically, the variance) *in predicting Y has been reduced by knowing X*. If, on one hand, this reduction has been great (as in Figure 10.3A), $\sigma^2_{est}$ will be small compared to $\sigma^2_Y$, and $r^2$ will be relatively large. If, on the other hand, little reduction has been achieved (as in Figure 10.3B), $\sigma^2_{est}$ will be about equal to $\sigma^2_Y$, and $r^2$ will be close to zero. In the example involving GPA and SAT scores, $r^2$ is equal to $(.65)^2$, or .42. So, in that example, 42% of the variance of the GPAs is said to be *explained by* the SAT scores. That is considered a rather high proportion of variance to be explained in one variable by a totally separate variable in behavioral research; clearly a fairly good job has been done of identifying a (linear) relationship between two variables that are not ostensibly measuring the same thing. Note that the expression *is explained by* does not mean *is caused by*, even though it sounds as if it should. Instead, we might say that some of the variance in the $Y$ variable is explained or *accounted for* by its *relationship with* the $X$ variable; unless random assignment has been conducted, the nature of that relationship cannot be deduced from the correlation.

It is useful to think of the total variance in the predicted variable ($\sigma^2_Y$) as the sum of two variances that are called the *explained variance* and the *unexplained variance*. Suppose you are predicting people's weight from their height, as in Figure 10.2. The vertical distance of each weight value from the regression line is the error of the estimate (i.e., the error of the prediction), but it can also be referred to as the portion of the weight value that is unexplained by the predictor variable (height, in this example). So, the variance of the errors, $\sigma^2_{est}$, is also called the unexplained variance. It follows then that the portion of the weight value that is *not* unexplained is the portion that *is* explained by the predictor variable, so the variance of the explained portions is called the explained variance. There is no standard symbol for the explained variance ($\sigma^2_Y - \sigma^2_{est}$).

### Estimated Standard Error From a Sample

The standard error of estimate computed from Formula 10.5 is a good descriptive statistic, which is appropriate when you are dealing with an entire population. (You may have inferred this from the fact that we used the Greek letter σ.) When you wish to draw inferences about the standard error of estimate in a population *from the data in a random sample*, you could correct for bias by using the following formula:

$$s_{est} = \sqrt{\frac{\sum (Y - Y')^2}{N - 2}}$$

Note that whereas $s_Y$ is based on $N - 1$degrees of freedom, $s_{est}$, like the Pearson $r$, is based on $N - 2$ degrees of freedom (*df*) (with the ordinary variance, you lose one *df* for the sample mean; with the variance around a regression line, you lose one *df* for the slope and another *df* for the $Y$-intercept that, together, define that regression line). However, this formula is clearly not convenient for calculation. Fortunately, you can greatly simplify the calculations by basing your formula on the Pearson $r$, as follows:

$$s_{est} = s_Y \sqrt{\frac{N - 1}{N - 2}\left(1 - r^2_{xy}\right)} \qquad \text{Formula 10.6}$$

where $s_Y$ is the unbiased SD of your $Y$ scores.

## The Connection Between Correlation and the *t* Test

Now that you have learned the concept of regression, it is possible to show its connection to the two-sample *t*-test procedure that you learned in Chapter 7. Turns out, you can treat the two-sample comparison as a form of correlation. That is, you can compute the correlation between a two-valued (i.e., *categorical*) variable (e.g., drug group versus placebo) and a continuous variable (e.g., some outcome). It can be useful to think of the two-sample *t* test as a special case of correlation, as we demonstrate in the next section. Note, however, that the two-sample *t* test is a commonly used procedure in the behavioral sciences; our goal in illustrating the similarity is to help you see that many of the procedures we describe in this book are variations on a few mathematical relationships that have been manipulated to allow researchers to analyze data that come in a variety of forms. Let us return once more to the problem of predicting weight from height (as depicted in Figure 10.2). This time, suppose that your data set contains five men and five women. By a strange coincidence, all of the women are the same height (i.e., 64 inches) and all of the men are the same height as each other (i.e., 69 inches). There is still variability in weight for each group, as you can see in the scatterplot in Figure 10.4. The data for Figure 10.4 are shown, and the Pearson *r* is calculated for the height/weight relationship (using the *z*-product method) in Table 10.1.

It should not be surprising that the linear correlation is high ($r = .894$) for these data (note the similarity between Figure 10.4 and Figure 10.3A), even though the *X* variable has only two values in this case.

What *is* surprising is that the correlation would be exactly the same if the *X* values were simply 1 for all women and 2 for all men (or vice versa). In fact, *any* two different values for *X* would yield the same magnitude for the Pearson *r*. (However, the *r* would have a negative

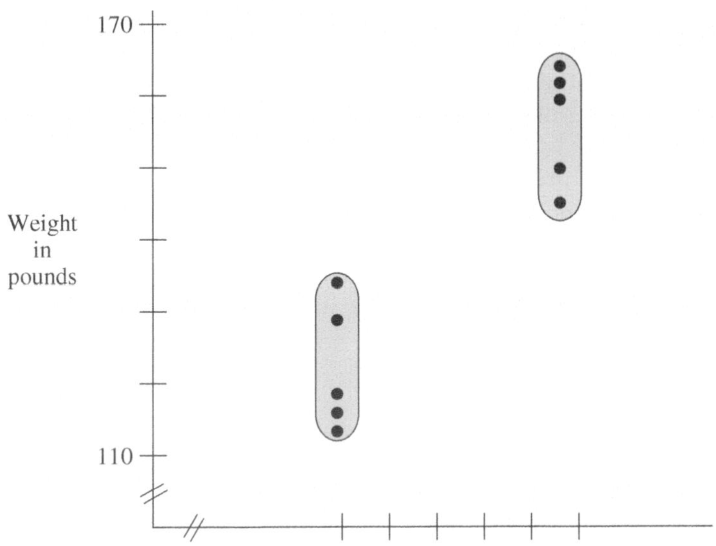


## Figure 10.4

**A Scatter Plot in Which One of the Variables Has Only Two Values**

## Table 10.1 **Height and Weight Data for Figure 10.1**

| Height (in.) $X$ | $z_X$ | Weight (lbs.) $Y$ | $z_Y$ | $z_X z_Y$ | |
|---|---|---|---|---|---|
| 64 | −1 | 116 | −1.31 | +1.31 | |
| 64 | −1 | 118 | −1.20 | +1.20 | $\overline{X}_1 = 123.8$ |
| 64 | −1 | 120 | −1.10 | +1.10 | $\sigma_1 = 8.635$ |
| 64 | −1 | 125 | −0.83 | +0.83 | $s_1 = 9.654$ |
| 64 | −1 | 140 | −0.03 | +0.03 | |
| 69 | +1 | 145 | +0.24 | +0.24 | |
| 69 | +1 | 150 | +0.51 | +0.51 | $\overline{X}_2 = 157.2$ |
| 69 | +1 | 162 | +1.15 | +1.15 | $\sigma_2 = 8.134$ |
| 69 | +1 | 164 | +1.26 | +1.26 | $s_2 = 9.094$ |
| 69 | +1 | 165 | +1.31 | +1.31 | |
| $\sum X = 665$ | $\sum z_X = 0$ | $\sum Y = 1405$ | $\sum z_Y = 0$ | $\sum z_X z_Y = +8.94$ | |
| $\overline{X} = 66.5$ | | $\overline{Y} = 140.5$ | | $r = \frac{\Sigma z_X z_Y}{N}$ | |
| $\sigma_X = 2.5$ | | $\sigma_Y = 18.69$ | | $= \frac{8.94}{10} = 0.894$ | |

sign if women were arbitrarily assigned the higher of the two $X$ values.) This phenomenon becomes much less surprising when you look carefully at the calculation of $r$ in terms of $z$-scores (see Table 10.1). You can see that whenever you convert a set of only two different numbers to $z$-scores, the $z$-score for the smaller number will be $-1$ and the $z$-score for the larger one will be $+1$, regardless of which two numbers you are working with.

### The Point-Biserial Correlation Coefficient

In our height/weight example, all of the women have one height and all of the men have another (i.e., height and *gender* are perfectly correlated). Therefore, the correlation between height and weight is also the correlation between *gender* and weight. You can correlate any variable with gender (as long as you have measured your DV on at least some individuals of each gender) by arbitrarily assigning an $X$ value to each gender (e.g., 1 and 2) and then correlating these $X$ values with your continuous variable as the $Y$ value for each individual. For instance, suppose you correlate gender with height. After assigning the $X$ value of 1 to females and 2 to males, you will be measuring the tendency for height to be higher whenever the gender value (1 versus 2) is higher.

The type of correlation just described is usually called a **point-biserial correlation**, symbolized as $\boldsymbol{r_{pb}}$ because one of the variables being correlated has only two levels, which are assigned arbitrary values. The main problem with $r_{pb}$ is that its sign will change if you reverse the assignment of the $X$ values (e.g., $r_{pb}$ for gender with height will become negative if males are assigned a value that is *lower* than the one assigned to the females). Therefore, we routinely ignore the sign of $r_{pb}$ and look at the means of the two groups to determine the direction of the relationship.

When squared, $r_{pb}$ is interpreted in the same way as any squared correlation coefficient; it represents the proportion of variability in the $Y$ variable that is explained by its relationship with the $X$ variable. Note that any time you can calculate an $r_{pb}$, you can also compute a $t$ value to compare the two group means instead. The information you get from the $r_{pb}$ complements the information you get from the $t$ value, as we will now demonstrate for the gender/weight example.

### The Relationship Between $r_{pb}$ and the $t$ Test

In the rightmost column of Table 10.1, we included the means and the biased and unbiased SDs for weight separately for each height (i.e., gender). Using the simplified $t$ formula for equal sample sizes (Formula 7.15), we can easily calculate $t$ for comparing the mean weights of the two groups:

$$t = \frac{\overline{X}_1 - \overline{X}_2}{\sqrt{\dfrac{s_1^2 + s_2^2}{n}}} = \frac{123.8 - 157.2}{\sqrt{\dfrac{93.2 + 82.7}{5}}} = \frac{-33.4}{\sqrt{35.18}} = \frac{-33.4}{5.93} = -5.63$$

This $t$ value shows that the difference in weight between the two genders is statistically significant. Recall that the $r_{pb}$ between height and weight (and therefore in this example, between gender and weight) was high—in fact, it was .894. However, even such a large $r$ should be tested for significance, which we will do by using the $t$ formula found in a footnote on page 191 in Chapter 9, substituting $r_{pb}$ for $r$. (We could just look up the critical value in Table C in the Appendix, but we want to make a point about the $t$ value.)

$$t = \frac{r_{pb}\sqrt{N - 2}}{\sqrt{1 - r_{pb}^2}} \qquad \text{Formula 10.7}$$

$$t = \frac{0.894\sqrt{8}}{\sqrt{1 - 0.8}}$$

$$t = \frac{2.5286}{0.4472}$$

$$t = 5.65,$$

where $N$ represents the total number of men and women, which is also the number of pairs on which $r_{pb}$ was calculated.

The slight difference in magnitude between these two $t$ values is due entirely to rounding off intermediate terms (e.g., the variance of each group). The sign of the $t$ value can be ignored in each case. The point is that when you test $r_{pb}$ for significance, you are testing the relationship between a (two-valued) independent variable and a continuous DV, which is equivalent to testing the difference of the means of the two groups on the DV using a $t$ test for two independent samples.

### The Relationship Between $r_{pb}$ and $d$ (Effect Size)

Recall that in Chapter 7, we introduced a measure of the relative separation of two sample means, called $d$. The ordinary (pooled-variance) two-sample $t$ test can be viewed as a test of the significance of $d$, where

$$t = d\sqrt{\frac{N_1N_2}{N_1 + N_2}}$$

In fact, both $r_{pb}$ and $d$ are alternative measures of effect size. We already solved the preceding formula for $d$ to create Formula 7.11 (page 137), which enables you to calculate $d$ from a published $t$ value (if you know the sample sizes):

$$d = t\sqrt{\frac{N_1 + N_2}{N_1N_2}} \qquad \text{Formula 7.11}$$

Similarly, you can modify the formula for testing the significance of $r_{pb}$ so that you can also calculate $r_{pb}$ from a published $t$ value:

$$r_{pb} = \sqrt{\frac{t^2}{t^2 + df}} \qquad \text{Formula 10.8}$$

where $df = N_1 + N_2 - 2$. If, in the formula for $r_{pb}$, you substitute the expression for $t$ in terms of $d$ and $N$ (where $N = N_1 + N_2$), you will get a formula (if you are really good with algebra) that relates $r_{pb}$ directly to $d$ (with an adjustment for sample size, $df/N$, that is only noticeable with small samples) and demonstrates how closely these two effect-size measures are related.

$$r_{pb} = \sqrt{\frac{d^2}{d^2 + 4(df/N)}} \qquad \text{Formula 10.9}$$

For the data in Table 10.1, $d$ equals $t\sqrt{(10/25)} = 5.65(0.632) = 3.57$—a very large value for an effect size. (We could also have calculated this value by dividing the difference between the male and female means by the square root of the pooled variance.) Of course, we should expect a large $d$ from already knowing that $r_{pb}$ is extremely large (i.e., close to 1.0). We demonstrate next that, although Formula 10.9 looks pretty strange, it does indeed produce the correct value for $r_{pb}$:

$$r_{pb} = \sqrt{\frac{3.57^2}{3.57^2 + 4(8/10)}} = \sqrt{\frac{12.75}{12.75 + 3.2}} = \sqrt{\frac{12.75}{15.95}} = \sqrt{0.8} = 0.894$$

Given these two choices for measuring effect size, the question is this: Under what circumstances would $r_{pb}$ be preferred, and when would $d$ be considered more desirable? As you will see in the next chapter, $d$ is the preferred effect-size measure for power analysis. Also, $d$

has a straightforward interpretation when dealing with roughly symmetric, bell-shaped curves; it gives you a clear sense of the degree to which the scores in one group overlap with the scores of the other group. This aspect of $d$ is described further in the next chapter as well.

### The Proportion of Variance Accounted for by a Grouping Variable

The advantage of $r_{pb}$ is that squaring $r_{pb}$, as with any correlation coefficient, gives the proportion of variance in one of your variables that is *accounted* for by the other. In the two-group case, you will, of course, always want to look at the proportion of variance in your DV that is accounted for by your grouping variable (e.g., experimental versus control group), rather than the other way around. As an example, we first compute the proportion of variance accounted for directly in the data of Table 10.1. First, we find the total variance in weight by squaring the (biased) SD at the bottom of the Weight column in Table 10.1: $\sigma_y^2 = 349.25$. The variance in weight that is *not* explained by gender is the average (a weighted average if the *n*s are not equal) of the variances *within* the two genders. Therefore, we square and average the σs of the two groups to obtain the unexplained variance: $(8.635^2 + 8.134^2)/2 = 140.725/2 = 70.36$. Now we can find the explained variance (i.e., the *variance accounted for*) by subtracting the unexplained variance (which is the same as $\sigma_{est}^2$) from the total:

$$\sigma_{explained}^2 = \sigma_{total}^2 - \sigma_{unexplained}^2 = 349.25 - 70.36 = 278.89$$

The proportion of the total variance in weight that is accounted for is therefore $278.89/349.25 = .799$, or approximately 80%. Note that $r_{pb}^2$ equals $.894^2 = .799$. This shows that we did not have to go through the trouble of dividing variances; squaring the point-biserial $r$ gives us the proportion of variance accounted for. Even when a large $t$ value is obtained that easily reaches statistical significance, it is possible that very little variance is being accounted for. This consideration is true of all procedures in inferential statistics. Statistical significance does *not* imply a *large* relationship or effect; it just suggests that the effect in the population is *unlikely to be zero*. A $t$ value does *not* offer a solution to this problem because it does not provide a ready measure of how large the relationship is (the effect size). So, it is desirable to convert values of $t$ obtained from the procedures in Chapter 7 into point-biserial correlation coefficients (or *d*s) and to report *both t* (with its $p$ value) and $r_{pb}$ (or $d$). A glance at the formula for $r_{pb}$ in terms of $t$ and $df$ reveals that any particular $t$ that is obtained with large samples (hence, many $df$) is less impressive (i.e., it will be associated with a smaller $r_{pb}$) than the same $t$ value obtained with smaller samples.

As an illustration, consider once again the caffeine study discussed in Chapter 7. It was found that caffeine did have a positive association with heart rates; a statistically significant value of $t$ was obtained. The question remains, however, as to *how strong* is the relationship between the presence or absence of that dosage of caffeine and heart rates. Recalling that $t = 2.91$ and $df = (22 + 20 - 2) = 40$, we have

$$\begin{aligned} r_{pb} &= \sqrt{\frac{(2.91)^2}{(2.91)^2 + 40}} \\ &= \sqrt{\frac{8.47}{48.47}} \\ &= \sqrt{0.175} \\ &= 0.42 \end{aligned}$$

Thus, there is a rather strong relationship between the presence or absence of caffeine and heart rates; $r_{pb}^2 = (0.42)^2$, so 17% of the variance in heart rates is explained by whether or not caffeine is present. You therefore know that caffeine is an important factor that accounts for differences among people in resting heart rate, though it is far from the only one—83% of the variance in heart rate is still not accounted for.

Let us switch gears for a moment and imagine that the DV was not heart rate but performance on an attention task. If it could be assumed that the amount of caffeine that had been given in the experiment had no harmful side effects, based on these positive results, a cautious recommendation that students take caffeine before a task requiring sustained attention would be justified.

Suppose instead that, using a very large sample, the statistically significant $t$ had yielded a value of $r_{pb}$ of .10. Here, only $(.10)^2$ or 1% of the variance in test scores is explained by the presence or absence of caffeine. You are therefore warned that this experimental finding, while statistically significant, is probably not of much practical value. You certainly would *not* recommend that caffeine actually be used because the probable effect on attention performance would be too small to matter. (The possibility does exist, however, that such a finding might be of theoretical importance in understanding brain functioning.)

## Estimating the Proportion of Variance Accounted for in the Population

As we have seen, squaring $r_{pb}$ produces a useful descriptive statistic. However, $r^2_{pb}$ tends to *over*estimate the proportion of variance that would be accounted for in the data for the entire population from which you drew your samples. Fortunately, the formula for $r^2_{pb}$ can easily be adjusted to remove nearly all of its *bias*.

It would be reasonable to expect that calculating $r^2_{pb}$ for the entire population would produce a quantity referred to as $\rho^2_{pb}$. However, the proportion of variance accounted for in the *population* is usually referred to as $\omega^2$ (the lowercase Greek letter omega, squared). To create a nearly unbiased estimate of $\omega^2$, we need only square Formula 10.8 for $r_{pb}$ and modify it slightly, as shown:

$$\text{est.}\omega^2 = \frac{t^2 - 1}{t^2 + df + 1} \qquad \text{Formula 10.10}$$

If we apply this formula to the caffeine study, we see that whereas 17.47% of the heart rate variance is accounted for by the presence of caffeine in the sample data, our best guess about the percent of variance that would be accounted for in the *population* is a bit less: 15.1%.

$$\text{est.}\omega^2 = \frac{2.91^2 - 1}{2.91^2 + 40 + 1} = \frac{7.468}{8.468 + 41} = \frac{7.468}{49.468} = 0.151$$

Note that when $t$ is less than 1, it does not make sense to use Formula 10.10 because variances cannot be negative. However, $t$ values less than 1 are so far from being statistically significant that it would generally be safer to assume that no (i.e., zero) variance is being accounted for in the population than to try to estimate that quantity.

### The Relationship Between $\omega^2$ and $d^2$

As just mentioned, if you were to calculate $r^2_{pb}$ on an entire population, you would refer to the result as $\omega^2$. And, as we explained in Chapter 7, calculating the effect-size measure $d$ for an entire population results in a quantity that we will be referring to as $\mathbf{d}$, the lowercase letter d printed in boldface. Note that although $d$ is a slightly biased estimator of $\mathbf{d}$, the bias is so small for fairly large sample sizes that you rarely see $d$ being corrected when it is used to estimate **d**. We will have much more to say about **d** when we discuss power in the next chapter.

Just as there is a simple relationship between $r_{pb}$ and $d$ for the data in your samples, there is a correspondingly simple relationship between $\omega^2$ and $\mathbf{d}^2$ in the population. To express this relationship as a formula, all that needs to be done is to square both sides of the equation relating $r_{pb}$ to $d$, drop out the $df/N$ term (there is no distinction between $df$ and $N$ when you are dealing with an entire population), and change the notation:

$$\omega^2 = \frac{\mathbf{d}^2}{\mathbf{d}^2 + 4} \qquad \text{Formula 10.11}$$

For example, when **d** is 2 (i.e., the means of the two populations are two SDs apart), the proportion of variance accounted for is .5 [i.e., 4/(4 + 4)]. That $\omega^2$ equals .5 tells us that the average variance within the populations is only half as large as the variance measured across the two populations. If the gender difference in height in an entire population were such that $\mathbf{d} = 2$, then you would know that the variance in height for all humans is twice as large as the variance measured separately for each gender. You can use the formula to verify the following examples for yourself: $\omega^2 = .2$ when $\mathbf{d} = 1$; $\omega^2 = .8$ when $\mathbf{d} = 4$.

### Publishing Effect-Size Estimates

As we mentioned in Chapter 7, supplementing the results of a $t$ test by reporting a corresponding confidence interval is especially helpful when you are dealing with familiar units (inches, kilograms, IQ points). When you are not, it is highly recommended to report a measure of effect size, such as $d$ (also known as Cohen's $d$), even though it has been calculated from sample data and has not been corrected for bias. Sometimes researchers use a correlational measure, such as $r_{pb}$ or $r^2_{pb}$, to represent effect size in the two-group case, but in our experience, the correlational measures are much more likely to be used in multi-group situations, such as those described in Chapter 13.

## Summary

To the extent that two continuous variables exhibit a linear relationship, we can use the value on one variable to predict a person's value on the other variable. The closer the points on a scatterplot of the two variables fall on a single straight (regression) line, the more accurate the predictions will be. However, the concept and mechanics of linear regression can also be useful for understanding the relationship between a continuous variable and a dichotomous (two-valued) variable.

### Linear Regression

When the goal is to obtain the predicted score $Y'$, given a raw score for $X$, the formula to use is:

$$Y' = b_{YX}X + a_{YX} \qquad \text{Formula 10.1}$$

where

$$b_{YX} = r_{XY}\frac{\sigma_Y}{\sigma_X} = r_{XY}\frac{s_Y}{s_X} \qquad \text{Formula 10.2}$$

$$a_{YX} = \overline{Y} - b_{YX}\overline{X} \qquad \text{Formula 10.3}$$

The standard error of estimate for predicting $Y$ from $X$, a measure of error in the prediction process, is equal to

$$\sigma_{est} = \sqrt{\frac{\sum (Y - Y')^2}{N}} = \sigma_Y \sqrt{1 - r_{XY}^2} \qquad \text{Formula 10.5}$$

This measure is a descriptive statistic. If you wish to estimate the standard error of estimate in a population from sample data, use the following formula (corrected for bias):

$$s_{est} = s_Y \sqrt{\frac{N-1}{N-2}\left(1 - r_{xy}^2\right)} \qquad \text{Formula 10.6}$$

### *Points to Remember*

1. The foregoing regression equations minimize the sum of squared errors in prediction, $\sum(Y - Y')^2$.
2. The closer $r$ is to zero, the greater the errors in prediction. If $r = \pm 1$, $\sigma_{est} = 0$ (prediction is perfect). If $r = 0$, $\sigma_{est}$ equals its maximum possible value $\sigma_Y$ (prediction is useless).
3. If $r$ is not statistically significant, linear regression should probably *not* be used.
4. The regression line always passes through the point $\overline{X}, \overline{Y}$.
5. The foregoing procedure is used only for purposes of predicting scores on $Y$ *from* $X$ ($Y$ is the criterion and $X$ is the predictor). Different formulas are needed to predict scores on $X$, and the simplest procedure in such cases is to change the designations (relabel $X$ as $Y$ and $Y$ as $X$) so that the foregoing notation may be used.

## The Connection Between Correlation and the *t* Test

An $r_{pb}$ is calculated like any other Pearson $r$, except that the two $X$ values may be assigned arbitrarily to two distinct groups. Therefore, the sign of the correlation is usually ignored.

### *Testing the Point-Biserial Correlation Coefficient for Significance*

The magnitude of $r_{pb}$ can be tested for significance with a $t$ test, like any other Pearson $r$:

$$t = \frac{r_{pb}\sqrt{N-2}}{\sqrt{1 - r_{pb}^2}} \qquad \text{Formula 10.7}$$

where $N$ is the total number of pairs, which equals the total number of cases in the two groups $= N_1 + N_2$.

### *Converting Significant Values of t to $r_{pb}$*

When a statistically significant value of $t$ in a test of the difference between two means is obtained, it is often helpful to convert it to $r_{pb}$ so as to determine the *strength* of the relationship. This is readily done as follows:

$$r_{pb} = \sqrt{\frac{t^2}{t^2 + df}} \qquad \text{Formula 10.8}$$

where

$$df = N_1 + N_2 - 2$$

***Comparing $r_{pb}$ to Another Measure of Effect Size, d***

Both $r_{pb}$ and $d$ are alternative measures of the strength of the relationship between a dichotomous grouping variable and a continuous DV. You can compute $r_{pb}$ directly from $d$ with the following formula:

$$r_{pb} = \sqrt{\frac{d^2}{d^2 + 4(df/N)}} \qquad \text{Formula 10.9}$$

**Estimating the Proportion of Variance Accounted for in the Population**

Squaring $r_{pb}$ gives you the proportion of variance accounted for in your data, but this is a *biased* estimate of that proportion in the population, which is called $\omega^2$. A nearly unbiased estimate of $\omega^2$ is given by the following formula, based on the $t$ value from the corresponding pooled-variance $t$ test comparing the means of the two groups:

$$\text{est.}\omega^2 = \frac{t^2 - 1}{t^2 + df + 1} \qquad \text{Formula 10.10}$$

The relationship between $\omega^2$ and Cohen's **d** for the population, two alternative measures of the population effect size, is a simple one, as shown in the following formula:

$$\omega^2 = \frac{\mathbf{d}^2}{\mathbf{d}^2 + 4} \qquad \text{Formula 10.11}$$

## Exercises

1. Use the data and your results from Exercise 1 in Chapter 9 to compute the linear regression equation for predicting the sophomore average ($Y$) from the number of cuts ($C$). What sophomore average would be predicted for a student who cut class eight times during the semester in question?
2. (a) Use the data and your results from Exercise 1 in Chapter 9 to compute the linear regression equation for predicting the sophomore average ($Y$) from the test score ($X$). (Advanced exercise: Compute the linear regression equation for predicting the test score [$X$] from the sophomore average [$Y$].) (b) Use the linear regression equation you calculated in part (a) to compute the predicted ($Y'$) sophomore average for each particular student. Then compute the error (or residual, $Y - Y'$) for each student and the squared error as well. (c) Using the results of part (b), compute the standard error of the estimate for predicting $Y$ from $X$ (i.e., the test score). Verify that this value ($\sigma_{est}$) is indeed equal to the value you would calculate by using Formula 10.5.
3. Compute the linear regression equation for predicting $Y$ from $X$ for each of the data sets in Exercise 3 in Chapter 9. What proportion of the variance in $Y$ is accounted for by $X$ in each data set? For each data set, use the appropriate formula to calculate the *unbiased* standard error of the estimate ($s_{est}$).
4. Use the $t$ value you calculated for Exercise 1 in Chapter 7 to find the point-biserial $r$ that corresponds to the difference in means between Turck and Kirk Halls.

Then use the appropriate $t$ formula to test the $r_{pb}$ you just computed for significance. How does this $t$ value compare with the original $t$ value you calculated for Exercise 1 in Chapter 7?

5. An industrial psychologist is investigating specific cafe locations of a local coffee chain, Spyhouse, to see if there is a difference in customer satisfaction between the original and newest locations. She obtained customer satisfaction scores from 41 customers who went to the original location of a local coffee chain and 31 customers who went to the chain's most recently opened location, with the following results (see Exercise 4 in Chapter 7): original location, M = 48.75 (SD = 9.0); newest location, M = 46.07 (SD = 10.0). First, calculate $d$ for these data. Then use the appropriate formula to calculate $r_{pb}$ directly from $d$. What proportion of variance in these data is accounted for by cafe location?

6. The following data come from Exercise 5 in Chapter 7. Calculate $r_{pb}$ for these data by assigning an $X$ value of 0 to Group 1 and an $X$ value of 1 to Group 2. What proportion of the variance in the scores is accounted for by group membership?

| Group 1 | Group 2 |
|---|---|
| 62 | 46 |
| 54 | 53 |
| 59 | 50 |
| 56 | 52 |
| 59 | 54 |

7. Convert each of the following statistically significant values of $t$, obtained from the $t$ test for the difference between two independent means, to $r_{pb}$.

| $t$ | $N_1$ | $N_2$ |
|---|---|---|
| 2.11 | 12 | 8 |
| 2.75 | 12 | 8 |
| 6.00 | 12 | 8 |
| 2.11 | 19 | 23 |
| 2.75 | 19 | 23 |
| 6.00 | 19 | 23 |
| 2.11 | 51 | 51 |
| 2.75 | 51 | 51 |
| 6.00 | 51 | 51 |

8. Another industrial psychologist asks a group of 9 assembly-line workers and 11 workers not on an assembly line (but doing similar work) to indicate how much they like their jobs on a 9-point scale (9 = like, 1 = dislike). The results are as follows:

| | | | | | | | | | | | |
|---|---|---|---|---|---|---|---|---|---|---|---|
| Assembly-line workers: | 4 | 4 | 4 | 2 | 1 | 5 | 3 | 4 | 3 | | |
| Other workers: | 6 | 5 | 7 | 5 | 3 | 7 | 6 | 8 | 7 | 3 | 3 |

(a) Test the null hypothesis that there is no relationship between the assembly line variable and job satisfaction by computing $r_{pb}$ and then using Table C to find the critical value. What should the psychologist decide? (b) Calculate the corresponding $t$ value using Formula 10.7, and then compute an estimate of omega squared ($\omega^2$) by using Formula 10.10. Does it look like there is a fairly large effect of job type on job satisfaction in the population?

## Thought Questions

1. In a research study using a sample of 30 participants, the Pearson $r$ correlation between $X$ and $Y$ is .09 and is *not* statistically significant. Should linear regression be used to predict scores on $Y$ given scores on $X$? Why or why not?
2. When using linear regression, does the researcher want the standard error of estimate to be small or large? Why? (a) If the standard error of estimate is large, what does this imply about the variability of scores on $Y$ for any given value of $X$? Why does this make predicting scores on $Y$ from scores on $X$ less accurate? (b) If the standard error of estimate is small, what does this imply about the variability of scores on $Y$ for any given value of $X$? Why does this make predicting scores on $Y$ from scores on $X$ more accurate?
3. Explain why each of the following statements is true: (a) The data from any research study that uses the $t$ test for the difference between two independent sample means can also be analyzed by computing a point-biserial correlation coefficient. (b) Any researcher using the $t$ test for the difference between two independent sample means should also compute a point-biserial correlation coefficient (or other measures of effect size) for the same data. (c) When computing a point-biserial correlation coefficient, it does not matter what two numbers you assign to the dichotomous variable.
4. (a) How do the variables in a research study that uses the ordinary Pearson $r$ correlation coefficient differ from the variables in a study that uses the point-biserial correlation coefficient? (b) Is the procedure for computing a point-biserial correlation coefficient different from the procedure for computing a Pearson $r$ correlation coefficient? Why or why not? (c) Is the procedure for testing a Pearson $r$ correlation coefficient for statistical significance different from the procedure for testing a point-biserial correlation coefficient for statistical significance? Why or why not? (d) In what way is the meaning of a squared point-biserial correlation coefficient the same as the meaning of any other squared Pearson $r$ correlation coefficient?
5. When using the Pearson $r$ correlation coefficient (and virtually all other statistical procedures), it is important to pay attention to the sign of a coefficient (plus or minus). Why, then, do we ignore the sign when interpreting a point-biserial correlation coefficient?

## Computer Exercises

1. Use your statistical package to perform a linear regression predicting the statistics quiz score from the academic interest score. Write out the formula for the raw-score regression line. What statistics quiz score would be predicted for a student who obtained a score of 20 on the academic interest questionnaire? A score of 40?
2. Perform a linear regression to predict the statistics quiz score from the student's self-reported procrastination level, and write out the raw-score regression formula. Use a scatter plot to help you interpret this result. Repeat the linear regression and scatter plot using procrastination to predict pre-quiz anxiety. (Advanced exercise: Redo these two regressions for just the psychology majors in Jackson's class.)
3. Use your statistical package to compute the Pearson $r$ between regular exercise and baseline anxiety for Jackson's students; calculate the $t$ value for testing $r$, using Formula 10.7, and compare it to the $t$ value you obtained for Computer Exercise 1 of Chapter 7.
4. Repeat Exercise 1 for baseline heart rate (compare to Chapter 7, Computer Exercise 2). How can you interpret the sign of the correlation for this and the previous exercise?
5. Create a new variable whose value is 1 for psychology majors and 2 for all other majors. Compute Pearson's $r$ between this new variable and the academic interest score. Interpret both the sign and the magnitude of this correlation.

## Bridge to SPSS

SPSS handles regression analyses completely separately from correlation. Moreover, to perform a simple linear regression (i.e., just one predictor with one criterion), SPSS employs the same module that it uses for multiple regression. Fortunately, it is easy to use that module for simple, bivariate linear regression and to interpret the output.

### Linear Regression

To compute any linear regression in SPSS, select **Regression** from the **Analyze** menu, and then choose **Linear** .... In the dialog box that opens (see Figure 10.5), move the variable that *you want to predict* (often called the criterion) to the area labeled *Dependent:* (there is room for only one variable in this space). Then move the variable that you want to use as your *predictor* to the area labeled *Independent(s)*; this space can accommodate more than one variable, in case you want to perform multiple regression. The **Options ...** button is used only for multiple regression, and the default choices from the **Statistics ...** button are all that you will need for simple linear regression. The choices available from the **Plots ...** button can help you to evaluate the assumptions (such as the linearity of the relation between the two variables) that underlie the use of linear regression, but are beyond the scope of this textbook. The **Save ...** button allows you to save the values predicted by the regression and/or the residuals (the difference between the actual value of the criterion and the predicted value), which could be useful for further analyses, but that is also beyond the scope of this book. The space labeled *Selection Variable:* allows you to perform your regression on a particular subgroup of your cases—e.g., only those who exercise regularly—but you can also do this by using **Select Cases ...** from the **Data** menu.

The default output for linear regression consists of six tables. The first two tables contain descriptive statistics due to the default selections from the **Statistics ...** button. The third table is really useful only for multiple regression, but it can be nice to have on your output

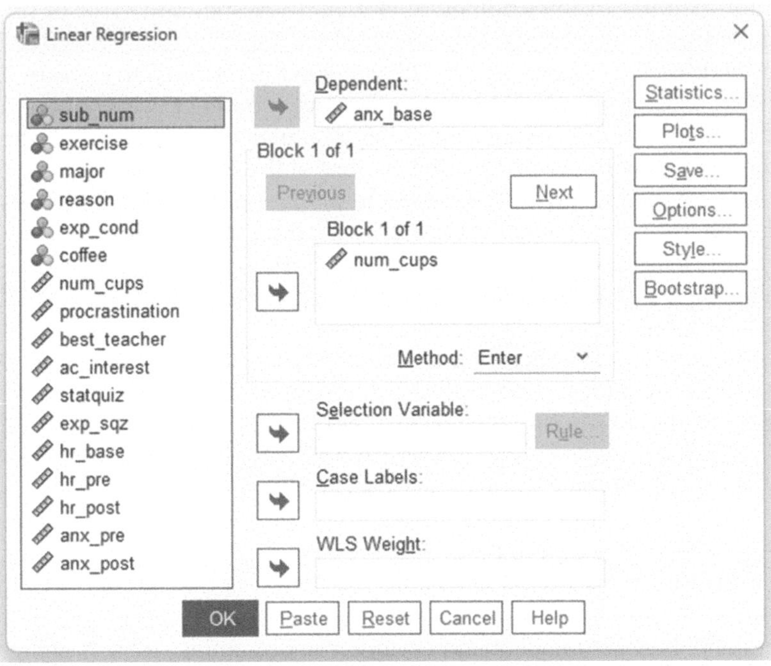


*Source*: Reprint courtesy of International Business Machines Corporation, © SPSS, Inc., an IBM company. SPSS was incorporated by IBM in October 2009

Figure 10.5

**The Dialog Box for Linear Regression**

# Figure 10.6

**Main Result Tables for Simple Linear Regression**

**Model Summary**

| Model | R | R Square | Adjusted R Square | Std. Error of the Estimate |
|---|---|---|---|---|
| 1 | .280[a] | .079 | .069 | 4.178 |

a. Predictors: (Constant), num_cups

**ANOVA[a]**

| Model | | Sum of Squares | df | Mean Square | F | Sig. |
|---|---|---|---|---|---|---|
| 1 | Regression | 145.962 | 1 | 145.962 | 8.362 | .005[b] |
| | Residual | 1710.548 | 98 | 17.455 | | |
| | Total | 1856.510 | 99 | | | |

a. Dependent Variable: anx_base

b. Predictors: (Constant), num_cups

**Coefficients[a]**

| Model | | Unstandardized Coefficients | | Standardized Coefficients | t | Sig. |
|---|---|---|---|---|---|---|
| | | B | Std. Error | Beta | | |
| 1 | (Constant) | 17.473 | .533 | | 32.790 | <.001 |
| | num_cups | 1.407 | .486 | .280 | 2.892 | .005 |

a. Dependent Variable: anx_base

because it tells you which variable was the predictor (IV) and which was the DV. The last three tables contain all the results you need to see from your regression (see Figure 10.6). The Model Summary table provides the Pearson $r$ between the IV and the DV (an uppercase R is used because with more than one predictor this value is the coefficient of *multiple* correlation), as well as the squared correlation (i.e., the proportion of variance explained), an adjusted squared correlation (similar to the estimate of $\omega^2$), and the standard error of the estimate.

The ANOVA table provides a significance test for the Pearson $r$ in terms of an $F$ ratio—a statistic that will be described in Chapter 13. The $p$ value (Sig.) is two-tailed by default and will allow you to decide if the correlation of the two variables is statistically significant. If the correlation is not significant, you may not want to make any use of the regression. Finally, it is the Coefficients table that provides you with the slope and intercept for the least-squares regression line, though the labels do not make this clear. The heading over the first two columns is *Unstandardized Coefficients* and the first column is labeled simply *B*. This is the column that gives you the slope and intercept of your regression equation. In the row labeled *(Constant)*, you will find the $Y$-intercept (17.473 for this example), and in the row under that is the slope of the regression line (1.407). These values are divided by the ones in the next column (*Std. Error*) to create $t$ values for testing both the intercept and the slope for significance. You will rarely be interested in the significance test for the intercept, and the $t$ value for the slope is exactly the same as the one you would get from the $t$ formula for testing $r$. In fact, squaring this $t$ value will give you the $F$ value in the ANOVA table, and the Sig. value in the ANOVA table (.005) will always be the same as the Sig. for your predictor in the Coefficients table when you are using only one predictor.

Given the slope and intercept in the Coefficients table, the regression equation for predicting baseline anxiety from the number of cups of coffee is: $Y = 1.407\ X + 17.473$. For example, if one of the students had actually consumed four cups of coffee, their baseline anxiety would be predicted to be $1.407 * 4 + 17.473 = 5.628 + 17.473 = 23.1$, a fairly high score.

### Point-Biserial Correlation

There is nothing special about computing a point-biserial correlation with SPSS; as long as a variable is Numeric, and not a *String* variable, it can be selected in the **Bivariate Correlations** dialog box as described in the previous chapter. And, as long as the variable contains at least two different values in the data sheet, it will produce a correlation coefficient when selected along with any other variable that meets the same criterion. However, whether the magnitude of that correlation will be meaningful depends on the nature of the variables involved. For example, if one of the variables has more than two arbitrary values (i.e., a categorical or qualitative variable), such as *undergrad major*, its correlation with any other variable will be meaningless. On the other hand, if a variable has only two arbitrary values, such as *exercise,* its correlation with any other variable that has quantitative levels (e.g., heart rate) will result in an interpretable correlation coefficient, which is usually referred to as a point-biserial correlation ($r_{pb}$). Just recall that the sign attached to this correlation can only be interpreted by knowing which level of the two-level variable was assigned the higher code number.

# Chapter 11
# Introduction to Power Analysis

## PREVIEW

### Introduction

What is the power of a statistical test? How is it related to a Type II error? What are the practical consequences of making a Type II error?

### Concepts of Power Analysis

How is the power of a statistical test related to the criterion of significance? To the size of the sample?

What is the population effect size, and how is it related to the power of a statistical test?

How large must the sample size be in order to use the procedures described in this chapter?

### Power Analysis for the Mean of a Single Population

What are the procedures for determining the power of statistical tests about the mean of a single population?

What are the procedures for determining the sample size required to have a specified power?

### Power Analysis for the Proportion of a Single Population

What are the procedures for determining the power of statistical tests about the proportion of a single population?

What are the procedures for determining the sample size required to have a specified power?

### Power Analysis for a Pearson *r*

What are the procedures for determining the power of statistical tests about the Pearson *r*?

What are the procedures for determining the sample size required to have a specified power?

### Power Analysis for the Difference Between Independent Means

What are the procedures for determining the power of statistical tests about the difference between two means, using independent samples?

What are the procedures for determining the sample size required to have a specified power?

### Power Analysis for the Difference Between the Means of Two Matched Populations

What are the procedures for determining the power of a matched-pairs *t* test? How do they differ from the procedures for independent samples?

How can the sample size be determined for a given level of power?

*Introductory Statistics for the Behavioral Sciences*, Eighth Edition. R. Brooke Lea and Barry H. Cohen.

Companion website: www.wiley.com/go/ISBS8e

**PREVIEW (*continued*)**

### Choosing a Value for d for a Power Analysis Involving Independent Means

How can you estimate the value of **d** from previously published results?

How can extreme possible values of **d** be used to make decisions about what sample sizes to use?

### Using Power Analysis Concepts to Interpret the Results of Null Hypothesis Tests

Why does a lack of statistical significance with small samples tell you little about whether the null hypothesis is true?

What can you conclude when statistically significant results are attained with smaller samples?

Why should you be extra cautious when interpreting the results of a significance test involving very large samples?

### The Null Hypothesis Testing Controversy Revisited

### Summary

### Exercises

### Thought Questions

### Computer Exercises

### Bridge to SPSS

## Introduction

Thus far, this text has been chiefly concerned with procedures for controlling the rate of Type I errors (incorrect rejections of the null hypothesis). This reflects a long-term emphasis among both behavioral researchers and statisticians. However, in hypothesis testing, there is a second important type of error to consider. You also run the risk that even if $H_0$ is false, you may fail to reject it. Incorrectly retaining a null hypothesis is called a *Type II error*. Of the cases in which the null hypothesis is *not* true, the proportion for which $H_0$ is nonetheless retained is the (conditional) probability of a Type II error, symbolized by $\beta$. The complement of this probability, $1 - \beta$, is the probability of getting a significant result when $H_0$ is false, and it is called the *power* of the statistical test.

To truly understand null hypothesis testing (NHT), you must also understand the factors that affect the rate of Type II errors and therefore statistical power.

When behavioral scientists test a null hypothesis, they almost always want to reject it; therefore, they want the power of the statistical test to be high rather than low. That is, they hope that there is a good chance that the statistical test will indicate that $H_0$ can be rejected. Despite its importance, this topic has received less emphasis in introductory statistical textbooks than it deserves. The unfortunate result of the ignorance about power is that research may be done in which, unknown to the investigator, power is low (and therefore the probability of a Type II error is high), a false null hypothesis is not rejected (i.e., a Type II error is actually made), and much time and effort are wasted. Worse, a promising line of research may be prematurely and mistakenly abandoned because the investigators do not know that they should have relatively little confidence concerning their failure to reject $H_0$.

This chapter deals first with the concepts involved in power analysis. Then, methods are presented for accomplishing the two major forms of power analysis that can be applied to the null hypothesis tests discussed thus far in this text.

## Concepts of Power Analysis

The material in this section, because it holds generally for a variety of statistical tests, is abstract and requires careful reading. In the ensuing sections, these ideas will be incorporated into concrete procedures for power and sample size analysis of five types of hypothesis tests. There are four major factors involved in power analysis:

1. ***The significance criterion, α.*** This, of course, is the familiar criterion for rejecting the null hypothesis. It equals the probability of a Type I error given that $H_0$ is true (usually .05). A little thought should convince you that the more stringent (the smaller) this criterion, other things being equal, the harder it is to reject $H_0$ even when it is *not* true, and therefore power is reduced.
2. ***The sample size, N.*** Whatever else the accuracy of a sample statistic may depend upon, it *always* depends on the size of the sample for which it has been determined. Thus, all the standard errors you have encountered in this book contain some function of *N* in the denominator. It follows that, other things being equal, error decreases and power increases as *N* increases.
3. ***The population "effect" size,* d.** Effect size is a critical concept for understanding power, and a very general one that applies to many statistical tests. When a null hypothesis about a population is false, it is false to some degree. (In other words, $H_0$ might be very wrong, or somewhat wrong, or only slightly wrong.) The parameter **d**, as described in Chapter 7, is a standardized measure of the *degree* to which the null hypothesis is false, or how large the "effect" is in the population. Within the framework of hypothesis testing, **d** can be looked at as a *specific* value that is an alternative to $H_0$. Specific alternative hypotheses, in contrast to such universal alternative hypotheses as $H_A$: $\mu_1 - \mu_2 \neq 0$ or $H_A$: $\rho \neq 0$, are what make power analyses possible. We will consider **d** in detail in subsequent sections.

   Other things being equal, power increases as **d**, the degree to which $H_0$ is false, increases. That is, you are more likely to reject a false $H_0$ if $H_0$ is very false. Given that power is a function of both **d** and *N*, it is generally true that the larger the **d**, the smaller the *N* that is required for significance to be obtained.
4. ***Power, or* 1 − β.** The fourth parameter is power, the probability of rejecting $H_0$ at the given significance criterion when $H_0$ is not true. It is equal to the complement of the probability of a Type II error. That is, power $= 1 - \beta$.

These four factors are mathematically related in such a way that any one of them is an exact function of the other three. We deal with the two most useful ones:

1. *Power determination.* Given that a statistical test is performed with a specified α and *N*, if you can specify (or at least estimate) the population state of affairs, **d**, the power of the statistical test can be determined.
2. *N determination.* For a given population state of affairs, **d**, if a statistical test using α is desired to have some specified level of power (say .80), the necessary sample size, *N*, can be determined.

One final foundational concept that will prove useful in statistical power analysis is **δ** (lowercase delta), where

$$\delta = \mathbf{d} \times f(N)$$

That is, $\delta$ is equal to **d** times a function of $N$. Thus, $\delta$ combines the population effect size and the sample size into a single index. The table from which power is read (Table D in the Appendix) is entered using $\delta$. As you will see, $\delta$ can be thought of as the expected value of your test statistic (e.g., $t$ value), and therefore, the larger $\delta$ is, the more likely it is that your test statistic will beat the critical value and become statistically significant. So, the larger $\delta$ is, the more power you have.

In the sections that follow, the general system described above for analyses of power determination and sample size determination will be implemented for five different statistical tests:

1. Tests of hypotheses about the mean of a single population (Chapter 6).
2. Tests of hypotheses about the proportion of a single population (Chapter 6).
3. Tests of the significance of a Pearson correlation coefficient (Chapter 9).
4. Tests of the difference between the means of two independent populations (Chapter 7).
5. Tests of the difference between the means of two matched populations (Chapter 7).

The procedures to be described are approximate and are not very accurate except for fairly large samples ($N$ at least 25 or 30). This is because the system we will be describing is based on the normal curve, which is only a good approximation to the $t$ distribution once $N$ is fairly large.

## Power Analysis for the Mean of a Single Population

In Chapter 5, we introduced NHT in an example involving the math SAT. The value of 500 was known to be the mean of the population, and the issue was whether the mean of individually tested students differed from that value. Thus, the null hypothesis is $H_0$: $\mu = \mu_0 = 500$.

The alternative hypothesis stated in Chapter 5 was merely that $H_A$: $\mu \neq 500$. However, power analysis is impossible unless a *specific* $H_A$ is stated.

### Power Determination

For the purposes of power analysis, let us assume that you suspect (or are interested in the possibility) that the population mean of individually tested participants is 25 points higher than 500, so that $H_A$: $\mu = 525$. Assume further that, to be cautious, a two-tailed .05 decision rule is to be used and that the sample size you have available is $N = 164$. Finally, assume that you know that the population standard deviation ($\sigma$) for the math SAT is 100. The power determination question can then be formulated as follows: If we perform a test at $\alpha = .05$ of $H_0$: $\mu = 500$ using a random sample of $N = 164$, and in fact, $\mu$ is 525, what is the probability that we will get a significant result and hence reject $H_0$?

The size of the effect postulated in the population, which is 25 points in *raw score* terms, must be expressed as a **d** value to accomplish the power analysis. For a test of the mean of a single population, **d** is expressed essentially as a $z$ score:

$$\mathbf{d} = \frac{\mu_1 - \mu_0}{\sigma}$$

In the present example,

$$\mathbf{d} = \frac{525 - 500}{100} = \frac{25}{100} = 0.25$$

Note that this is *not* a statement about *sample* results. It expresses, as an alternative hypothesis, the *population* state of affairs. That is, $\mu_1$ is postulated to be $0.25\sigma$ away from $\mu_0$, the value specified by $H_0$. Note also that the sign of **d** is ignored; an effect size of $-0.25$ (i.e., $\mu_1 = -475$) would lead to the same results for power.

Having obtained the measure of effect size, the next step is to obtain $\delta$. In the previous section, we pointed out that $\delta = \mathbf{d} \times f(N)$. For the test of the mean of a single population, the specific function of $N$ required is $\sqrt{N}$, so that

$$\delta = \mathbf{d}\sqrt{N} \qquad \text{Formula 11.1}$$

In the present example,

$$\delta = 0.25\sqrt{164} = 3.2$$

Entering Table D in the Appendix with $\delta = 3.2$ and $\alpha = .05$, the power is found to be .89. Thus, if the mean of the population of individually tested participants is 25 points away from 500, the probability of rejecting $H_0$ in this situation is .89 (or the probability of a Type II error is .11). If the mean of the population of individually tested participants is more than 25 points away from 500, the power will be greater than .89. Conversely, if the population mean is less than 25 points away from 500, the power will be less than .89.

It is important to understand that power analysis is based entirely on population values, whether we know them or whether we are just hypothesizing about them. The above analysis could well take place before the data were gathered to determine what the power would be under the specified $\alpha$, **d**, and $N$. Or it could take place after an experiment was completed to determine the power the statistical test most likely *had,* given $\alpha$, **d**, and $N$, using your own experimental results to estimate **d**, as we will explain shortly.

If the power for a reasonably postulated **d** were equal to a low value like .2 or .3, and the results of the experiment turned out not to be statistically significant, the nonsignificant result would hardly be surprising, since the a priori probability of obtaining significance was so small (and the probability of a Type II error was so high). However, a power of .90 associated with a nonsignificant result tends to suggest that the actual **d** is not likely to be as large as postulated.

In order to compute **d** in the above example, it was necessary to posit a specific value of $\mu_1$ (525) and also to know $\sigma$ (100). This can often be difficult when the unit is not a familiar one. (For example, extensive a priori data may not be available for a new psychological test.) For such situations, it is useful to specify conventional values corresponding to "small," "medium," and "large" values of **d**, which, although arbitrary, are reasonable and widely accepted (in much the same way as the .05 decision rule). For the test of the mean of a single population, the values suggested by J. Cohen (1988) are:

$$\textit{small}: \mathbf{d} = .20$$
$$\textit{medium}: \mathbf{d} = .50$$
$$\textit{large}: \mathbf{d} = .80$$

If you are unable to posit specific values of $\mu_1$ or $\sigma$, you can select the value of **d** corresponding to how large you believe the effect size in the population to be. However, do *not* use the above conventional values if you can specify **d** values that are appropriate to the specific problem or field of research in which the statistical test occurs, for conventional values are only reasonable approximations. As you get to know a substantive research area, the need for reliance on these conventions should diminish. Later in this chapter, we discuss how previously published research results can help you to estimate **d** for your study.

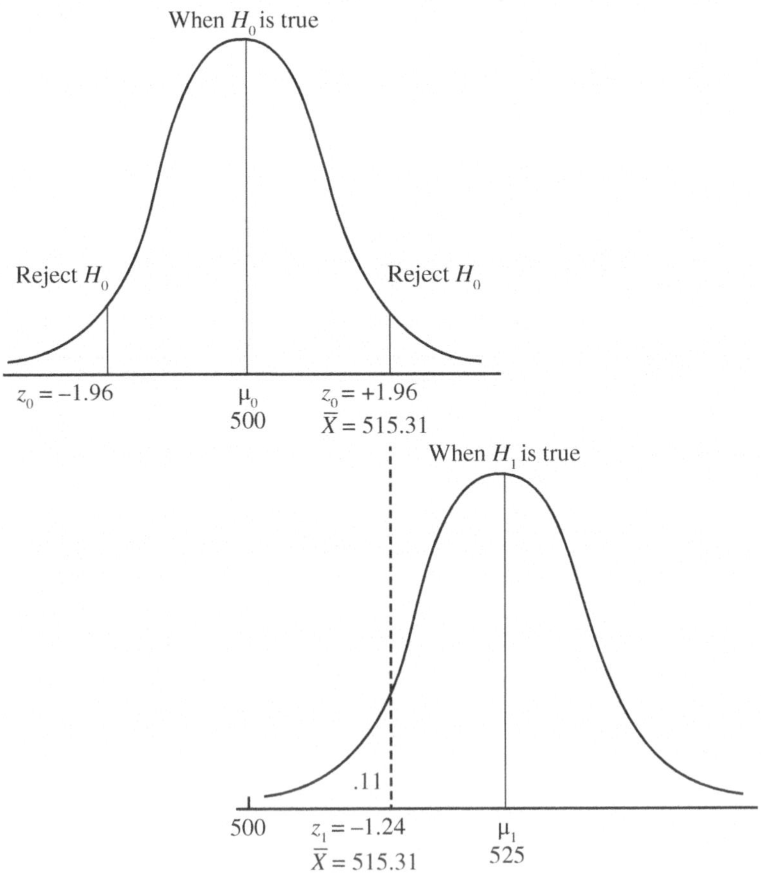


## Figure 11.1

**Sampling Distribution of Means ($N$ = 164) Assuming $\mu_0$ = 500 and $\mu_1$ = 525**

To clarify the general logic of power analysis, let us reconsider the preceding problem in terms of the concepts introduced in Chapter 5. The null hypothesis states that $\mu_0 = 500$ and the standard error of the mean is equal to 7.81 ($\sigma/\sqrt{N}$ is equal to 100/12.81), so the sampling distribution of means postulated by $H_0$ can be illustrated by the graph shown at the *top* in Figure 11.1. Assume that the true value of $\mu$ is equal to 525, for which the sampling distribution of means is as shown by the graph at the *bottom* in Figure 11.1. How likely is it that you will retain the incorrect $\mu_0$ of 500?

Because we are using the normal curve as an approximation to the appropriate $t$ distribution, the critical value ($z_0$) is equal to $\pm 1.96$. If a sample mean is obtained that (when expressed as a $z$ value) is numerically less than 1.96, the incorrect $H_0$ will be retained. This $z_0$ value can readily be converted to a critical value of the sample mean:

$$z_0 = \frac{\overline{X} - \mu_0}{s_{\overline{X}}}$$

$$1.96 = \frac{\overline{X} - 500}{7.81}$$

$$\overline{X} = (1.96)(7.81) + 500$$

$$\overline{X} = 515.31$$

The incorrect $H_0$ will be rejected if the sample mean is greater than 515.31. To see how likely this is to occur when the population mean is actually equal to 525, consider only the bottom curve in Figure 11.1, and convert the mean value of 515.31 to a $z$ value:

$$z_1 = \frac{\overline{X} - \mu_1}{\sigma_{\overline{X}}}$$

$$z_1 = \frac{515.31 - 525}{7.81} = \frac{9.69}{7.81} = -1.24$$

The percent of the area under the normal curve to the right of this $z$ value can be found by using Table A: At $z = -1.24$, it is about 89% (39.25% + 50.00%). Therefore, when $\mu = 525$, about 89% of the sample means that one can obtain would be greater than 515.31 and thus significant. So, the probability of rejecting $H_0$ when $\mu = 525$, and hence the power of this test, is .89. Of course, it was much easier to find the power for this problem by calculating $\delta$ and then looking up the power in Table D. However, we would like you to know how the values in Table D arose. Because the test is two-tailed and the analysis symmetrical, the preceding steps would lead to the same conclusion if the true $\mu$ were equal to 475, 25 points *below* $H_0$. Thus, we can conclude that, under these conditions, if $\mu = 525$ *or* 475, the probability of rejecting $H_0$ ($\mu_0 = 500$) is .89.

If the null hypothesis is so wrong that there is virtually no chance of a Type II error (i.e., if the effect size is extremely large), power will be virtually perfect (approaching 1.0). This does *not* indicate that the research is necessarily good, for the researcher may be testing a proposition so obviously wrong that the research would be pointless. (For example, the null hypothesis might be that the mean weekly hours of training of Olympic athletes is 6!) If, however, the null hypothesis is only slightly wrong (the researcher is studying a small-size effect), a large sample size will be needed in order for the research to have acceptable power.

## Sample Size Determination

We now turn to what is arguably the main purpose of statistical power analysis, sample size determination. Here, $\alpha$, $\mathbf{d}$, and the desired power are specified, and you wish to determine the necessary sample size. This form of analysis is particularly useful in experimental planning because it provides the only rational method for making the crucially important decision about the size of $N$.

First, we must consider the concept of "desired power." Your first inclination might be to set power very high, such as .99 or .999. But, as your intuition might suggest, the quest for near certainty is likely to result in the requirement of a very large $N$, usually far beyond your resources. This is similar to the drawback of choosing a very small significance criterion, such as .0001: Although a Type I error is very unlikely with such a criterion, you are also very unlikely to obtain statistical significance unless the effect size and/or sample size is unrealistically large. Therefore, just as it is prudent to seek less than certainty in minimizing Type I errors by being content with a significance criterion such as .05 or .01, a prudent power value is also in order insofar as Type II errors are concerned.

Although a researcher is free to set any power value that makes sense to her, we suggest the value .80 (which makes the probability of a Type II error equal to .20) when a conventional standard is desired. Why is this suggested value larger than the customary .05 value for the probability of a Type I error? In most instances in science, it is seen as less desirable to mistakenly reject a true $H_0$ (which leads to falsely positive, and possibly even embarrassing, claims) than to mistakenly fail to reject $H_0$ (which leads only to the failure to find something and no claim at all: "The evidence is insufficient to warrant the conclusion that …"). The bottom line is that, if desired power was conventionally set at .95 (making $\beta = .05$), most studies would demand larger samples than most investigators could muster. We repeat, however, that the .80 value should be used only as a general standard, with investigators quite free to make their own decisions based on the availability of participants, the cost of running the study, and other pertinent factors.

Let us return to the individual SAT test study, but now with a different purpose. Instead of assuming that $N$ is to be 164, suppose that you are now interested in the following question: "If I test at $\alpha = .05$ the null hypothesis that $\mu = 500$ when in fact $\mu = 525$ or 475 (and $\sigma = 100$), how large must $N$ be for me to have a .80 probability of rejecting $H_0$ (i.e., to have power = .80)?" To answer this question, the first step is to obtain two values: **d**, determined by the same methods as in the preceding section and equal to $(\mu_1 - \mu_0)/\sigma$ or 0.25 in this case, and the desired power, specified as .80. Next, $\delta$ is obtained by entering Table E in the Appendix in the row for desired power = .80 and the column for $\alpha$ (two-tailed) = .05; $\delta$ is found to be 2.80. (You could find the power value of .80 in the body of Table D and then see that it corresponds to a $\delta$ of 2.8, but Table D will not always contain the nice even values of power that you will usually want to look up, so we created Table E as a convenience.) Finally, for this test of the mean of a single population, $N$ is found as follows:

$$N = \left(\frac{\delta}{\mathbf{d}}\right)^2 \qquad \text{Formula 11.2}$$

In the present example,

$$N = \left(\frac{2.80}{0.25}\right)^2$$
$$= (11.2)^2$$
$$= 125$$

To have power = .80 in the present situation, the sample must have 125 cases in it. Note that this is consistent with the previous result where we saw that, other things (**d**, $\alpha$) being equal, $N = 164$ resulted in power = .89. To illustrate the consequences of demanding very high power, consider what happens in this problem if desired power is set at .999. From Table E in the Appendix, $\delta = 5.05$. Substituting,

$$N = \left(\frac{\delta}{\mathbf{d}}\right)^2 = \left(\frac{5.05}{0.25}\right)^2 = (20.2)^2 = 408$$

In this problem, to go from .80 to .999 power requires increasing $N$ from 125 to 408. This is more than most psychological researchers can easily manage. Of course, if data are easily obtained (e.g., a brief online survey), or if the cost of making a Type II error is great, no objection can be raised about such a high power demand.

## Power Analysis for the Proportion of a Single Population

The structure of the power analysis procedures remains the same as before; only the details need to be adjusted when dealing with proportions. The null hypothesis in question is that the population proportion, $\pi$, is equal to some specified value. That is, $H_0$: $\pi = \pi_0$. $H_0$ is tested against some specific alternative, $H_A$: $\pi = \pi_1$.

### Power Determination

Let us return to the worried politician in Chapter 6, who wants to forecast the results of an upcoming two-person election by obtaining a random sample of $N = 400$ voters and testing the null hypothesis that the proportion favoring her is $\pi_0 = .50$. Suppose that she has not yet conducted the survey, but she thinks that she is separated from her opponent by about .08. That is, she expects the vote to be .54 to .46 or .46 to .54. (Her expectations are stated in both directions because a two-tailed significance test is intended.) The question can be summarized as follows: If a statistical test is performed at $\alpha = .05$ of $H_0$: $\pi = .50$ against the specific alternative $H_A$: $\pi = .54$ (or .46) with $N = 400$, what is the power of this statistical test?

For the test of a proportion from a single population, the effect size, **d**, and $\delta$ are defined as follows:

$$\mathbf{d} = \frac{\pi_1 - \pi_0}{\sqrt{\pi_0(1 - \pi_0)}}$$

$$\delta = \mathbf{d}\sqrt{N}$$

For the data of this problem,

$$\begin{aligned}\mathbf{d} &= \frac{0.54 - 0.50}{\sqrt{0.50(1 - 0.50)}} \\ &= \frac{0.04}{0.50} \\ &= 0.08 \\ \delta &= 0.08\sqrt{400} \\ &= 1.60\end{aligned}$$

Entering Table D with $\delta = 1.60$ and $\alpha = .05$, the power is found to be .36. So, our worried politician has something else to worry about: As she has planned the study, she has only about one chance in three of coming to a positive conclusion (rejecting the null hypothesis that $\pi = .50$) if the race is as close as she thinks ($H_A$: $\pi = .54$ or .46). If the poll had already been conducted as described and the results were not statistically significant, she should consider the results inconclusive. Even if $\pi$ is as far from .50 as .54 or .46, the probability of a Type II error ($\beta$) is .64.

### Sample Size Determination

Let us now invert the problem to one of determining $N$, given the same $\mathbf{d} = 0.08$, $\alpha = .05$ and specifying the desired power as .80. The formula for $N$ for the test of a proportion from a single population, like the previous formula for $\delta$, is the same as in the case of the mean of a single population:

$$N = \left(\frac{\delta}{\mathbf{d}}\right)^2$$

The value of $\delta$ for the joint specification of power = .80, and $\alpha$ (two-tailed) = .05 is found from Table E (again) to be 2.80. Therefore,

$$\begin{aligned}N &= \left(\frac{2.80}{0.08}\right)^2 \\ &= (35)^2 \\ &= 1{,}225\end{aligned}$$

This sample size is considerably larger than the $N = 400$ which yielded a power of .36 under these conditions. It is hardly a coincidence that in surveys conducted both for political polling and for market and advertising research, sample sizes typically run about 1,000 to 1,500.

If conventional values for effect size for this test are needed, the conventional **d** values for correlation (see next section) can be used: small, .10; medium, .30; large, .50.

## Power Analysis for a Pearson *r*

Recall from Chapter 9 that for testing Pearson's $r$, the null hypothesis is usually $H_0$: $\rho = 0$. For the purpose of power analysis, the alternative hypothesis is that the population value is some *specific* value other than zero: $H_A$: $\rho = \rho_1$. For example, you might be testing an $r$ for significance when expecting that the population value is .30. In this case, $H_A$: $\rho = .30$. (Again, for two-tailed tests, the value is taken as either .30 or −.30.)

### Power Determination

When dealing with a Pearson $r$, you do not need a formula to define the effect size, because $r$ is already in the form of an effect-size measure; the value for **d** is simply the value of $\rho$ specified by $H_A$ (.30 in this example). To find $\delta$, the appropriate function of $N$ to be combined with **d** is $\sqrt{(N-1)}$.

$$\delta = \mathbf{d}\sqrt{N-1} = \rho_1\sqrt{N-1} \qquad \text{Formula 11.3}$$

Let us consider again the study in Chapter 9 in which we measured the degree of linear relationship between GPA and SAT scores in a random sample of 25 college students. Suppose we had planned a two-tailed statistical test with $\alpha = .01$, and we had expected the population $\rho$ to be .43; thus $\rho_1 = .43$, whereas $H_0$: $\rho = 0$. What is the power of this test? We can go directly to

$$\delta = .43\sqrt{25-1} = 2.10$$

Entering Table D for $\delta = 2.10$ and $\alpha = .01$, power is found to be .32. One chance in three of finding significance seems hardly worth the effort. We may then reconsider the stringency of our significance criterion and check $\alpha = .05$; the power of .56 with this more lenient criterion may still be disappointing. If so, we could plan to increase our sample size (as shown in the next section). However, if $\rho$ was actually equal to the obtained sample $r$ of .65 (see Table 9.2), $\delta$ would be about 3.2, with $N = 25$, and power would be as high as .73, even for $\alpha = .01$.

If the investigator has difficulty in formulating an alternative hypothetical value for $\rho_1$, the following conventional values are offered (J. Cohen, 1988): small, .10; medium, .30; large, .50. The value of .50 may not seem "large," but over most of the range of behavioral science where correlation is used, correlations between different variables (e.g., not two measures of anxiety) do not often get much larger than that.

### Sample Size Determination

Returning to the SAT/GPA study, we can ask what $N$ is necessary for a test at $\alpha$ (two-tailed) = .05, assuming $\mathbf{d} = \rho_1 = .30$, in order to have power of (let us say) .75. The value of $N$ is just one more than for the other two one-sample tests previously described:

$$N = \left(\frac{\delta}{\mathbf{d}}\right)^2 + 1 = \left(\frac{\delta}{\rho_1}\right)^2 + 1 \qquad \text{Formula 11.4}$$

$\delta$ is found from Table E for power = .75, $\alpha = .05$, to be 2.63, so

$$N = \left(\frac{2.63}{0.30}\right)^2 + 1$$
$$= (8.77)^2 + 1$$
$$= 78$$

To have a .75 chance (i.e., three-to-one odds) of finding $r$ to be significant if $\rho_1 = .30$ (or $-.30$), the researcher needs a sample of 78 cases.

## Power Analysis for the Difference Between Independent Means

The next significance test whose power analysis we consider is the test of the difference between the means of two independently drawn random samples. This is probably the most frequently performed test in the behavioral sciences, so we will go into a good deal of detail for this case.

As stated in Chapter 7, the null hypothesis most frequently tested is $H_0$: $\mu_1 - \mu_2 = 0$ (often written as $\mu_1 = \mu_2$). For power analysis, a *specific* alternative hypothesis is needed. We will write this as $H_A$: $\mu_1 - \mu_2 = \theta$ (theta), where $\theta$ is the difference between the means expressed in *raw* units. To obtain **d**, the standard measure of effect size, this difference must be standardized. This is done using the standard deviation of the population, $\sigma$ (which is a single value since we assume that $\sigma_1 = \sigma_2$ for the two populations). Thus, the value of **d** in this instance is conceptually the same as for the test of a single population mean, namely

$$\mathbf{d} = \frac{\text{alternative-hypothetical } \mu_1 - \mu_0}{\sigma} = \frac{\theta}{\sigma}$$

Thus, the effect size in the two-group case can be looked upon as the difference in means expressed in units of $\sigma$. Again, the direction of the difference (the sign of $\theta$) is ignored in two-tailed tests.

This device of "standardizing" the difference between two means is generally useful, and not only for purposes of power analysis. Frequently in behavioral science, there is no sure sense of how large one unit of a raw score really is. How large *is* a point? Well, they come in different sizes: An IQ point comes about 15 to the standard deviation ($\sigma_{IQ}$), while an SAT point is much smaller, coming 100 to the standard deviation ($\sigma_{SAT}$). By always using $\sigma$ as the unit of measurement, we achieve comparability from one measure to another, as was the case with $z$ scores.

This device helps us with regard to other related problems. One occurs whenever we have not had much experience with a measure (e.g., a new test of attention), and we have little if any basis for estimating the population $\sigma$, which the foregoing formula for **d** requires. Paradoxically enough, we can use this $\sigma$ as our unit, despite the fact that it is unknown, by thinking directly in terms of **d**. Thus, a **d** of 0.25 indicates a difference between two population means (whose exact value we do not know) equal to $0.25\sigma$ (whose exact value we also do not know). It is as if our ignorance cancels out, and in a most useful sense $\mathbf{d} = 0.25$ is always the same size difference whether we are talking about IQ, height, socioeconomic status, or a brand-new measure of "tree planting desire."

How large *is* a **d** of 0.25, or any other? Here, as before, it is possible (and sometimes necessary) to appeal to some conventions, and they are the same as in the test of the mean of a single population: small, .20; medium, .50; large, .80. In this framework, a **d** of 0.25 would be characterized as a "smallish" difference. To gain a more concrete feeling with respect to the size of **d**, it may help to think about **d** in terms of the overlap of the two distributions involved. When **d** is 2.0, the mean of one population is two SDs away from the mean of the other; only a little more than 2% of the population with the lower mean, for instance, will extend above the mean of the "higher" population. When **d** is merely "large" (see Figure 11.2A), about 21% of the lower population exceeds the mean of the higher population, and when **d** is "small," more than 42% of the lower population is above the mean of the higher one (see Figure 11.2B). By contrast, with a **d** of 4.0, there would be so little overlap between the two distributions that relatively few scores from the lower population would surpass *any* of the scores from the higher population (see Figure 11.2C).

### Power Determination

For the comparison of two means, the value of $\delta$ is equal to

$$\delta = \mathbf{d}\sqrt{\frac{n}{2}} \qquad \text{Formula 11.5}$$

where

$n$ = size of *each* of the two samples (thus $2n$ cases are needed in all)

Compare this formula to Formula 7.9A for finding $t$ in terms of $d$ (the effect size in your samples) and $n$ (reproduced here):

$$t = d\sqrt{\frac{n}{2}} \qquad \text{Formula 7.9A}$$

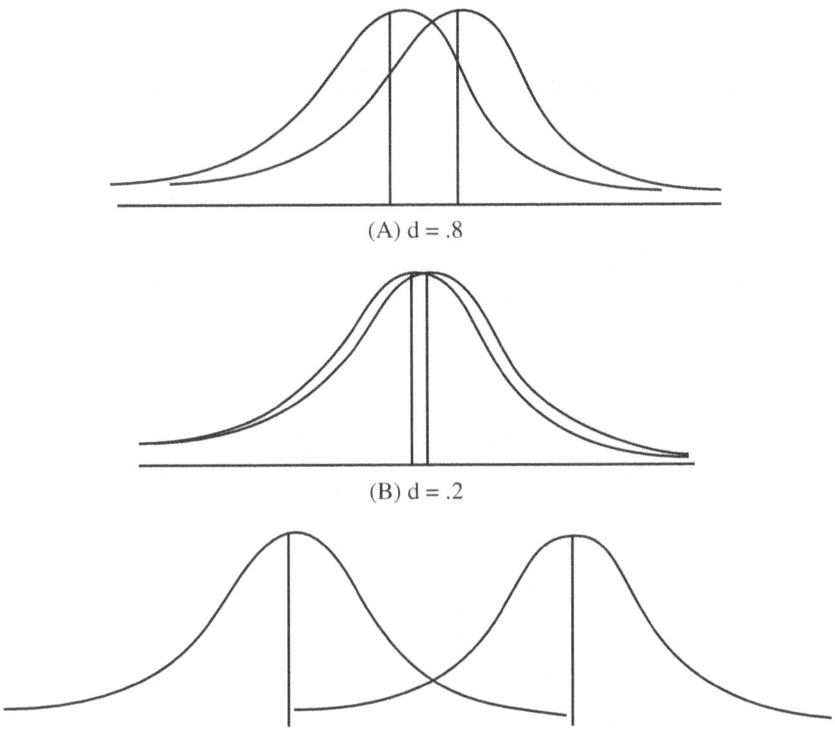


# Figure 11.2

**Overlap of Populations as a Function of Effect Size**

Over many exact replications of a particular experiment, the values of $d$ will average out (approximately) to **d**, the effect size in the population. And the $t$s for these replications should, in the long run, average out to (approximately) $\delta$ (delta), which can be thought of as the *expected t value* for an experiment with a particular value for **d** and a fixed size of $n$ for each of the two samples.

As an illustration, consider the caffeine study in Chapter 7. Given two groups, each with $n = 40$ cases, one of which is given a dose of caffeine and the other a placebo, we will test at $\alpha = .05$ (two-tailed) the difference between their mean heart rates following the treatment. If between the populations there is a mean difference of medium size, operationally defined as **d** = .50, what is the power of the test? With **d** given on the basis of convention, there is no need to compute it and hence no need to estimate $\mu_1$, $\mu_2$, or $\sigma$. Thus, given a fixed $n$, all we need to calculate is:

$$\delta = .50\sqrt{\frac{40}{2}} = 2.24$$

Entering Table D in the column for $\alpha = .05$, we find that for $\delta = 2.2$ power = .59 and for $\delta = 2.3$ power = .63. Given that 2.24 is about halfway between 2.2 and 2.3, it is reasonable to say that our power is about halfway between .59 and .63, or approximately .61. Thus, for medium differences (as defined here) and samples of 40 cases each, we have only a .61 probability of rejecting $H_0$ for a two-tailed test using $\alpha = .05$.

Ordinarily, in research where variables are to be manipulated, such as the caffeine experiment described here, it is possible to arrange the experiment so that the total available pool of participants is divided equally among the groups. This division is usually optimal. But it is not always possible or desirable to have equal sample sizes. For example, it would be a mistake to reduce the larger of two available samples of convenience to make it equal to the smaller. Instead, to determine power when $N_1 \neq N_2$, we can use the same device we employed in Chapter 7; we can compute the *harmonic mean* of the two $N$s and then use this obtained value for $n$ in the formula for $\delta$. As an example, consider a psychiatric hospital study where there are resources to place a random sample of 50 newly admitted schizophrenic patients into a special experimental treatment program. After one month, their progress is compared with a control group of 150 cases receiving regular treatment for the same period. It is planned to rate each patient on an improvement scale 30 days following admission and to test for the significance of the difference in means for rated improvement, using a two-tailed test where $\alpha = .05$. If it is anticipated that $\mathbf{d} = 0.40$ (between "small" and "medium"), what is the power of the test?

Because $N_1 \neq N_2$, first compute:

$$\overline{X}_{\text{harmonic}} = \frac{2N_1N_2}{N_1 + N_2} = \frac{2(50)(150)}{50 + 150} = 75$$

Then enter this value as $n$ in the formula for $\delta$:

$$\delta = 0.40\sqrt{\frac{75}{2}} = 2.45$$

Interpolating between the values for $\delta = 2.4$ and $\delta = 2.5$ in Table D in the $\alpha = .05$ column yields power $= .69$ (i.e., midway between .67 and .71).

Note that a total of $N_1 + N_2 = 200$ cases will be studied. But since $N_1 \neq N_2$, the resulting power is the same as if two samples of 75 cases each, or 150 cases, were to be studied. There is a smaller effective $n$, and hence less power, when the sample sizes are not equal. Equal sample sizes usually yield the optimal level of power for a given total $N$. Were it possible to divide the total of 200 cases into two equal samples of 100, $\delta$ would then equal $0.40\sqrt{(100/2)} = 0.40 \times 7.07 = 2.83$, and the resulting power at $\alpha = .05$ obtained from Table D by interpolation is about .81.

### Sample Size Determination

The sample size needed for *each* independent sample when two means are to be compared can be found for specified values of $\alpha$, $\mathbf{d}$, and the desired power, as was shown for the other statistical tests in this chapter. The combination of $\alpha$ and power is expressed as $\delta$ by looking in Table E, and the values for $\delta$ and $\mathbf{d}$ are substituted in the equation appropriate for this statistical test, which is:

$$n = 2\left(\frac{\delta}{\mathbf{d}}\right)^2 \qquad \text{Formula 11.6}$$

For example, it was found in the Chapter 7 caffeine study example that, for $\alpha = .05$ and $\mathbf{d} = 0.50$, the plan to use samples of $n = 40$ each resulted in power $= .61$. If power $= .90$ is desired, what $n$ per sample is needed?

For a two-tailed test where $\alpha = .05$ and power $= .90$, Table E yields $\delta = 3.24$. Substituting these values in Formula 11.6, we obtain:

$$n = 2\left(\frac{3.24}{0.50}\right)^2$$

$$= 84 \text{ cases in each group, or 168 cases in all}$$

Again, we see that large desired power values demand large sample sizes. Were the more modest power of .80 specified, $\delta$ would equal 2.80 (Table E) and $n$ would equal $2 \times (2.80/0.50)^2$, or about 63, a substantially smaller number.

Suppose you want to be extra careful to avoid a Type I error, so you decide to use $\alpha = .01$, instead of the more common $\alpha = .05$. What do you suppose will happen to your estimate of power? For a given effect size and sample size, reducing the Type I error rate will increase the Type II error rate, and therefore reduce power. Making alpha smaller makes the critical value you need to beat larger, and therefore you will tend to attain significance less often. To keep power at the level you want with a smaller $\alpha$, but the same effect size, you will need to use a larger sample. This can be made clear with a numerical example.

You have seen in Table E that for $\alpha = .05$, and power $= .8$, the required $\delta$ is 2.80. Now, look in the column for $\alpha = .01$ and you will see that to maintain power at .8, you will have to have a $\delta$ of 3.42. If the effect size you expect is the same as in the previous example (**d** = .50), the $n$ you will need for .8 power is $2 \times (3.42/0.50)^2 = 2 \times (6.84)^2 = 2 \times 46.8 = 93.6$, so 94 participants per group. Compare this number to the sample size of 63, when $\alpha$ was fixed at .05. Conversely, if you keep your $\delta$ as 2.80 (e.g., sample size of 63 per group, with the effect size at .5), but $\alpha$ at .01, Table D shows you that your power will be reduced from .80 to .59. For any given value for $\delta$, you can see in Table D that as you move to the right in that row (i.e., to smaller values for $\alpha$), the amount of power goes down considerably.

## Power Analysis for the Difference Between the Means of Two Matched Populations

The last significance test whose power analysis we will consider is the test of the difference between *dependent* or *correlated* means—that is, the test that is applied to the data from a matched-pairs design. The formulas and procedures for power analysis, in this case, differ from the preceding one (i.e., the difference between two independent means) by the inclusion of one additional term that depends on the degree to which the two sets of scores are matched in the population, as expressed by the correlation coefficient $\rho$.

### Power Determination

Given that **d** is defined as in the case of two independent means, the value of $\delta$ for a matched (or repeated-measures) design is given by:

$$\delta = \mathbf{d}\sqrt{\frac{1}{1-\rho}}\sqrt{\frac{n}{2}} \qquad \text{Formula 11.7A}$$

where $\rho$ is the correlation between the two sets of scores in the population. Note that as $\rho$ gets closer to 1.0, the denominator of the first term gets smaller, making the first term larger. When $\rho$ is .75, the term based on $\rho$ becomes 2.0, and $\delta$ therefore becomes twice as large as it would be for an otherwise similar experiment without matching. The power boost from this degree of matching is substantial; for a population correlation as high as .75, it is equivalent to the boost you would get by quadrupling your sample sizes. It is instructive to express Formula 11.7A in terms of Formula 11.5:

$$\delta_{\text{matched}} = \delta_{\text{independent}}\sqrt{1/(1-\rho)} \qquad \text{Formula 11.7B}$$

where $\delta_{\text{independent}}$ is given by Formula 11.5.

You can see from Formula 11.7B that a $\rho$ of zero would leave $\delta_{\text{independent}}$ unchanged—that is, $\delta_{\text{matched}}$ would equal $\delta_{\text{independent}}$ in that case. Conversely, a $\rho$ close to +1 would cause $\delta_{\text{matched}}$ to be much larger than $\delta_{\text{independent}}$. A negative $\rho$, which is unlikely when you are making an effort to match participants, would actually *reduce* your $\delta$ and therefore your power. To illustrate, let us return to the data from the Stroop experiment as shown in Table 7.2 of Chapter 7. Suppose that you expect a large effect size (**d** = .8), and a population correlation ($\rho$) of .5, for the 10 participants in the study. You plan to perform a two-tailed matched-pairs $t$ test at the .05 level. Given these values,

$$\delta = .8\sqrt{\frac{1}{1-.5}}\sqrt{\frac{10}{2}} = .8\sqrt{2}\sqrt{5} = (.8)(1.414)(2.236) = 2.53$$

From Table D, we can see that a $\delta$ of 2.53 corresponds to a power of about .72 for a two-tailed significance test using .05 for alpha. This is a barely adequate level of power for conducting an experiment, and if a lower value of **d** or ρ were expected, the sample size would have to be increased to maintain even this minimal level of power.

### Sample Size Determination

The formula for finding the sample size required for a given level of power, with values specified for **d** and ρ, is the same as the one for independent samples multiplied by a factor of (1 − ρ). So, if you expect a matching correlation of .9 in the population, you will only need one-tenth the number of participants (1 − .9) to achieve the level of power that would result from no matching at all.

For the heart rate comparison, suppose you want power to be .8 for a .05, two-tailed test (δ = 2.8, from Table D), and you expect a medium effect size (**d** = .5) and a ρ of only .3. The number of pairs of scores that would be required is given by the following formula:

$$n = 2(1 - \rho)\left(\frac{\delta}{\mathbf{d}}\right)^2 \qquad \text{Formula 11.8}$$

For this example,

$$n = 2(1 - .3)\left(\frac{2.8}{.5}\right)^2 = 1.4(31.36) = 43.9$$

Therefore, 44 participants would each have to be measured twice in order to attain the desired level of power.

## Choosing a Value for d for a Power Analysis Involving Independent Means

Students often feel uneasy when they first discover that power analysis is only as accurate as your estimate of **d**. Unless you are replicating an experiment for which you already have the results, it is not obvious how you can make a reasonable guess about the **d** that corresponds to your new experiment. Let us now consider the relatively simple case in which you *are* contemplating the replication of a published result.

### Estimating d From Previous Research

Suppose you read a journal article reporting that a group of participants listening to music composed by Mozart produced significantly higher scores on a test of spatial ability than a control group who listened to popular (e.g., "soft rock") music. Further suppose that the chief result was reported as "$t(34) = 2.4$, $p < .05$." If you know that the two groups were the same size, then you know that $n$ was 18 ($2n - 2 = 34$), and you can then find $d$ from $t$ and $n$ by using Formula 7.9B:

$$d = t\sqrt{\frac{2}{n}} = 2.4\sqrt{\frac{2}{18}} = 2.4\left(\frac{1}{3}\right) = .8$$

If you were to replicate this study exactly, it would be reasonable to use $d$ from the previous study as the estimated **d** for your new study when calculating power for different possible sample sizes. Even if you planned to use a different spatial ability measure whose mean and standard deviation were quite different from those in the previous study, it would not be unreasonable to suppose that **d** (which is in standardized units) would be fairly similar for both studies.

At what point is your new study so different from a previously published study that you can no longer use $d$ from the previous study as an estimate of **d** for your new study? This is a judgment call that requires a thorough knowledge of your research domain. It is the same kind of judgment that underlies the increasingly popular statistical procedure known as *meta-analysis* (discussed further in the next chapter). In some forms of meta-analysis, the *d*s for any number of similar studies are averaged together, with greater weight being given to studies with larger samples, to arrive at an improved estimate of **d** (with a slight correction for bias, *d* is known as Hedge's *g*). This (weighted) average *d* or *g* can then be tested for significance using the combined *N* of all the studies involved. This method creates one large "meta" study, which greatly increases power relative to any one individual study, and thus makes a Type II error much less likely.

If the two samples are not equal in size in a previous study but you are given both $N_1$ and $N_2$, you can still calculate *d* from *t* by using the unequal-*N* formula for *d* (based on the harmonic mean) given in Chapter 7, and reproduced here:

$$d = t\sqrt{\frac{N_1 + N_2}{N_1 N_2}} \qquad \text{Formula 7.11}$$

Suppose that the *N*s for the Mozart study were as follows: $N_1 = 30$ and $N_2 = 6$ (degrees of freedom [*df*] is still 34). In order for *t* to be 2.4, *d* would have to be:

$$d = 2.4\sqrt{\frac{36}{180}} = 2.4\sqrt{\frac{1}{5}} = 1.07$$

We showed earlier that as your two *N*s diverge, your power decreases. Here we are showing a related principle: When your *N*s are unequal, you need a larger effect size to have the same power you would have had with the same total *N*, but equal sample sizes.

### Trying Extreme Values for d

One possible approach to sample-size determination begins with deciding on the smallest effect size for which you would like to have adequate power. For instance, suppose you feel that for your experiment there would be little point in having a barely adequate level of power, say, .7, for an effect size of only .1. Given that power = .7 corresponds to a δ of 2.48 for a .05, two-tailed test, this constraint corresponds to a size for each sample of

$$n = 2\left(\frac{\delta}{\mathbf{d}}\right)^2 = 2\left(\frac{2.48}{.1}\right)^2 = 2(24.8)^2 = 2(615) = 1{,}230$$

Thus, you would not consider running your experiment with more than about 1,200 participants in each group, because that would give you adequate power for an effect size too small to be of interest.

Conversely, you can begin by deciding on the largest value that is reasonably possible for your effect size. Suppose you think it unlikely that **d** is greater than 1.0. Again using the value of .7 for barely adequate power, your target sample size would be:

$$n = 2\left(\frac{2.48}{1}\right)^2 = 2(6.15) = 12.3$$

This calculation tells you that there would be little point in running your experiment with *fewer* than 13 participants per group, because in that case, you would expect your power to be less than adequate (i.e., below .7), even with **d** as large as 1.0.

## Using Power Analysis Concepts to Interpret the Results of Null Hypothesis Tests

Whether you ever perform a power analysis before running an experiment, it is important to understand the factors that affect the power of a statistical test. This can help you interpret the results of null hypothesis tests correctly and avoid some serious misconceptions.

### Lack of Statistical Significance Does Not Imply That the Null Hypothesis Is Likely to Be True

The smaller your samples (all else being equal), the less power you will have to detect a given effect size in the population. In many common situations, power is surprisingly low. For instance, if your two-group experiment involves 32 participants in each group, your power will be only about .5 whenever you test medium-size effects at the .05 level. Here's the calculation that demonstrates that:

$$\delta = .5\sqrt{\frac{32}{2}} = .5(4) = 2.0$$

Looking up δ in Table D, you can see that you will reject $H_0$ (correctly) 52% of the time, but you will retain $H_0$ (and therefore commit a Type II error) nearly half of the time. So, when you see a study involving small samples and $H_0$ is *not* rejected, you should be aware that $H_0$ could easily be false, nonetheless. Unfortunately, even when the true effect size is considerably more than zero, there is a good chance that the null hypothesis will not be rejected if the samples are small. All that a lack of statistical significance can suggest when moderate sample sizes are used is that the true effect size is not likely to be very large.

### Attaining Statistical Significance When the Samples Are Small

For a given value of $t$, the *smaller* $n$ is, the larger $d$ has to be. And a larger $d$ for your samples suggests a larger **d** in the population. However, when using small samples, $d$ could accidentally be much larger or much smaller than the **d** of which it is just an estimate because the standard deviation for $d$ is large when the $n$s are small. By itself, a large $t$ value only suggests that **d** is not very likely to be zero, but a large $t$ obtained with small $n$s tells you that $d$ is fairly large and that **d** therefore is not likely to be very small. But note that there are strong advantages to using large samples if doing so is practical and affordable (e.g., obtaining online responses to a brief questionnaire). Using large samples will produce a relatively narrow confidence interval when estimating **d**.

### Large Samples Tend to Produce More Accurate Results But Can Lead to Misleading Conclusions

The larger your samples, the more accurate is $d$ as an estimate of **d**. Because it is the magnitude of **d** that you would really like to know (not just whether it is unlikely to be zero, which is all you can conclude from NHT), larger samples are always more informative. However, there is usually some economic cost that comes with increasing your sample size (e.g., paying participants, paying research assistants), and the additional information may not be worth the cost. Moreover, very large samples make it easy to obtain statistically significant $t$ values even when $d$ (and, therefore, probably **d**) is rather small. So, you should not be overly impressed with statistically significant results that are obtained from large samples; in such cases, you need to pay extra attention to effect-size estimates.

Also, remember that using repeated measures or matched samples can be similar to using larger samples without matching, in terms of attaining statistical significance with experimental treatments that are quite weak in their effects. Note that *statistical* significance is less likely to imply *practical* significance when the samples are very large or the correlation between matched sets of scores is very high.

## The Null Hypothesis Testing Controversy Revisited

As we mentioned at the end of Chapter 5, there is much confusion concerning what NHT does and does not tell you about the state of affairs in a population, and there has been much debate in the literature on this topic, especially in recent years. (We repeat our suggestion here that a good way to learn more about these issues is by reading a comprehensive review article, such as Nickerson [2000]). However, even though there does seem to be some increased reporting of confidence intervals and effect-size estimates, NHT, with its reporting of $p$ values and decisions about $H_0$ still dominates the results sections of articles published in many branches of behavioral science research.

There is one major advantage of NHT that would be difficult to achieve with any new system: the widespread agreement on the .05 level as the largest, routinely acceptable value for alpha. This fixed criterion of significance, coupled with the usual constraints on sample sizes (behavioral researchers rarely employ very large samples), tends to make it likely that *statistically* significant results will be associated with effect sizes that are large enough to be interesting. Conversely, whereas NHT screens out 95% of studies in which the null hypothesis is actually true (i.e., with $\alpha = .05$, 95% of the null experiments will fail to attain statistical significance), it also screens out about 93% of very small effect sizes (e.g., $\mathbf{d} \leq .1$) when using small samples (e.g., about 30 participants in each of the two groups being compared). However, it is important to note that if the null hypothesis were true for the majority of studies conducted in a particular area of research, screening out 95% of them would not be very helpful; the absolute number of Type I errors committed could still get rather large. If there is reason to expect that the null hypothesis is often true in a research domain, studies in that domain ideally should be required to use a smaller value for alpha.

Technically, failing to reject $H_0$ when **d** is anything more than zero is a Type II error, but given all of the small effects that behavioral researchers may explore, many of these Type II errors save us the trouble of dealing with effects too small to be practical or even worthy of further exploration. For example, consider the "Mozart effect": the finding that listening to music composed by Mozart improves performance on a spatial ability test (Rauscher, Shaw, & Ky, 1993). It is not really interesting if Mozart's music produces an extremely tiny, temporary effect on spatial ability (especially when compared to popular music, as in the original study). Did anyone think that listening to Mozart's music was likely to reduce spatial ability? And would anyone be surprised if the music of other classical composers also produced such an effect? So, in many cases, especially when dealing with purely exploratory studies, for which the true effect size might easily be quite small, some Type II errors can actually be helpful.

However, (effect) size isn't everything. Sometimes even a tiny effect can be very interesting indeed, especially if it represents a phenomenon that was thought to be impossible (e.g., learning to control directly and voluntarily the function of your kidneys). It is often important to look carefully at estimates of effect size when a result is statistically significant, but it is also important to consider the theoretical significance of a study. When you are dealing with a tiny but important effect, it may indeed be worth the expense of gathering the large samples you will need to obtain adequate power, unless, of course, you can achieve a large power gain with matching or repeated measures.

## Summary

Power is the *probability of getting a significant result* in a statistical test for which $H_0$ is *not* true. Power is equal to $1 - \beta$, where $\beta$ = the probability of a Type II error.

### The Importance of Power

If power is not known, a researcher may waste a great deal of time by conducting an experiment that has little chance to produce significance even if $H_0$ is false. Perhaps worse, she may abandon a promising line of research because she does not know that she should have relatively little confidence concerning her failure to reject $H_0$ (i.e., the probability of a Type II error is large).

### The Four Major Parameters of Power Analysis

1. *The significance criterion*, $\alpha$. A Type II error is more likely as $\alpha$ gets smaller because you would fail to reject $H_0$ more often. Thus, power decreases as $\alpha$ decreases.
2. *The sample size, N*. Larger samples yield better estimates of population parameters and make it more likely that you will reject $H_0$ when it is correct to do so. Thus, power increases as $N$ increases.
3. *The population "effect" size*, **d**. You are less likely to fail to reject a false $H_0$ if $H_0$ is "very wrong"—that is, if there is a large effect size in the population. Thus, power increases as **d** increases.
4. Power, $1 - \beta$.

Any one of these four parameters can be determined mathematically from the other three.

### General Procedures for the Two Most Common Forms of Power Analysis

1. Power determination.
   (a) Compute or posit the value of **d**.
   (b) Compute $\delta$ (delta), which combines the $N$ that you are considering and **d**.
   (c) Obtain power from the appropriate table, according to your desired $\alpha$.
2. Sample size determination.
   (a) Specify desired power. (If a conventional value is needed, we recommend .80 for most purposes.)
   (b) Compute or posit the value of **d**.
   (c) Obtain $\delta$ from the appropriate table, according to your desired $\alpha$.
   (d) Compute $N$.

If you have no basis for estimating **d**, or the population values from which **d** is computed, you can use the appropriate conventional values of **d**, as devised by J. Cohen (1988), for the statistical test in question. Also, note that using Tables D or E to look up a value of delta or power leads to approximate results for power or required sample sizes, respectively, because these tables are based on the normal distribution. However, the obtained values become more accurate for larger samples and are a good approximation for sample sizes of 30 or more.

### Specific Formulas for d, δ, and *N*

The exact form of the required formulas depend on the statistical test that is being performed.

1. Power Analysis for the mean of a single population:

   (a) $\mathbf{d} = \frac{\mu_1 - \mu_0}{\sigma}$

   where

   $\mu_0$ = value of $\mu$ specified by $H_0$
   $\mu_1$ = value of $\mu$ specified by $H_A$ (a *specific* $H_A$ is always necessary in power analysis)
   $\sigma$ = population standard deviation

   (b) $\delta = \mathbf{d}\sqrt{N}$ Formula 11.1

   (c) $N = \left(\frac{\delta}{\mathbf{d}}\right)^2$ Formula 11.2

   (d) Conventional values of **d**: small, .20; medium, .50; large, .80

2. Power Analysis for the proportion of a single population:

(a) $\mathbf{d} = \dfrac{\pi_1 - \pi_0}{\sqrt{\pi_0(1 - \pi_0)}}$

where

$\pi_0$ = value of $\pi$ specified by $H_0$
$\pi_1$ = value of $\pi$ specified by $H_A$

(b) $\delta = \mathbf{d}\sqrt{N}$

(c) $N = \left(\dfrac{\delta}{\mathbf{d}}\right)^2$

(d) Conventional values of **d**: small, .10; medium, .30; large, .50

3. Power Analysis for a Pearson $r$ (includes $r_{pb}$)

(a) $\mathbf{d} = \rho_1$, the value specified by $H_A$

(b) $\delta = \mathbf{d}\sqrt{(N - 1)}$ Formula 11.3

(c) $N = \left(\dfrac{\delta}{\rho_1}\right)^2 + 1$ Formula 11.4

(d) Conventional values of **d**: small, .10; medium, .30; large, .50

4. Power Analysis for the difference between independent means

(a) $\mathbf{d} = \dfrac{\theta}{\sigma}$

where

$\theta$ = value of $\mu_1 - \mu_2$ specified by $H_A$
$\sigma$ = population standard deviation (a single value because it is assumed that $\sigma_1 = \sigma_2$)

(b) $\delta = \mathbf{d}\sqrt{\dfrac{n}{2}}$ Formula 11.5

where
$n$ is the size of *each* sample (thus, $2n$ cases have been used in the experiment)

If the sample sizes are unequal, use the harmonic mean:

$$\bar{X}_{\text{harmonic}} = \frac{2N_1N_2}{N_1 + N_2} \quad \text{Formula 7.10}$$

$$n = 2\left(\frac{\delta}{\mathbf{d}}\right)^2 \quad \text{Formula 11.6}$$

where
$n$ is the size of *each* sample (so $2n$ cases will be needed in all)

(a) Conventional values of **d**: small, .20; medium, .50; large, .80

5. Power Analysis for the difference between *dependent* (i.e., correlated) means

(a) $\mathbf{d} = \dfrac{\theta}{\sigma}$ where $\theta$ is the difference of the population means according to the null hypothesis

(b) $\delta = \mathbf{d}\sqrt{\dfrac{1}{1 - \rho}}\sqrt{\dfrac{n}{2}}$ Formula 11.7A

where $\rho$ represents the linear correlation of the two sets of scores in the population

(c) $n = 2(1 - \rho)\left(\frac{\delta}{\mathbf{d}}\right)^2$ Formula 11.8

6. Estimating **d** from published research
   (a) You can calculate $d$ from a published $t$ value if you are given the two sample sizes:

$$d = t\sqrt{\frac{N_1 + N_2}{N_1 N_2}} \qquad \text{Formula 7.11}$$

   (b) If the sample sizes are equal, this formula reduces to

$$d = t\sqrt{\frac{2}{n}} \qquad \text{Formula 7.9B}$$

   where $n$ is the size of each of the two samples.

## Exercises

1. Suppose that the students at Bigbrain University are planning to test whether the mean math SAT score for their school is higher than the national average ($\mu$) of 500 (assume that $\sigma = 100$).
   (a) If they believe that their mean is 520, and they plan to sample 25 students, what is the power of their statistical test at the .05 level, two-tailed? What would the power be for a one-tailed test? Explain why power is higher for the one-tailed test.
   (b) Recalculate the one- and two-tailed power values in part (a) assuming that the Bigbrain students expect their average to be 50 points higher than the national average. Explain why these power values are higher than the ones you calculated in part (a).
   (c) Redo part (a) assuming that a sample of 100 Bigbrain students is being planned. Explain why these power values are higher than the ones you calculated in part (a).
   (d) Given the expected effect size in part (a), what sample size would be needed to obtain a power of .8 for a two-tailed .05 test? For a two-tailed .01 test?
   (e) Repeat part (d) given the expected effect size in part (b).
2. Calculate the power for the tests you conducted for Turck and Kirk Halls in Exercise 1 of Chapter 6. (Use the $t$ values you calculated for that exercise as the values for $\delta$ in your power calculation.)
3. (a) Exercise 5 in Chapter 6 was based on an artist who needed more than 50% of the voters for a new mascot at a certain midwestern college to choose her highland cow design over the current mascot. If it wins, her logo design will be prominently featured throughout the school. To find out how her campaign is going, she plans to obtain a random sample of 81 voters and see how many plan to vote for the highland cow, rather than the lochness monster. She is willing to posit a specific alternative hypothesis of 60% (or 40%) and wishes to use the .01 criterion of significance. Compute the power of her statistical test. How would you evaluate her plan for predicting her chances of winning the vote?
   (b) Suppose the artist decides to switch to the .05 criterion of significance (but that the other values are not changed). Will this improve the power of the statistical test to a satisfactory level?
   (c) The artist finally resigns herself to doing more work and obtaining a larger sample. She wants power to be .75. How large a sample does she need (using the .05 criterion)?

4. (a) A personality theorist expects that two particular traits have a linear correlation on the order of .40 in the population she plans to sample. She wishes to test the null hypothesis that the correlation between the two traits is exactly zero, using the .05 criterion of significance and a random sample of 65 subjects. Is the power of this statistical test satisfactory? Explain.
   (b) How large a sample would the theorist in part (a) need to obtain power = .9 with alpha set at .01 for a two-tailed test? For a one-tailed test?
5. If the sample $r$ you calculated in Exercise 2 of Chapter 9 was equal to $\rho$ for the population, how large a sample would you need to obtain power = .7 for a two-tailed test at the .05 level?
6. (a) Calculate $d$ for the comparison of the means of Turck and Kirk Halls (if you did not already calculate that value for Exercise 1 of Chapter 7). Use $d$ as your estimate of $\mathbf{d}$ and compute the sample sizes you would need for power = .75, for a .05 two-tailed test, comparing the two dormitories.
   (b) Given the sizes of the samples from Turck and Kirk Halls, how large would $\mathbf{d}$ have to be to obtain power = .85, for a .01, two-tailed test?
7. (a) Calculate $d$ from the pooled-variance $t$ value you computed to solve Exercise 2 in Chapter 7. Use that value for $d$ as your estimate of $\mathbf{d}$ and compute the sample sizes you would need for power = .9, for a .05 two-tailed test, assuming that you plan to use two equal-size samples to compare Turck and Dupre Halls.
   (b) Given the sizes of the samples in Exercise 2 of Chapter 7, how large would $\mathbf{d}$ have to be to obtain power = .8, for a .05, two-tailed test?
8. Calculate the missing values in the following table, assuming that you are comparing two independent, equal-size samples with a two-tailed $t$ test. (Note: Small = .2, Medium = .5, Large = .8.)

| | | | | ***n* in *each* sample needed for power of .85 if:** | |
|---|---|---|---|---|---|
| **Proposed *n* in each sample** | **Effect size** | **Power $\alpha = .05$** | **Power $\alpha = .01$** | **$\alpha = .05$** | **$\alpha = .01$** |
| 30 | Small | _____ | _____ | | |
| 100 | Small | _____ | _____ | _____ | _____ |
| 30 | Medium | _____ | _____ | | |
| 100 | Medium | _____ | _____ | _____ | _____ |
| 30 | Large | _____ | _____ | | |
| 100 | Large | _____ | _____ | _____ | _____ |

9. Suppose that you plan to match students between Turck and Kirk Halls so that the population correlation corresponds to .4. Given the $\mathbf{d}$ you found in part (b) of Exercise 6 and the sizes of the samples from Turck and Kirk Halls, how much power would you have for a two-tailed matched-pairs $t$ test at the .01 level?
10. Recalculate the required sample sizes for part (a) of Exercise 7 if the two populations are matched with $\rho = .6$.
11. (a) Calculate the power your test would have in Exercise 7 of Chapter 7 if the two sets of scores were not correlated at all in the population (use $d$ for those data as $\mathbf{d}$ in your power calculation).
    (b) Calculate the power your test would have in Exercise 7 of Chapter 7 if the two sets of scores had a population correlation of $\rho = .5$ (again, use $d$ for those data as $\mathbf{d}$ in your power calculation).

## Thought Questions

1. What is the power of a statistical test? Why is it important when planning a research study?
2. "Using the .05 criterion of significance, the probability of a Type I error is .05. Using the .001 criterion of significance, the probability of a Type I error is .001 (much smaller). Therefore, researchers should always use the .001 criterion of significance." Use the concept of statistical power to explain why the last sentence in the quotation is not a good idea.
3. A researcher argues as follows: "I performed a correlational study with a very large sample. Because my sample was so large, the correlation of $r = .07$ which I obtained was statistically significant. But this correlation is too small to be of any practical value, as there is very little (linear) relationship between the two variables that I studied. I therefore warn other researchers *not* to use large samples so as to avoid getting such misleading results." (a) Use the concept of statistical power to explain why the last sentence in the preceding quotation is not entirely true. (b) What is the advantage of obtaining the largest sample that a researcher can afford?
4. In the following table, decide which of the numbered statements belongs in each box:

| Sample size | You retained $H_0$ | You rejected $H_0$ |
|---|---|---|
| Very small | | |
| Very large | | |

    1. You are studying an effect that is likely to be quite large.
    2. Caution is indicated because you are studying an effect that may be quite small.
    3. Power is likely to be so low, and the probability of a Type II error so large, that no conclusions of any kind should be drawn.
    4. Power is likely to be high, so it may well be that $H_0$ is (approximately) true.

## Computer Exercises

1. Suppose that Jackson is planning to compare the regular exercisers with the nonexercisers in their class. Use the power analysis function in your statistical package, or use a power calculator on the web, to answer the following questions. Assume that alpha is .05 and that each test is two-tailed.
    (a) Use the $t$ value you computed for the exercise status difference in baseline anxiety (first computer exercise of Chapter 7) as your value for delta ($\delta$) to determine the power of that test.
    (b) If the population effect size for another dependent variable (DV) was only .3, how much power would Jackson have for that comparison?
    (c) If Jackson had equal-size groups of frequent and infrequent exercisers, how large would each have to be to attain power = .8, given the effect size in part (b)?
    (d) Repeat part (b) for $\mathbf{d} = .7$.
2. Solve this problem using the same methods and assumptions that you used for Exercise 1.
    (a) Use the $t$ value you computed to compare the students in the "impossible to solve" condition with those in the "easy to solve" condition in terms of postquiz heart rates (see Computer Exercise 3 of Chapter 7) as your value for $\delta$ (delta) to determine the power of that test.
    (b) Had the population effect size for this DV been medium in size ($\mathbf{d} = .5$), how many students would Jackson have needed to have in each condition to attain power = .7?

## Bridge to SPSS

In this chapter, we simplified power analysis by using the normal distribution as an approximation of the *t* distribution; that is why we needed only two brief tables (Tables D and E) to complete our analyses. Unfortunately, as the samples you are dealing with get smaller, the normal approximation becomes less accurate. In fact, the distributions that are needed for an accurate power analysis when comparing the means of two fairly small samples are not the family of *t* distributions but an even more complex, nonsymmetrical family known as the **noncentral *t* distribution** (NCTD). For large samples, the NCTD, like the ordinary *t* distribution, begins to resemble the normal distribution fairly well, but in the past, for greater accuracy, researchers used printed tables and graphs to estimate the power of a proposed study (see J. Cohen, 1988). In recent years, free, easy-to-use power calculator programs have become available on the Web. SPSS has been surprisingly late in following this trend, but in the last few years, SPSS has finally added power analysis to its Analyze menu, and it is a welcome addition. In fact, Power Analysis is now the first choice on the Analyze menu.

We will begin our description of power analysis by SPSS with the common problem of deciding how large your samples should be to have an acceptable chance of attaining significance for a two-group comparison. Let us suppose that you expect regular exercise to have the effect of reducing an individual's anxiety. You are aware of the comparison in Jackson's study showing that regular exercisers report significantly less anxiety than their fellow students who don't do much exercise, but you are concerned that it is just an observational result—it is possible that greater anxiety causes people to exercise less, not that exercise causes a reduction in anxiety. So, you plan a longitudinal study, in which students are randomly assigned either to follow a regular exercise routine for two months or to refrain from exercise during the same period. Assuming that the two groups are equivalent in baseline anxiety, you want to know how many students you will need in each group to have a good chance of statistical significance.

To use SPSS to run the power analysis, first make the following menu selection: **Analyze/Power Analysis/Means/Independent-Samples *T* Test.** A dialog box will open with the default settings shown in Figure 11.3. Immediately under **Test Assumptions**, there is a drop-down menu labeled *Estimate:* that has only two choices. The default choice, *Sample size*, is the one we need for this first example. In the *Single power value:* space, type in the amount of power you want to have; that is, the probability that your results will attain statistical significance, knowing that the higher the requested power, the larger the required sample sizes will be. As we mentioned earlier in this chapter, a power of .80 is generally considered a good compromise. For most experimental studies, there is no reason to change the default value for *Group size ratio:* (i.e., 1).

Next, you will see another two-choice drop-down menu, labeled *Specify:*. In this case, you will probably want to change from the default option, which asks you to hypothesize separately about the population means and standard deviations, to the other option: *Effect Size*. This is where you have to make the best educated guess you can about the effect size you can expect to be created by your intervention (or that already exists in the population if you are selecting from pre-existing groups). For this illustration, we will use one of the effect size measures found for Jackson's exercise/anxiety test, as shown in Figure 7.4. To be on the cautious side, we will use the point estimate of Hedges' correction of Cohen's *d*, and type in .433 in the space labeled *Value:*. Finally, under Test Direction, we will leave the defaults in place: a two-sided test at the .05 level.

The results of the choices just described are shown in Figure 11.4. From this table, you can see that you would need to have 85 students in each group to have 80% power with a smallish effect size of 0.433. This SPSS procedure makes it very easy to explore different effect sizes, different levels of desired power, and different one- or two-tailed alpha levels, to see how many participants you might need.

Sometimes, it is interesting to know the power that your study actually had, if the population effect size were the same as the effect size you found in your samples. This value is called the *Observed Power*, and it can be obtained by choosing the other option in the

## Figure 11.3

**The Dialog Box for the Power Analysis of an Independent-Samples *t* Test**

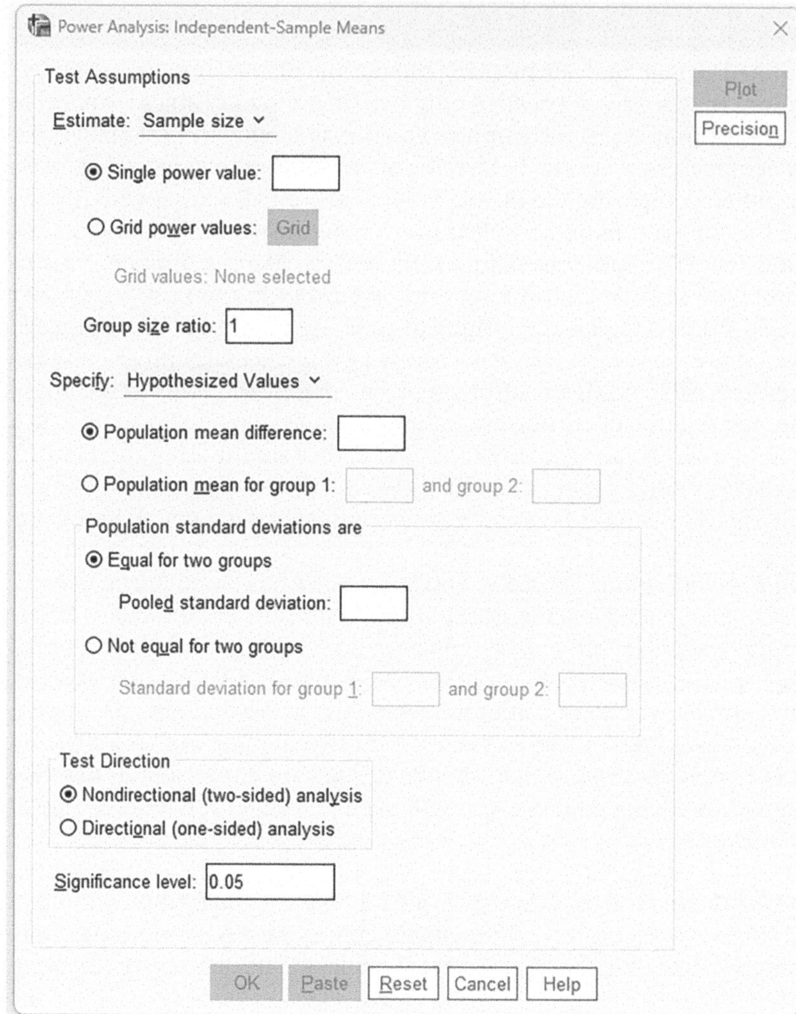


*Source*: Reprint courtesy of International Business Machines Corporation, © SPSS, Inc., an IBM company. SPSS was incorporated by IBM in October 2009

## Figure 11.4

**The Results Table for a Power Analysis to Determine the Sample Sizes Needed for an Independent-Samples *t* Test**

**Power Analysis Table**

| | | | | Test Assumptions | | |
|---|---|---|---|---|---|---|
| | N1 | N2 | Actual Power[b] | Power | Effect Size[c] | Sig. |
| Test for Mean Difference[a] | 85 | 85 | .801 | .8 | .433 | .05 |

a. Two-sided test.

b. Based on noncentral t-distribution.

c. Group variances are assumed to be equal.

*Estimate:* menu, which is *Power*. If you choose this option, the dialog box will ask you to enter the sizes of the two samples. Following the example from Jackson's study, we entered 57 and 43, and left the *Effect Size* (.433) in place (you would type that value in, of course, if this were your first use of this dialog box). In Figure 11.5, you can see that if you knew that the population effect size was about .43 and you were stuck with sample sizes of 57 and 43, it would be barely worth running the study. You would have a better than 50/50 chance of

**Power Analysis Table**

| | Power[b] | Test Assumptions | | | |
|---|---|---|---|---|---|
| | | N1 | N2 | Effect Size[c] | Sig. |
| Test for Mean Difference[a] | .565 | 57 | 43 | .433 | .05 |

a. Two-sided test.

b. Based on noncentral t-distribution.

c. Group variances are assumed to be equal.

## Figure 11.5

**The Results Table for a Power Analysis to Determine the Observed Power for an Independent-Samples *t* Test Already Conducted**

getting significance, but your probability of attaining significance would be less than .6. An observed power of .5 would mean that your $t$ value was exactly equal to the critical $t$ value, so your $p$ value would be .05, right at the border of statistical significance.

Note that for a given total $N$ of the two samples (e.g., 100), power is maximized when the two groups are the same size. When we changed the sample sizes from 57 and 43 to 50 and 50, the observed power turned out to be .573, instead of .565, as in Figure 11.5.

# Chapter 12
# Beyond Traditional Null Hypothesis Testing

## PREVIEW



## More on Criticisms of NHT (and Some Rebuttals)

Near the end of Chapter 5, and again at the end of Chapter 11, we mentioned that there have been sharp criticisms of NHT, since its inception, from quantitative psychologists. Before we consider some of the alternative methods proposed for data analysis, we will continue to discuss some of the major criticisms of NHT.

### The Null Hypothesis is Never (or Very Rarely) True

One persistent criticism of NHT is that the null hypothesis is never exactly true in psychological research so, strictly speaking, there cannot be any Type I errors. It is extremely unlikely, for instance, that the mean depression score for an entire population of those taking


*Introductory Statistics for the Behavioral Sciences*, Eighth Edition. R. Brooke Lea and Barry H. Cohen.



Companion website: www.wiley.com/go/ISBS8e

a new antidepressant will be *exactly* equal to the population mean of a placebo control group. However, psychologists do sometimes study interventions that in reality have very little effect, and they often fail to reject the null hypothesis in those cases, unless they are using unusually large sample sizes. This means that Type II errors can be quite common when tiny effects are being studied without using large sample sizes. Tiny effects can be important, especially in the medical arena, or when testing an effect with dramatic theoretical consequences, but for routine experimental psychology studies, missing a very tiny effect does not have dire consequences and may actually be helpful. Rejecting the null for a very tiny effect may lead others to waste their time studying an effect that is too small to be worth pursuing.

### With Large Enough Samples, the Null Will Always Be Rejected (Unless it is Exactly True)

As you learned in the previous chapter, unless an effect is actually zero, power increases as the sample sizes increase, so no matter how tiny the effect is, you can reject the null by using large enough samples. This argument builds on the one in the previous paragraph: If the null is virtually never true, why bother testing your intervention to see if it has zero effectiveness? Later in this chapter, we will discuss why large samples can be worth the trouble, but at this point, we need to consider that it is extremely rare to reject the null without stating and interpreting the direction of your results. For example, in a two-group comparison of means, the direction of the results is simple: Which group had the larger mean? We almost always care about the direction of our results and not just whether the null can be rejected. You will recall that the two-tailed test is the default for psychologists performing a $t$ test. That applies to the critical value for significance, but we almost never stop by merely rejecting the null—we declare the results to be statistically significant in one direction or the other.

The importance of the description in the previous paragraph is that whereas large enough samples virtually guarantee rejection of the null, they cannot guarantee that the results will come out in the direction you are predicting (e.g., the mean of the experimental group is larger than the mean of the control group). In fact, NHT helps to control Type III errors, which occur when you reject the null, but your results are in the wrong direction—e.g., the phonics method for reading is better than the visual/holistic method in the entire population, but in your study, the visual/holistic group performed significantly better than the phonics group due only to chance factors, such as sampling error. Type III errors are rare (less than half of your alpha level), but if you did not use NHT and just reported the direction of your results, directional errors would be pretty common when the effect sizes are small. So, as you can see, NHT can play a useful role with respect to tiny effects.

### The $p$ Value Does Not Tell You If the Null Hypothesis is Likely to be True (but Bayes' Statistics Can!)

#### *Disease Detection Analogy*

Imagine that an all-purpose disease detection device has been invented that can tell whether a person is harboring a particular disease that the device is set for, even though the person is exhibiting no symptoms of illness at all. The device is not perfect, but you can set how sensitive it is for detecting disease. Because we want to control worrisome false alarms, we will set the device so that only 5% of totally healthy people are labeled as having the disease being tested for. For this hypothetical device, the 5% false alarm setting results in 20% of those with the disease to be falsely labeled healthy (the disease has been "missed"). Further reducing the percentage of false alarms would result in less sensitivity and therefore more missed diseases. The percentages of false alarms (Type I errors) and misses (Type II errors) initially selected for this analogy were chosen to resemble a common situation that occurs with null hypothesis testing.

As a pilot experiment, this device is used to determine which of the people randomly tested for a relatively rare new Coronavirus (CV-25) have that virus in their system. Those labeled as having that virus are all sent to the same new hospital wing that has been set up for treating the virus. After testing a large sample for the CV virus, the new hospital wing contains 100 people awaiting treatment for a virus that some may not even have. Now we can ask the question this analogy has been building towards: How many of the 100 people in this ward are actually free of the CV-25 virus, having been falsely identified as having it? Are you tempted to say that five are virus-free because 5% of virus-free people tested in this experiment are being sent to this ward? If so, you would be in good company – many people assume that the answer is 5% -- but you would be wrong. That would mean that you are looking at the percentages in this analogy in a backwards way.

The answer to the question about how many in that ward are actually virus-free is not knowable given the information you have because we did not tell you the total number of people who were tested with this device to arrive at 100 people testing positive for CV-25. Even more importantly, you do not know how many of those tested actually had the CV-25 virus. For instance, if 500 people were tested by the device and 400 were totally healthy and only 100 really had the CV-25 virus, then 20 people would have been sent to that ward for treatment they did not need (5% of 400) along with 80 of those infected with CV-25 (because 20% of the 100 infected would have been falsely declared healthy, while 80% are correctly labeled as having the virus). Because of the very high percentage of healthy people being declared infected in this example, 20% (20 out of 100) of those in the ward are actually virus-free and are there by mistake (a lot more than just 5%).

Now consider a very different example, perhaps with a different, much more common, virus being tested for (e.g., the Epstein–Barr virus, which about 95% of all adults will eventually have in their bodies). A total of 144 people are tested, of which only 20 are virus-free, while 124 are infected. In this case, only one healthy person is sent to the treatment ward (5% of 20) and 99 of the infected (80% of 124, with a little rounding off ) are sent. This time, only 1 of the 100 people in the ward is actually free of the virus, or just 1%. We know the percentage of virus-free people who will be labeled infected, because we set that in the device at 5%, just as we traditionally set the alpha level of null hypothesis testing to .05. If we want to know the reverse percentage—that is, the percentage of people labeled infected who are really virus-free—we need to use a mathematical equation known as Bayes's theorem. That is what we have been doing informally in the above example.

### Connecting the Analogy to *NHT*

The purpose of our disease-detection analogy is to make our understanding of NHT more concrete. Let us define a "null" study as one for which the population effect is actually zero. This will correspond to a virus-free person in the analogy. Obtaining statistical significance for a null study, then, is analogous to judging a virus-free person to be infected. The focus of NHT is to limit the number of null studies that are declared significant (i.e., Type I errors), just as our hypothetical device is trying to minimize false alarms so as to minimize the need for follow-up procedures and needless worry (of course, the setting for false alarms should be adjusted according to how likely the virus being tested for could cause serious disease). Note that the Type I error rate, like the proportion of virus-free people judged infected, is a *conditional* probability. It is not the probability that a statistical test will be found significant, it is the probability that a statistical test will be significant, *given that the null hypothesis is true* (in psychological research, the null hypothesis is almost always a zero effect in the population).

If we use S to represent significance and N to represent the null being true, this can be written as P(S | N), where the vertical line can be read as "given that" and the whole statement would be read as "the probability of S given N". This is the Type I error rate, or alpha, that we decide on. The more interesting question is: What proportion of significant results are really Type I errors (i.e., What is the probability that the null is true, given that significance was found?) This can be represented as P(N | S). (In the analogy, we were interested in the proportion of people in the hospital ward who did not need to be there.) To find P(N | S), we need to know the value we set for P(S | N)—i.e., alpha, but *we also need to know two other important probabilities*. If we use H to represent the probability that the null is not true (some alternative Hypothesis is true), we need to know P(S | H), which is the power of the test, and we also need to know how often the null is true and how often it is not, represented by P(N) and P(H), respectively (these two probabilities must add up to 1.0, of course). These are all the probabilities we need in order to apply Bayes's theorem to find P(N | S).

### Bayes's Theorem

For this example, Bayes's theorem looks like this:

$$P(N \mid S) = \frac{[P(S \mid N) \times P(N)]}{[P(S \mid N) \times P(N) + P(S \mid H) \times P(H)]} \qquad \text{Formula 12.1}$$

P = probability of
S = significance
N = true Ho
H = true alternative hypothesis
| = given that

If you multiply the numerator of the above formula by the total number of studies performed, you get the number of Type I errors that will be made. Multiplied by the number of studies, the denominator gives you the total number of tests that will come out statistically significant (the number of Type I errors plus the number of correct rejections of the null hypothesis). The ratio then gives you the proportion of significant results that are really Type I errors. Then, you would have the tool you need to answer the following question: If the last 100 studies you read all found significant results just barely at the .05 level, how many of them are likely to be Type I errors? Unfortunately, many behavioral researchers erroneously guess about 5%. By now, you should know better. You usually have no way to know how many of those studies were "null" (i.e., testing zero population effects). As an exercise, use an alpha of .05, pick a value of P(N), noting that P(H) = 1 − P(N), and a value for power, P(S | H), and use Bayes's Theorem (i.e., Formula 12.1) to find out just what proportion of significant results are really Type I errors in that case. Try some extreme values to see what happens.

How often would you guess that psychological researchers test completely ineffective experimental manipulations? Or, compare pre-existing populations that have identical means on the variable of interest? Would you say about 5%, or 50%, or less than 1%? It likely differs from area to area within psychology. If one is testing various paranormal phenomena (e.g., clairvoyance; telekinesis), then it is easy to imagine that a high percentage of the studies involve a zero effect. In that case, most, if not all, significant results would be Type I errors. If, however, one is testing minor variations on an intervention that has worked well many times in the past, there may be no null studies, and no chances for Type I errors.

If you have solid information concerning the likelihood of various effect sizes (including the null hypothesis of zero) being true for your study, you can use a statistical method that can be more informative than NHT: Bayesian statistics. In this way, Bayesian statistics makes it possible to improve the accuracy of your conclusions by using data from prior research. The ultimate goal is to have a truly cumulative science in which the results of each study improve the accuracy of the next similar study. Unlike ordinary NHT, which, by itself, tells you nothing about how likely it is that the null hypothesis is true, Bayesian statistics allows you to estimate the relative likelihood of the null being true as compared to an alternative hypothesis being true. We will go into further detail concerning Bayesian statistics in the appendix to this chapter.

## Improving NHT with Robust Statistics

For each statistical test we have described in this book, we have detailed the assumptions required for the test, and particularly for your *p* value, to be valid. One assumption that underlies all of the parametric tests in this book (but not ordinal/ranked statistics) is that your dependent variable (DV) is normally distributed in the population from which you are sampling. Unfortunately, many variables that psychologists study have strongly skewed distributions. For example, reaction time (RT) has a lower limit defined by how quickly the organism can physically respond, but sometimes participants are inattentive, causing you to end up with some unusually long RTs, creating a positively skewed distribution. According to the Central Limit Theorem, the means of large samples will follow a fairly normal distribution (ND) even if the individual scores do not, but the stronger the skewing, the larger the samples you need to normalize a badly skewed distribution. In the next four subsections, we will describe different ways of dealing with violations of the ND assumption. These statistical methods are considered *robust* with respect to the normality assumption because they tend to give accurate $p$ values even if your original distribution is far from being "normal."

First, to make the distribution of your scores look more like the ND, you can use a data transformation. Second, in some data situations, it is considered legitimate to actually delete or replace extreme scores (referred to as *outliers*) using accepted algorithms. Third, you can use the distribution of your own data as an estimate of the population distribution (see #3 Bootstrapping). Finally, you can avoid any distributional assumptions by using nonparametric statistics. We will not go into too much detail about these methods here, but you should be aware that these possibilities are acceptable solutions for dealing with data from distinctly non-NDs.

1. *Data transformations*. Suppose that six of your data points looked like this: 1, 4, 4, 9, 9, 36. This looks like a fairly strong positive skew. But if you were to take the square root of each score, they would look like this: 1, 2, 2, 3, 3, 6, much less skewed. You could then use the square roots of your scores for your statistical tests. One of the most common transformations for strongly skewed data, such as RTs, is to take the logarithm of each score. Data transformations can complicate the interpretation of your results somewhat, but it may be worth the greater accuracy of your test by bringing your data closer to a ND.
2. *Deleting/Replacing outliers*. Extreme scores are sometimes the result of the inattention of a participant, a misunderstanding of instructions, or some other mistake (including a typo when entering scores into a data sheet); if you know the cause (ideally before you look at the data, e.g., knowing that a participant fell asleep during the experiment—hey, it happens!), it is considered legitimate to delete those scores. This is called data cleaning and should always be done before running your statistical tests. Note that it is not legitimate to delete scores just for being extreme—e.g., one of your participants may have an unusually good memory when you are conducting a memory experiment. You need some sort of evidence to conclude that a participant, or data point, is not part of the distribution you are trying to understand. For example, if you are studying the effects of mindfulness on the ability to solve difficult anagrams, you would like to generalize your results to the population of people who are actually trying to solve the anagrams. If you have evidence that a participant was not really trying to solve the puzzles (they started scrolling TikTok during the experiment), then their data are not relevant to the population you would like to understand. However, in some situations, such as RTs, some extreme scores (long RTs) not related to your experimental conditions are to be expected, and in such cases, it is reasonable to set a rule, such as deleting any scores that are more than three standard deviations from the mean.

   When extreme scores on both ends of a distribution are expected from past research, one could adopt a method known as *trimming*, in which a fixed percentage of the largest and smallest data points are automatically deleted before analyzing the data. For example, if your sample size is 50 and you plan to perform 10% trimming, the five largest and five smallest values would be deleted, leaving you with an $N$ of 40. (Note that 10% trimming implies that 10% is trimmed from each end of the distribution.) The mean calculated on the remaining values is then referred to as a trimmed mean and its associated (trimmed) variance must be adjusted accordingly. A related procedure is called *Winsorizing* your data; it involves replacing each of your outliers with the most extreme score that is *not* considered an outlier. Statistics performed on trimmed or Winsorized means are best handled by sophisticated statistical modules, as can be found in R.

3. *Bootstrapping*. Rather than assuming that the values in your data were sampled from a normally distributed population, you can instead assume that the distribution of your data values is a reasonably accurate reflection of the distribution in the population from which they were sampled. A simple example is finding a confidence interval (CI) for the population mean based on your sample data.

   On the one hand, if you assume that your data come from a normal population distribution, you can use the methods of Chapter 6, and calculate the standard error of the mean (SEM) by dividing your sample standard deviation by the square root of your sample size (Formula 6.1, page 111). The limits of the 95% CI are approximately two SEMs above and below your sample mean for large sample sizes (however the .05 critical $t$ gets considerably larger than 2.0 for very small samples). On the other hand, if you do not want to make any assumptions about the distribution in the population, you can use the *bootstrapping* method to create a sampling distribution of the mean from the data you have.

   Let us say that you have only six data points (e.g., 1, 4, 4, 9, 9, 36), so N = 6. You take a random sample of size six from your data to create a new sample. How can the new sample be different from the original one if both are the same size? The answer is to use *sampling with replacement*. Let's say your first random selection is "1". The "1" is not deleted from the original data set, so your next random selection could also be "1". In fact, though it is very unlikely, your first random sample could consist of six 1s, or, perhaps, six 36s. In this example, you would calculate the mean of your first random sample (e.g., 1 or 36, or something in between) and then draw a second, separate random sample and calculate its mean, as well.

   This process is repeated until you have at least 100 sample means, though computer software makes it easy to create a thousand means or more. These sample means will form a distribution, which, in this example would go from 1 to 36, with a mean of about 10.5, and very low probabilities for values near the extremes of 1 and 36. The main point of the bootstrapping in this case is to calculate the standard deviation of all those means to get an SEM that is derived from bootstrapping your data and not from a formula that assumes a normal population distribution. The bootstrapped SEM (bSEM) could be considerably different from the standard deviation of your original six scores divided by the square root of six. The bSEM is then used to create a CI around the mean of the sample as an estimate of the population mean.

   The CI created by bootstrapping is used in the same way as the CI you would create by using Formula 6.4. For instance, if you create a 95% CI for the population mean, you know that any hypothesized population mean that falls outside of the interval would be rejected as a null hypothesis at the .05 level, and any population mean within the interval would not be. A 99% CI would allow you to know immediately which population means could be rejected at the .01 level and which could not. Thus, CIs are more informative than NHT alone, which is why they are often recommended as supplements to the $p$ values you get from NHT. Some NHT critics have even proposed that CIs should replace NHT, as discussed in a later section.

   The current versions of many popular statistical software packages, such as SPSS and RStudio, now include an option to replace some of the traditional statistical tests based on the ND assumption (e.g., the two independent samples $t$ test) with tests based on bootstrapping, usually in the form of a CI for the population value of interest (e.g., the difference of two population means).

4. *Using nonparametric statistics*. You can avoid making *any* assumptions about the distribution of your data in the population by using nonparametric (sometimes called distribution-free) tests. As described in Chapter 8, ordinal tests are based on replacing your original scores by ranking them and therefore do not require any assumption about the underlying population distribution. Returning to the example in which the six scores were 1, 4, 4, 9, 9, 36, replacing them with ranks transforms

them to: 1, 2, 3, 4, 5, 6, thus completely eliminating the skewing. When reporting descriptive measures in the context of nonparametric statistics, the median should be used in place of the mean and the interquartile (or semi-interquartile) range instead of the standard deviation. Later in this textbook, you will be introduced to nonparametric tests that are suitable for data that cannot even be ranked, because they involve qualitatively different categories (e.g., different religious denominations).

The methods described above tend to not only avoid the problem of non-normal distributions, they also tend to reduce the issue involved with comparing two sample means that are associated with very different variances. However, all of these methods still rest on the assumption that participants are selected randomly and independently of each other, except for matching participants in pairs, in which case each pair must be independent with respect to all of the other pairs.

## p Hacking, HARKing, and the "File Drawer Problem"

The improvements just described can make the *p* values found by NHT more accurate, and the explanations of the NHT criticisms laid out earlier in this chapter can help researchers avoid common mistakes, but there are some problems surrounding the use of NHT that are particularly hard to deal with because they involve what are called "researcher degrees of freedom", which are ways that researchers can increase their chances of obtaining statistically significant results without properly controlling their probability of committing Type I errors.

### p Hacking

For example, although *t* tests are fairly robust (i.e., relatively unaffected) with respect to dealing with skewed distributions, if one is working with samples that are not very large and exhibit extreme skewing, researchers are advised to use one of the methods for improving NHT just described. However, each of those methods will produce a somewhat different *p* value. If someone were to try all of those methods, it is possible that only one of those *p* values would fall below .05. Then, if that person were to report only that result and not mention all of the methods that failed to reach significance, they would be engaging in an improper practice known as *p hacking*. Trying different ways to analyze the same data in order to get a significant *p* value, which is the definition of *p* hacking (Simmons et al., 2011), can lead to results that greatly exaggerate what is really a tiny (or even zero) effect in the population (the only way to get a smaller *p* value for the same data is to find a way to make the effect size in the sample appear larger).

Insisting on the use of .05 as the cutoff for significance is a way to ensure that when the null hypothesis is true for a whole set of studies, no more than about 5% of them will reach significance. All of the "null" studies (there is no real effect in the population) that reach significance will be Type I errors, but at least 95% of the null studies will not make the cutoff. However, analyzing the same data from a null study in many ways increases the probability of getting a significant result; if everyone did that when using an alpha of .05, the percentage of Type I errors would likely climb above 5%. If more Type I errors are made and then published as real findings, the chances increase that other researchers will see a false significant result and waste time trying to replicate it or expand on it, not knowing that there is no real population effect. Attempts to replicate a finding that was really just a Type I error often leads to a failure to replicate and that contributes to what has been called the "replication crisis" in social science and medical research. We will discuss the replication crisis shortly.

### HARKing

The previous example, using different methods to compensate for a skewed distribution is just one avenue for what is sometimes called "data fishing." As experimental designs become more complicated, there is a sharp increase in the way that the resulting complex data set can be analyzed, leading to a higher probability of Type I errors getting made and published. Related to *p hacking*, another unfortunate research practice that tends to lead to an increase in Type I errors is called *HARKing*, which stands for Hypothesizing After the Results are Known. In the worst case of *HARKing*, researchers try analyzing their data in every way imaginable (e.g., looking at various subsets of their samples, or subscales of their questionnaires) and when they find a significant result, no matter how unexpected, they write the introduction to their article in such a way that it seems to anticipate that surprise result, and then write a hypothesis that predicts that result. As in the case of *p hacking*, testing many possible results and only reporting the significant ones, increases the possibility of committing and publishing Type I errors. Possible solutions to these problems include preregistering data analysis plans and will be discussed shortly.

### The File-Drawer Problem

There is a clear bias on the part of journal editors with respect to which manuscripts submitted to them will ultimately be published. The bias we are referring to, known as *publication bias*, favors studies whose main results are statistically significant. This bias is not unreasonable given that, compared to significant results, it is harder to ignore the possibility that nonsignificant results are due entirely to chance factors involved in random sampling. Being well aware of this bias, researchers tend not to submit articles based on nonsignificant results for publication, unless for some reason it is particularly interesting to demonstrate a lack of effect (e.g., testing a controversial theory in a large, rigorous study). Consequently, many studies with negative (i.e., nonsignificant) results are simply stored in a researcher's (real or metaphorical) file drawer and never made public.

To understand the problem created by not sharing negative results publicly, imagine the following situation: 20 different experimenters, unknown to each other, independently test the same manipulation, such as one of the many versions of the so-called Mozart effect (e.g., playing music by Mozart while participants solve certain cognitive problems, in order to improve their performance), each using the .05 level. Assuming the null hypothesis is true in every case, the probability is, nonetheless, better than 50% that at least one of these experimenters will find a significant result (i.e., $p < .05$). Suppose that only one study turns out to be significant and that one is submitted to a journal and eventually published. Thus, a Type I error has been published. If the other 19 studies were *not* placed in file drawers, but rather on a searchable website, it could become known that 19 similar studies did not find a significant effect. This knowledge would, as it should, cast doubt on the one published study.

Unfortunately, as of this writing, there is no one comprehensive searchable database dedicated to making negative psychological findings available. However, there is a less official form of publication known as *gray literature* (Paez, 2017). Items in the gray (often spelled as grey) literature have not been commercially published but may be accessible on the internet and may include results that did not attain statistical significance. Examples of gray literature include unpublished theses and dissertations, papers and posters presented at conferences, some government reports, posted preprints of articles submitted for publication, and papers posted in various archival services. Before conducting your own study, it makes sense to search the gray literature for similar studies that may have been tried. However, because many articles in the gray literature have not been subjected to any peer review, or only minimal review (e.g., posters presented at regional conferences) extra caution must be exercised in evaluating their results. Many of these "gray" articles can be found on http://scholar.google.com or through an online searchable database (e.g., http://apa.org/pubs/databases/psycextra) available through your college's library.

A particularly promising repository for psychology articles that have not yet been published (and may never be) is http://PsyArXiv.com, established by the Center for Open Science as a free and open-access website for manuscripts that need not have significant results. A description posted on their blog (http://blog.psyarxiv.com/2016/12/08/psyarxiv-press-release) states that: "PsyArXiv welcomes contributions from all areas of psychology, and hosts papers under review, working papers, and manuscripts that might be difficult to publish in traditional venues, such as replications of previous work or failures to replicate." An important feature of this venue is that articles posted there can be commented on by other researchers, thus providing potentially useful feedback for ongoing studies and the design of follow-up studies.

## The Replication Crisis in Psychological Research and Possible Solutions

It has long been observed that when researchers try to replicate a particularly novel or controversial finding that has been published in a prominent journal, their results sometimes fail to reach statistical significance and may not even come close (Nelson et al., 2022). There are many possible reasons for these failures of replication. Here is a nonexhaustive list of some common ones: (a) The replication may have involved too small a sample and thus too little power to detect the likely effect size; (b) The replication was more conceptual than exact, meaning that some possibly critical features of the experiment were altered (e.g., an easier or more difficult task was employed); (c) Related to the previous reason, the crucial manipulation in the experiment may be a subtle one (e.g., producing some type of emotional state or expectations in the participants) and those performing the replication lacked the skill or detailed knowledge to create an effective manipulation; and (d) The population being sampled for the replication may differ considerably from that in the original study in their cultural traditions or values, or there may have been important societal changes that, over time, have changed the attitudes of most people in the population.

Because some psychological functions are more easily affected by the experimental and/or the social environment of the participants, it can be expected that research in social psychology, for instance, may be harder to replicate than research involving more basic sensory and physiological functions. However, as mentioned in the preceding section, publication bias, combined with *p hacking*, can affect any type of research, and may be contributing to an abundance of Type I errors and exaggerated effect sizes published even in highly selective journals.

Due to a growing concern over replication failures, the Center for Open Science (www.cos.io) was founded in 2013, which then created the Open Science Framework (OSF) to foster greater communication among researchers, as well as to increase the transparency of their methods and analyses. The OSF's first major initiative was the Reproducibility Project, which was aimed at assessing the replicability of scientific studies, and the first subject of this project was psychological research. The project employed multiple labs and investigators to attempt direct (i.e., as exact as possible) replications of the principal findings of 100 different studies published in three major psychology journals. The results, reported in 2015, were disturbing, leading many in the field to declare that psychology has a "replication crisis." The average effect size in the replications was only half as large (e.g., 0.2 rather than 0.4) as it was in the original studies. Moreover, although the replications used considerably larger samples to increase power, only 36% of the replications produced statistically significant results (Open Science Collaboration, 2015).

### The Problem of Type I-ish Errors

We have been focusing so far on the problem of having too many Type I errors published. Type I errors are easy to understand; there is no effect in the population, but significant results were nonetheless attained. However, as we described at the beginning of this chapter, it is hard to imagine that any two-group comparison, whether between pre-existing groups or between two groups treated differently in an experiment, involves absolutely no effect in the population, except in unusual cases (e.g., testing paranormal phenomena). But it is not hard to imagine that the true effect being studied is quite small (e.g., some of the possible Mozart effect studies). If a researcher, using fairly small samples, studies a phenomenon that has a tiny effect in the population and yet finds a large effect in their samples, thus obtaining significant results, a Type I error has not been committed. But something is wrong here. Quite accidentally, the effect size in the sample is much larger than the effect size in the population, leading to a misleading result. Someone trying to replicate that result with similar sample sizes is unlikely to get significant results. Although there is no official name for this severe overestimation of the true effect size, we like to refer to it as a Type I-ish error (Cohen, B.H., 2017). It has an effect on reproducibility that is very similar to a Type I error and is likely to be much more common.

We think it is safe to say that although there may be few genuine Type I errors in the published psychological literature, there is likely to be a much larger amount of Type I-ish errors. Just how prevalent Type I-ish errors are is hard to say, especially because there is no exact criterion for a Type I-ish error—that is, there is no way to say just how tiny the true population effect has to be, or how much it needs to be overestimated in someone's samples, in order to qualify as a Type I-ish error. However, it looks like the Reproducibility Project just described uncovered a surprising number of likely Type I-ish (and perhaps even some Type I) errors.

### The Preregistration Solution

There is a logical and straightforward solution to the problem of Type I error inflation due to *p hacking*, data fishing, and *HARKing—preregistration* (Nosek & Lindsay, 2018). From the Association for Psychological Science (https://www.psychologicalscience.org/observer/research-preregistration-101): "Preregistering a research project involves creating a permanent record of your study plans before you look at the data. The plan is stored in a date-stamped, uneditable file in a secure online archive. You can give others (e.g., potential reviewers) access to the preregistered plan, and you can do so while maintaining your and the reviewers' anonymity. The main purpose of preregistration is to make clear which hypotheses and analyses were specified a priori and which were more exploratory and driven by the data."

Exploratory analyses are permitted, but fellow researchers will tend to give more weight to analyses that were planned at the outset of the study, knowing that they were not the product of *p hacking* or *HARKing*, and therefore less likely to be overestimating the true population effect size. One convenient place to post the plan for a study, including the analyses to be performed, before the data are collected, is PsyArXiv, part of the OSF, and mentioned in a previous section.

Some journals provide peer review for a preregistered study, called a Registered Report, which includes a guarantee of publication regardless of whether the planned analyses attain statistical significance provided that the plan has been approved by the relevant reviewers and editors of the journal. This procedure plays an important role in reducing the effects of publication bias in favor of results being statistically significant. Because the reviewers and editors had decided in advance that the study was well worth performing, they would have no problem publishing negative results.

### Meta-Analysis

Suppose that a search for both published and unpublished studies of the Mozart effect yields ten studies, of which three obtained significant results and seven fell short. This does not look good for the Mozart effect until it is noticed that all ten studies found results in the same direction—that is, all showed some improvement in performance relative to a control condition (e.g., the playing of modern instrumental music). If the null hypothesis were true that the Mozart effect is actually zero, it is unlikely that all ten results would come out in the same direction. That is like flipping ten heads (or ten tails) in a row with a fair coin. Now it is looking like the effect is not zero. But we still do not know just how large the effect might be in the population, because we have ten different estimates from our ten studies. This is a case in which a meta-analysis would be appropriate and could be quite helpful. Although its details can get complex, the concept of meta-analysis is easy to understand. The goal is to combine the ten studies into one much larger "meta" study.

Once you have determined which studies are similar enough to be combined (more on that later), the next step is to calculate the same measure of effect size for each study. For a two-group comparison, Cohen's $d$ can be used, but more commonly a slightly corrected-for-bias version of $d$, known as Hedge's $g$ is used. Alternatively, the point-biserial $r$, described in Chapter 10 can be used. Then, the effect sizes of the included studies are averaged together, using an appropriately weighted average, to determine the effect size of the meta-study. Returning to our example of Mozart-effect studies, the size of the meta-study is found by adding the $N$s of the ten individual studies. Now that you have one large two-group study, you can calculate a $t$ value for the *meta*-study, and test it for significance with the $df$ based on your much larger sample sizes.

Although it is encouraging to be able to reject the null hypothesis for the Mozart effect as tested in the ten studies, the more important question you can now answer is: How large is the Mozart effect in the population? You have an effect-size estimate that is based on ten studies, which, combined, are likely to have much larger $N$s for the two groups than any one Mozart study. However, it would be even more informative to construct a 95% CI around the average effect size. With the large $N$s of the meta-analysis, you should be able to find a fairly narrow range of plausible effect sizes in the population, which can tell you whether this line of research is worth pursuing further (see Confidence Intervals for Effect Sizes, in the next section).

For the purposes of this textbook, we have described a very simplified version of meta-analysis. There are some complex computations involved in combining effect sizes, but the really tricky part of meta-analysis is deciding which studies are similar enough to be combined. Continuing with the Mozart example, differences among the studies considered may include the age range of the participants, the condition (e.g., type of music) used for the control group, the details of the task performed, or even the particular selection of Mozart music played during the task performance. Judgment is involved and reasonable people may disagree as to which studies can be meaningfully combined. Fortunately, the use of standardized effect-size measures at least reduces the problem of task performance being measured on different scales.

## Alternatives to NHT (The "New" Statistics)

### Interpreting Effect Sizes

In recent years, some psychologists have been using online services (e.g., Amazon Mechanical Turk) to recruit large numbers of diverse participants relatively inexpensively. Although very large sample sizes do not guarantee that the results will come out in the direction you hypothesized (as previously mentioned), your statistical test is likely to be significant even if the effect size is fairly small. In that case, statistical significance can be a bit misleading, if you take the term "significant" too literally. Even a small effect can be interesting and important, but you need to be open about reporting the effect size that you found, how it compares to what you expected, and what it means for your theory and its possible practical implications. Of course, this can be important no matter how large or small the effect size is.

Those who are critical of NHT point out that by itself your obtained $p$ value tells you only about statistical significance, but nothing about the size of the effect. It has been suggested that, as an alternative to NHT, the effect size should be reported and interpreted, and that $p$ values not be included at all. This suggestion may be reasonable when dealing with variables that have a clear quantitative relationship with each other, but when your independent variable has just two levels, so that, for instance, one group is instructed to think about their own death (a mortality salience manipulation) and the other (control) group is told to think about a difficult exam, and the DV is their inclination to vote for a conservative versus a liberal leader measured on a 1–10 scale, the effect size is not likely to be easy to predict or interpret. Nonetheless, it is a good practice to always report your effect size along with your $p$ value when presenting NHT results. In fact, in some cases, it is helpful to calculate a CI for the population effect size based on the results of your samples.

### Interpreting Confidence Intervals

Just as some NHT critics favor replacing NHT with reports and interpretations of effect sizes, others emphasize replacing NHT with reports of CIs (as mentioned in the previous section), especially when looking at the difference of two group means, and then discussing the implications of your interval estimates. Although reporting a CI gives you more information than just reporting a *p* value, the usefulness of the CI can depend on the way you are measuring your DV. To illustrate this point, we will return to the caffeine example in Chapter 7. The DV was heart rate in beats per minute (bpm) and the mean for the caffeine group was 81 while the mean for the control group was only 68. The difference of the two sample means, 13, is our best estimate of the difference of the caffeine and placebo population means. However, with fairly small samples (*N*s of 22 and 20 in this case), and a good deal of person-to-person variability (SDs of 12 and 16.8), we know that there is a lot of room for error in our estimate. In fact, the 95% CI we calculated in that example ranges approximately from a mean population difference of 4 bpm to 22 bpm.

Knowing the CI gives us some confidence that the difference of the population is not zero, because zero is not within the limits of the CI, but it also lets us know that there is a wide range of plausible values for the difference of the population means. Because heart rate is a much-studied variable, the practical, and even clinical, implications of these differences are meaningful. A difference of 4 bpm is not very consequential, but a difference of 22 bpm might be. Suppose, however, that the means of 81 and 68 for the two groups represent the average number of targets participants in the two groups detected in an attention task created for this study. Clearly, the caffeine group detected more targets, but is 13 more targets a large effect or not so much? When the DV is not well known from previous research, it is helpful to convert the mean difference to a more standard measure that is easier to interpret. The most common standardized measure of a difference between two sample means is Cohen's *d*, which we calculated for the caffeine study in Chapter 7 and, at 0.9, it was rather large. It is also our best single (i.e., point) estimate of the population effect size for this study, symbolized as **d**.

Having an estimate of the effect size for the population is helpful; in this case it tells us that there seems to be a fairly good separation of the two population distributions (not a great deal of overlap). However, the same issue that led us to calculate a CI for the difference in population means applies to the effect size, as well. There could be a good deal of error with respect to our estimate and there may be a fairly large range of reasonable values for the true population effect size.

### Confidence Intervals for Effect Sizes

To calculate an accurate CI for **d**, the effect size for the difference between two population means, centered around your sample estimate, *d*, requires the use of a noncentral *t* distribution (a more complex, nonsymmetric variation of the familiar *t* distribution.) Fortunately, when your sample sizes are fairly large, you can get a reasonable approximation of that CI by simply dividing the upper and lower limits of your usual CI for the difference of population means (as found by Formula 7.7) by an appropriate estimate of the standard deviation of each population (because **d** is the difference of population means divided by their common population standard deviation). Assuming homogeneity of variance, the standard deviation of both populations is the same, symbolized as $\sigma$. The best estimate of $\sigma$ from a two-group study is the square root of the pooled variance—that is, $s_p$. Returning to the caffeine study example, we have already calculated $s_p$ in Chapter 7, because we needed it to obtain the sample effect size. (We divided the difference of the sample means, 13, by $s_p$, which turned out to be 14.48, and obtained $d = 0.9$). This time we need to divide the limits of the 95% CI for the difference of the population means by $s_p$. Those limits were 4 and 22 bpm. Dividing each by 14.48, we find that the 95% CI for the effect size ranges from 0.276 to 1.52. Of course, the *d* we calculated, 0.9, is in the middle of the CI. Because an effect size of zero does not fall within the 95% CI, we know that the null hypothesis can be rejected at the .05 level, with a two-tailed test.

While the effect size estimate from the caffeine study is fairly large (the sample means are almost a standard deviation apart), the 95% CI tells us that there is a reasonable chance that the true population effect size is fairly small (less than 0.3) and also a reasonable chance that the effect size is quite large (more than 1.5). This is the drawback of using small sample sizes. If it is important to have a fairly accurate estimate of the effect of some experimental condition or clinical treatment or educational intervention, etc., you need to use large samples, or meta-analysis as described previously. Finding an accurate CI for an effect size when dealing with small samples is best left to statistical software, because of the difficulty of dealing with the noncentral $t$ distribution.

## Summary

This chapter covers some advanced methods and concepts that involve improvements to, as well as replacements for, traditional null hypothesis testing.

### Some Major Criticisms of NHT

It has been argued that it is virtually impossible for the null hypothesis to be true for any real-world study, so why bother testing if the null is true? However, using NHT prevents us from wasting time studying effects so tiny that they do not reach significance with ordinary sample sizes. Unless the effects have great theoretical or practical importance, it is useful that these studies result in Type II errors. It is true that by using large enough sample sizes even trivially tiny effects can produce significant results, but the results are not guaranteed to come out in the predicted direction. So, if the tiny effect is important enough, it could be worthwhile using very large samples, if feasible, to find out the true direction of the results in the population.

Another criticism is that the $p$ value of NHT tells you if your data are likely if the null is true, but not if the null is likely to be true given your data. Bayes's theorem can help you to reverse the $p$ value, given some assumptions. A disease detection analogy was used to make the concept of Bayesian statistics more concrete.

### Improving NHT with Robust Statistics

The $p$ values of traditional NHT are only accurate if some assumptions are met; one of the most important assumptions is that the variable you are measuring has a ND in the population. Outliers and/or a strong skewing of the distribution can throw off the accuracy of the $p$ value. Fortunately, there are statistical methods that are robust (not easily affected) with respect to the ND assumption. These include: 1) mathematically transforming the values of your measured variable (e.g., taking the logarithm or square root); 2) trimming (i.e., deleting) your data by a fixed percentage on both the low and high end of your sample, or Winsorizing your data (replacing the most extreme values on both ends with a value that falls short of being extreme) and then using modified formulas to find more accurate $p$ values; 3) creating an empirical sampling distribution by taking (usually thousands) of samples of the same size (i.e., the size of your sample), with replacement, from the data you collected, and then basing your $p$ value on the distribution you created by repeated resampling (a method called bootstrapping); and 4) using nonparametric statistics such as those already described in Chapter 8 and others that will be detailed in Chapters 17 and 18.

### *p* Hacking, HARKing, and the "File Drawer Problem"

Analyzing your data in many different ways (e.g., both parametric and nonparametric methods; with outliers and without outliers; analyzing different subgroups of participants; combining variables in different ways) and then reporting only the statistically significant outcomes is commonly called $p$ hacking and it can greatly increase the chances of committing a Type I error or hugely overestimating the true population effect size. A related method,

HARKing, involves writing your hypotheses to match whichever results happen to reach significance. The file drawer problem occurs when only significant results are published, even though similar studies may have been performed many times with nonsignificant results, and were therefore not submitted for publication.

### The Replication Crisis in Psychological Research and Possible Solutions

A replication crisis in psychological research was declared when a group of researchers tried to replicate a large number of published results and, more often than not, found smaller, nonsignificant effect sizes as compared to the original study. Unfortunately, it is difficult to replicate a study exactly and many seemingly minor differences can account for failures to obtain results similar to the original study. However, due to *p* hacking and other practices mentioned in the previous section, there is also an increased chance that the published studies that were the target of replications were based on Type I or Type I-ish (greatly overestimated effect sizes) errors. Possible solutions include preregistering a study, so that the researchers are committed to specific analyses, before seeing the data, thus preventing *p* hacking and HARKing. Moreover, if a journal commits to publishing the article regardless of whether the results attain significance, a publication bias that favors significant results, is thus reduced. Another solution is to combine many relatively small studies into one larger meta-study by averaging their effect sizes and using the increased power of a much larger sample size. However, it can be difficult to decide which studies are similar enough to be combined, and, unfortunately, many of the smaller studies can be hard to find because their results were not significant and therefore were not published in a traditional way.

### Alternatives to NHT (The "New" Statistics)

The simplest form of an effect size is the difference of two sample means. If the outcome variable is well known and measured on a standard scale, such as weight in pounds (or kilograms), the difference can be immediately evaluated. For example, if, after three months, the mean weight of a group assigned to the Mediterranean diet is ten pounds less than the mean of a group on the Paleo diet that is meaningful—a one-pound difference, not so much. A 95% CI for that weight difference (e.g., 2–18 pounds) would also be meaningful and would put that mean difference in perspective.

However, if the measured variable is how many words are recalled from a list created for the study, a standardized effect-size measure is needed, such as Cohen's *d*. A mean difference of ten words recalled would not be immediately interpretable, but a *d* of 0.7 or only 0.1, would be. Moreover, just as a CI would put a difference of means in perspective, the same function would be served by constructing a CI for the effect size.

## Thought Questions

1. What do you think are the two or three most serious mistakes people make when drawing conclusions from NHT?
2. Which criticism(s) of NHT do you think are the least justifiable?
3. In what kind of situation does the use of Bayesian statistics make the most sense?
4. Describe an experiment in which a very skewed distribution and/or one with extreme outliers might be expected. Describe two different ways to reduce the effects of the skewing or the extreme scores.
5. What are some of the real-world consequences of the widespread use of *p hacking* and similar bad practices?
6. Describe some variables for which you do not need standardized effect sizes to interpret the size of the effect. Describe some situations in which standardized effect sizes would definitely be needed to evaluate the size of the effect.

## Appendix: A Brief Introduction to the Use of Bayesian Statistics

Although *p* values can be useful, they only tell you how unlikely your data (or data even more extreme) would be if the null hypothesis were true. The smaller the *p* value, the more comfortable you can be dismissing the null hypothesis as a plausible explanation for your results. However, it would be even more useful to know the probability that the null is true given your data. Using the symbol D for your data and N for the null hypothesis, the *p* value can be written as P(D | N) (the probability of obtaining your data given that the Ho is true), while what you would really like to know is P(N | D) (the probability that the Ho is true, given your data). You can find P(N | D) from P(D | N) using Formula 12.1 (substituting D for S), as long as you know P(N)—that is, how likely is it that the null hypothesis is really true for your study? In a two-group experiment, that translates to how likely it is that the intervention or manipulation you are using has zero effect in the population, and any difference you see between your two samples is just accidental (i.e., due to random sampling). When the researcher is dealing with a slight variation on an intervention that has worked before, it is likely that P(N) will be quite low (perhaps, fairly close to zero). By contrast, when dealing with a treatment that is completely novel, and even counterintuitive, P(N) may be relatively large.

### Prior and Posterior Probabilities

In Bayesian statistics, which seeks to quantify P(N | D), P(N) is referred to as the *prior probability*, or more simply as a "prior." It is your (hopefully educated) guess about the relative probabilities of N and H before you see your data. As an example, suppose you have no idea how likely the null hypothesis is for your experiment, so you decide that you might as well assume that P(N) = P(H) = .5. (This implies that half the studies like yours have no effect at all, which does not seem very likely in psychological research, but it makes for an easy example.) You will also have to have a specific alternative hypothesis (H), in order to calculate P(D | H). Now suppose that your *p* value [i.e., P(D | N)] for your statistical test equals .02, but P(D | H) is larger because the data are considerably closer to the value you hypothesized. Assuming that P(D | H) = .4, we can insert these values into Formula 12.1, modified to substitute D for S:

$$P(N \mid D) = \frac{[P(D \mid N) \times P(N)]}{[P(D \mid N) \times P(N) + P(D \mid H) \times P(H)]}$$

$$= \frac{.02 \times .5}{(.02 \times .5) + (.4 \times .5)} = \frac{.01}{.01 + .2}$$

Thus, P(N | D) = .01/.21 = .0476. This is called a *posterior probability* because it is the probability of N after being adjusted according to the data you obtained from your study. In this particular case, the probability of the null being true given your data is about .05. Your data have led you to update your assumption about the null being true for your study from your initial guess of .5 to the much lower value of .05. Calculate P(N | D) for yourself if the *p* value were .3 and the P(D | H) were only .1. In this case, the probability of the null being true given your data is higher than your initial guess—the probability of N being true is a rather high .75. The influence of your data was to increase the apparent probability that the null hypothesis is actually true.

Thus far, we have been dealing with the prior probabilities as single values for the probability of a particular N and a particular H. It is often useful to express N and H as population effect sizes; for a two-group comparison, **d** (the difference of the two population means divided by their common standard deviation) is the most common measure (this is the value that is estimated by Cohen's *d*). The null hypothesis is usually that the population effect size is zero (i.e., **d** = 0), but as illustrated in the mortality salience example in the earlier subsection on *Interpreting Effect Sizes*, predicting a specific effect size for H is difficult in many areas of psychology, in which researchers do not posit quantitative relationships between their independent variables (IVs) and DVs. As in the case of power analysis, the best you can usually do is to use the effect-size estimate from the most similar published study you can find.

### Prior Distributions

We began the previous subsection by assuming that N and H were equally likely. This was an example of an *uninformative prior distribution*, sometimes referred to as a neutral or objective prior because it is not based on any previous knowledge of how likely N or H might be. If, in addition to N, there were two competing hypotheses, H1 and H2, the uninformative prior, as in the two-hypothesis case, would be a uniform (or flat) distribution, meaning that each hypothetical value would be assigned the same prior probability (e.g., $1/3$, $1/3$, $1/3$) and this principle could be extended for any number of discrete possibilities. However, totally flat (uniform) priors are rarely used for a full range of hypotheses, because we would not expect very large effect sizes to be as likely as tiny effect sizes. More commonly used as a neutral prior, is a hypothetical probability distribution of effect sizes with its highest peak at zero. Therefore, the height (i.e., probability density) of most neutral prior distributions decreases as effect sizes differ more from zero (or some other hypothesized null value). A prior distribution centered on zero is sometimes called a *skeptical* prior, because you are not really expecting any effect to occur in either direction.

Suppose that prior research has shown that a particular experimental intervention is associated with a fairly small effect size of .2, but you have designed a study that you believe will increase that effect size. In that case, your prior distribution would be informed by the previous research; you do not expect the effect of your study to be zero. If your new intervention is no better than the previous one, you have a good reason to expect an effect size of .2. Therefore, your prior distribution could be a normal distribution (ND) centered on .2 instead of zero. This would be called an *informative prior*. If from past experience, you expect a good deal of variability in outcomes, you might propose that your ND prior, though centered on .2, has quite a large standard deviation, allowing for a large range of possible effect sizes. That is an example of a *weakly informative* prior distribution. By contrast, a relatively narrow distribution centered on .2 would be a *strongly informative* prior.

### Posterior Distributions and Cumulative Science

In a previous subsection, we expressed Bayes's theorem in terms of P(N) and P(H), the prior probabilities, P(D | N), the probability of obtaining your data (or a value more extreme) if the null were true (i.e., your *p* value), and P(D | H), what your *p* value would be if H were true. We inserted these values in order to obtain P(N | D), the posterior probability of the null being true, which we expect to be at least a bit different from the prior probability, given that it has been adjusted by the results of our study. Ideally, the posterior probability of N (i.e., the null hypothesis) for one study could serve as the prior probability of the next similar study. And, this chain could continue from one study to the next in a series of studies, leading to more and more informative results.

When using a prior distribution to represent the probabilities of different possible effect sizes, you can still combine that distribution with the data you collected, but without two discrete values, you cannot use Bayes's theorem as expressed in Formula 12.1. Fortunately, computer software now makes it possible to do the complex calculations needed to work with distributions and can compute a posterior distribution for you.

### Reporting the Results of a Bayesian Test

There are a number of ways to express the results of a Bayesian test, but the most common and easiest to explain seems to be the *Bayes factor* (BF). This requires positing a specific value for H that differs from your null hypothesis value. Then, you can calculate both P(D | N) and P(D | H). The ratio of P(D | H) to P(D | N) is the BF (your best friend!). This ratio tells you how much more likely it is that the data you have (often in terms of a sample mean) comes from a distribution centered on H rather than N, which is something you would really like to know. A BF of 1.0 means that H and N are equally likely and a BF of less than 1.0 is disappointing, in that it means that the null hypothesis is more likely, given your data, than your hypothesized value. A BF between 1 and 3 is considered a negligible amount of evidence in favor of H, but as BF increases between 3 and 10, the BF suggests a reasonable amount of evidence for H. A BF greater than 10 is considered strong evidence for H because it means that H is 10 times more likely to give you the data you collected than N would be. Analogous to a CI, a Bayesian analysis allows you to construct what is called a *credible interval*, based on the posterior distribution. If the boundaries of your credible interval contain 95% of the posterior distribution, you can then assert that the probability is .95 that the population value you are looking for is in that interval.

### Advantages and Disadvantages of NHT and Bayesian Statistics

An advantage of NHT is that it gives you a single number, the *p* value, that allows you, by convention, to draw a conclusion about whether or not the null hypothesis is true for your study (e.g., whether the group means you are comparing would really be different if you had data for the entire population). Moreover, there is widespread agreement on comparing the *p* value to .05 (the most common value for alpha) to make that decision. By now, however, you know the main drawbacks of the *p* value: (a) that it doesn't really tell you the probability of the null being true (just the probability of obtaining data at least as extreme as yours *if* the null is true); (b) that reliance on a *p* value can easily lead to erroneous conclusions, such as thinking that a *p* value larger than .05 is good evidence that the null is true, or that there really is a big difference between data that yield a *p* value of .04 and data that lead to a *p* value of .06; and (c) that there are many ways researchers can play with their data to get the *p* value below .05.

Bayesian statistics avoid most of the problems associated with NHT and *p* values. A Bayesian procedure does provide an estimate of the null hypothesis being exactly true for your study based on the data you collected. Another advantage is that you do not have to decide ahead of time how much data you will collect because it is permissible to look at your results at any point and then collect more data if desired. Bayesian statistics focuses on evaluating the probabilities of different hypothesized values (e.g., differences between two population means), rather than making a yes or no decision concerning whether a result is statistically significant. Finally, Bayesian statistics have the advantage of creating a cumulative science, in which previous relevant results can inform the analysis of subsequent studies.

The main disadvantage of Bayesian statistics is that the results of the analysis rely on the prior distribution that is used. Because of a lack of cumulative research and replications in the field of psychology, it is often difficult to construct an informative prior based on previous studies. Therefore, Bayesian analyses often employ so-called "neutral" priors. However, there is no widespread agreement on what constitutes the most appropriate neutral prior for any given study. In fact, no prior distribution can be entirely neutral, because choosing a prior always involves some subjective judgment. The complications concerning the use of Bayesian statistics explain in part why these methods have not displaced NHT to any great degree, though there seems to be an increase in the number of published studies that include Bayesian statistics, to supplement the reported *p* values of NHT.

# Part III
# Analysis of Variance Methods

# Chapter 13
# One-Way Analysis of Variance

## PREVIEW

### Introduction

We wish to draw inferences about the differences among more than two population means. Why is it a poor idea to perform numerous $t$ tests between the various pairs of means?

What is the experimentwise error rate?

In what way is one-way analysis of variance similar to the $t$ test for the difference between two means? How does it differ?

### The General Logic of ANOVA

If we wish to draw inferences about population means; why are we analyzing variances?

What is the within-group (or error) variance estimate? What is the between-group variance estimate?

What is the $F$ ratio?

### Computational Procedures

What are the procedures for drawing inferences about any number of population means by computing a one-way analysis of variance?

What are sums of squares? Mean squares?

### Testing the $F$ Ratio for Statistical Significance

What is the correct statistical model to use for testing an ANOVA?

How do we test the $F$ ratio for statistical significance?

### Calculating the One-Way ANOVA From Means and Standard Deviations

How can you calculate the $F$ ratio for an ANOVA if you do not have the raw data but only a table of means, standard deviations, and sample sizes? Why might you need to do this?

### Comparing the One-Way ANOVA With the $t$ Test

What is the relationship between the $t$ value and the $F$ ratio in the two-group case?

What is the relationship between the pooled variance of the $t$ test and $MS_W$ in ANOVA?

### A Simplified ANOVA Formula for Equal Sample Sizes

How does the equal-$n$ formula show that the magnitude of the $F$ ratio depends on the variance of the sample means, the variance within the samples, and the size of the samples?

### Effect Size for the One-Way ANOVA

How can the formula for $F$ be separated into a measure of effect size and a measure of sample size?

Why is it desirable to convert a statistically significant value of $F$ into a correlational measure that shows the strength of the relationship between the independent and dependent variables? How is this done? And how can this measure be corrected for bias?

*Introductory Statistics for the Behavioral Sciences*, Eighth Edition. R. Brooke Lea and Barry H. Cohen.

Companion website: www.wiley.com/go/ISBS8e

## Introduction

Chapter 7 provided techniques for testing the significance of the difference between two means. These techniques enable you to determine the effect of a single independent variable (e.g., the presence or absence of a particular dosage of caffeine) on the mean of a dependent variable (DV) (e.g., mathematics test scores) when there are *two* samples of interest, such as an experimental group and a control group.

Suppose that you would like to determine whether *different dosages* of caffeine affect performance on a test of manual dexterity. As before, there is one independent variable (amount of caffeine), but this time you wish to include the following *five* conditions:

Sample 1: very large dose of caffeine

Sample 2: large dose of caffeine

Sample 3: moderate dose of caffeine

Sample 4: small dose of caffeine

Sample 5: placebo (no caffeine)

As usual, you would like the probability of a Type I error in this experiment to be .05 or less. Therefore, it would *not* be correct to perform 10 separate $t$ tests for the difference between two means (i.e., first test $H_0$: $\mu_1 = \mu_2$; then test $H_0$: $\mu_1 = \mu_3$; etc.) and then compare each obtained $t$ value to the appropriate critical $t$ value for $\alpha = .05$.

The more statistical tests you perform on a given data set, the more likely it is that some will be statistically significant purely by chance. When $\alpha = .05$, the probability that one $t$ test will yield statistical significance when $H_0$ actually is true is .05. This is the probability of committing a Type I error. However, if you run 20 $t$ tests when $H_0$ is always true, an average of one of them ($.05 \times 20 = 1.0$) will be statistically significant just on the basis of chance, so it is likely that you will commit at least one Type I error somewhere along the line. Similarly, if you run 10 separate *t* tests in order to test the null hypothesis for every possible pair of different caffeine dosages, the probability that you will commit at least one Type I error is clearly *greater* than the desired .05. (It can be shown that this probability is closer to .3.) The greater the number of sample means among which you perform pairwise significance tests, the more likely you are to make at least one Type I error. In statistical terminology, the rate of occurrence of *any* Type I errors over a *series* of individual but related statistical tests is called the **experimentwise error rate, $\alpha_{EW}$**.

Fortunately, there is a procedure for testing differences among three or more means for statistical significance simultaneously, which overcomes this difficulty. This procedure is the *analysis of variance*, abbreviated as ANOVA. The null hypothesis tested by ANOVA is that the means of the populations from which the samples were randomly drawn are *all* equal. For example, the null hypothesis in the caffeine experiment is:

$$H_0: \mu_1 = \mu_2 = \mu_3 = \mu_4 = \mu_5$$

The alternative hypothesis merely states that $H_0$ taken as a whole is *not* true. There are many ways in which it may be false: It may be that $\mu_1 \neq \mu_2$, or that $\mu_3 \neq \mu_5$, or both, or that all five population means are unequal, or only two are equal, and so forth. The rejection of the ANOVA $H_0$ tells us only that *some* inequality exists. In the next chapter, we will describe separate procedures for testing pairwise mean differences following an ANOVA. Such procedures enable us to add greater specificity to the meaning of rejecting the ANOVA $H_0$.

The caffeine experiment is an example in which both the independent (amount of caffeine) and dependent (heart rate) variables are measured quantitatively. More commonly, the independent variable in an ANOVA has qualitative levels (e.g., persuasive arguments are attributed either to a fellow student, a faculty member, or a member of the school administration), and that is the type of ANOVA on which this chapter is based. The type of ANOVA in which the independent variable (often called the **factor**) has quantitative levels is described only briefly near the end of the chapter.

## The General Logic of ANOVA

It may seem paradoxical to test a null hypothesis about *means* by testing *variances*, so the general logic of ANOVA will be discussed before proceeding to the computational procedures. The ANOVA procedure is based on a mathematical proof that the sample data can be made to yield two independent estimates of the population variance:

1. **Within-group (or "error") variance estimate.** This estimate is based on how different each of the scores in a given sample (or group) is from other scores in the same group.
2. **Between-group variance estimate.** This estimate is based on how different the *means* of the various samples (or *groups*) are from one another, as well as on the sizes of those groups.

If the samples all come from the same normally distributed population (or from normally distributed populations with equal means and variances), it can be proved mathematically that the between-group variance estimate and the within-group variance estimate will be about equal to each other and equal to the common population variance ($\sigma^2$). The larger the between-group variance estimate is in comparison to the within-group variance estimate, the more likely it is that the samples do *not* come from populations with equal means.

As an illustration, consider the hypothetical data in Table 13.1A. There are five groups, each of which listened to music by a different classical composer while completing a spatial ability test. Group 1 listened to Mozart, Group 2 to Chopin, Group 3 to Bach, Group 4 to Schubert, and Group 5 to Beethoven. Each group contains five people, so there are 25 participants in all, and the entries in the table represent the score of each person on the spatial ability test. There is some between-group variation and some within-group variation. The means of the five groups may differ because the composer of the music does have an effect on test scores

## Table 13.1 Two Versions of the Music Experiment (Hypothetical Data)

***A. Within-group variation and between-group variation are about equal***

| | 1 | 2 | 3 | 4 | 5 |
|---|---|---|---|---|---|
| | 16 | 16 | 16 | 16 | 14 |
| | 16 | 14 | 14 | 14 | 13 |
| | 14 | 13 | 12 | 12 | 13 |
| | 13 | 13 | 10 | 12 | 10 |
| | 12 | 10 | 10 | 12 | 10 |
| Sample means: | 14.2 | 13.2 | 12.4 | 13.2 | 12.0 |
| Standard deviations: | 1.789 | 2.168 | 2.608 | 1.789 | 1.871 |

→ within-group variation (within each group's column)

← *between-group variation* → (across the sample means)

*grand mean = 13.0*

***B. Between-group variation is much greater than within-group variation***

| | 1 | 2 | 3 | 4 | 5 |
|---|---|---|---|---|---|
| | 16 | 23 | 26 | 11 | 10 |
| | 16 | 21 | 24 | 9 | 9 |
| | 14 | 20 | 22 | 7 | 9 |
| | 13 | 20 | 20 | 7 | 6 |
| | 12 | 17 | 20 | 7 | 6 |
| Sample means: | 14.2 | 20.2 | 22.4 | 8.2 | 8.0 |
| Standard deviations: | 1.789 | 2.168 | 2.608 | 1.789 | 1.871 |

→ within-group variation (within each group's column)

← *between-group variation* → (across the sample means)

*grand mean = 14.6*

or merely because they are affected by the presence of sampling error (i.e., the accident of which cases happened to be in each sample), or for both reasons. Whether to retain or reject the null hypothesis that the samples all come from the same population is decided by applying ANOVA procedures to the data.

What if the composer of the music *does* have an effect on spatial ability, at least temporarily? Suppose that listening to one particular composer (e.g., Mozart) causes a real increase in test scores. Ideally, this implies that this composition by Mozart produces about the same helpful effect on all of the participants in the Mozart group. If that is true, then the variability *within* that group will barely be affected by this effect, because adding a constant to all scores or subtracting a constant from all scores does *not* change the variance or standard deviation. Thus, the overall within-group variance estimate based on all the groups will also be little affected. The mean of the Mozart group, however, can be expected to be higher than the mean of the other groups. Therefore, the *between*-group variance estimate will increase. Similarly, if several different classical composers have different effects on test scores (e.g., some composers increase test scores more than others), the within-group variance estimate will not necessarily increase, but the between-group variance estimate will certainly get larger. The greater the differences among the group means, the larger will be the between-group variance estimate.

As an illustration, consider the data in Table 13.1B. Imagine that the groups in this version of the experiment heard very different *types* of music (e.g., country, jazz, classical, Hip-Hop, and heavy metal). (In reality, these data were obtained from the data in Table 13.1A as follows: Scores in Group 1 were not changed; 7 was added to each score in Group 2; 10 was added to each score in Group 3; 5 was subtracted from each score in Group 4; and 4 was subtracted from each score in Group 5.) The variance of each of the groups in Table 13.1B, and hence the within-group variance estimate, is equal to that of Table 13.1A. However, the between-group variance estimate is much greater in the case of Table 13.1B. Thus, it is more likely that the five samples of scores in Table 13.1B do *not* come from the same population.

To clarify further the meaning of between-group variation and within-group variation, consider the score of the first person (let's call him Evan) in Group 1 of Table 13.1A, which is equal to 16. The total deviation of Evan's score from the grand mean is 3.0 (i.e., 16.0 − 13.0). However, Evan's score also deviates by 1.8 points from the mean of his group (16.0 − 14.2 = 1.8). This part of his deviation is "error" in the sense that it cannot be explained by the composer variable because all people in Evan's group (Group 1) listened to the same music. It simply reflects the basic person-to-person variance of the spatial ability scores.

The difference between the mean of Group 1 and the grand mean (14.2 − 13.0) is equal to 1.2 points. This part reflects the amount of Evan's deviation that can be explained by the composer he was listening to (Mozart). Of course, the within-group variation of 1.8 and the between-group variation of 1.2 sum to 3.0, the total variation of Evan's score from the grand mean. The three deviation scores involving just this one person are:

1. Total deviation = 3.0
2. Within-group (from own group) deviation (*error*) = 1.8
3. Between-group (from own group to grand mean) deviation = 1.2

The total variance of the spatial ability scores, the within-group variance estimate, and the between-group variance estimate are a function of the magnitude of these deviations, respectively, across everyone in the experiment. As you may recall, variances are computed by squaring all the deviations from the mean first, then summing them, and then dividing by the appropriate number of degrees of freedom (*df*). However, for some purposes, it is easier to put off dividing by the *df*s and deal instead with just the sums of the squared deviations (the *SS*s for short). For instance, it is useful that the total sum of squared deviations from the grand mean (the total sum of squares, or $SS_T$) is equal to the sum of the between-group sum of squares ($SS_{bet}$) and the within-group sum of squares ($SS_W$). (The proof is given in the appendix at the end of this chapter.) This does not hold true for the variances.

It is important to understand that the between-group variance estimate is influenced by *both* the effects of the different types of music (if any) *and* error variance. (The sample means will tend to vary because the individual scores tend to vary, even if the independent variable is *not* having any effect; since individual variation in spatial ability is not the object of study in this experiment, it is called "error variance.") Actually, there are two distinct sources of error variance; thus, the group means will be affected by not only the experimental treatment but also by individual differences in spatial ability as well as error in the spatial ability scores due to imprecision of measurement. The within-group variance estimate, however, reflects solely the error variance (which includes both individual differences and measurement error). The effect due solely to the different types of music can therefore be isolated by computing the following *F* ratio:

$$F = \frac{\text{treatment variance} + \text{error variance}}{\text{error variance}} = \frac{\text{between-group variance estimate}}{\text{within-group variance estimate}}$$

If $H_0$ is true, there is no treatment variance. The between-group and within-group variance estimates will therefore be approximately equal on average so that *F* will usually be near 1.0. The more *F* is greater than 1.0, the more sure you can be that music type does have an effect on test scores. You reject $H_0$ when *F* is so large as to have a probability of .05 or less of occurring if $H_0$ is true.

Values of *F* less than 1.0 would indicate that $H_0$ should be retained because the between-group variance estimate is smaller than the within-group variance estimate. Since the value of *F* expected if $H_0$ is true is about 1.0, one might conceive of a *significantly small* value of *F* (e.g., $F = 0.2$). Such a result would mean that the sample means are *closer together* than you would expect given the person-to-person variability! There would be no obvious explanation for such a result other than chance or perhaps failure of one of the assumptions underlying the *F* test. Thus, the *F* associated with ANOVA can be thought of as a one-tailed test; the $H_0$ is retained no matter how small the *F* ratio gets, and can only be rejected when *F* exceeds the critical value in the positive direction.

## Computational Procedures

Although we do not expect that you will be using these formulas to calculate ANOVAs on research data you collect in the future, we feel that calculating a few ANOVAs by hand will give you a deeper appreciation of both the numerator and denominator of an ANOVA *F* ratio than you could ever get by writing code or clicking on the menu of a statistical program and then inspecting the results.

### Sums of Squares

The first step in the calculation of any ANOVA is to compute the **sum of squares between groups** (symbolized by $SS_{\text{Bet}}$), **the sum of squares within groups** (symbolized by $SS_{\text{W}}$), and the **total sum of squares** (symbolized by $SS_{\text{T}}$). As we hope you recall, *a sum of squares* (SS) is just the numerator of the variance formula, as defined in Chapter 3.

1. ***Total sum of squares*** ($SS_{\text{T}}$)**.** The definition formula for the total sum of squares is:

$$SS_{\text{T}} = \sum (X - \overline{X}_G)^2 \qquad \text{Formula 13.1A}$$

where

$\overline{X}_G$ = grand mean of all observations in the experiment (sometimes symbolized by X with two horizontal bars over it)
$\sum$ = summation across all *observations*

In Table 13.1A, for example,

$$\begin{aligned} SS_{\text{T}} &= (16-13)^2 + (16-13)^2 + (14-13)^2 + \cdots + (13-13)^2 + (10-13)^2 + (10-13)^2 \\ &= 3^2 + 3^2 + 1^2 + \cdots + 0^2 + (-3)^2 + (-3)^2 \\ &= 100 \end{aligned}$$

As is the case with any variance (see Chapter 3), however, it is also possible to compute $SS_{\text{T}}$ by using a shortcut computing formula:

$$SS_{\text{T}} = \sum X^2 - N_{\text{T}}\overline{X}_G^2 \qquad \text{Formula 13.1B}$$

where

$N_{\text{T}}$ = total number of observations in the experiment

For the data in Table 13.1A,

$$\begin{aligned} \sum X^2 &= 16^2 + 16^2 + 14^2 + \cdots 13^2 + 10^2 + 10^2 \\ &= 4{,}325 \\ N_{\text{T}}\overline{X}_G^2 &= 25(13)^2 \times 25 \times 169 = 4,225 \\ SS_{\text{T}} &= 4{,}325 - 4{,}225 = 100 \end{aligned}$$

2. ***Sum of squares between groups*** ($SS_{\text{Bet}}$)**.** The definition formula for the sum of squares between groups is

$$SS_{\text{Bet}} = \sum_{i=1}^{k} N_i(\overline{X}_i - \overline{X}_G)^2 \qquad \text{Formula 13.2A}$$

where

$N_i$ = number of scores in group $i$
$\overline{X}_i$ = mean of group $i$
$\overline{X}_G$ = grand mean
$k$ = the number of groups
$\sum_{i=1}^{k}$ = summation across all the *groups* (*not* the subjects)

As we have seen, the between-groups sum of squares is based on the differences between the mean of each group and the grand mean. The multiplication by $N_i$ in the formula ensures that the squared difference of the group mean from the grand mean is effectively counted once for each person in the group. (Samples need not be of equal size.)

In Table 13.1A,

$$\begin{aligned} SS_{\text{Bet}} &= 5(14.2 - 13.0)^2 + 5(13.2 - 13.0)^2 \\ &\quad + 5\,(12.4 - 13.0)^2 + 5(13.2 - 13.0)^2 + 5(12.0 - 13.0)^2 \\ &= 5(1.2)^2 + 5(0.2)^2 + 5(-\,0.6)^2 + 5(0.2)^2 + 5(-\,1.0)^2 \\ &= 14.4 \end{aligned}$$

When all of the samples are the same size, the formula for $SS_{\text{Bet}}$ can be simplified to:

$$SS_{\text{Bet}} = n\sum_{i=1}^{k}(\overline{X}_i - \overline{X}_G)^2 \qquad \text{Formula 13.2B}$$

where

$n$ = number of scores in *each* of the $k$ groups

This is the same as finding the *SS* for the $k$ sample means and then multiplying it by the common sample size.

3. ***Sum of squares within groups ($SS_W$).*** The definition formula for the *SS* within groups is:

$$SS_W = \sum^{n_1}(X - \overline{X}_1)^2 + \sum^{n_2}(X - \overline{X}_2)^2 + \cdots + \sum^{n_k}(X - \overline{X}_k)^2 \qquad \text{Formula 13.3}$$

where

$\overline{X}_1$ = mean of first group, $\overline{X}_2$ = mean of second group, and so on
$\sum$ = summation across the $n_i$ cases of the group in question (i.e., the $i$th group)

This is the same as calculating the *SS* separately for the scores in each group and then adding all of these SSs together.

In Table 13.1A,

$$\begin{aligned} SS_W &= (16 - 14.2)^2 + (16 - 14.2)^2 + (14 - 14.2)^2 + (13 - 14.2)^2 \\ &\quad + (12 - 14.2)^2 + (16 - 13.2)^2 + (14 - 13.2)^2 + (13 - 13.2)^2 \\ &\quad + \cdots + (10 - 12.0)^2 + (10 - 12.0)^2 \\ &= 85.6 \end{aligned}$$

That is, the difference between each score and the mean of the group containing the score is squared. (In this example, there is a total of 25 squared differences, but we are showing only the first 8 and the last 2.) Once this has been done for every score, the results are summed. This is the most tedious of the three sums of squares to compute, and it is possible to save some work by making use of the fact that

$$SS_T = SS_{\text{Bet}} + SS_W \qquad \text{Formula 13.4}$$

That is, the sum of squares between groups and the sum of squares within groups must add up to the total sum of squares. (See the Appendix at the end of this chapter for a proof.) Therefore, the within-group sum of squares can readily be computed as follows:

$$SS_W = SS_T - SS_{Bet} = 100 - 14.4 = 85.6$$

You should be careful in using this shortcut, however, because it provides no check on computational errors. If $SS_W$ is found directly, the foregoing formula can be used as a check on the accuracy of computation. (In a later section, we will show you how to skip the calculation of $SS_W$ and instead calculate $MS_W$ directly from the standard deviations of your groups.)

### Mean Squares

Next, $SS_{Bet}$ and $SS_W$ are each divided by the appropriate degrees of freedom. The values thus obtained are called *mean squares* (MSs) and are estimates of the population variance. The degrees of freedom between groups (symbolized by $df_{Bet}$) are equal to:

$$df_{Bet} = k - 1 \qquad \text{Formula 13.5}$$

where

$k$ = number of groups

The degrees of freedom within groups (symbolized by $df_W$) are equal to:

$$df_W = N_T - k \qquad \text{Formula 13.6}$$

where

$N_T$ = total number of observations

This is equivalent to obtaining the degrees of freedom for each group separately ($N_i - 1$) and then adding the degrees of freedom across all groups. The *total df*, helpful as a check on the calculations of *df*, is equal to $N_T - 1$.

In Table 13.1A, $df_{Bet} = (5 - 1) = 4$; $df_W = (25 - 5) = 20$. The total $df = N_T - 1 = 24$, which is in fact equal to 4 + 20.

The MS between groups (symbolized by $MS_{Bet}$) and the MS within groups (symbolized by $MS_W$) are equal to:

$$MS_{Bet} = \frac{SS_{Bet}}{df_B} \qquad \text{Formula 13.7}$$

$$MS_W = \frac{SS_W}{df_W} \qquad \text{Formula 13.8}$$

Thus, for Table 13.1A,

$$MS_{Bet} = \frac{14.4}{4} = 3.60$$

$$MS_W = \frac{85.6}{20} = 4.28$$

For Table 13.1B, $MS_W$ is the same, but $MS_{Bet} = 884.4/4 = 221.1$. (We suggest that you calculate $MS_{Bet}$ for Table 13.1B, using Formula 13.2B, as a check on your comprehension of this chapter thus far.)

## Testing the *F* Ratio for Statistical Significance

Before we calculate and test our *F* ratio for significance, let us review the ANOVA procedure in terms of the three-step method of null hypothesis testing introduced in Chapter 7.

**Step 1: State the null and alternative hypotheses and the significance criterion,** $\alpha$: Because we are dealing with five distinct populations (corresponding either to five different composers or five types of music), the null hypothesis is stated in terms of the equality of those five means: $H_0$: $\mu_1 = \mu_2 = \mu_3 = \mu_4 = \mu_5$. The only easy way to state the alternative hypothesis is as follows: $H_A$: not $(\mu_1 = \mu_2 = \mu_3 = \mu_4 = \mu_5)$. As usual, we will set alpha at the maximum value ordinarily allowed: $\alpha = .05$.

**Step 2: Calculate the appropriate test statistic:** Having computed the mean squares (*MS*s), it is easy to compute the *F* ratio (we are pursuing two examples at once, so we will calculate both *F* ratios), as follows:

$$F = \frac{MS_{\text{Bet}}}{MS_{\text{W}}} \qquad \text{Formula 13.9}$$

For Table 13.1A,

$$F = \frac{3.60}{4.28} = 0.84$$

For Table 13.1B,

$$F = \frac{221.1}{4.28} = 51.66$$

**Step 3: Find the critical value and make a statistical decision with respect to your null hypothesis:** As noted earlier, we encounter a new statistical model when we wish to test the ANOVA $H_0$ for statistical significance: the *F distributions*. Here again, the correct model (i.e., the null hypothesis distribution) enables us to determine the probability of obtaining the results observed in the samples if $H_0$ is true.

Just as there are different *t* distributions for different degrees of freedom, so are there different *F* distributions for all combinations of different $df_{\text{Bet}}$ and $df_{\text{W}}$. Although the exact shape of an *F* distribution depends on $df_{\text{Bet}}$ *and* $df_{\text{W}}$, all of the *F* distributions used in psychological research are positively skewed. Two are illustrated in Figure 13.1.

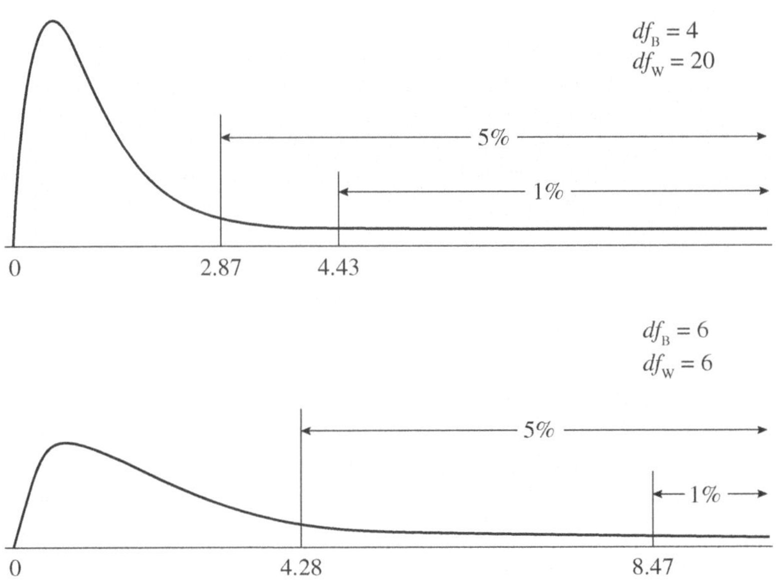


Figure 13.1

***F* Distributions for $df_{\text{Bet}} = 4$ and $df_{\text{W}} = 20$, and for $df_{\text{Bet}} = 6$ and $df_{\text{W}} = 6$**

**Table 13.2** **Summary of the One-Way ANOVA of the Second Music Experiment (from Table 13.1B)**

| Source of variation | *SS* | *df* | *MS* | *F* | *p* |
|---|---|---|---|---|---|
| Between groups | 884.4 | 4 | 221.1 | 51.66 | < .01 |
| Within groups (error) | 85.6 | 20 | 4.28 | | |
| Total | 970.0 | 24 | | | |

To obtain the minimum value of $F$ needed to reject $H_0$, you can refer to Table F in the Appendix of this book. Check the *column* corresponding to $df_{\text{Bet}}$ (the degrees of freedom in the *numerator* of the $F$ ratio) and the *row* corresponding to $df_W$ (the degrees of freedom in the *denominator* of the $F$ ratio). The critical $F$ values for 4 $df$ (numerator) and 20 $df$ (denominator) are as follows:

| | | *df*-numerator | | | | | |
|---|---|---|---|---|---|---|---|
| | | 1 | 2 | 3 | 4 | 5 | ... |
| *df*-denominator | 1 | | | | | | |
| | 2 | | | | | | |
| | 3 | | | | | | |
| | ... | | | | | | |
| | 20 | | | | 2.87<br>**4.43** | | |
| | ... | | | | | | |

The critical values for 20 $df$ (numerator) and 4 $df$ (denominator) are *not* the same; be sure to enter the table in the proper place. The smaller of the two values is the .05 criterion; the larger (in boldface) is the .01 criterion. In the case of Table 13.1A, the obtained value of 0.84 is less than the critical value of 2.87. Therefore, using $\alpha = .05$, you retain $H_0$ and conclude that there is not sufficient reason to believe that there are any differences among the five population means. (Because the computed value of $F$ is less than 1.0, you could have reached this conclusion without bothering to consult the table in this instance; because an ANOVA is essentially a one-tailed test, an $F$ ratio of 1.0 or less can never be statistically significant.)

For the experiment shown in Table 13.1B, the calculated $F$ value of 51.66 easily exceeds the critical value for the .05 level (2.87) and even the .01 level (4.43). We present these statistically significant results in the form of a summary table.

### The ANOVA Summary Table

It is customary to summarize the results of an analysis of variance in a table such as the one shown in Table 13.2. Note that the between-groups source of variation is listed first and that the value of $F$ (and sometimes $p$ as well) is entered at the extreme right of the between-groups row. The denominator of the $F$ ratio (i.e., $MS_W$ for this type of ANOVA) is often called the *error term*, for reasons already explained previously under "The General Logic of ANOVA."

## Calculating the One-Way ANOVA From Means and Standard Deviations

Traditionally, when statistics texts describe the calculation of a $t$ test, they do so in terms of the means, standard deviations, and sample sizes of the two groups. However, when these same texts describe the calculation of an ANOVA, the emphasis is usually placed on proceeding directly from the raw data (i.e., individual scores) to the *SS* components of the ANOVA, without stopping to calculate the SD or variance of each separate group. (That is the approach we have taken in this chapter thus far.) There is certainly some computational efficiency to this approach, but in the age of handheld calculators and statistical computer

**Table 13.3** **Data From the Caffeine Experiment of Chapter 7**

| | Caffeine group | Control group |
|---|---|---|
| $X$ | 81 | 68 |
| $S$ | 12.0 | 16.8 |
| $N$ | 22 | 20 |

programs, computational efficiency is no longer an important advantage. From a pedagogical standpoint, there is much understanding to be gained by describing the calculation of an ANOVA from the means and SDs, in a way that is parallel to the usual calculation of the *t* test. Moreover, it is not inconceivable that you may someday want to calculate an ANOVA from a published table of the means, SDs, and *n*s for a study in which the *F* ratio you are most interested in is not presented. Therefore, in this section, we will start with the type of data summary table that is often published for the results of an experiment and show you how to calculate the ANOVA from those data.

A particularly convenient table to use as an example for this section is the one that summarized the data for the caffeine experiment in Chapter 7, which is reproduced here as Table 13.3.

Before we can use Formula 13.2A for $SS_{Bet}$, we need to calculate the grand mean for these data. If the two groups were the same size, $\overline{X}_G$ would be the simple average of the two group means, which is 74.5 in this case. However, when the *n*s are not equal, as in Table 13.3, a weighted average of the two means must be used. The grand mean for these data is:

$$\overline{X}_G = \frac{22(81) + 20(68)}{22 + 20} = \frac{3142}{42} = 74.8$$

Now we can calculate $SS_{Bet}$:

$$\begin{aligned} SS_{Bet} &= n_1\left(\overline{X}_1 - \overline{X}_G\right)^2 + n_2\left(\overline{X}_2 - \overline{X}_G\right)^2 \\ &= 22(81 - 74.8)^2 + 20(68 - 74.8)^2 = 845.68 + 924.8 = 1770.48 \end{aligned}$$

To find $SS_W$ when given SDs but not the raw data, you have two options. The first is based on the fact that MSw is nothing more than an average of the sample variances; when the sample means are not equal, then MSw is a *weighted average* of the sample variances. You may recall from Chapter 7 that we computed the weighted average of two sample variances, which we called the *pooled variance* (Formula 7.1 page 130; we expand upon the section on comparing ANOVA with the *t* test, below).

$$s^2_{\text{pooled}} = \frac{(N_1 - 1)s_1^2 + (N_2 - 1)s_2^2}{N_1 + N_2 - 2} \qquad \text{Formula 7.1}$$

With ANOVA, we simply need to extend Formula 7.1 so that it will work with any number of variances:

$$MS_W = \frac{\sum_{j=1}^{k}\left((n_j - 1)\hat{s}_j^2\right)}{N_T - k} \qquad \text{Formula 13.10}$$

Once you have computed MSw directly, we can derive SSw by multiplication. Recall Formula 13.8:

$$MS_W = \frac{SS_W}{df_W} \qquad \text{Formula 13.8}$$

If we multiply both sides of this equation by $df_W$, then we can solve for $SS_W$: $MS_W * df_W = SS_W$.

The second option for deriving $SS_W$ from SDs is simply a variation. First, square each $s$ to obtain the variance and then multiply each $s^2$ by its $df$. (Recall that $s^2 = SS/df$, so $SS = s^2 df$, where $df = N - 1$ for that group.) In this example, $SS_{W - \text{caffeine}} = 21(12)^2 = 3{,}024$, and $SS_{W - \text{control}} = 19(16.8)^2 = 5{,}362.56$. Therefore, $SS_W = 3{,}024 + 5{,}362.56 = 8{,}386.6$.

Next, we calculate the two $MS$s for the ANOVA by dividing each $SS$ by the appropriate $df$ term: $df_{Bet} = k - 1 = 2 - 1 = 1$; $df_W = N_T - k = 42 - 2 = 40$. Therefore, $MS_{Bet} = SS_{Bet}/1 = 1{,}770.48$, and $MS_W = 8{,}386.6/40 = 209.66$. Finally, we calculate the $F$ ratio: $MS_{Bet}/MS_W = 1{,}770.48/209.66 = 8.44$.

## Comparing the One-Way ANOVA With the *t* Test

When you compare the means of only two groups, you have a choice. You can test the difference of the means with a $t$ test or a one-way ANOVA. Although both procedures lead to the same $p$ value, it is instructive to see how closely connected these two seemingly diverse statistical methods are. That is why we chose a two-group example for our calculation in the previous section. If you take the $t$ value calculated for these same data in Chapter 7 on page 130 ($t = 2.91$) and square it, you will get (within rounding error) the $F$ ratio just calculated for those data. In fact, $F$ always equals $t^2$ in the two-group case. This also holds true for the corresponding critical values. The critical $t$ that we found in Chapter 7 for this example was 2.02, and the critical $F$ for the corresponding ANOVA is $F_{.05}\ (1, 40)$, which is 4.08, which equals $2.02^2$.

Comparing the $t$ and $F$ calculations for the same data reveals another important connection. Consider once again the calculation of the standard error of the difference for the $t$ test of these data in Chapter 7:

$$s_{\overline{X}_1 - \overline{X}_2} = \sqrt{209.664\left(\frac{1}{22} + \frac{1}{20}\right)}$$

Note that the value for $s^2_{\text{pooled}}$ in this formula is 209.664, which rounds off to the value we calculated for the denominator of the $F$ ratio, $MS_W$. As mentioned in the previous section, $MS_W$ is just a new label for the pooled variance when more than two groups are involved; it is a weighted average of the variances of all the groups in an ANOVA. When you add the $SS$s from all the groups to create $SS_W$ and then divide by the sum of the $df$s from all the groups (i.e., $df_W$), what you are actually finding is the weighted average of the variances, also known as the pooled variance.

## A Simplified ANOVA Formula for Equal Sample Sizes

When all of the samples are the same size in a one-way ANOVA, the $F$ ratio formula can be reduced to a simple expression that is easy to understand as well as to use. If we divide the expression for $SS_{Bet}$ when the $n$s are equal (shown previously) by $df_{Bet}$, we get

$$MS_{Bet} = \frac{n\sum(\overline{X}_i - \overline{X}_G)^2}{df_{Bet}} = n\frac{\sum(\overline{X}_i - \overline{X}_G)^2}{df_{Bet}}$$

Notice that the part of this formula that is multiplied by $n$ is really a formula for the variance of the sample *means*, so we can simplify the formula to $ns^2_{\overline{X}}$. When all the $n$s are equal, $MS_W$ has a simple form as well; it is just the ordinary average of the sample variances:

$$MS_W = \frac{\sum s_i^2}{k}$$

Putting the two halves of the *F* ratio together, we come up with a convenient formula for the (fairly common) case when all the samples are the same size:

$$F = \frac{ns_{\bar{X}}^2}{\frac{\sum s_i^2}{k}} \qquad \text{Formula 13.11}$$

From this formula we can see at a glance that the magnitude of *F* depends entirely on three important factors: how far apart the sample means are from each other (the larger the spread, the larger is $s_{\bar{X}}^2$); how large the samples are ($n$ is the size of *each* sample); and how much individuals vary from each other within each group. Because all of the $n$s are equal in Table 13.1, we can illustrate the use of this simplified formula by applying it to the music/spatial ability data. We begin by finding the (unbiased) variance of the five sample means in Table 13.1B. For 14.2, 20.2, 22.4, 8.2, and 8.0, $s_{\bar{X}}^2 = 44.22$.[1] $MS_{\text{Bet}}$ equals $n\,s_{\bar{X}}^2$, so $MS_{\text{Bet}} = (5)(44.22) = 221.1$. (Note that this value agrees with the entry for $MS_{\text{Bet}}$ in Table 13.2.)

Next, we find $MS_{\text{W}}$ by squaring the (unbiased) *SD*s for the five groups and averaging them. (The *SD*s are the same in both the A and B sections of Table 13.1.) $MS_{\text{W}}$ equals $(1.789^2 + 2.168^2 + 2.608^2 + 1.789^2 + 1.871^2)/5 = 21.4/5 = 4.28$. (Again, this value agrees with Table 13.2.) Dividing $MS_{\text{Bet}}$ by $MS_{\text{W}}$ yields the same *F* ratio that we calculated previously with the more tedious raw-score formula.

## Effect Size for the One-Way ANOVA

Just as we were able to write a *t* formula in terms of the effect size in the sample data ($d$) and the size of each sample ($n$), we can begin with the equal-$n$ formula for *F*:

$$F = n \times s_{\bar{X}}^2 / MS_{\text{W}}$$

and then separate it into an effect size term and a sample size term. If we remove $n$ from the above formula, we can define a new quantity we will call $f^2$ as $f^2 = s_{\bar{X}}^2/MS_{\text{W}}$, then the one-way ANOVA formula can be written as $F = nf^2$.

If we do not square it, $f$ is an effect size measure comparable to $d$ for the two-sample $t$ test. (In fact, in the two-group case, $f$ is equal to $d$ divided by the square root of 2.) Just as $d$ is an estimate of **d** in the population, $f$ can be viewed as an estimate of a corresponding population quantity that J. Cohen (1988) referred to as **f** (a lower-case f in **boldface**). Although estimates of **f** are sometimes used as the basis of power analysis for one-way ANOVA, they are not routinely reported as effect-size measures in journal articles. The commonly reported measures of effect size for ANOVA are correlational measures related to the square of the point-biserial $r$ in the two-group case.

The analysis of variance procedure begins (at least conceptually) by dividing $SS_{\text{total}}$ into two components: $SS_{\text{Bet}}$ and $SS_{\text{W}}$. These components can be thought of, in regression terms, as the variability explained and unexplained, respectively. Therefore, dividing $SS_{\text{Bet}}$ by $SS_{\text{T}}$ gives the proportion of variance explained in your DV by the ANOVA factor. Although this proportion comes from sample data, it is most commonly symbolized by a Greek letter (usually reserved for population values) and referred to as $\eta^2$ (eta squared); it can be found from the following simple formula:

$$\eta^2 = \frac{SS_{\text{Bet}}}{SS_{\text{T}}} \qquad \text{Formula 13.12A}$$

In the two-group case, $\eta^2$ will always be exactly equal to $r_{\text{pb}}^2$.

[1] What makes this formula so easy to use is the ready availability of scientific or statistical calculators that give you the unbiased standard deviation at the touch of a button after the scores are entered. This SD is then squared to obtain the (unbiased) variance.

And just as $r^2_{pb}$ can be calculated from a $t$ value without access to the raw data, $\eta^2$ can be calculated from your $F$ ratio by the following formula:

$$\eta^2 = \frac{df_{Bet}F}{df_{Bet}F + df_W} \qquad \text{Formula 13.12B}$$

For example, the $F$ ratio for testing the group differences in Table 13.1B was 51.66, with $df_{Bet} = 4$ and $df_W = 20$. Such a large $F$ ratio associated with moderate sample sizes tells us that a large proportion of variance has been accounted for. We can confirm this by using Formula 13.12B:

$$\eta^2 = \frac{4(51.66)}{4(51.66) + 20} = \frac{206.64}{226.64} = .91$$

Such a large value for $\eta^2$ will not often occur in behavioral research, unless you are performing something like a manipulation check (e.g., did the participants who were moved to tears by watching a sad movie clip report more sadness than those participants who were laughing at a comedy clip?). Note that you will get the same value for $\eta^2$ if you divide $SS_{Bet}$ by $SS_T$, as given in Table 13.2: $\eta^2 = 884.4/970 = .91$.

### Interpreting the Size of Eta Squared

Clearly, .9 is a very large value for $\eta^2$, given that the largest possible value is 1.0, but what is less obvious is that a value of .25—that is, one-fourth of the total variation in the dependent variable (DV) accounted for by the independent variable (IV)—is also considered very large in psychological research. In fact, according to conventional guidelines any $\eta^2$ larger than .14 is considered to be large. What then is considered a small effect size for ANOVA? Without a doubt, an $\eta^2$ that is less than .01 (i.e., only 1% of the variance accounted for) is considered so small that for many studies it is viewed as a negligible effect that is not worth pursuing further (unless, perhaps, it is a very interesting or important effect). An $\eta^2$ between .01 and .06 is still considered small, but worth considering, whereas an $\eta^2$ between .06 and .14 is viewed as a medium-sized effect. Of course, these are just rough, fairly arbitrary, guidelines but they can be useful for performing a power analysis, when you have no other basis for predicting the effect size of a study you are planning.

Another way to think about effect size is that it is inversely related to the amount of overlap among the population distributions being sampled, while being completely unrelated to the size of the samples. By itself, $\eta^2$ just describes the overlap of the scores among your different samples, but it is also your best basis for guessing about the overlap among the corresponding populations, after correcting for its bias (see next section). A very large effect size suggests, for instance, that rather few scores in the group with the lowest mean will be as high as even just the middle of the group with the highest mean. Conversely, a very small effect size suggests a great deal of overlap from group to group. It is critical to remember, however, that even when there is a great deal of overlap among the populations and this is reflected in your samples, it is still possible to obtain a large $F$ ratio with a correspondingly small $p$ value, if the samples are very large. A small $p$ value does suggest that the population effect size is not zero, but, by itself, even a very tiny $p$ value does not tell you anything about the size of the effect in the population or even in your samples. That is why it is so important to supplement your test statistic (e.g., $t$ or $F$) with an effect size measure that is independent of the sample sizes. Simply deciding whether you think the population effect size is zero or not, which is all that you are doing when you decide whether to reject the null hypothesis, is usually not very informative.

### Unbiased Estimate of the Amount of Variance Accounted for in the Population (Omega Squared)

Just as in the two-group case, the proportion of variance accounted for in the population is designated by $\omega^2$. And, like $r^2_{pb}$, $\eta^2$ is a biased (over)estimate of $\omega^2$. More than one way has been recommended to correct this bias, but the simplest correction is to subtract 1 from the calculated $F$ ratio before using Formula 13.12B for $\eta^2$, as follows:

$$\text{est.}\omega^2 = \frac{df_{\text{Bet}}(F - 1)}{df_{\text{Bet}}(F - 1) + df_{\text{W}}} \qquad \text{Formula 13.13}$$

One advantage of this correction formula is that you can see at a glance that it is pointless to estimate $\omega^2$ if your $F$ ratio is less than 1.0 ($\omega^2$ cannot be negative). Whenever $F$ is less than 1, the separation between the sample means is *less* than you would expect from sampling error, so the data are suggesting that the effect in the population, and therefore $\omega^2$, is zero.

## Some Comments on the Use of ANOVA

The general considerations we included in Chapter 7 under the heading "The $t$ Test for Two Sample Means" apply equally well to the ANOVA, with minor modifications, as described next.

### Underlying Assumptions

The procedures discussed in this chapter assume that observations are *independent* both within and between all of the samples. The value of any observation should *not* be related in any way to that of any other observation, as would be the case if all of the samples in your ANOVA were truly selected at random from an infinite, or at least very large, population. (For experimental studies, the usual procedure is to employ random assignment to all conditions in the study from one large convenience sample, in the manner discussed for the two-sample $t$ test in Chapter 7.) It is also assumed that the populations are normally distributed. Fortunately, ANOVA is fairly robust with regard to the latter assumption, and will yield accurate results even if the population distributions depart considerably from normality (unless you also have small sample sizes or the distributions for different groups are skewed oppositely).

In addition, the use of $MS_{\text{W}}$ as the error term requires that you assume that all of the populations have the same amount of (within-group) variance. As in the case of the $t$ test, homogeneity of variance (HOV) can be tested, and if it looks like you cannot assume HOV, an adjusted form of ANOVA (see B. H. Cohen, 2008) can be performed. However, when researchers are dealing with sample sizes and variances that both vary a great deal or small samples that are very skewed, they are more likely to apply either an appropriate data transformation or a nonparametric test, such as the *Kruskal–Wallis* test, described at the end of this chapter.

### Publishing Your Results

In behavioral research journals, the syntax for reporting an $F$ ratio is as follows (as applied to the results presented in Table 13.2): "The different types of music produced large differences in spatial ability and yielded a significant one-way ANOVA on the group means, $F(4,20) = 51.66$, $MS_{\text{W}} = 4.28$, $p < .01$, $\eta^2 = .91$." Note that in most journals MSw is written as MSE (for MS error), and that $df_{\text{Bet}}$ and $df_{\text{W}}$ are always given, in that order, in parentheses after $F$. However, $p$ may be given exactly (except when four or more zeroes are required to the right of the decimal point), or in terms of an even smaller alpha (e.g., $p < .001$), if applicable. As was pointed out in previous chapters, statistical significance (even at an unusually small alpha level) does not necessarily imply a strong relationship between the IV and DV. It is therefore desirable, and becoming increasingly common, to report a distance measure of effect size (e.g., $g$ of $f$) or strength of relationship measure (e.g., estimated $\omega^2$) in addition to the test of significance. (Instead of estimated $\omega^2$, $\eta^2$ is often reported without correction although in the example given here, $F$ is so large that it made little difference; $\eta^2$ was .91, while est. $\omega^2$ was .89.) However, effect-size measures of any type are generally not reported if the results fail to reach significance at the .05 level.

### Following Up on a Significant ANOVA

An investigator is rarely satisfied with knowing that a set of three or more population means is not equal, yet this is all that the rejection of the ANOVA null hypothesis signifies when $k$ is greater than 2. Instead, the researcher would usually like to conclude that certain particular population means are larger (or smaller) than others. Therefore, it is usually necessary to make more specific comparisons among the $k$ means. Depending on the structure and purpose of the experiment, these *multiple comparisons* may take many forms and are subject to many considerations.

Sometimes the investigator wishes to make all $k(k - 1)/2$ of the possible pairwise mean comparisons. At other times, only some of the possible comparisons are of interest (e.g., when several experimental groups are each to be compared with a single control group). Another important issue is whether a given comparison was planned in advance of the research (a priori) or was suggested by seeing the actual results (post hoc).

The topic of multiple comparisons is a rather complex one, which is why we devote the entire next chapter to it.

### ANOVA When the Independent Variable Has Quantitative Levels

In the type of ANOVA we have been presenting in this chapter, the dependent variable has been measured on an interval or ratio scale (ANOVA for ordinal, or ranked, DVs are discussed a little later in this chapter), and the independent variable has categorical levels, often in the form of qualitatively different experimental treatments. However, it is possible, and sometimes very desirable, for the levels of the ANOVA factor to be in the form of different amounts of the same treatment—for example, different dosages of a drug (usually including zero, in the form of a placebo); different numbers of letters in an anagram to be solved; different numbers of repetitions of words in a list to be recalled. When the factor has such levels, the ANOVA procedure of this chapter does not capitalize on an important potential source of regularity in the data: The DV, on average, may increase linearly, or in some other regular way, as a function of the level (i.e., amount) of the factor. For instance, if performance on a task (the DV) improves to a greater degree according to the amount of caffeine participants are given (the IV), the data are said to exhibit an increasing (possibly linear) trend. Even if the performance starts to decline for the largest doses, producing a simple curve (often referred to as a quadratic trend), a modified ANOVA procedure that is sensitive to such trends can produce considerably larger $F$ ratios (and therefore more power) than the ordinary one-way ANOVA. The greater the number of levels the IV has, the greater is the potential advantage of an ANOVA based on trends, as compared to the ANOVA described in this chapter. The error term of a trend-based ANOVA is the same as the $MS_W$ described in this chapter, but the $MS_{Bet}$ is calculated differently, and generally more efficiently, for each possible trend. Because trends in the data fall under the general heading of *multiple comparisons*, they are dealt with, albeit briefly, in the next chapter.

### Computational Practice for the One-Way ANOVA: Three Unequal-Sized Groups

Suppose that the dean of a small college wishes to survey student attitudes about participating in determining college curricula and requirements. She circulates a five-item questionnaire in which students respond to each item on a Likert scale that goes from 1 (no interest in student participation) to 5 (very interested). Therefore, total scores on this questionnaire can range from 5 to 25. Although the students do not give their names, they do provide such demographic items as their college major, year in school, and gender.

Looking only at the data she has collected so far for seniors, the dean notices that student attitudes seem to differ according to their principal areas of study. She decides to arrange the scores of the seniors according to whether they are majoring in a natural science,

**Table 13.4** **ANOVA of Attitudes of Students From Three Areas of Study at a Small College to Student Participation in Determining College Curricula**

| Areas of study | | | |
|---|---|---|---|
| **Natural sciences** | **Social sciences** | **Humanities** | |
| 18 | 17 | 69 | $F = \frac{85.10}{16.74} = 5.08$ |
| 15 | 22 | 9 | |
| 12 | 5 | 12 | |
| 12 | 15 | 11 | |
| 9 | 12 | 11 | |
| 10 | 20 | 8 | |
| 12 | 14 | 13 | |
| 20 | 15 | 14 | |
| | 20 | 7 | |
| | 21 | | |

a social science, or one of the humanities. The (hypothetical) data, broken down by these three areas, are shown in Table 13.4. We leave it as an exercise for the reader to calculate the *SS* components using Formulas 13.2A and 13.3, and therefore present the results of the ANOVA in Table 13.4 only in terms of the *MS*s and *F* ratio. Because the critical *F* at the .05 level for 2 and 24 *df* is 2.064, the null hypothesis for these data ($\mu_1 = \mu_2 = \mu_3$) can be rejected, and measures of effect size should be calculated. For this example, the estimate of $\omega^2$ comes out to .254, which, according to the guidelines described in a preceding section, is considered large. For additional practice, we suggest that you use both Formula 13.12A and Formula 13.12B to calculate eta squared from the data in Table 13.4 in order to demonstrate that (a) both formulas yield the same value for $\eta^2$ and (b) the value for $\eta^2$ is somewhat larger than the value for $\omega^2$.

## A Nonparametric Alternative to the One-Way ANOVA: The Kruskal–Wallis *H* Test

Anytime you are dealing with more than two groups but it seems more appropriate to perform the rank-sum test than a *t* test to compare any two of the groups, you can compare all of your groups simultaneously with a distribution-free test based on ranks, called the **Kruskal–Wallis (K–W) *H* test**. The K–W *H* test is an extension of the rank-sum test (described in Chapter 8) to the multigroup case, just as the one-way ANOVA is an extension of the *t* test. As you might guess, when *k* (the number of groups) is equal to 2, the rank-sum test and the Kruskal–Wallis *H* test yield the same *p* values. Moreover, the K–W *H* test resembles the one-way ANOVA in several important ways. However, because statistics based on ranks are simple functions of *N*, the computation of the *H* statistic is much more basic than is the computation of *F*. In fact, *H* is a function of just the sum of squares between groups ($SS_{\text{Bet}}$), except that $SS_{\text{Bet}}$ is computed on the ranks instead of the scores.

To illustrate the calculation of the *H* statistic, we have applied the K–W *H* test to the data in Table 13.4. The ranks for the data are shown next to the original scores for each group in Table 13.5. Those ranks were obtained by taking the total of ($N =$)27 observations and ranking them together from the lowest score of all (rank = 1) to the highest score among all three groups (rank = 27); then we placed those ranks next to their corresponding scores, organized into $k = 3$ groups, as in the original table (i.e., Table 13.4). Tied scores were assigned mean ranks in the usual manner. For example, there are two scores of 9 (one from natural sciences and one from humanities) that should occupy ranks 5 and 6, so each has been assigned the rank of 5.5 (the mean of 5 and 6). The next highest score, 10 (which is not tied), therefore received a rank of 7.

**Table 13.5** **The Kruskal–Wallis *H* Test for $k = 3$ Independent Samples**

| Natural (Group sciences 1) | | Social sciences (Group 2) | | Humanities (Group 3) | |
|---|---|---|---|---|---|
| *X* | *R* | *X* | *R* | *X* | *R* |
| 15 | 19 | 17 | 21 | 6 | 2 |
| 18 | 22 | 22 | 27 | 9 | 5.5 |
| 12 | 12 | 5 | 1 | 12 | 12 |
| 12 | 12 | 15 | 19 | 11 | 8.5 |
| 9 | 5.5 | 12 | 12 | 11 | 8.5 |
| 10 | 7 | 20 | 24 | 8 | 4 |
| 12 | 12 | 14 | 16.5 | 13 | 15 |
| 20 | 24 | 15 | 19 | 14 | 16.5 |
| | | 20 | 24 | 7 | 3 |
| | | 21 | 26 | | |
| | $T_1 = 113.5$ | | $T_2 = 189.5$ | | $T_3 = 75$ |
| | $N_1 = 8$ | | $N_2 = 10$ | | $N_3 = 9$ |

$$N_T = N_1 + N_2 + N_3 = 27$$

$$\sum R = N_T(N_T + 1)/2 = (27)(28)/2 = 378$$

$$= T_1 + T_2 + T_3 = 113.5 + 189.5 + 75 = 378$$

The sum of the ranks for each of the three groups (using the symbol $T_i$ to represent the total of the ranks for any one group) is shown in Table 13.5: $T_1 = 113.5$; $T_2 = 189.5$; and $T_3 = 75$. (To guard against errors in ranking or summing, the sum of all *T* values should be verified as equal to $\sum R$ or $N_T[N_T + 1]/2$, as shown at the bottom of the table.) The next step is to compute the sum of squares between groups, $SS_{Bet}$, for the ranks; the following formula is particularly convenient in that it is based on the sum of the ranks for each group:

$$SS_{Bet} = \frac{T_1^2}{N_1} + \frac{T_2^2}{N_2} + \cdots + \frac{T_k^2}{N_k} - \frac{N_T(N_T + 1)^2}{4} \qquad \text{Formula 13.14}$$

Substituting the *T* and *N* values from Table 13.5 yields:

$$SS_{Bet} = \frac{(113.5)^2}{8} + \frac{(189.5)^2}{10} + \frac{(75)^2}{9} - \frac{27(28)^2}{4}$$
$$= 5826.31 - 5292.00$$
$$= 534.31$$

The final step is to compute the Kruskal–Wallis *H* statistic:

$$H = \frac{12SS_{Bet}}{N_T(N_T + 1)} \qquad \text{Formula 13.15}$$

The value of 12 in the numerator is a constant, and it arises from the fact that the variance of a full set of *N* untied ranks is equal to $(N + 1)^2/12$. In our example,

$$H = \frac{12(534.31)}{(27)(28)} = 8.48$$

Under the null hypothesis that the locations of the *k* populations are identical, *H* follows, approximately, a distribution known as the chi-square (symbolized as $\chi^2$) distribution with $k - 1$ degrees of freedom. Thus, the computed *H* value of 8.48 can be compared to a critical value found in Table H of the Appendix. The critical value in Table H that corresponds to the .05 criterion of significance for 2 *df* is 5.99. Because 8.48 is larger than 5.99, you can reject the null hypothesis that the three populations have equal locations. The chi-square distribution is discussed in some detail in Chapter 17, in which it becomes the main tool for null hypothesis testing with categorical data.

Like the normal approximation for the rank-sum test, the chi-square approximation used to test *H* for statistical significance is a good one unless the size of any group is very small. If there are no more than three groups, each group should have at least five cases in it; otherwise the Kruskal–Wallis *H* test should not be used. With more than three groups, as few as two cases in a group are sufficient. The existence of tied ranks tends to make the *H* test conservative (smaller and less likely to yield statistical significance), but this tendency is slight unless ties are very long and numerous. If necessary, it is possible to correct for ties and to perform this test for very small samples using special tables (see Siegel & Castellan [1988]). Finally, as compared to the analogous parametric *F* test in a one-way ANOVA, the *H* test has a power efficiency ranging from 90% for small samples to 95% for large samples.

### A Measure of the Strength of the Relationship: $\eta^2_R$

A measure of the strength of the relationship between membership in one of the *k* groups and rank on the dependent variable may be obtained by computing a form of eta squared. This is the same index that was used to measure effect size for the one-way ANOVA, but it is now applied to the ranks rather than the original scores. Just as $\eta^2$ for the scores was a function of *F* and the degrees of freedom, so $\eta^2_R$ may be found from *H*, *k*, and $N_T$:

$$\eta^2_R = \frac{H - k + 1}{N_T - k} \qquad \text{Formula 13.16}$$

where

$k$ = number of groups
$N_T$ = total number of observations

For example, the ranked attitude scores in Table 13.5 yielded an *H* of 8.48. Therefore, for these data,

$$\eta^2_R = \frac{(8.48 - 3) + 1}{27 - 3} = \frac{6.48}{24} = 0.27$$

For the continuous score data in Table 13.4, $\eta^2$ was equal to .298 (i.e., $SS_{Bet}/SS_T$ = 170.21/572)—a very similar (and rather large) value. This similarity will usually be the case, with the nonparametric eta squared usually being a bit smaller. However, sometimes a few outliers so increase the error variance in a one-way ANOVA that the nonparametric eta squared is actually larger, and the K–W *H* test may attain significance when the ANOVA does not. (Just as the statistical significance of $\eta^2$ is tested by the *F* ratio, so the significance of $\eta^2_R$ is tested by *H*. So, as with the other measures of strength of relationship that we have discussed, there is no need to conduct an additional significance test for $\eta^2_R$.)

## Summary

Analysis of variance (ANOVA) permits null hypotheses to be tested that involve the means of three or more samples (groups). One-way ANOVA deals with one independent variable made up of membership in one of *k* groups or levels. Total variance is partitioned into two sources: *between-group variance* and *within-group (error) variance*. These are compared by using the *F* ratio to determine whether the independent variable can be said to have an effect on the dependent variable. A significant *F* ratio in ANOVA indicates that it is reasonable to decide that not all *k* population means are equal. Follow-up procedures can be used to specify which pairs of means differ significantly, as detailed in the next chapter

### Meaning of Symbols

| Symbol | General meaning |
|---|---|
| $k$ | Number of groups (or the last group) |
| $N_T$ | Total number of observations |
| $N_i$ | Number of observations in group $i$ |
| $\overline{X}_G$ | Grand mean |
| $\overline{X}_i$ | Mean of group $i$ |
| $X_i$ | A score in group $i$ |

### Definition and Computing Formulas

1. Total sum of squares

$$\text{Definition: } SS_T = \sum (X - \overline{X}_G)^2 \qquad \text{Formula 13.1A}$$

$$\text{Computing formula: } SS_T = \sum X^2 - N_T\overline{X}_G^2 \qquad \text{Formula 13.1B}$$

2. Between-groups sum of squares

$$\text{Definition: } SS_{Bet} = \sum_{i=1}^{k} N_i(\overline{X}_i - \overline{X}_G)^2 \qquad \text{Formula 13.2A}$$

$$\text{For equal } ns \text{ only: } SS_{Bet} = n\sum_{i=1}^{k} (\overline{X}_i - \overline{X}_G)^2 \qquad \text{Formula 13.2B}$$

where $n$ is the size of *each* sample

3. Within-groups sum of squares

Definition:

$$SS_W = \sum^{n_1} (X - \overline{X}_1)^2 + \sum^{n_2} (X - \overline{X}_2)^2 + \cdots + \sum^{n_k} (X - \overline{X}_k)^2 \qquad \text{Formula 13.3}$$

Shortcut computing formula: $SS_W = SS_T - SS_{Bet}$

### Steps in One-Way Analysis of Variance

1. Compute $SS_T$, $SS_{Bet}$, and $SS_W$.
2. Compute the degrees of freedom:

$$df_{Bet} = k - 1 \qquad \text{Formula 13.5}$$

$$df_W = N - k \qquad \text{Formula 13.6}$$

3. Compute the two mean squares ($MS$s):

$$MS_{Bet} = \frac{SS_{Bet}}{df_{Bet}} \qquad \text{Formula 13.7}$$

$$MS_W = \frac{SS_W}{df_W} \qquad \text{Formula 13.8}$$

If all of the *n*s are equal:

$$MS_{Bet} = ns^2_{\bar{X}} \qquad \text{(Top of Formula 13.11)}$$

where $s^2_{\bar{X}}$ is the unbiased variance of the sample means, and

$$MS_W = \frac{\sum s_i^2}{k}$$

where $s_i^2$ is the unbiased variance within group *i*.

4. Compute the *F* ratio:

$$F = \frac{MS_{Bet}}{MS_W} \qquad \text{Formula 13.9}$$

5. Obtain the (.05 or .01) critical value from the *F* table ($dfs = k - 1, N - k$) and test for significance. If the computed *F* is equal to or greater than the tabled *F*, reject $H_0$ (that all population means are equal) in favor of $H_A$ (that $H_0$ is not true). Otherwise, retain $H_0$.
6. Especially if *F is statistically significant*, a corresponding effect-size measure should be computed. The most commonly used strength-of-relationship measure is the proportion of variance accounted for as determined by the following formula:

$$\eta^2 = \frac{SS_{Bet}}{SS_T} \qquad \text{Formula 13.12A}$$

Eta squared can also be calculated from the *F* ratio using this formula:

$$\eta^2 = \frac{df_{Bet}F}{df_{Bet}F + df_W} \qquad \text{Formula 13.12B}$$

A less biased estimate of the variance accounted for in the population requires the following adjustment:

$$\text{est.}\omega^2 = \frac{df_{Bet}(F-1)}{df_{Bet}(F-1) + df_W} \qquad \text{Formula 13.13}$$

### The Kruskal–Wallis *H* Test

1. Parametric analog: *F* test of one-way ANOVA

   Power efficiency: Approximately 90–95%

2. Computational procedures: Rank *all* scores from 1 (smallest) to *N* (largest), ignoring which group they are in. In case of ties, follow the usual procedure of assigning the mean of the ranks in question to each of the tied scores. Then find the sum of ranks separately for each group, and compute the sum of squares between groups ($SS_{Bet}$) for the ranks, using the following simple formula, which was specially designed for ranked data:

$$SS_{Bet} = \frac{T_1^2}{N_1} + \frac{T_2^2}{N_2} + \cdots + \frac{T_k^2}{N_k} - \frac{N_T(N_T+1)^2}{4} \qquad \text{Formula 13.14}$$

where

$T_1$ = sum of ranks in Group 1
$T_2$ = sum of ranks in Group 2
$T_k$ = sum of ranks in Group $k$
$N_1$ = number of observations in Group 1
$N_2$ = number of observations in Group 2
$N_k$ = number of observations in Group $k$
$k$ = number of groups
$N_T$ = total number of observations

Once $SS_{Bet}$ has been obtained, compute

$$H = \frac{12SS_{Bet}}{N_T(N_T + 1)} \qquad \text{Formula 13.15}$$

The value of $H$ is then referred to the $\chi^2$ table (Table H) with $k - 1$ degrees of freedom. If $H$ exceeds the tabled value, reject the null hypothesis that all populations have equal locations. Otherwise retain $H_0$. Do *not* use this procedure if there are only two or three groups *and* any group has fewer than five cases.

3. Measure of strength of relationship: If $H$ is statistically significant, a measure of the strength of relationship between group membership and rank on the dependent variable may be obtained by computing eta squared as applied to ranks ($\eta^2_R$):

$$\eta^2_R = \frac{H - k + 1}{N_T - k} \qquad \text{Formula 13.16}$$

where

$k$ = number of groups
$N_T$ = total number of observations

## Exercises

1. Following are two separate (hypothetical) sets of data that are somewhat exaggerated to help clarify the procedures underlying analysis of variance. In each case, the experimenter is interested in conditions that affect the number of errors made by participants on a simple clerical task that must be performed quickly. Group 1 listens to upbeat dance music while performing the tasks; Group 2 listens to soothing, New Age music; and Group 3 listens to white noise (e.g., the steady hum of a machine in the background).

| | **Experiment 1** | | | **Experiment 2** | | |
|---|---|---|---|---|---|---|
| | **Group 1** | **Group 2** | **Group 3** | **Group 1** | **Group 2** | **Group 3** |
| | 1 | 8 | 6 | 4 | 10 | 10 |
| | 3 | 7 | 5 | 0 | 0 | 0 |
| | 2 | 5 | 3 | 5 | 9 | 8 |
| | 1 | 4 | 3 | 0 | 1 | 1 |
| | 3 | 6 | 3 | 1 | 10 | 1 |
| $\overline{X}$ | 2 | 6 | 4 | 2 | 6 | 4 |

For Experiment 1: [$\overline{X}_G = 4$, $SS_T = 62$]; Experiment 2: [$\overline{X}_G = 4$, $SS_T = 250$]

(a) By inspection, in which case would you guess that the difference among groups is more likely to be statistically significant? Why?

(b) Carry out the analysis of variance for Experiment 1. Are the results significant at the .05 level? Calculate eta squared.
(c) Carry out the analysis of variance for Experiment 2. Are these results significant at the .05 level? Calculate eta squared.
(d) Briefly describe why the *F* ratio and eta squared statistics differ dramatically for the two experiments even though the sample means (i.e., 2, 6, and 4) are the same in both cases.
(e) Calculate an unbiased estimate of $\omega^2$ for whichever experiment is significant.

2. Using the means and standard deviations you have already calculated for the data from the four dormitories (i.e., Turck, Kirk, Dupre, and Doty Halls), perform an analysis of variance to decide whether you can reject the null hypothesis that the four samples come from populations with identical means. Calculate eta squared for this $\overline{X}$ comparison, and comment on the proportion of variance accounted for by the different dormitories. [Turck: $\overline{X} = 13$, $s = 4.36$; Kirk: $\overline{X} = 9.2$, $s = 4.16$; Dupre: $\overline{X} = 7$, $s = 2.58$; Doty: $\overline{X} = 11$, $s = 4.47$]

3. For each of the following two experiments, calculate the means and variances for each group first, and then use those statistics to perform the ANOVA.
(a) Can you reject the null hypothesis for Experiment 1? Display your results in an ANOVA summary table, and state the significance of your *F* ratio in a sentence, using the proper format. Include the value for eta squared. $[\overline{X}_G = 12.17]$.

| Experiment 1 | | |
|---|---|---|
| **Group 1** | **Group 2** | **Group 3** |
| 17 | 15 | 9 |
| 12 | 11 | 21 |
| 3 | 4 | 3 |
| 10 | 26 | 7 |
| 1 | 18 | 20 |
| 14 | 23 | |
| 5 | | |
| $\overline{X} = 8.86$ | $\overline{X} = 16.17$ | $\overline{X} = 12$ |
| $s = 35.81$ | $s = 64.6$ | $s = 65$ |

(b) Can you reject the null hypothesis for Experiment 2? Display your results in an ANOVA summary table, and state the significance of your *F* ratio in a sentence, using the proper format. Include an estimate of omega squared.

| Experiment 2 | | |
|---|---|---|
| **Group 1** | **Group 2** | **Group 3** |
| 1 | 4 | 26 |
| 10 | 17 | 21 |
| 3 | 11 | 15 |
| 12 | 20 | 9 |
| 3 | 14 | 23 |
| 5 | 18 | |
| 7 | | |
| $\overline{X} = 5.86$ | $\overline{X} = 14$ | $\overline{X} = 18.8$ |
| $s = 16.14$ | $s = 34$ | $s = 46.2$ |

4. Calculate the one-way ANOVA and determine its statistical significance for the following data:

| | Group | | | | |
|---|---|---|---|---|---|
| | 1 | 2 | 3 | 4 | 5 |
| $\overline{X}$ | 23 | 30 | 34 | 29 | 26 |
| $S$ | 6.5 | 7.2 | 7.0 | 5.8 | 6.0 |
| $N$ | 12 | 15 | 14 | 12 | 15 |

5. For each of the two (separate) experiments that follow, perform the Kruskal–Wallis $H$ test. If an experiment yields statistically significant results, compute the appropriate eta squared.

| (a) Experiment 1 | | | (b) Experiment 2 | | |
|---|---|---|---|---|---|
| Group 1 | Group 2 | Group 3 | Group 1 | Group 2 | Group 3 |
| 36 | 43 | 26 | 36 | 33 | 43 |
| 8 | 33 | 28 | 33 | 6 | 36 |
| 33 | 6 | 19 | 22 | 26 | 28 |
| 14 | 11 | 36 | 14 | 19 | 38 |
| 26 | 22 | 46 | 28 | 28 | 43 |
| 43 | 24 | 9 | 11 | 8 | 46 |
| 28 | 28 | 31 | 26 | 30 | 42 |
| 35 | 36 | 38 | 31 | 9 | 48 |
| 42 | 40 | | 35 | 24 | |
| 30 | | | 36 | | |
| 48 | | | 40 | | |

6. The following data table comes from Exercise 4 in Chapter 8, in which a psychology professor used three different methods of instruction in three small classes. In Chapter 8, you were asked to compare each pair of groups with the nonparametric rank-sum test. This time we would like you to: (a) test the null hypothesis that all three methods of instruction have the same effect on examination scores, using the Kruskal–Wallis $H$ test, and compute $\eta_R^2$; (b) repeat the analysis by computing the ordinary one-way ANOVA and $\eta^2$; (c) compare the results of parts a and b.

| Group 1 | Group 2 | Group 3 |
|---|---|---|
| 94 | 60 | 86 |
| 97 | 97 | 42 |
| 100 | 96 | 61 |
| 72 | 57 | 73 |
| 99 | 93 | 40 |
| 96 | 90 | 63 |
| 98 | 92 | 87 |
| 97 | | 65 |
| | | 67 |

7. Consider the following hypothetical results, as published in a hypothetical journal: "The participants had been divided equally among three experimental conditions—negative feedback, positive feedback, and no feedback—and the mean scores for the three groups at the end of the study were 11.1, 14.7, and 12.4, respectively. Moreover, the group differences were found to be statistically significant, $F(2, 57) = 3.69$, $p < .05$." (a) Find the value of $\eta^2$ for these data. How would you describe the size of this effect? (b) Estimate $\omega^2$ for the population represented by these groups.

## Thought Questions

1. A researcher wishes to test the hypothesis that the means of five populations are different. Why would it be a bad idea to test the researcher's hypothesis by computing 10 separate *t* tests (first test population 1 versus population 2, then test population 1 versus population 3, then test population 1 versus population 4, etc.)?
2. If there are only two groups (i.e., the researcher wishes to test the hypothesis that the means of two populations are different), will a one-way ANOVA yield the same results as the *t* test for two independent sample means? Explain.
3. If the results of a one-way ANOVA are statistically significant, why is it desirable to compute a measure of effect size?
4. Give an example of when the Kruskal–Wallis *H* test should be used instead of an ordinary one-way ANOVA.

## Computer Exercises

1. Perform one-way ANOVAs to test whether the different quiz conditions (last question is easy, moderate, difficult, or impossible) had a significant effect on postquiz anxiety and postquiz heart rate. Request descriptive statistics and draw a graph of the sample means, with the levels of the independent variable on the horizontal axis. Use the sample means to explain the results of your ANOVA.
2. Using college major as the independent variable, perform one-way ANOVAs to test for significant differences in both the academic interest score and the nonexperimental statistics quiz. Request descriptive statistics and a homogeneity of variance (HOV) test. Use the sample means to explain the results of each ANOVA. Do the standard deviations of the samples seem consistent with the results of the HOV test in each case?
3. Create a grouping variable from the number of previous outstanding teachers (Group 1 = none; Group 2 = one or two; Group 3 = three or more), and perform one-way ANOVAs on both the academic interest score and the nonexperimental statistics quiz. In each case, explain your results in terms of the means of the three groups.
4. Repeat Exercise 2 using the Kruskal–Wallis *H* test, and compare these results to the corresponding ANOVA results.

## Bridge to SPSS

There are two ways to perform a one-way ANOVA in SPSS: by selecting **One-Way ANOVA ...** from the **Analyze/Compare Means and Proportions** menu or by selecting **Univariate** from the **Analyze/General Linear Model** (GLM) menu. The latter method, because it allows for a variety of complex ANOVA procedures, lacks some of the useful features found in the more specific **One-Way ANOVA** module. So, we will describe the Compare Means method here and in the next chapter, and postpone a description of the GLM method until Chapter 15.

One advantage of the **One-Way ANOVA** dialog box not shared by its GLM counterpart is that you can move any number of variables into the *Dependent List:* area (but, of course, only one IV into the *Factor* slot), and after you click **OK**, separate one-way ANOVAs will be performed for each DV in the list. You do not have to specify the levels of your factor; every different value of your factor variable will be assumed to indicate a different group of cases (see Figure 13.2). For our example, we selected *statquiz* and *exp_sqz* for our DVs and *reason* for our factor (IV). Note that the checkbox for effect size estimates is checked by default.

The results of each one-way ANOVA requested are presented in a summary table much like the one in Table 13.2, except that the exact *p* value is displayed under the heading "Sig." Unlike the *t* test procedure, descriptive statistics and HOV tests are not reported

## Figure 13.2

**The Dialog Box for One-Way ANOVA**

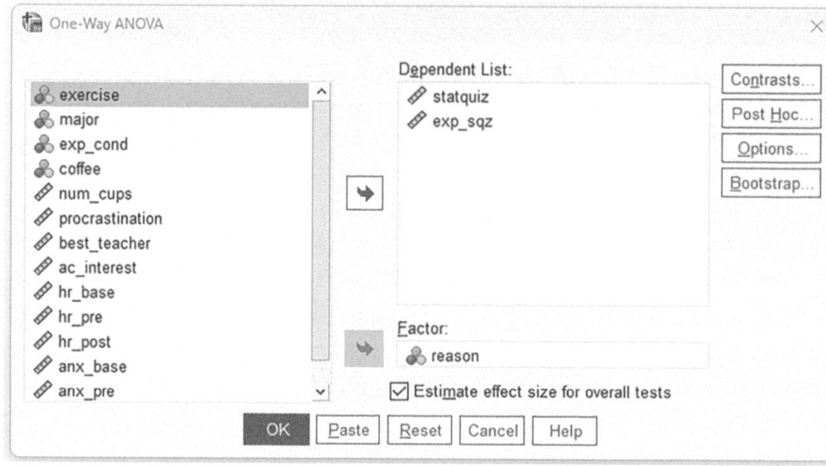


*Source*: Reprint courtesy of International Business Machines Corporation, © SPSS, Inc., an IBM company. SPSS was incorporated by IBM in October 2009.

automatically for a one-way ANOVA; they must be requested by checking the appropriate choices after clicking on the **Options...** button. The functions of the **Contrasts ...** and **Post Hoc** ....buttons, will be described in the next chapter.

If you select *Descriptive* and *Homogeneity of variance* test, those two results tables will appear first (see Figure 13.3). In the HOV table, SPSS gives you more choices than you need, but if none of the *p* values for a DV is less than .05, you can accept the results for that DV in the *ANOVA* table below. However, if any of the variations of Levene's test is significant at the .05 level, it is suggested that you use one of the modified ANOVA procedures available by clicking the **Options...** button. Unfortunately, those procedures, cousins of the separate-variances *t* test, are too advanced to be covered in this textbook. The *Descriptives* table is needed to make sense of your results in the ANOVA table, as you will soon see.

The *ANOVA* table contains the one-way ANOVA summary for each DV in your *Dependent:* list. You can see that the results for the stats quiz fall short of significance, but the results for the experimental quiz are significant even at the .01 level. So, we know we can reject the null hypothesis that, in the population, one's reason for taking a statistics course is not related to performance on the experimental stats quiz. But the significant F ratio does not by itself tell us which reason is associated with the highest performance and which is the lowest. That is why looking at the *Descriptives* table is important. We can see from that table that those taking stats due to a personal interest get the highest scores on average, while those who take the course based on an advisor's recommendation score the lowest. Moreover, because the two quizzes are highly correlated, it is not surprising that the means for the stats quiz are in the same order as the means for the experimental quiz.

Note that declaring that the population stats quiz means associated with the three reasons are not the same does not automatically imply that each mean is significantly different from each of the others. Determining which pairs of means differ significantly is the topic of the next chapter. It is also important to recall that statistical significance does not guarantee that the effect size you are dealing with is large. That is why it is important to look at the last of the results tables, *ANOVA Effect Sizes*. We won't explain all four of the choices in that table, but the first one, Eta-squared, is the same as the $\eta^2$ described earlier in this chapter and that can be calculated with Formula 13.12A or 13.12B. Because we have the ANOVA summary table in the output we can use Formula 13.12A to calculate $\eta^2$ for exp_sqz. We simply divide the SS Between Groups (52.04) by the SS Total (442.11) to obtain .1177, which rounds off to the value in the table.

### The Kruskal–Wallis *H* Test

As we mentioned in Chapter 8, SPSS provides continuity for long-term users through its "Legacy Dialogs." If you are using one of the latest versions of SPSS, we recommend using the Legacy Dialogs for your nonparametric tests.

**Descriptives**

| | | N | Mean | Std. Deviation | Std. Error | 95% Confidence Interval for Mean Lower Bound | 95% Confidence Interval for Mean Upper Bound | Minimum | Maximum |
|---|---|---|---|---|---|---|---|---|---|
| statquiz | Program | 41 | 6.85 | 1.711 | .267 | 6.31 | 7.39 | 1 | 9 |
| | Personal | 29 | 7.38 | 1.635 | .304 | 6.76 | 8.00 | 3 | 9 |
| | Advisor | 30 | 6.37 | 1.650 | .301 | 5.75 | 6.98 | 3 | 10 |
| | Total | 100 | 6.86 | 1.700 | .170 | 6.52 | 7.20 | 1 | 10 |
| exp_sqz | Program | 41 | 6.63 | 1.907 | .298 | 6.03 | 7.24 | 1 | 10 |
| | Personal | 29 | 7.90 | 1.896 | .352 | 7.18 | 8.62 | 4 | 11 |
| | Advisor | 30 | 6.07 | 2.227 | .407 | 5.23 | 6.90 | 2 | 11 |
| | Total | 100 | 6.83 | 2.113 | .211 | 6.41 | 7.25 | 1 | 11 |

**Tests of Homogeneity of Variances**

| | | Levene Statistic | df1 | df2 | Sig. |
|---|---|---|---|---|---|
| statquiz | Based on Mean | .068 | 2 | 97 | .935 |
| | Based on Median | .243 | 2 | 97 | .785 |
| | Based on Median and with adjusted df | .243 | 2 | 92.302 | .785 |
| | Based on trimmed mean | .206 | 2 | 97 | .814 |
| exp_sqz | Based on Mean | 1.444 | 2 | 97 | .241 |
| | Based on Median | 1.810 | 2 | 97 | .169 |
| | Based on Median and with adjusted df | 1.810 | 2 | 90.700 | .170 |
| | Based on trimmed mean | 1.508 | 2 | 97 | .227 |

**ANOVA**

| | | Sum of Squares | df | Mean Square | F | Sig. |
|---|---|---|---|---|---|---|
| statquiz | Between Groups | 15.124 | 2 | 7.562 | 2.707 | .072 |
| | Within Groups | 270.916 | 97 | 2.793 | | |
| | Total | 286.040 | 99 | | | |
| exp_sqz | Between Groups | 52.041 | 2 | 26.021 | 6.471 | .002 |
| | Within Groups | 390.069 | 97 | 4.021 | | |
| | Total | 442.110 | 99 | | | |

**ANOVA Effect Sizes[a,b]**

| | | Point Estimate | 95% Confidence Interval Lower | 95% Confidence Interval Upper |
|---|---|---|---|---|
| statquiz | Eta-squared | 053 | .000 | .147 |
| | Epsilon-squared | .033 | −.021 | .129 |
| | Omega-squared Fixed-effect | .033 | −.020 | .128 |
| | Omega-squared Random-effect | .017 | −.010 | .068 |
| exp_sqz | Eta-squared | .118 | −.017 | .232 |
| | Epsilon-squared | .100 | −.003 | .217 |
| | Omega-squared Fixed-effect | .099 | −.003 | .215 |
| | Omega-squared Random-effect | .052 | −.001 | .120 |

a. Eta-squared and Epsilon-squared are estimated based on the fixed-effect model.

b. Negative but less biased estimates are retained, not rounded to zero.

Figure 13.3

**Results from a One-Way ANOVA**

From the **Analyze/Nonparametric Tests/Legacy Dialogs** menu, select **K Independent Samples ....** It is the first (and default) choice in the dialog box under *Test Type* that you will want to use, and it is appropriately labeled **Kruskal–Wallis *H*.** The main difference between this dialog box (titled **Tests for Several Independent Samples**) and the one for two independent samples (see Chapter 8) is that you are asked to **Define Range ...** after moving over the *Grouping Variable:*, instead of to **Define Groups ...** (see Figure 13.4). You must enter integers to define the minimum and maximum values of interest for your grouping variable (IV). Those integers entered, and all integer values between them, will each define a different group in the analysis. You can select Descriptive after clicking the **Options...** button, if you want to see the usual quantitative statistics. For the K–W test example, we are using the same IV and DVs as we did for the one-way ANOVA, so we can compare the parametric and nonparametric results.

By default, the Kruskal–Wallis procedure creates two results tables in your output, the first of which contains the sample size and mean rank for each group (see Figure 13.5). The second box contains only the $H$ statistic, the formula for which was given earlier in this chapter, along with its $df$ (one less than the number of groups being compared) and its approximate $p$ value, labeled **Asymp. Sig**. Note that for a small $N$, it may be desirable to find the exact $p$ value for $H$ by selecting *Exact*, after clicking the **Exact...** button, but for larger $N$ it makes very little difference. SPSS does not have an option to report an effect-size measure for this test, but you can calculate an appropriate version of eta-squared from Formula 13.16.

If you compare the $p$ values for the one-way ANOVA with the K–W test on the two DVs, you will see that the results are similar for *exp-sqz*: .002 and .005, respectively. It is typical for the K–W test to have only a little less power than the one-way ANOVA on the same variables. However, there is a noteworthy discrepancy in $p$ values for the *statquiz*: .072 (ANOVA) versus .008 (K–W). This can be explained by the fact that *statquiz* has a strong negative skew (most of the scores are above 6 with a few scores that are low) and ranking eliminates the skewing. The ANOVA works best with normal distributions, whereas ordinal tests do not require that assumption. Strong skewing and outliers are reasons to use nonparametric statistics, especially when the samples are fairly small.

The relation between the Kruskal–Wallis and the rank-sum tests is perfectly analogous to the relation between the one-way ANOVA and the independent-samples $t$ test. If you run

## Figure 13.4

**The Dialog Box for the Kruskal–Wallis *H* Test**

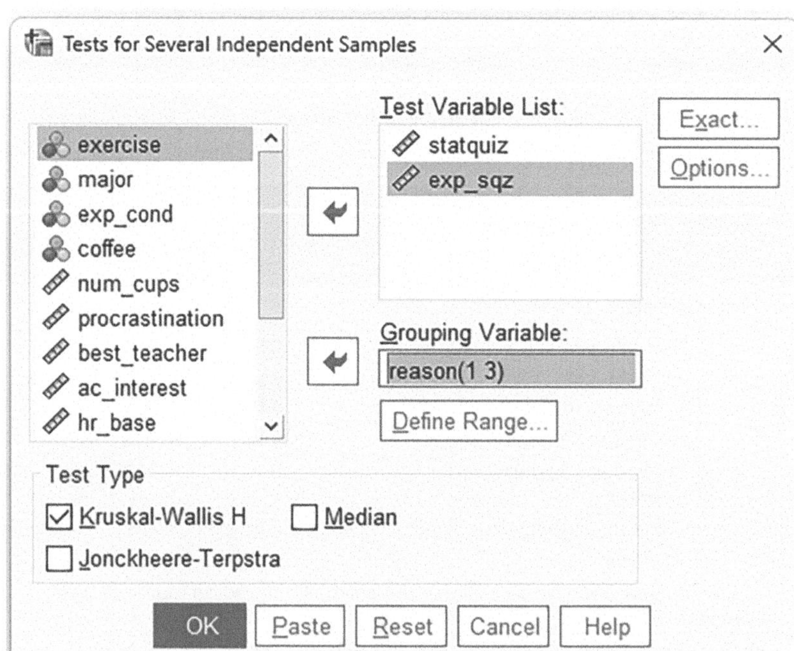


*Source*: Reprint courtesy of International Business Machines Corporation, © SPSS, Inc., an IBM company. SPSS was incorporated by IBM in October 2009.

**Ranks**

| | reason | N | Mean Rank |
|---|---|---|---|
| statquiz | Program | 41 | 51.56 |
| | Personal | 29 | 61.34 |
| | Advisor | 30 | 38.57 |
| | Total | 100 | |
| exp_sqz | Program | 41 | 47.65 |
| | Personal | 29 | 64.41 |
| | Advisor | 30 | 40.95 |
| | Total | 100 | |

**Tests Statistics[a,b]**

| | statquiz | exp_sqz |
|---|---|---|
| Kruskal-Wallis H | 9.732 | 10.593 |
| df | 2 | 2 |
| Asymp. Sig. | .008 | .005 |

a. Kruskal Wallis Test
b. Grouping Variable: reason

## Figure 13.5

**Results from a Kruskal–Wallis *H* Test**

the Kruskal–Wallis *H* test on just two groups, the *p* value you will obtain is the same as the *p* value from the rank-sum test on the same two groups. To follow up a significant K–W test by testing each pair of groups, you can use the K–W test or the rank-sum test for each pair.

## Appendix: Proof That the Total Sum of Squares Is Equal to the Sum of the Between-Group and the Within-Group Sum of Squares

We begin with the mathematical statement that the total deviation of any score from the grand mean is equal to the sum of two parts: the score's deviation from the mean of its own group plus the deviation of the mean of its group from the grand mean:

$$X - \overline{X}_G = (X - \overline{X}_i) + (X_i - \overline{X}_G)$$

Squaring both sides of the equation and summing over *all* people yields:

$$\sum (X - \overline{X}_G)^2 = \sum (X - \overline{X}_i)^2 + \sum (\overline{X}_i - \overline{X}_G)^2 + 2\sum (X - \overline{X}_1)(\overline{X}_i - \overline{X}_G)$$

For any one group, $(X_i - X_G)$ is a constant, and the sum of deviations about the group mean must equal zero. The last term is therefore always equal to zero, leaving

$$\sum (X - \overline{X}_G)^2 = \sum (X - \overline{X}_i)^2 + \sum (\overline{X}_i - \overline{X}_G)^2$$

For ANOVA purposes, the sums of squares are then divided by the appropriate degrees of freedom to yield estimates of the population variance, which are then compared by the *F* ratio.

# Chapter 14
# Multiple Comparisons

## PREVIEW

### Introduction

What is the purpose of pairwise multiple comparisons?

### Fisher's Protected *t* Tests and the Least Significant Difference (LSD)

How is the error term used by Fisher's protected *t* tests an improvement over the one used by the ordinary *t* test for comparing two independent means? How can these tests be simplified when all of the samples are the same size?

### Tukey's Honestly Significant Difference (HSD)

What do Fisher's protected *t* tests "protect" against, and under what conditions does that protection break down?

How does Tukey's HSD test use the studentized range statistic to provide greater protection against Type I errors than Fisher's protected *t* tests?

What is meant by the statement that the HSD test is overly "conservative" when dealing with a three-group study?

How can you use HSD when your samples differ slightly in size?

### Other Multiple Comparison Procedures

Why did the Newman–Keuls test become so popular for multiple comparisons, and why has its use decreased in recent years?

Which multiple comparison procedure is recommended for which number of groups?

### Planned and Complex Comparisons

How does the Bonferroni correction provide greater power when you are planning to test a small subset of the comparisons that are possible? Why is the Bonferroni test overly conservative if you want to test all of the possible pairwise comparisons?

What is a complex comparison, and how can it lead to greater power if you predict the pattern of sample means correctly?

### Nonparametric Multiple Comparisons: The Protected Rank-Sum Test

What is the appropriate way to follow up a significant Kruskal–Wallis *H* test with pairwise comparisons?

Under what conditions would a Bonferroni correction be recommended?

### Summary

### Exercises

### Thought Questions

### Computer Exercises

### Bridge to SPSS

*Introductory Statistics for the Behavioral Sciences*, Eighth Edition. R. Brooke Lea and Barry H. Cohen.

Companion website: www.wiley.com/go/ISBS8e

## Introduction

To introduce the need for *multiple comparisons*, we now return to the analysis of variance (ANOVA) we presented for computational practice in Table 13.4 of the previous chapter. For that ANOVA, the dean was able to reject the null hypothesis and conclude that attitudes toward student participation in campus governance differ depending on whether seniors at her college are majoring in a natural science, a social science, or one of the humanities. However, it is hard to imagine that anyone would be satisfied with that general conclusion and not want to probe further for more specific conclusions. For instance, do all three groups differ significantly from each other, or perhaps the natural science majors differ significantly from each of the other two groups, but the difference between social science and humanities majors is not significant? It is important to understand that rejecting the null hypothesis in a three-group experiment does *not* automatically allow you to conclude that all three population means are different from each other (i.e., $\mu_1 \neq \mu_2 \neq \mu_3$). There are three other possibilities that are consistent with rejecting the null hypothesis: $\mu_1 = \mu_2 \neq \mu_3$; $\mu_1 \neq \mu_2 = \mu_3$; and $\mu_1 \neq \mu_3 = \mu_2$. To decide which of the four possibilities (including $\mu_1 \neq \mu_2 \neq \mu_3$) is true requires what are called, in this context, *pairwise* multiple comparisons. Pairwise comparisons can be performed with the ordinary two-sample *t* tests you learned about in Chapter 7, but they can be slightly improved, as you will see next. We reproduced Table 13.4 here as Table 14.1, and added means and standard deviations, to help us illustrate the procedure for performing pairwise comparisons on those data.

Now, you might be thinking: If I must follow my ANOVA with *t* tests anyway, why can I not just skip the ANOVA and proceed directly to the *t* tests? This is not a silly question to ask in the three-group case when there are only three pairwise comparisons to perform (e.g., natural versus social sciences, and each of these versus humanities); the problem with performing multiple *t* tests is not obvious in this simple case. However, suppose that the dean has collected enough data for her to conduct a one-way ANOVA on the means from *eight* different college majors. As mentioned in the previous chapter, the number of possible pairwise comparisons is given by the formula $k(k-1)/2$, so in the eight-group case there would be $(8 \times 7)/2 = 28$ *t* tests to perform. If you were to use .05 (i.e., 1 in 20) as your alpha, the chances would be better than 50% of obtaining at least one Type I error even if all eight of the corresponding population means were exactly the same. In fact, the *experimentwise* alpha ($\alpha_{EW}$) in this case would be at least .6. The advantage of performing an ANOVA at the .05 level first is that it will screen out about 95% of those experiments in which all of the population means are equal, and follow-up *t* tests will be performed only on the 5% that accidentally reach significance. For this reason, Fisher, who was the first to formalize the ANOVA procedure back in the 1920s, devised the following **protected *t* test** procedure.

## Fisher's Protected *t* Tests and the Least Significant Difference (LSD)

According to Fisher's original recommendation, if *and only if* an ANOVA *F* test has resulted in the rejection of the overall null hypothesis that all *k* means are equal, any (or all) of the paired means may be compared by *t* tests using the usual (.05) decision rule for each pair. If it is reasonable to assume homogeneity of variance (HOV) among all of the populations represented in the ANOVA, these follow-up *t* tests can be slightly improved relative to ordinary two-group *t* tests. An ordinary *t* test between any $\overline{X}_i$ and $\overline{X}_j$ (where the subscript *i* represents any one of the samples in the ANOVA, and the subscript *j* represents any *other* sample) would be based on only the $N_i$ and $N_j$ observations involved, and it would only have degrees of freedom $(df) = N_i + N_j - 2$. However, by making the HOV assumption, you can take advantage of the more stable estimate of the population variance provided by the ANOVA's $MS_W$, which is based on $df_W = N_T - k$. Specifically, you can use $MS_W$ (sometimes referred to as the *omnibus* error term) in place of the pooled variance in each of your (protected) follow-up *t* tests, and therefore you can base your critical *t* value on the $df_W$ from the ANOVA. Because $df_W$ in a multigroup study will always be larger than the *df* for just two of the groups, the critical *t* value will be smaller for these follow-up *t* tests, giving them a

**Table 14.1** **ANOVA of Attitudes of Students From Three Areas of Study at a Small College to Student Participation in Determining College Curricula**

| Areas of study | | | |
|---|---|---|---|
| **Natural sciences** | **Social sciences** | **Humanities** | |
| 15 | 17 | 6 | $F = \frac{85.10}{16.74} = 5.08$ |
| 18 | 22 | 9 | |
| 12 | 5 | 12 | |
| 12 | 15 | 11 | |
| 9 | 12 | 11 | |
| 10 | 20 | 8 | |
| 12 | 14 | 13 | |
| 20 | 15 | 14 | |
| | 20 | 7 | |
| | 21 | | |
| $\overline{X}_1 = 13.5$ | $\overline{X}_2 = 16.1$ | $\overline{X}_3 = 10.11$ | |
| $S_1 = 3.85$ | $s_2 = 5.13$ | $S_3 = 2.76$ | |
| $N_1 = 8$ | $N_2 = 10$ | $N_3 = 9$ | $N_T = 27$ |

better chance at statistical significance (i.e., more power) than ordinary *t* tests (unless the sample sizes are so large that your critical *t* values are close to critical *z*'s even for an ordinary *t* test). Note that by requiring the ANOVA *F* to be significant, we protect the follow-up *t* tests from the large experimentwise Type I error rate that would otherwise occur. At the same time, these *t* tests are more powerful in detecting real population mean differences, if HOV can be assumed because then they are based on more *df*.

Next, we illustrate the protected *t* procedure using the results of the ANOVA on student attitudes, as shown in Table 14.1. Because the *F* test for the ANOVA was indeed significant at $\alpha = .05$, we are allowed to proceed to compare the three pairs of means by *t* tests. The null hypothesis for any pair of means, $\overline{X}_i$ and $\overline{X}_j$, can be tested by

$$t = \frac{\overline{X}_i - \overline{X}_j}{\sqrt{MS_W\left(\frac{1}{N_i} + \frac{1}{N_j}\right)}} \qquad \text{Formula 14.1}$$

Here, *df* for the critical *t* equals $N_T - k$, and $MS_W$ is taken from the one-way ANOVA results (the denominator of the *F* ratio given in Table 14.1). Note that Formula 14.1 is just the ordinary formula for the two-group *t* test, with $MS_W$ serving as the pooled-variance estimate.

Table 14.2 uses the means, group sizes, and $MS_W$ of Table 14.1 to illustrate the computation of these *t*'s. For $df = 24$, the critical value for *t* at $\alpha = .05$ (two-tailed) is 2.064. Only the difference between the means of Groups 2 and 3 leads to a *t* value that is larger than 2.064, so only this pairwise difference can be declared significant. We can now get more specific in our rejection of the overall ANOVA null hypothesis; rather than merely stating that it is not the case that all three population means are equal to each other, we can assert that, in particular, $\mu_2 \neq \mu_3$. At the same time, we have to retain the null hypotheses that $\mu_1 = \mu_2$ and $\mu_1 = \mu_3$.

When all groups are of the same size, $N_i = N_j$, the denominator of the foregoing *t* test becomes a constant for all the pairwise comparisons. It then becomes possible to greatly simplify the protected *t* test procedure. There must be some difference between the two means that is just large enough so that when it is divided by the (constant) denominator, it yields a *t* value exactly equal to the critical *t* needed for significance at a given alpha level. This particular difference between the means is called the **LSD**. When it is inserted into the *t* formula given previously, we get

**Table 14.2 Protected *t* Tests Among the Three Mean Attitude Scores Following a Significant ANOVA *F***

| Means | $\overline{X}_1 = 13.50$, $N_1 = 8$ | $\overline{X}_2 = 16.10$, $N_2 = 10$ | $\overline{X}_3 = 10.11$, $N_3 = 9$ |
|---|---|---|---|

$$MS_W = 16.74$$

$\overline{X}_1$ versus $\overline{X}_2$:

$$t = \frac{13.50 - 16.10}{\sqrt{16.74\left(\frac{1}{8} + \frac{1}{10}\right)}} = \frac{-2.60}{1.94} = -1.34$$

$\overline{X}_1$ versus $\overline{X}_3$:

$$t = \frac{13.50 - 10.11}{\sqrt{16.74\left(\frac{1}{8} + \frac{1}{9}\right)}} = \frac{-3.39}{1.99} = 1.70$$

$\overline{X}_2$ versus $\overline{X}_3$:

$$t = \frac{16.10 - 10.11}{\sqrt{16.74\left(\frac{1}{10} + \frac{1}{9}\right)}} = \frac{-5.99}{1.88} = 3.19$$

$df$ for $t$'s $= df_W = N_T - k = 24$

$$t_{crit} = \frac{\text{LSD}}{\sqrt{MS_W\left(\frac{2}{n}\right)}} = \frac{\text{LSD}}{\sqrt{\frac{2MS_W}{n}}}$$

where $n$ is the size of each sample.

Solving for LSD, we obtain the following easy-to-use formula:

$$\text{LSD} = t_{crit}\sqrt{\frac{2MS_W}{n}} \qquad \text{Formula 14.2}$$

The value of $t_{crit}$ is chosen for $df = N_T - k$ and the two-tailed $\alpha$ of the decision rule. Assume that all of the sample sizes were 9 ($= n$) in the preceding example, and the usual $\alpha = .05$ decision rule is to be used. Then the $df$ would still be 24 (i.e., 27 − 3), and the critical $t$ value would still be equal to 2.064. If we assume that $MS_W$ is still 16.74, the size of LSD would be:

$$\text{LSD} = 2.052\sqrt{16.74\left(\frac{2}{9}\right)} = 2.052(1.93) = 3.96$$

Finding LSD saves you the work of performing all of the follow-up $t$ tests. Instead, you need only look at the amount of difference for each pair of means. In this example, all pairs of means differing by at least about 4 attitude scale points would be significantly different at $\alpha = .05$ (two-tailed). Looking at the numerators of the $t$ tests in Table 14.2, we see differences of −2.6, −3.39, and −5.99; only the last of these—the difference between Groups 2 and 3—is larger in magnitude than LSD, so that is the only difference that would attain significance according to the LSD test. Of course, these conclusions are consistent with the conclusions of the individual $t$ tests (in the equal-$n$ condition, the conclusions would have to always be consistent).

Although the protected $t$ procedure provides good control and balance of Type I and Type II errors when you are dealing with only three population means, the protection afforded by requiring $F$ to be significant drops unacceptably as $k$ increases. Some other procedure should be used (which will be described shortly) when $k$ is more than 3, especially when strict control of the experiment-wise Type I error rate is desired. Also, it is possible (though highly unlikely) to find $F$ significant and none of the pairwise $t$'s significant, in which case the proper conclusion is that the $k$ means are not all equal but the data do not justify any further specification (*not* that all paired means are equal, which would be a contradiction). Much more often, one of the $t$ tests is significant when the ANOVA is not, which is why requiring a significant ANOVA before performing the $t$ tests makes a real difference in reducing Type I errors.

### Confidence Intervals for the Protected *t* Test

In the previous section, we tested the null hypothesis for each pair of means using three protected $t$ tests. An alternative procedure would be to calculate three confidence intervals (CIs), as discussed in Chapters 6 and 7. This has the advantage of specifying all reasonably likely values of the difference between the two population means in each comparison.

To use the CI procedure for the protected $t$ test, the ANOVA $F$ test must have resulted in the rejection of the overall null hypothesis that all $k$ means are equal. As we saw in Chapter 7, the CI for the difference between two population means is given by Formula 7.7, reproduced here:

$$\left[(\overline{X}_1 - \overline{X}_2) - t_{\text{crit}} s_{\overline{X}_1 - \overline{X}_2}\right] \leq \mu_1 - \mu_2 \leq \left[(\overline{X}_1 - \overline{X}_2) + t_{\text{crit}} s_{\overline{X}_1 - \overline{X}_2}\right] \qquad \text{Formula 7.7}$$

The critical value of $t$ for a 95% CI is the same as for the corresponding .05, two-tailed $t$ test, which is 2.064 in this example. For each comparison, the standard error of the difference is equal to $\sqrt{MS_W(1/N_i + 1/N_j)}$ Since $MS_W = 16.74$, $N_1 = 8$, $N_2 = 10$, and $N_3 = 9$, the three standard errors of the difference that we need are:

1. Group 1 versus Group 2 : $\sqrt{16.74(1/8 + 1/10)} = 1.94$
2. Group 1 versus Group 3 : $\sqrt{16.74(1/8 + 1/9)} = 1.99$
3. Group 2 versus Group 3 : $\sqrt{16.74(1/10 + 1/9)} = 1.88$

Therefore, the three 95% CIs are:

1. Group 1 versus Group 2:

$$-2.60 - (2.064)(1.94) \leq \mu_1 - \mu_2 \leq -2.60 + (2.064)(1.94)$$
$$-6.60 \leq \mu_1 - \mu_2 \leq 1.40$$

2. Group 1 versus Group 3:

$$3.39 - (2.064)(1.94) \leq \mu_1 - \mu_3 \leq 3.39 + (2.064)(1.94)$$
$$-0.72 \leq \mu_1 - \mu_3 \leq 7.50$$

3. Group 2 versus Group 3:

$$5.99 - (2.064)(1.88) \leq \mu_2 - \mu_3 \leq 5.99 + (2.064)(1.88)$$
$$2.11 \leq \mu_2 - \mu_3 \leq 9.87$$

As would be expected from the results in Table 14.2, only for the comparison between Groups 2 and 3 does zero fall outside the CI. Here again, CIs provide us with important additional information about the probable difference between the population means in each comparison.

## Tukey's Honestly Significant Difference (HSD)

The problem with using Fisher's procedure when you are dealing with more than three groups is that you are "protected" only in experiments for which the *complete* null hypothesis is true (e.g., for four groups, $H_0$: $\mu_1 = \mu_2 = \mu_3 = \mu_4$). As the number of groups increases, there is an increasing number of ways that the null hypothesis can be *partially* true.

For five groups, a worst-case scenario for Fisher's procedure would be an experiment for which $\mu_1 = \mu_2 = \mu_3 = \mu_4 \neq \mu_5$. Perhaps four antidepressant drugs are being tested against each other and against a placebo, and the drugs have identical effects, all of which exceed the placebo effect. If the ANOVA involving the five means reaches statistical significance (this would not be surprising, given the difference of the placebo from the drugs), then, according to Fisher, it is acceptable to test each of the six possible pairs of drugs for significance, in addition to testing each drug against the placebo. But in this example, all of the drug-to-drug comparisons are "null." If any accidentally reach significance, you have committed a Type I error within the experiment, even though you were correct to reject the (complete) null hypothesis of the ANOVA. The more "null" pairwise comparisons you are permitted to perform, the larger the experimentwise alpha becomes for your study.

We cannot know how many "partial-null" experiments are being conducted (or how "partial" they are), but the more that are tested, the more often $\alpha_{EW}$ will rise above .05. J. W. Tukey (the same statistician who brought us the boxplot) understood this problem and was able to find a distribution that could keep $\alpha_{EW}$ below .05, regardless of how many groups are involved in the study and how complete the null hypothesis might be. This distribution is known as the *studentized range*.

### Studentized Range Statistic

Tukey recognized that when you draw, say, five samples from the same population and compare the smallest sample mean to the largest, the difference is likely to be considerably larger than if you had only drawn two sample means and compared them. Therefore, his **studentized range statistic** (symbolized by $q$) grows appropriately larger as the total number of samples becomes larger. Also, the statistic is "studentized," like the $t$ distributions, in that it gets smaller as the size of each sample gets larger, although it changes very little after the size of the samples rises above 40 or so.

It is not unreasonable to think of $q$ as a $t$ statistic that has been adjusted for the total number of groups being compared. We can see the extent of this adjustment by looking at Table G in the Appendix, which is a table of the critical values of the studentized range statistic for alpha = .05, two-tailed. An excerpt of this table is reprinted here as Table 14.3.

**Table 14.3 Selected Critical Values of Tukey's Studentized Range Statistic**

| | Number of groups | | | | | | |
|---|---|---|---|---|---|---|---|
| $df_W$ | 2 | 3 | 4 | 5 | 6 | 7 | 8 |
| ... | ... | ... | ... | ... | ... | ... | ... |
| 4 | 3.93 | 5.04 | 5.76 | 6.29 | 6.71 | 7.05 | 7.35 |
| ... | ... | ... | ... | ... | ... | ... | ... |
| 8 | 3.26 | 4.04 | 4.53 | 4.89 | 5.17 | 5.40 | 5.60 |
| ... | ... | ... | ... | ... | ... | ... | ... |
| 12 | 3.08 | 3.77 | 4.20 | 4.51 | 4.75 | 4.95 | 5.12 |
| ... | ... | ... | ... | ... | ... | ... | ... |
| 16 | 3.00 | 3.65 | 4.05 | 4.33 | 4.56 | 4.74 | 4.90 |
| ... | ... | ... | ... | ... | ... | ... | ... |
| 40 | 2.86 | 3.44 | 3.79 | 4.04 | 4.23 | 4.39 | 4.52 |
| ... | ... | ... | ... | ... | ... | ... | ... |
| ∞ | 2.77 | 3.31 | 3.63 | 3.86 | 4.03 | 4.17 | 4.29 |

Notice that the value of $q$ increases as you move to the right in any of the rows in Table 14.3. However, by the time you are dealing with six groups, adding another group or two does not have nearly the impact on $q$ that it does when you are dealing with only two or three groups.

Although increasing the size of the groups (i.e., going down the columns in Table 14.3) decreases the size of $q$, these decreases get smaller as the samples become large (as is the case with $t$). It may seem odd that $q$ is larger than $t$ even for only two groups (e.g., at infinity, $q$ for two groups equals 2.77, while the critical $t$ equals 1.96). It should seem odd because this is not an actual adjustment in the test statistic but rather an artifact of how Tukey decided to present his table and his formula, as we show next.

### Using Tukey's HSD Formula

Tukey's original formula merely substituted $q_{crit}$ for $t_{crit}$ in Fisher's formula and called the minimal difference for significance the **Honestly Significant Difference (HSD)**:

$$\text{HSD} = q'_{crit}\sqrt{\frac{2MS_W}{n}}$$

where $q'_{crit}$ refers to a value from Tukey's original $q$ table. However, Tukey decided to simplify his formula.

We can imagine that he first separated it, like this,

$$\text{HSD} = q'_{crit}\sqrt{2}\sqrt{\frac{MS_W}{n}},$$

and then multiplied his original $q'$ values by the square root of 2 to obtain critical values of $q$. The resulting formula is:

$$\text{HSD} = q_{crit}\sqrt{\frac{MS_W}{n}} \qquad \text{Formula 14.3}$$

where $q_{crit}$ is a value from Table G and equals $\sqrt{2}$ times $q'_{crit}$. Because the first column of Table G is for two groups, and therefore requires no adjustment from the critical $t$ value, it actually consists of the .05, two-tailed critical values of $t$, each multiplied by the square root of 2 (e.g., $2.77 = 1.96\sqrt{2}$). Unfortunately, dropping the number 2 from the formula for LSD to create HSD, originally intended just to simplify its calculation (at a time before there were handheld calculators and personal computers), serves these days only to confuse students of statistics, who may think that the formulas for LSD and HSD are more different than they actually are. Let us see what happens when we apply Tukey's formula to the data in Table 14.1.

For three groups and $df = 24$, the critical $q$ from Table G is 3.53. For the purpose of comparison with LSD, we will once again use 9 for the common $n$. Therefore,

$$\text{HSD} = 3.53\sqrt{\frac{16.74}{9}} = 3.53(1.364) = 4.81$$

Note that HSD, which equals 4.81, is considerably larger than LSD, which equals 3.96. This leads to larger CIs for Tukey's than Fisher's test, as we demonstrate next.

### Confidence Intervals for Tukey's HSD Test

It could hardly be any easier to find CIs for the difference of any two population means represented by samples in your study after you have calculated HSD because HSD is already designed to tell you the difference between two means that is just on the borderline of significance. The formula for the CI for any two means in your study, according to Tukey's test, is:

$$\mu_i - \mu_j = \overline{X}_i - \overline{X}_j \pm \text{HSD} \qquad \text{Formula 14.4}$$

For example, the CI for the difference between the social sciences and the humanities, as represented by Groups 2 and 3, respectively, in Table 14.1 is:

$$\mu_i - \mu_j = 5.99 \pm 4.81$$

So, the 95% CI extends from +1.18 to +10.80.

Note that this CI is wider than the one we found for these same two groups based on Fisher's procedure (+2.13 to +9.85). This will always be the case when there are more than two groups in the study because Tukey's test uses a larger critical value.

Of course, the percentage of the CI depends on the alpha you used to look up the $q$ value that you used to calculate HSD. If you found $q$ from our Table G, which is only for $\alpha = .05$, then Formula 14.4 will give you 95% CIs. We have not included $q$ values for $\alpha = .01$, because Tukey's test is already quite conservative with $\alpha = .05$, but that is what you would need to construct 99% CIs that correspond to Tukey's test.

### Tukey's HSD for Unequal Sample Sizes: The Harmonic Mean

Both the LSD and HSD formulas assume that all of the samples have the same size ($n$). If your sample sizes differ only slightly, as when equal groups were planned but a few data points have been lost, it is acceptable to calculate the harmonic mean of your actual sample sizes and use that value as $n$ in the formula for LSD or HSD. As a crude approximation, we have been using the arithmetic mean of the sample sizes in Table 14.1 (i.e., 9) to calculate LSD and HSD. However, it is more accurate to use the harmonic mean. In Chapter 7, we presented a formula (7.10) for the harmonic mean of two numbers; the harmonic mean of any number of sample sizes can be expressed as follows:

$$n_h = \frac{k}{\sum \frac{1}{n_i}} \qquad \text{Formula 14.5}$$

For the sample sizes in our present example, the harmonic mean ($n_h$) is:

$$n_h = \frac{3}{\frac{1}{8} + \frac{1}{9} + \frac{1}{10}} = \frac{3}{0.33611} = 8.93$$

As you can see, with small differences among your sample sizes, the harmonic mean does not differ much from the arithmetic mean, but the harmonic mean is always more accurate in these cases, and it becomes increasingly discrepant from the arithmetic mean when your numbers are more spread out proportionally. However, note that when your sample sizes differ considerably, Tukey's HSD test loses its accuracy and is therefore not recommended.

### Comparing HSD to LSD

Although HSD (4.81) is considerably larger than LSD (3.96), our pairwise conclusions do not change. The difference between Group 2 and Group 3, which is 5.99, is still statistically significant. Given that HSD can never be smaller than LSD, you cannot have a difference that exceeds HSD but not LSD. However, the reverse is certainly possible (e.g., a mean difference in this example of 4.2 would be significant by the LSD but not the HSD procedure). Although both procedures usually lead to the same set of conclusions, there is no doubt that Tukey's HSD is the more *conservative* of the two, which means that it does a better job of controlling Type I errors than Fisher's LSD. In fact, HSD keeps $\alpha_{\text{EW}}$ below .05 for any number of groups, which makes Tukey's test *overly* conservative, especially in the case of three groups, for which Fisher's procedure maintains $\alpha_{EW}$ appropriately at the value used to look up the critical $t$. Therefore, Fisher's protected $t$ tests should be used whenever you are dealing with only three groups, and the LSD formula can be used in particular whenever the samples are the same size or nearly so. Fisher's system is too *liberal*, however, when dealing with more than three groups: The larger the number of groups, the more Fisher's procedure allows $\alpha_{EW}$ to increase.

### Follow-Up Tests for Published Results

The results of the ANOVA performed on the data in Table 14.1, if published in a journal that follows the guidelines of the *APA Publication Manual* (APA, 2020), would be expressed something like this: "The mean attitudes toward student participation in campus governance were found to differ among students majoring in a natural science ($M$ = 13.5), a social science ($M$ = 16.1), or one of the humanities ($M$ = 10.11), and this difference attained statistical significance, $F(2, 24) = 5.08$, $p < .05$." Sometimes an effect-size measure is included, and the error term (i.e., denominator of the $F$ ratio) is also given. Suppose, however, that the error term is *not* given, and the authors do not report the results of follow-up tests at all (unlikely) or report the $p$ values associated with pairwise comparisons according to Tukey's procedure without reporting the size of HSD (fairly likely). Do you have enough information to perform Fisher's protected $t$ tests instead? You do, as long as you know the size of each sample ($n$s = 8, 10, and 9, respectively, for this example). Then you could calculate $SS_{Bet}$ using Formula 12.2A, and divide it by $df_{Bet}$ to obtain $MS_{Bet}$, which was 85.1. Because you know that $F$ equals $MS_{Bet}/MS_W$, and you were given $F$ in the report (5.08), and you calculated $MS_{Bet}$, you can find $MS_W$ with a little algebra: $MS_W = 85.1/5.08 = 16.75$ (which is within rounding error of the $MS_W$ we obtained from the raw data). Once you have $MS_W$, you can perform your own follow-up $t$ tests; for instance, in the three-group case, you could use the more powerful, and perfectly acceptable Fisher procedure if, say, the follow-up tests were given according to Tukey's procedure. For more than three groups, Fisher's protected $t$ tests are not considered acceptable. There have been other follow-up procedures that are more *powerful* than Tukey's test (i.e., less conservative) and were thought to be acceptable, but modern methods have shown that they are not. We describe one of these formerly popular alternatives next.

## Other Multiple Comparison Procedures

Although Tukey's HSD test is acceptable for any number of groups (as long as they are all about the same size), it is more conservative and therefore less powerful than is desirable (i.e., it is less likely to detect small effects than it could be). This realization has led many statisticians to devise alternative (and sometimes highly complex) multiple comparison procedures designed to keep $\alpha_{EW}$ at a predetermined rate while maximizing power. One such test, which uses the studentized range statistic in a way that makes it more powerful than HSD, is called the Newman–Keuls (N–K) test. This test was the most popular multiple comparisons method in the behavioral sciences a few decades ago, but its use has been abandoned in recent years. This rejection of a once popular test is due to the results of computer simulation studies that have demonstrated that the N–K test was not keeping $\alpha_{EW}$ fixed, as was previously thought. Rather, the N–K test was gaining most of its extra power by letting $\alpha_{EW}$ increase as the number of groups increased. Although the N–K test is not as liberal as LSD when many groups are involved, it is no longer considered acceptably conservative.

In the past few decades, a number of multiple comparison procedures have been shown to keep $\alpha_{EW}$ at the desired level without being overly conservative. Unfortunately, most of these methods are unreasonably tedious to calculate by hand, and only some of them have been included in major statistical software packages. The SPSS package does offer several more modern tests for multiple comparisons, but because these newer tests are both difficult to explain and rarely used, we will not cover them in this textbook.

### Which Multiple Comparison Test Should I Use?

Our recommendation is clear. For multiple pairwise comparisons, use Fisher's protected *t* tests (LSD, if all *n*s are equal) when dealing with only three groups, and use the Tukey HSD test with more than three groups. However, if you are dealing with more than three samples and their sizes differ considerably, it is not legitimate to use either of the tests mentioned above. You may have to use an even more conservative general-purpose comparison test, like the Bonferroni or Scheffé tests described in the next section.

## Planned and Complex Comparisons

Whenever we have used the expression *multiple comparisons*, we were referring only to **post hoc pairwise comparisons**. The Latin term post hoc means "after the fact." In this context, it implies that no decision was made about which particular *t* tests to compute before looking at the data or performing the ANOVA. If you inspect your sample means and then test the difference between the largest and the smallest, you need the same $\alpha_{EW}$ protection that you would if you had conducted all of the possible *t* tests. Selecting specific pairs of means for comparison based on your theoretical research questions *before* seeing your actual data can give you a boost in power much like the added power that you can derive from planning a one- rather than two-tailed test. You can get "credit" for predicting the statistical significance of a few particular **planned comparisons**.

### The Bonferroni Correction

If you want to keep $\alpha_{EW}$ at .05, for instance, you should not use .05 as your alpha for each of the possible pairwise comparisons (the alpha for any one comparison can be symbolized as $\alpha_{pc}$, for alpha per comparison). But if you plan on testing relatively few of the possible comparisons, you do not need to increase your critical value as much as you would for Tukey's HSD. The simplest and most common procedure for finding the appropriate alpha for each of several planned (also called "a priori") comparisons is based on a formula for the maximum accumulation of probabilities, derived by the Italian mathematician Carlo Bonferroni in the 1930s. Applied to the problem of multiple comparisons, Bonferroni's "inequality" can be stated as follows: $c \times \alpha_{pc} \leq \alpha_{EW}$, where $c$ is the number of comparisons being performed.

For example, if you use an alpha of .01 for each of five null comparisons, the chance that one or more of the five tests will turn out to be statistically significant (i.e., $\alpha_{EW}$) will not be greater than $c \times \alpha_{pc} = 5 \times .01 = .05$. This fact leads to a simple rule for adjusting alpha for each comparison, based on the number of comparisons planned and the experimentwise value for alpha that you do not want to exceed. The formula for the *Bonferroni correction* (or "adjustment") is

$$\alpha_{pc} = \frac{\alpha_{EW}}{c} \qquad \text{Formula 14.6}$$

To illustrate the use of the Bonferroni correction, let us return to the example mentioned at the beginning of the HSD section—that is, the study of four similar drugs and a placebo. If you plan to test each drug against the placebo, but you do not plan to compare one drug to another, you are planning 4 comparisons out of a total of 10 (i.e., $5 \times 4/2$). To keep $\alpha_{EW}$ from rising above .05, you should use the following alpha for each of the 4 drug-to-placebo comparisons: $\alpha_{pc} = .05/4 = .0125$.

Before the ready availability of statistical software, special tables were needed to find critical *t* values corresponding to the various possible values of $\alpha_{pc}$. If you are using statistical software to conduct your planned *t* tests, you will get an exact *p* value for each of your comparisons (the two-tailed *p* value is usually the default option). For this example, any *t* test producing a $p$ value less than .0125 would be declared significant, but $H_0$ would not be rejected whenever $p$ was .0125 or larger.

For the purpose of comparison with other procedures, note that the critical $t$ for $df_W$ = 20 and $\alpha_{pc}$ = .0125 equals 2.744. This is a rather drastic increase over the critical $t$ for the .05 level, which equals 2.086. The Bonferroni correction is very conservative. That is, it tends to keep $\alpha_{EW}$ well below .05. But if you are planning relatively few comparisons, the Bonferroni test gives you more power than Tukey's HSD. You can see from Table G that $q$ for five groups and $df_W$ = 20 is 4.23. This is equivalent to a critical $t$ of $4.23/\sqrt{2}$, which equals 2.99, and is even larger than the Bonferroni-corrected critical $t$ of 2.744.

The Bonferroni correction is too conservative, however, to be used as a post hoc test when you are performing all of the possible $t$ tests. In the five-group case, a total of 10 pairwise comparisons is possible. So $\alpha_{pc}$ would equal .005, which corresponds to a Bonferroni-corrected critical $t$ ($df_W$ = 20) of 3.15. This is even larger than the equivalent $t$ for the HSD test (2.99), which is already a bit more conservative than is necessary. A reasonable compromise would be to use the Bonferroni correction to test a few planned pairwise comparisons (assuming you have a solid basis on which to plan those tests) and then use Tukey's HSD (basing $q$ on the total number of groups) to test the rest of the pairs. There are some modern procedures that modify the Bonferroni correction to make it suitable for testing all of your comparisons, such as the Hochberg (1988) sequential test and the *false discovery rate*, but these tests are too complex to be included here.

### Complex Comparisons

Selecting a few pairs of conditions for planned tests is one way to increase power. A more sophisticated (and increasingly popular) way is to create a **complex comparison**, which involves more than two conditions in one test.

For example, if your study included a drug, an herbal remedy, and a placebo, you could subtract the average of the herbal remedy and placebo means from the drug mean to create a single difference or *contrast*. This contrast can then be tested for significance as a planned comparison. Such planned contrasts allow you to get "credit" for correctly predicting the pattern, or relative spacing, of the means. If it turns out that the sample mean for the herbal remedy is much closer to the placebo mean than the drug mean, the $F$ ratio for testing our contrast (the average of herbal and placebo versus drug) will be larger than if the herbal and drug means were close together and very different from the placebo. (Conversely, planning to average the drug and herbal means for comparison with the placebo would imply that you are predicting that the effect of the herbal remedy will be comparable to the drug and that both are clearly better than the placebo.) If the pattern of the sample means fits well with your prediction, a planned comparison can attain significance, legitimately, even when the ANOVA would not. With few planned comparisons, it is reasonable to test each at the .05 level. When testing more than a few such comparisons, a Bonferroni correction on alpha may be called for.

Although it is not commonly done, when trying to follow up the results of a significant one-way ANOVA, you could devise and test a complex comparison based on a pattern you have seen in your data. For such post hoc complex contrasts, the Scheffé test is widely recommended to keep $\alpha_{EW}$ under control. The Scheffé test is so conservative that when the ANOVA is not significant, you can be sure that you will not be able to create any comparison that will be found significant by Scheffé's test (see B. H. Cohen, 2013).

### Trend Components

When the factor in your one-way ANOVA has quantitative levels, the types of complex comparisons you are likely to want to test are the simplest trends—usually just the linear and quadratic trends, and sometimes the cubic trend—rarely any more complex trend. The fact that the levels are quantitative automatically justifies the test of the linear trend as a planned comparison. The $SS_{Bet}$ for the linear trend is based on how close the means for the levels come to being on a single straight line; the closer the means of the dependent variable

(DV) come to exhibiting a linear trend, the more similar $SS_{Bet}$ for the trend will be to the $SS_{Bet}$ for the one-way ANOVA. The added power of testing the linear trend, as with any complex comparison, comes from the fact that the *MS* for the linear trend is found by dividing the *SS* for the linear trend by 1, whereas the $SS_{Bet}$ for the ANOVA must be divided by one less than the number of levels of the factor. The greater the number of quantitative factor levels, the greater the potential power advantage of testing any one particular trend (or other complex comparisons).

If the means of your factor levels tend to show one reversal in trend (e.g., the linear trend starts out steeply upward and then levels off or even begins to slope downward), testing your means for a quadratic trend may give you a larger *F* ratio (and therefore a better chance for statistical significance) than testing for a linear trend, though both may prove to be significant. Some extra justification is generally needed to test the quadratic trend as a planned comparison, but it often makes sense that the independent variable (IV) levels, if extended far enough, will eventually reach an optimum point, after which further increases will have decreasing effects on the DV. If the data tend to exhibit a second reversal, this will contribute to the amount of the cubic trend, and so on for higher trend components.

## Nonparametric Multiple Comparisons: The Protected Rank-Sum Test

We have seen that when the overall (often called *omnibus*) *F* test in an ANOVA is statistically significant and there are only three groups, it is then acceptable to perform the three possible (protected) *t* tests, each at the .05 level, in order to draw more specific conclusions about which of the population means can be considered different from which other ones. Similarly, if and only if your Kruskal–Wallis *H* is statistically significant for a three-group study, a method called the **protected rank-sum test** may be used to determine which pairs of populations differ significantly in location. As with the protected *t* test, you may perform the three possible pairwise rank-sum tests using an alpha of .05 for each, because the tests are protected against a larger experimentwise Type I error rate by the precondition that *H* be significant. Note that in the nonparametric case, there is no omnibus error term to use; completely separate (two-group) rank-sum tests, just like the ones described in Chapter 8, are performed. Also, note that there is no nonparametric version of Tukey's test. When *k* is larger than 3, a Bonferroni adjustment of the alpha for each follow-up rank-sum test is recommended.

To illustrate the use of rank-sum tests following a significant *H* test, we turn once again to the data in Table 14.1. Since *H* was statistically significant (see Chapter 13, in which *H* was found to be 8.48, $p < .05$), any or all of the three pairs of groups (1 versus 2, 1 versus 3, or 2 versus 3) may be compared by using the protected rank-sum test. In order to do so, however, *the scores must be reranked for each test* (rather than ranking all three groups together, as shown in Table 13.5). For example, if Groups 1 and 2 are being compared, the third group is ignored. The third score in Group 2 is ranked 1 (the lowest score), the fifth score in Group 1 is now ranked 2, the score right below this one is now ranked 3, and so on until the highest score in these two groups (i.e., the second score in Group 2) receives a rank of 18. Thus $N_1 = 8$, $N_2 = 10$, and $N_T = 18$ for this comparison. After reranking it is found that $T_1 = 59.5$ (and $T_2 = 111.5$), $T_E = (8)(19)/2 = 76$, and $\sigma_T = \sqrt{(8 \times 10 \times 19)/12} = 11.25$. Therefore, using the formula for the rank-sum test given previously, we obtain:

$$z = \frac{T_1 - T_E}{\sigma_T} = \frac{59.5 - 76}{11.25} = -1.47$$

This *z* value is not statistically significant. When Group 1 and Group 3 are compared, a nonsignificant *z* value (1.73) is also obtained. But when Group 2 and Group 3 are compared, a statistically significant *z* value of 2.69 results. Therefore, the conclusion is that the locations of these three populations are not the same because the location (or median) of Population 2 is greater than that of Population 3. However, the data do not warrant concluding that the location of Population 1 differs from either of the other two.

## Summary

When the $F$ ratio for a one-way ANOVA is statistically significant, it is likely that the researcher will want to run multiple comparisons on the data to determine which of the population means can be said to differ from one another. A general procedure for post hoc pairwise comparisons, which can be applied whether the sample sizes are equal or not even similar, is the Fisher protected $t$ test. It is "protected" only if the null hypothesis for the one-way ANOVA can be rejected. Unfortunately, it is not adequately conservative when dealing with more than three groups.

### Fisher's Protected *t* Tests

For any (or every) pair of groups, compute as follows:

$$t = \frac{\overline{X}_i - \overline{X}_j}{\sqrt{MS_W\left(\frac{1}{N_i} + \frac{1}{N_j}\right)}}, \qquad \text{Formula 14.1}$$

$$df = N_T - k$$

where

$\overline{X}_i$ = mean of group $i$
$\overline{X}_j$ = mean of group $j$
$MS_W$ = within-groups mean square
$N_i$ = number of observations in group $i$
$N_j$ = number of observations in group $j$

If all of the sample sizes are equal, Fisher's procedure can be simplified by calculating LSD. Any pair of means that differs by more than LSD can be said to differ significantly. The formula is

$$\text{LSD} = t_{\text{crit}}\sqrt{\frac{2MS_W}{n}} \qquad \text{Formula 14.2}$$

where $n$ is the size of *each* sample and $t_{\text{crit}}$ is two-tailed and is based on $df_W$ and the alpha used to test the one-way ANOVA. Fisher's procedure is not recommended for following up an ANOVA that involves more than three groups, because it allows $\alpha_{EW}$ to increase as $k$ increases, for $k > 3$.

### Tukey's HSD Test

This test is appropriate for any number of samples, as long as the samples are all the same size, or nearly so. (In the latter case, the harmonic mean of the sample sizes is used as $n$; see Formula 14.5.) It is not necessary to perform a one-way ANOVA before applying this test, but Tukey's test will rarely find a pair of means to differ significantly when the ANOVA would not be significant. The difference between each pair of means is compared to

$$\text{HSD} = q_{\text{crit}}\sqrt{\frac{MS_W}{n}} \qquad \text{Formula 14.3}$$

where $n$ is the size of each sample and $q_{\text{crit}}$ is a value from Table G, based on $k$ groups and $df_W$. The HSD test is more conservative than desirable, but it is simpler and better known than most alternative multiple comparison procedures. You can obtain a CI for the difference between any two population means represented in your study by using the following formula, based on Tukey's test:

$$\mu_i - \mu_j = \overline{X}_i - \overline{X}_j \pm \text{HSD} \qquad \text{Formula 14.4}$$

### The Harmonic Mean for $k$ Sample Sizes That Are Not All Equal

$$n_h = \frac{k}{\sum \frac{1}{n_i}} \qquad \text{Formula 14.5}$$

### The Bonferroni Correction

If a relatively small number of comparisons are planned in advance, each one can be tested using the following value for alpha:

$$\alpha_{pc} = \frac{\alpha_{EW}}{c} \qquad \text{Formula 14.6}$$

where $c$ is the number of planned comparisons and $\alpha_{EW}$ is the largest acceptable value for the experiment-wise alpha. This test is too conservative to be used as a post hoc test—for example, to test all possible pairwise comparisons following an ANOVA.

## Exercises

1. Calculate LSD for the two experiments in Exercise 1 of the previous chapter. For which experiment is the calculation of LSD justified? Determine which pairs of means differ significantly in the experiment for which the calculation of LSD is justified.
2. Calculate protected $t$ tests to compare all possible pairs of the four dormitories (e.g., Turck, Kirk, etc.), using the error term for Exercise 2 of the previous chapter. Which pairs differ significantly at the .05 level? Give two reasons why it would not be appropriate to calculate LSD for the data from the four dormitories.
3. Calculate both LSD and HSD for the two experiments in Exercise 3 of the previous chapter, using the harmonic mean of the three sample sizes as your value for $n$. What conclusions can you draw from each experiment? Use your values for LSD and HSD to compare the relative statistical power of these two procedures.
4. For your convenience, the data from Exercise 4 from the previous chapter are reprinted in the following table:

| | Group | | | | |
|---|---|---|---|---|---|
| | 1 | 2 | 3 | 4 | 5 |
| $X$ | 23 | 30 | 34 | 29 | 26 |
| $s$ | 6.5 | 7.2 | 7.0 | 5.8 | 6.0 |
| $N$ | 12 | 15 | 14 | 12 | 15 |

Using the harmonic mean of all the sample sizes as your value for $n$, calculate HSD for these data, and determine which pairs of groups differ significantly.

5. A study using five samples finds that the mean of each sample is as follows: Sample 1 = 6.7; Sample 2 = 14.2; Sample 3 = 13.8; Sample 4 = 10.4; Sample 5 = 15.8. Suppose that HSD (with $q$ based on $\alpha = .05$) for this study is 5.3. (a) Which population means should be regarded as different from each other at the .05 level? (b) Find the 95% CIs for comparing Sample 5 to each of the others.
6. For this exercise, we will use the sample means in Table 14.4; imagine that the size of each sample ($n$) is 7, but you do not know $MS_W$. Fortunately, you are given the results of the ANOVA: $F(4, 30) = 5.53$.
   (a) Find the error term for the ANOVA and use it to calculate Tukey's HSD.
   (b) Which types of music differ significantly from which other types, according to Tukey's HSD test?

**Table 14.4** **Means for the Music Experiment Data in Table 13.1B**

| Bluegrass | Jazz | Classical | New Age | Heavy metal |
|---|---|---|---|---|
| 14.2 | 20.2 | 22.4 | 8.2 | 8.0 |

7. In Exercise 7 of the previous chapter, the means for the negative feedback, positive feedback, and no feedback groups were $\overline{X}_{\text{neg}} = 11.1$, $\overline{X}_{\text{pos}} = 14.7$, and $\overline{X}_{\text{no}} = 12.4$, and the $F$ ratio for the one-way ANOVA was equal to 3.69. Given that the ANOVA was significant at the .05 level and that there were 20 participants in each group: (a) Find the value of LSD for this study. (b) Which pair(s) of means differ significantly, according to the LSD test?

## Thought Questions

1. An $F$ test from an ANOVA using five samples is statistically significant. The researcher now wishes to determine which pairs of the five population means differ significantly from each other. Why are multiple comparisons procedures necessary rather than just performing all of the possible $t$ tests?
2. If a multiple comparisons procedure involves the use of $t$ tests, why is it that a researcher cannot skip the ANOVA and just do these $t$ tests?
3. (a) When should you use Tukey's HSD test rather than Fisher's protected $t$ tests? (b) When should you use Fisher's protected $t$ tests rather than Tukey's HSD test? (a) What is the difference between post hoc comparisons and a priori comparisons? (b) What advantage results from using a priori comparisons?
4. Give an example of when the Bonferroni correction should be used.

## Computer Exercises

1. Redo the one-way ANOVAs requested in Computer Exercise 1 of the previous chapter, selecting both LSD and Tukey from the list of possible post hoc tests in your statistical software for each ANOVA. For postquiz anxiety, which pairs of quiz conditions (last question easy, moderate, difficult, or impossible) differ significantly from each other, according to each multiple comparison procedure? Answer this question again for postquiz heart rate. Compare LSD and Tukey with respect to power, using the results you found in this exercise to illustrate the differences.
2. Redo the one-way ANOVAs requested in Computer Exercise 2 of the previous chapter, selecting both Tukey and Bonferroni from the list of possible post hoc tests in your statistical software for each ANOVA. What is the problem with using HSD to make comparisons between groups in this exercise? Use the results for this exercise to compare the power of Tukey and Bonferroni for testing all possible pairs of means.
3. Redo the one-way ANOVAs requested in Computer Exercise 3 of the previous chapter, selecting LSD for pairwise comparisons. Also, use your statistical software to perform ordinary $t$ tests for each pair of math groups for each DV. Explain the differences between the LSD test and ordinary $t$ tests as applied to the data in this exercise.
4. (a) Use the Kruskal–Wallis procedure to test whether the different quiz conditions (last question easy, moderate, difficult, or impossible) had a significant effect on postquiz anxiety scores. Regardless of the significance of your test, perform all of the possible pairwise comparisons as rank-sum tests. Which of these pairwise tests would be significant after a Bonferroni adjustment of your alpha for each comparison?

   (b) Repeat part (a) for the postquiz heart rates.

## Bridge to SPSS

When performing a one-way ANOVA, you can obtain a wide selection of multiple-comparison procedures by clicking on the **Post Hoc** ... button from the initial dialog box. The LSD test is checked by default; we checked Tukey and Bonferroni for this example as well (see Figure 14.1). If you can assume that all of the population variances are equal (e.g., the HOV test in the previous chapter was not significant at the .05 level), there are 14 choices for multiple comparisons, only a few of which we have discussed in this chapter. The choice labeled "Tukey" corresponds to the HSD test we have described; "Tukey's-b" refers to a less-used modification that we did not discuss. "S–N–K" refers to the Newman–Keuls test, which we discussed briefly and did not recommend. (The "S" in S–N–K stands for Student and is sometimes added to the name as a reminder that this test is based on the "studentized" range statistic.)

For each post hoc test selected, SPSS will give you an appropriately adjusted and interpretable $p$ value for each possible pairwise comparison, assuming that your goal is to keep the experimentwise alpha at .05. For instance, if you choose "Tukey," your output will not present you with the size of HSD. Rather, $t$ tests will be performed for each possible pair of levels of your factor, and the exact $p$ level will be given for each, adjusted according to the studentized range statistic. As an example, if the difference between a pair of means is exactly equal to HSD, its $p$ value (labeled "Sig.") will be given as .05, because it is just on the borderline of significance for Tukey's test. For LSD, no adjustment is made to the $p$ values, so it is up to you to note whether the ANOVA is significant and to decide whether to use LSD if dealing with more than three groups (not recommended). For a pair of means that yields a $p$ value of .05 for "Tukey," the $p$ value for LSD will be smaller (i.e., more "significant"), because the latter is not adjusted.

If Bonferroni is selected, the $p$ value from the LSD test is simply multiplied by the total number of possible pairwise comparisons. For example, if there are four conditions in your one-way ANOVA (and therefore six possible pairs), and the $p$ value for comparing a particular pair by the LSD test is .01, the Bonferroni $p$ for that pair will be given as $6 \times .01 = .06$, and therefore that pair will not differ significantly at the .05 level, when using the Bonferroni correction. (Note that this method yields the same conclusions as dividing .05 by 6 to get a Bonferroni-adjusted alpha of .00833. Because .01 is not less than .00833, a pair associated with that $p$ value will not differ significantly when the Bonferroni procedure is used.) The other post hoc choices are beyond the level of this text, but they are rarely used, anyway.

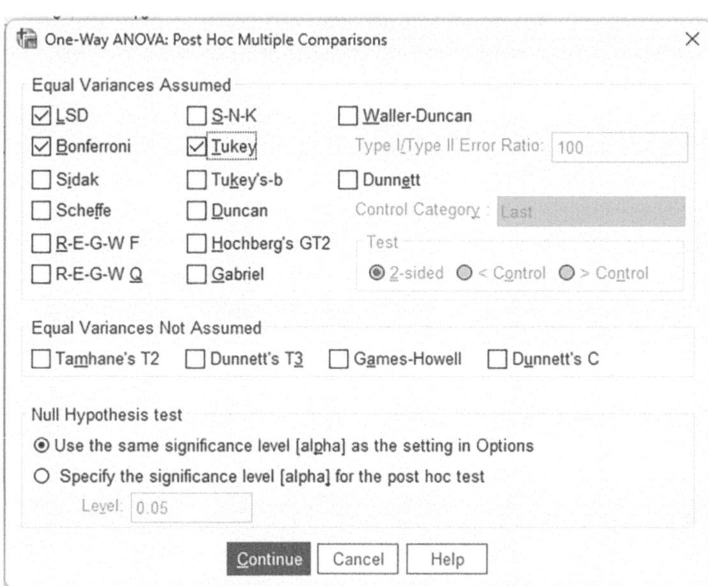


Figure 14.1

**The One-Way ANOVA: Post Hoc Multiple Comparisons Dialog Box**

*Source*: Reprint courtesy of International Business Machines Corporation, © SPSS, Inc., an IBM company. SPSS was incorporated by IBM in October 2009

# Figure 14.2

**The Results from Selecting LSD, Tukey, and Bonferroni as Multiple Comparisons in a One-Way ANOVA with Three Groups**

**Multiple Comparisons**

Dependent Variable: statquiz

| | (I) reason | (J) reason | Mean Difference (I-J) | Std. Error | Sig. | 95% Confidence Interval Lower Bound | Upper Bound |
|---|---|---|---|---|---|---|---|
| Tukey HSD | Program | Personal | −.526 | .405 | .401 | −1.49 | .44 |
| | | Advisor | .487 | .402 | .448 | −.47 | 1.44 |
| | Personal | Program | .526 | .405 | .401 | −.44 | 1.49 |
| | | Advisor | 1.013 | .435 | .057 | −.02 | 2.05 |
| | Advisor | Program | −.487 | .402 | .448 | −1.44 | .47 |
| | | Personal | −1.013 | .435 | .057 | −2.05 | .02 |
| LSD | Program | Personal | −.526 | .405 | .198 | −1.33 | .28 |
| | | Advisor | .487 | .402 | .228 | −.31 | 1.28 |
| | Personal | Program | .526 | .405 | .198 | −.28 | 1.33 |
| | | Advisor | 1.013* | .435 | .022 | .15 | 1.88 |
| | Advisor | Program | −.487 | .402 | .228 | −1.28 | .31 |
| | | Personal | −1.013* | .435 | 022 | −1.88 | −.15 |
| Bonferroni | Program | Personal | −.526 | .405 | .594 | −1.51 | .46 |
| | | Advisor | .487 | .402 | .684 | −.49 | 1.47 |
| | Personal | Program | .526 | .405 | .594 | −.46 | 1.51 |
| | | Advisor | 1.013 | .435 | .066 | −.05 | 2.07 |
| | Advisor | Program | −.487 | .402 | .684 | −1.47 | .49 |
| | | Personal | −1.013 | .435 | .066 | −2.07 | .05 |

*. The mean difference is significant at the 0.05 level.

To obtain the results in Figure 14.2, we reran one of the ANOVAs from the previous chapter: the IV was *reason* and the DV was *statquiz*. The Multiple Comparisons table that was produced is based on the selections shown in Figure 14.1. Looking at Figure 14.2, you will see that for LSD, there is one significant comparison: Personal versus Advisor ($p = .022$). However, in the previous chapter, we found that the one-way ANOVA for all three groups was not significant ($p = .072$). This illustrates the need to check the significance of the one-way ANOVA before accepting the LSD results. Note that the Bonferroni $p$ value (.066) for this comparison is three times larger than the $p$ for LSD; that is because, with three groups, there are three different pairwise comparisons that can be tested. You can also see that Tukey's HSD test is more conservative than LSD, but less conservative than Bonferroni. The Tukey $p$ value for Personal versus Advisor is nearly significant at .057, which is greater than .022 (LSD), but less than .066 (Bonferroni).

If you click on the **Contrasts ...** button and then check the *Polynomial* box, you can select from the *Degree:* menu whether you want to test only for a Linear trend (the default), or higher trends in addition. You can also create a complex contrast by choosing Coefficients: in the same order as the levels of your factor. For example, if your factor is *reason*, entering the coefficients −1, 2, and −1 in that order (the coefficients must add up to zero), will compare the average of the Program and Advisor with the Personal Interest group. A more complete discussion of complex comparisons and how to choose the needed coefficients is beyond the scope of this text (see B. H. Cohen, 2008).

# Chapter 15

# Introduction to Factorial Design: Two-Way Analysis of Variance

## PREVIEW

### Introduction

When should you use a factorial design instead of a one-way analysis of variance? In what way are these two procedures similar? What are some of the important differences between them?

What is meant by the interaction between two variables?

### Computational Procedures

What are the procedures for computing a two-way (factorial) analysis of variance?

Why do we calculate the sums of squares for the main effects before the *SS* for the interaction?

How do we test the *F* ratios for statistical significance?

What are the multiple comparison procedures in two-way analysis of variance?

What is meant by the term *familywise alpha*, and how does it differ from experiment-wise alpha?

### The Meaning of Interaction

What is meant by a zero interaction versus some interaction between two variables?

What are the two major types of interaction for a $2 \times 2$ factorial design?

Why are we cautious about interpreting the results of the main effects when the interaction is significant?

### Following Up a Significant Interaction

What are simple main effects, and when is it appropriate to test them?

How can tests of simple main effects lead to more specific conclusions from a two-way ANOVA?

How do you follow up a significant simple main effect that involves three or more levels of a factor?

### Measuring Effect Size in a Factorial ANOVA

How does a partial eta squared differ from an ordinary eta squared? When would each be preferred?

### Summary

### Exercises

### Thought Questions

### Computer Exercises

### Bridge to SPSS

*Introductory Statistics for the Behavioral Sciences*, Eighth Edition. R. Brooke Lea and Barry H. Cohen.

Companion website: www.wiley.com/go/ISBS8e

## Introduction

The one-way analysis of variance (ANOVA) presented in Chapter 13 is used to investigate the relationship of a *single* independent variable to a dependent variable, where the independent variable has two or more levels (i.e., groups). For example, the music experiment in Chapter 13 dealt with the effect of five different types of music (five levels of the independent variable) on performance on a spatial ability test (the dependent variable).

The **factorial design** is used to study the relationship of *two or more* independent variables (called *factors*) to a dependent variable, where each factor has two or more levels. Suppose you are interested in the relationship between four different dosages of caffeine (four levels: large, moderate, small, and zero) and fatigue (two levels: high and low) to scores on a test of manual dexterity. There are several hypotheses of interest: Different dosages of caffeine may affect test scores differently; test performance may differ between fatigued and nonfatigued participants; certain caffeine dosages may affect test scores more for one fatigue level than for the other. These hypotheses may be evaluated in a single statistical framework by using a factorial design. This example would be called a "two-way" ANOVA because there are two independent variables. It could also be labeled as a $4 \times 2$ factorial design because there are four levels of the first independent variable and two levels of the second independent variable. Each combination of levels of the two factors is called a **cell**, so there are eight possible cells in a $4 \times 2$ factorial design. If all of the possible cells are tested in a study, the design is said to be *completely crossed*. When we refer to a factorial design, we always mean a completely crossed factorial design. (It is rare to see a factorial design with any empty cells because that situation greatly complicates the analysis.)

The logic of the factorial design begins with the logic of the simpler one-way design. The total sum of squares is partitioned into within-group (error) sum of squares and between-group sum of squares. In the factorial design, however, the between-group sum of squares (which could just as well be called the between-cells *SS*) is itself partitioned into several parts: variation due to the first factor, variation due to the second factor, and variation due to the joint effects of the two factors called the **interaction** (see Figure 15.1). One example of an interaction effect would be if the highest dosage of caffeine improved test scores for highly fatigued participants but *not* for participants who are low in fatigue, while other dosages had the same effect on test scores for both fatigue levels. (The interaction, a particularly important feature of factorial design, is discussed in detail later in this chapter.)

It can be shown algebraically that the total sum of squares is equal to the sum of these various parts: (a) the within-group (error) sum of squares, (b) the sum of squares due to Factor 1, (c) the sum of squares due to Factor 2, and (d) the sum of squares due to the interaction of Factors 1 and 2. Thus, a factorial design makes it possible to break down the total variability into *several* meaningful parts. That is, it permits several possible explanations as to why people are different on the dependent variable. As was pointed out in earlier chapters, explaining variation—why people differ from one another—is the *raison d'être* of the behavioral scientist.

Figure 15.1

**Partitioning of Variation in a Two-Way Factorial Design**

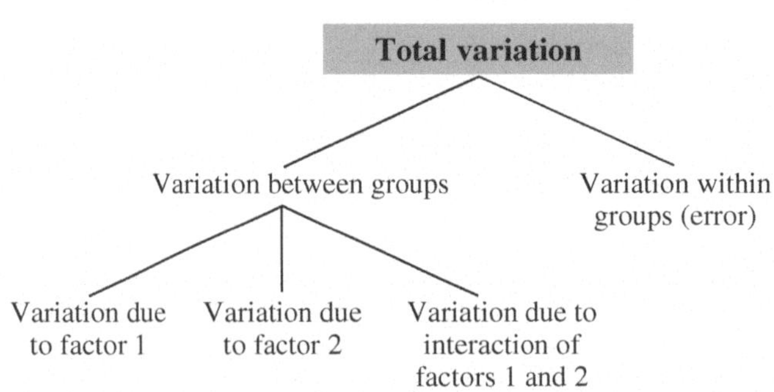

## Computational Procedures

The outline of the raw-score computational procedure for the two-way factorial design is as follows:

1. (a) Compute $SS_T$.
   (b) Compute $SS_{Bet}$.
   (c) Subtract $SS_{Bet}$ from $SS_T$ to obtain $SS_W$ (error).
   (d) Compute $SS_1$ (the $SS_{Bet}$ just for Factor 1).
   (e) Compute $SS_2$ (the $SS_{Bet}$ just for Factor 2).
   (f) Subtract $SS_1$ and $SS_2$ from $SS_{Bet}$ to obtain the sum of squares for the interaction of Factors 1 and 2 (called $SS_{inter}$ or $SS_{1\times 2}$).
2. Convert the sums of squares in steps (c), (d), (e), and (f) to mean squares (*MS*s) by dividing each one by the appropriate number of degrees of freedom.
3. (a) Test the *main effect* for Factor 1 ($MS_1$) for statistical significance by computing the appropriate *F* ratio (i.e., divide $MS_1$ by $MS_W$).
   (b) Test the main effect for Factor 2 ($MS_2$) for statistical significance by computing the *F* ratio, as in (a).
   (c) Test the interaction of the two factors ($MS_{1\times 2}$) for statistical significance by computing the *F* ratio, as in (a) and (b).

The two-way factorial design permits you to test three null hypotheses—one concerning the effect of Factor 1, one concerning the effect of Factor 2, and one concerning the *joint effect* of Factor 1 and Factor 2—in a single statistical framework. The separate effects of Factors 1 and 2 are referred to as the **main effects** of those factors.

As an illustration, consider the hypothetical results for the music experiment shown previously in Table 13.1A. There are five observations in each *cell*; for example, the scores of 16, 16, 14, 13, and 12 are the test scores of five participants with no musical training who completed a spatial ability test while listening to Mozart. Just as in the one-way design, the within-group variance estimate is based on the variability *within* each of the eight cells. Variation due to the composer factor is reflected by the variability across the four *column* means, while variation due to the musical background factor is reflected by the variability (i.e., difference) of the two *row* means (row and column means are both referred to as *marginal means*). (Note that the data for the participants with no musical training in Table 15.1 come from Table 13.1A, after deleting Group 4.)

**Table 15.1** **Scores on a 20-Item Spatial Ability Test as a Function of Classical Composer and Musical Background (4 × 2 Factorial Design)**

| | Musical composer (Factor 1) | | | | | | | | |
|---|---|---|---|---|---|---|---|---|---|
| | Mozart | | Chopin | | Bach | | Beethoven | | Row means |
| | 16 | | 16 | | 16 | | 14 | | |
| | 16 | $\overline{X}$ = 14.2 | 14 | $\overline{X}$ = 13.2 | 14 | $\overline{X}$ = 12.4 | 13 | $\overline{X}$ = 12.0 | 12.95 |
| No musical training | 14 | | 13 | | 12 | | 13 | | |
| | 13 | $s$ = 1.79 | 13 | $s$ = 2.17 | 10 | $s$ = 2.61 | 10 | $s$ = 1.87 | |
| | 12 | | 10 | | 10 | | 10 | | |
| Musical background (Factor 2) | | | | | | | | | |
| | 16 | | 17 | | 14 | | 16 | | |
| | 18 | $\overline{X}$ = 17.4 | 10 | $\overline{X}$ = 13.2 | 14 | $\overline{X}$ = 12.2 | 10 | $\overline{X}$ = 12.8 | 13.90 |
| Professional musician | 20 | | 13 | | 11 | | 13 | | |
| | 15 | $s$ = 1.95 | 12 | $s$ = 2. 95 | 10 | $s$ = 1.79 | 14 | $s$ = 2.39 | |
| | 18 | | 14 | | 12 | | 11 | | |
| Column means | 15.8 | | 13.2 | | 12.3 | | 12.4 | | Grand mean = $\overline{X}_G$ = 13.425 |

**Sums of Squares**

1. ***Total sum of squares ($SS_T$).*** The total *SS* is computed in the same way as in Chapter 13, using Formula 13.1B, repeated here:

$$SS_T = \sum X^2 - N_T\overline{X}_G^2 \qquad \text{Formula 13.1B}$$

where

$N_T$ = total number of observations
$\sum$ = summation across all observations

In Table 15.1,

$$\begin{aligned}\sum X^2 &= 16^2 + 16^2 + 14^2 + \cdots + 13^2 + 14^2 + 11^2\\ &= 7{,}467\\ N_T\overline{X}_G^2 &= 40(13.425)^2 = 7{,}209.2\\ SS_T &= 7.467 - 7.209.2\\ &= 257.8\end{aligned}$$

(See the note on using your calculator to obtain $SS_T$ in the corresponding part of Chapter 13.)

2. ***Sum of squares between groups ($SS_{Bet}$).*** We can ignore for a moment the fact that this is a factorial design, treat the data in Table 15.1 as coming from eight groups, and find $SS_{Bet}$ as in Chapter 13. Because this is a *balanced* ANOVA design[1] (i.e., all of the groups, or cells, are the same size), we could use the equal-*N* version of the $SS_{Bet}$ formula (Formula 13.2B), but to further reduce computational effort, we use an even more convenient version of that formula as shown here:

$$SS_{Bet} = n\sum_{i=1}^{k}\overline{X}_i^2 - N_T\overline{X}_G^2 \qquad \text{Formula 15.1}$$

where *n* is the size of *each* cell, and *k* is the number of cells.

For the data in Table 15.1, $k = 8$ and $n = 5$. Therefore,

$$\begin{aligned}n\sum_{i=1}^{k}\overline{X}_i^2 &= 5\left(14.2^2 + 13.2^2 + 12.4^2 + 17.4^2 + 13.2^2 + 12.2^2 + 12.8^2\right)\\ &= 5(1{,}463.32)\\ &= 7{,}316.6\end{aligned}$$

We have already calculated $N_T\overline{X}_G^2$ as part of the formula for $SS_T$, so

$$SS_{bet} = 7{,}316.6 - 7{,}209.2 = 107.4$$

3. ***Sum of squares within groups (error) ($SS_W$).*** The within-groups *SS* may be found by Formula 13.3 or, more easily, by subtraction as shown in Chapter 13 and here:

$$\begin{aligned}SS_W &= SS_T - SS_{Bet}\\ &= 257.8 - 107.4\\ &= 150.4\end{aligned}$$

[1] The calculation of factorial ANOVAs that are *not* balanced is a rather complex topic, which is beyond the scope of an introductory statistics text.

Later, we will divide $SS_W$ by $df_W$ to obtain the usual denominator of our *F* ratios, $MS_W$. However, $MS_W$ can be found directly from the unbiased standard deviations (SDs) given in Table 15.1. You may recall from Chapter 13 that in the equal-*n* one-way ANOVA, $MS_W$ is the average of all the group variances. Similarly, in a balanced factorial design, $MS_W$ is just the average of the *cell* variances. If you square all of the *s*s (i.e., SDs) in Table 15.1, sum them, and then divide by 8, you will get 4.7061. This agrees, within rounding error, with $SS_W/df_W$, as you will soon see.

Computing $MS_W$ directly from the SDs can act as a check on the accuracy of your other calculations, but it has a more important function. If for any reason you need to calculate a factorial ANOVA from a table of means and SDs, and you do *not* have the individual scores, you will not be able to calculate $SS_T$ directly. Therefore, you will not have the option of finding $SS_W$ by subtraction. But that is not a problem because you can always square and average the SDs to find $MS_W$ directly—provided, of course, that the SDs have been reported, and you are dealing with a balanced design.

Before we proceed to divide $SS_{Bet}$ into smaller, more specific pieces, some additional notation will help prevent confusion. Let us arbitrarily define Factor 1 as the *column factor* and represent the number of columns by *c*. Similarly, we will define Factor 2 as the *row factor* and use *r* to represent the number of rows. Thus, the number of cells in a two-way design can be written as *rc*, and the total $N(N_T)$ therefore equals *rcn* (For the design in Table 15.1, $N_T = rcn = 2 \times 4 \times 5 = 40$).

4. ***Sum of squares for Factor 1 ($SS_1$).*** We have defined the column factor as Factor 1. In this example, the columns vary by musical composer. The *SS* for the composer factor, *which ignores differences in musical background*, is computed using a slight variation of Formula 13.2B:

$$SS_1 = rn\sum_{i=1}^{c}\overline{X}_i^2 - N_T\overline{X}_G^2 \qquad \text{Formula 15.2A}$$

where $\overline{X}_i$ is the mean of any *column* and *rn* is the number of scores in any column (i.e., the number of rows times the size of each cell), which replaces *n* in Formula 13.2B.

If the composer has a substantial effect on test scores (ignoring musical background), the means of the *columns* of Table 15.1 should show considerable variability. The value of $N_T\overline{X}_G^2$ has already been found to be equal to 7,209.2. Then,

$$\begin{aligned} SS_1 &= 10(15.8^2 + 13.2^2 + 12.3^2 + 12.4^2) - 7{,}209.2 \\ &= 7{,}289.3 - 7{,}209.2 \\ &= 80.1 \end{aligned}$$

5. ***Sum of squares for Factor 2 ($SS_2$).*** In this example, musical background is the row factor and hence Factor 2. The *SS* for the background factor, *which ignores differences in musical composer*, is computed by the same formula used for $SS_1$, except for changing *rn* to *cn*:

$$SS_2 = cn\sum_{i=1}^{r}\overline{X}_i^2 - N_T\overline{X}_G^2 \qquad \text{Formula 15.2B}$$

where $\overline{X}_i$ is the mean of any row and *cn* is the number of scores in any row (i.e., the number of columns times the size of each cell). If musical background has an effect on test scores (ignoring composer), the means of the rows of Table 15.1 should differ at least somewhat.

$$\begin{aligned} SS_2 &= 20(12.95^2 + 13.90^2) - 7{,}209.2 \\ &= 7{,}218.2 - 7{,}209.2 \\ &= 9.0 \end{aligned}$$

6. ***Sum of squares for interaction ($SS_{1\times2}$).*** The interaction *SS* is part of the variability of the eight cells and is obtained by subtraction:

$$SS_{1\times2} = SS_{Bet} - SS_1 - SS_2 \qquad \text{Formula 15.3}$$

For the data in Table 15.1,

$$SS_{1\times 2} = 107.4 - 80.1 - 9.0$$
$$= 18.3$$

### Mean Squares

The next step is to convert each sum of squares to an estimate of the population variance, or *MS*. This is done by dividing by the appropriate degrees of freedom (*df*), as shown in the following table. The way that the *df* are partitioned in a two-way ANOVA is illustrated in Figure 15.2.

| Source | Degrees of freedom | Computation for Table 15.1 |
|---|---|---|
| Total | $N_T - 1$ | $df_T = 40 - 1 = 39$ |
| Within groups | $N_T - k$ | $df_W = 40 - 8 = 32$ |
| Between groups | $k - 1$ | $df_{Bet} = 8 - 1 = 7$ |
| Factor 1 | $c$ (i.e., number of levels of factor 1) $- 1$ | $df_1 = 4 - 1 = 3$ |
| Factor 2 | $r$ (i.e., number of levels of factor 2) $- 1$ | $df_2 = 2 - 1 = 1$ |
| Interaction | $df_1 \times df_2$ (i.e., $[r - 1] \times [c - 1]$) | $df_{1\times 2} = 3 \times 1 = 3$ |

Note: $N_T$ = total number of observations
$k$ = number of cells = $rc$

The first three values in the preceding table are the same as in the case of the one-way design (Chapter 13). The total *df* equal $N_T - 1$, or one less than the total number of observations. The within-group *df* is equal to $N_T - k$, where $k$ equals the number of cells (or groups). This is equivalent to obtaining the *df* for each cell (one less than the number of observations in the cell, or 4 in this example) and summing over all cells ($4 \times 8 = 32$). The between-group *df* equals one less than the number of cells (i.e., $rc - 1$).

It is also necessary, however, to partition the between-group *df* in the same way as the between-group variance is partitioned. The *df* for Factor 1 is one less than the number of levels of Factor 1. In this example, Factor 1 has four levels (the four different composers), so the *df* are $4 - 1 = 3$. Similarly, the *df* for Factor 2, musical background, are $2 - 1 = 1$. The *df* for the interaction of composer and background are found by *multiplying* the *df* for each factor ($3 \times 1 = 3$).

The mean squares (*MS*s) are then found by dividing each *SS*, as usual, by the corresponding *df*:

Mean square within groups (i.e., cells):

$$MS_W = \frac{SS_W}{df_W} = \frac{150.4}{32} = 4.70$$

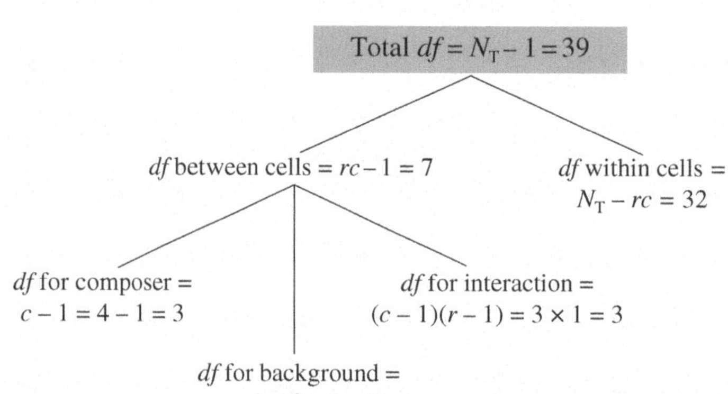


## Figure 15.2

**Partitioning of Degrees of Freedom in the Music Experiment**

Mean square for composer:

$$MS_1 = \frac{SS_1}{df_1} = \frac{80.1}{3} = 26.7$$

Mean square for background:

$$MS_2 = \frac{SS_2}{df_2} = \frac{9.0}{1} = 9.0$$

Mean square for interaction:

$$MS_{1\times 2} = \frac{SS_{1\times 2}}{df_{1\times 2}} = \frac{18.3}{3} = 6.1$$

### *F* Ratios and Tests of Significance

As with the one-way ANOVA, the appropriate statistical model to use with factorial designs is the *F* distribution. The first null hypothesis to be tested is that regardless of which of the four classical composers they listen to, the populations will have equal means. The *MS* for the composer factor is divided by the *MS* within, yielding the following *F* ratio:

$$F = \frac{MS_1}{MS_W} = \frac{26.7}{4.70} = 5.68$$

The critical value from the *F* table for 3 *df* in the numerator and 32 *df* in the denominator and $\alpha = .05$ is 2.90. Since the computed *F* of 5.68 is greater than this value, you reject $H_0$ and conclude that the participants in the four composer conditions do *not* come from populations with equal means—that is, different musical composers *do* have an effect on spatial ability test scores.

The second null hypothesis to be tested is that the musicians and novices come from populations with equal means for spatial ability (we refer to the participants who lack musical training as *novices*, for convenience). The *MS* for the background factor is divided by the *MS* within, yielding the following *F* ratio:

$$F = \frac{MS_2}{MS_W} = \frac{9.0}{4.70} = 1.915$$

The critical value from the *F* table for 1 *df* in the numerator and 32 *df* in the denominator and $\alpha = .05$ is 4.15. (Note that the critical values for the various *F* tests differ if the *df*s differ because they come from different *F* distributions.) The computed value of 1.915 is less than the critical value, so you retain $H_0$ and conclude that there is not sufficient evidence to believe that musical training is associated with differences in spatial ability.

The third null hypothesis to be tested is that the interaction effect is zero:

$$F = \frac{MS_{1\times 2}}{MS_W} = \frac{6.1}{4.70} = 1.30$$

This computed value of *F* is less than the critical value obtained from the *F* table of 2.90 for 3 and 32 *df* and $\alpha = .05$. So, you retain $H_0$ and conclude that there is not sufficient evidence to reject the null hypothesis of no interaction effect.

### ANOVA Summary Table

The results of the foregoing ANOVA are summarized in Table 15.2. Note that the factors are identified by name for the convenience of the reader. Also, as was the case with the one-way ANOVA, within-group variation (error) is listed last. No *F* value is listed for error because error is used as the denominator of the various *F* ratios and is not itself the subject of a statistical test.

**Table 15.2** Summary of Two-Way ANOVA of Music Experiment

| Source | *SS* | *df* | *MS* | *F* |
|---|---|---|---|---|
| Composer | 80.1 | 3 | 26.7 | 5.68 |
| Music background | 9.0 | 1 | 9.0 | 1.92 |
| Composer × background | 18.3 | 3 | 6.1 | 1.30 |
| Error (i.e., within cells) | 150.4 | 32 | 4.7 | |

### Multiple Comparisons Following a Factorial ANOVA

We saw in Chapter 13 how we could specify the meaning of a significant $F$ ratio by means of a multiple comparison procedure. In a two-way factorial design, there are potentially three $F$ tests that may require further specification as to which means differ from which others, and the same types of procedures may be followed.

When the $F$ for interaction is *not* significant, it is meaningful to follow up each of the independent variables (factors) as if it were from a one-way ANOVA. If the $F$ for a factor is *not* significant, there is nothing to follow up. Doing so is not only unnecessary but improper because the resulting tests would not be protected. Of course, if a significant factor has only two levels, there is nothing to follow up because there is only one difference, so it must be significant. But a significant $F$ for a factor of three or more levels requires specification just as in the one-way design.

In the foregoing example, the interaction $F$ is not significant but the $F$ for composer is. So, you may proceed to perform follow-up tests among the column means corresponding to the four different classical composers. Table 15.1 gives these as 15.8, 13.2, 12.3, and 12.4. All of the levels have the same $N_i = rn = 10$, and more than three levels are being compared, so Tukey's Honestly Significant Difference (HSD) test is recommended. Here, Formula 14.3 has been modified to show the appropriate "$n$":

$$\text{HSD} = q\sqrt{\frac{MS_W}{rn}}$$

where $q$ (from Table G) is based on four groups (i.e., columns), $df_W$ is from the ANOVA, and $rn$ is the number of scores in each group (i.e., column).

Filling in the relevant values in the formula, we find that

$$\text{HSD} = 3.84\sqrt{\frac{4.70}{10}} = 3.84(0.686) = 2.63$$

Thus, *two* column means must differ by at least 2.63 points on the spatial ability test to be declared statistically significant at the .05 level. By this criterion, the mean for the Mozart condition (15.8) is significantly larger than the means for both the Bach (12.3) and Beethoven (12.4) conditions (and very nearly significantly larger than the Chopin mean [13.2]), thus further refining the "Mozart effect" we first mentioned in Chapter 13. There are no significant differences among the other three composers.

Alternatively, a confidence interval (CI) may be established for each pairwise comparison, using Formula 14.4:

$$\mu_i - \mu_j = \overline{X}_1 - \overline{X}_2 \pm \text{HSD}$$

Thus, the 95% CI for the difference between Column 1 (Mozart) and Column 3 (Bach) is

$$[(15.8 - 12.3) - 2.63] \le \mu_1 - \mu_3 \le [(15.8 - 12.3) + 2.63]$$
$$+0.87 \le \mu_1 - \mu_3 \le +6.13$$

The value of zero falls outside (below) this interval, indicating that this comparison is statistically significant. In addition, the CI specifies all reasonably likely values of the difference between the two population means. CIs for the other composers may be obtained in a similar fashion.

### Familywise Alpha

The second factor in the music experiment, music background, was not significant in this example. So, no follow-up tests would be justified. It should also be noted that even if the music background factor *were* significant, no follow-up tests would be possible because the factor has only two levels. It is when both main effects in a two-way ANOVA are significant and both have more than two levels that an interesting question arises as to how to perform multiple comparisons for *both* factors and still keep the experimentwise alpha ($\alpha_{EW}$) at .05. Suppose, for instance, that music background had three levels (e.g., novice, music student, music teacher) and led to a significant *F* ratio. Because there are only three levels, it would be appropriate to use the LSD test to make comparisons among them. However, if the .05 level is used to find the critical *t* for the LSD test and the .05 level is used again to find the critical *q* for Tukey's HSD test on the composer factor, are we not using the .05 level *twice* to follow up on the *same* experiment? And would not that raise $\alpha_{EW}$ above .05? The answer is yes! But rather than try to protect $\alpha_{EW}$, the traditional approach is to invoke a related concept called the **familywise alpha** ($\boldsymbol{\alpha_{FW}}$) and direct our efforts toward controlling $\alpha_{FW}$.

It is important to note that it is entirely acceptable to test each of the three *F* ratios in a two-way ANOVA at the .05 level without using a Bonferroni correction or any other adjustment for multiple tests within the same experiment. That is because all three of the *F* ratios are automatically considered planned comparisons in a two-way ANOVA. Moreover, the set of post hoc comparisons that would be needed to follow up on any one of these *F* ratios is thought of as representing a separate *family* of tests, and it is considered acceptable to perform all of the tests in such a family if the corresponding *F* ratio is significant at the traditional .05 level. For a set of three possible post hoc comparisons, the significance of the *F* ratio is considered adequate protection to perform all three tests using $\alpha = .05$ for each (e.g., the LSD test). If the set of possible comparisons is greater than three, you would need a procedure like Tukey's HSD to keep the alpha for the whole family of tests (i.e., $\alpha_{FW}$) down to .05. Thus, if all three *F* ratios in a two-way ANOVA are significant at the .05 level, you are permitted to perform three sets of post hoc tests, as long as you keep the overall (i.e., familywise) alpha for each set to the .05 level. However, when the *F* ratio for the interaction of the two factors is significant, you may lose interest in comparing column or row means.

When the *F* for interaction *is* significant, it indicates that the factors operate *jointly*. Under these circumstances, our interest may be drawn to differences between *cells*—that is, specific combinations of the two factors. In the previous example, the *F* for interaction was not significant. So, the focus was on the two main effects. But an alternative set of results for this experiment, illustrating a large interaction, is posited in Table 15.3B. The way follow-up tests are conducted for a two-way ANOVA is quite different when the interaction of the factors is found

## Table 15.3 Cell Means Illustrating Some Interaction and Zero Interaction

| Music background | Mozart | Chopin | Bach | Beethoven | Row means |
|---|---|---|---|---|---|
| | A. Cell means from Table 14.1 (some interaction) | | | | |
| Novices | 14.2 | 13.2 | 12.4 | 12.0 | 12.95 |
| Musicians | 17.4 | 13.2 | 12.2 | 12.8 | 13.90 |
| $\overline{X}_{\text{musician}} - \overline{X}_{\text{novice}}$ | 3.2 | 0.0 | −0.2 | 0.8 | 0.95 |
| | B. Cell means illustrating zero interaction | | | | |
| Novices | 14.2 | 13.2 | 12.4 | 12.0 | 12.95 |
| Musicians | 12.6 | 11.6 | 10.8 | 10.4 | 11.35 |
| $\overline{X}_{\text{musician}} - \overline{X}_{\text{novice}}$ | 1.6 | 1.6 | 1.6 | 1.6 | 1.60 |

to be statistically significant. We describe procedures for further specifying a significant interaction after we discuss several ways to interpret the meaning of an interaction.

## The Meaning of Interaction

Interaction refers to the *joint* effect of two or more factors on the dependent variable. As an illustration, consider once again the music experiment. The interaction effect of composer and music background refers to the unique effect of particular combinations of levels of the two factors, such as the musicians listening to Mozart or the novices listening to Bach, and *not* to the sum of the separate effects of the two factors. The interaction is the joint effect *over and above* the sum of the separate effects.

### Comparing Zero to Some Interaction

The interaction in the music experiment is highlighted in Table 15.3A, which displays the cell means from Table 15.1, along with the differences between the two background groups for each composer. The fact that these differences (3.2, 0, −0.2, 0.8) are not all the same tells you that the amount of interaction is not zero. If, however, the professional musicians always scored exactly the same number of points higher (or lower) in spatial ability than the musical novices (say 1.6 points higher, as in Table 15.3B), regardless of the classical composer being listened to, the interaction would indeed be equal to zero. Also, note that a zero amount of interaction implies that the difference between mean test scores for any two composers would be the same for both musicians and novices (see Table 15.3B). If the mean differences for each composer were considerably different from one another, there could be a significant interaction effect; the composer factor would affect test scores differently for participants with different musical backgrounds. (Cell means that produce a significant interaction effect are shown in Table 15.4.) The interaction effect for the data in Table 15.3A was found to be *not* significant, so the differences among the observed mean differences of 3.2, 0, −0.2, and 0.8 are likely enough to have occurred by chance (random sampling error) that there is a lack of sufficient evidence to believe that there is an interaction effect in the population from which the samples in this experiment were drawn.

A graphic illustration of the examples in Table 15.3 is shown in Figure 15.3. When the interaction effect is zero (Figure 15.3B), the line connecting the points corresponding to the cell means for musicians follows the same pattern as the corresponding line for novices. That is, the two lines are parallel, which means that the distance between the two lines is the same at all points. When there is some amount of interaction (Figure 15.3A), musical background differences are different from one composer to another. From Figure 15.3A, you can see that whereas musicians produced the higher means while listening to Mozart and Beethoven, novices had the higher mean during the Bach condition (but these differences did not attain statistical significance).

Figure 15.3

**Graphic Representation of the Data in Table 15.3**

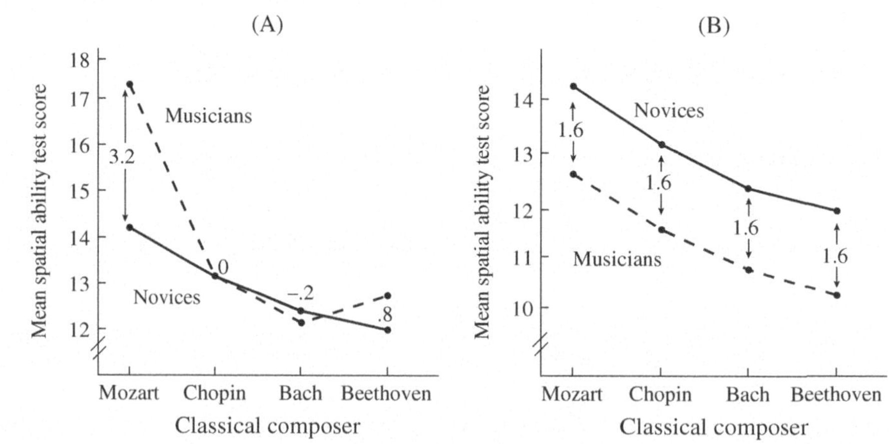

### Different Types of Interactions

The (main) effects of the two factors in a two-way ANOVA are independent of each other. That is, either one could be significant while the other is not. Indeed, one factor could have a very large effect while the other has none at all. Or both effects, or neither effect, could be significant.

The size of the interaction in a two-way ANOVA is completely independent of the sizes of the main effects. Both main effects could be very large, while the amount of interaction is zero. Or there could be a large interaction with no main effects at all. Different relative amounts and directions of the two main effects can combine with different amounts of interaction to produce a variety of different patterns when the cell means are graphed, as illustrated in Figure 15.4.

In order to categorize two major types of cell-mean patterns, we will deal with an example of the simplest two-way ANOVA design: the 2 × 2 ANOVA. That is, each factor has only two levels. The first factor is the time of day the drug was administered, and the two levels are morning and evening. The second factor is drug type: The pill that is given at the beginning of the experiment contains caffeine for half of the participants who took the drug in the morning and half of the participants who took the drug in the evening, but only sugar (i.e., it is a placebo) for the other half of the participants. The dependent variable is the score each participant obtains on a test of manual dexterity. For this experiment, two different kinds of interaction are shown in Figure 15.4.

In Figure 15.4A, those who took the drug in the evening obtained higher average dexterity scores than those who took the drug in the morning in the placebo condition. But when caffeine is administered, it is those who took the drug in the morning who achieve higher scores. This "reversal" effect leads to a large amount of interaction while virtually obliterating the main effects (e.g., when you average the morning and evening participants together for each drug condition, you get about the same mean for caffeine as you do for placebo). This pattern is often referred to as a *disordinal interaction* because the effect of one factor changes its direction (order) depending on the levels of the other factor. For example, the time-of-day difference changes its direction from the placebo to the caffeine condition.

There is a fair amount of interaction in Figure 15.4B (the two lines are far from being parallel), but this time there is no reversal. Because the order of the effect for one factor does not change with different levels of the other factor (e.g., the time-of-day difference favors those that took the drug in the evening in both drug conditions), this pattern is called an *ordinal interaction*. With this pattern, it is not uncommon to find that both main effects are significant, as well as the interaction.

Notice that a large interaction can cause the main effects to be misleading. You could easily get a significant time-of-day effect for the data depicted in Figure 15.4B, but it would be due almost entirely to the time-of-day difference that is produced by the caffeine condition. Similarly, if you attained a significant drug effect for these data, it would be due mostly to the

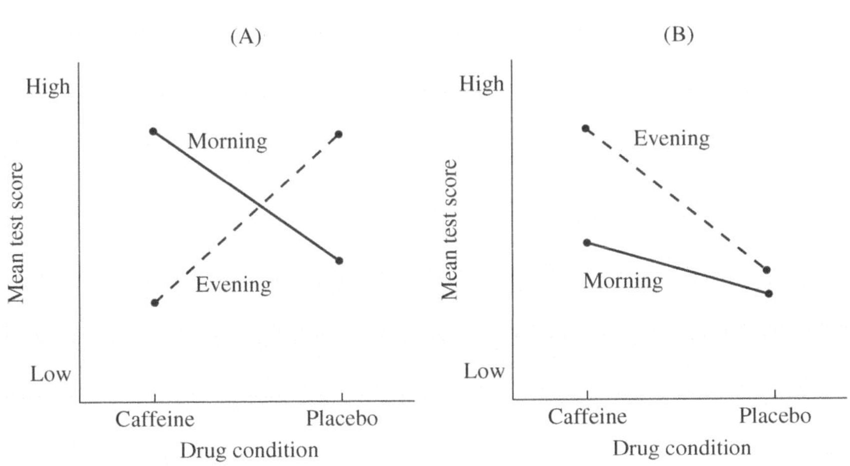


Figure 15.4

**Two Kinds of Interaction Patterns for Caffeine and Time of Day on Test Scores (2 × 2 Factorial Design)**

effect of caffeine on those who took the drug in the evening, as there is not much of a caffeine/placebo difference for those who took the drug in the morning. The main effects are even more misleading in Figure 15.4A. Reporting that both main effects were small and not significant would seem to suggest that caffeine is no different from placebo in its effect on manual dexterity and that there are no time-of-day differences. However, a quick look at Figure 15.4A shows that this is not the case. A large interaction, especially one that is statistically significant, is an indication that you should probably *not* focus your follow-up tests on the main effects (e.g., comparing pairs of row or column means), whether they are significant or not. Following up a significant interaction usually involves tests that are more specific; such comparison procedures are described in the next section.

## Following Up on a Significant Interaction

A common approach to further specifying the source of a significant interaction is to test what are called **simple main effects**. A simple main effect is the effect of one factor at only one level of the other factor. This is easier to understand in terms of a graph of cell means. For example, in Figure 15.4A, each line is a simple main effect. The line that represents those who took the drug in the morning is the main effect of drug *only for those who took it in the morning*; hence, it is a *simple* main effect. The same is true for the line representing those who took the drug in the evening. In addition, the cell means for those who were given caffeine either in the morning or the evening represent the (simple) main effect of time of day *only for the caffeine condition*. Similarly, there is a simple main effect for time-of-day differences just for the placebo condition. You could highlight the simple main effects of time of day by regraphing the cell means of Figure 15.4A so that the two times of day would be specified along the horizontal axis. Then each line would represent the simple main effect of time of day for one of the drug conditions.

### Significant Interactions in a 2 × 2 ANOVA

In the 2 × 2 case, testing the simple main effects is just like testing cell-to-cell comparisons, and you can proceed by using Fisher's protected $t$ test procedure. The cell-to-cell tests are protected if the $F$ ratio for the interaction attains significance, as would be likely for the data shown in Figure 15.4A. However, not all of the possible cell-to-cell comparisons represent simple main effects. Of the six possible pairs of cells (i.e., 4 × 3/2), two pairs make little sense to test. If you were to compare taking caffeine in the morning with taking the placebo in the evening, or vice versa, and obtain a significant difference, you could not say which of the two factors (time of day or drug versus placebo) is responsible, or whether both are. However, it would be meaningful to compare either the two times of day separately for both the placebo and caffeine conditions or the two drug conditions for each time of day. Given a balanced two-way design, you could use the LSD formula to test either pair of differences of cell means:

$$\text{LSD} = t_{\text{crit}}\sqrt{\frac{2MS_W}{n}},$$

where $n$ is the size of each cell. However, if you want to test all four of the meaningful pairwise comparisons, the more conservative approach would be to use Tukey's HSD.

In Figure 15.4A, it looks like all four of the simple main effects could easily be significant. However, the situation would likely be different for the pattern represented in Figure 15.4B. Again, the interaction could turn out to be significant. But whereas you could expect the simple main effect of drug condition to be significant for those who took the drug in the evening, the relative flatness of the line representing those who took the drug in the morning suggests that this simple main effect might fail to attain significance. Similarly, it looks like the simple main effect of time of day could easily be significant for the caffeine but not the placebo condition.

The results you can obtain from testing simple main effects provide a more nuanced and specific story concerning the effects of your two factors, a story that can have practical implications. For example, if the simple effects of drug in Figure 15.4A were significant for both times of day (albeit in different directions), the clear recommendation for workers would be that a cup of coffee or two could improve performance if taken in the morning but any source of caffeine should be avoided in the evening. However, the results in Figure 15.4B seem to suggest that there are benefits of having caffeine in the evening, whereas the results for having caffeine in the morning are less clear.

In general, the larger the interaction, the more the simple main effects differ from each other. A statistically significant interaction implies that the simple main effects differ from each other significantly. Specifying the source of a significant interaction becomes more complicated when one or both factors have more than two levels because there are more options. The most common approach is to test simple main effects as an intermediate step before proceeding to cell-to-cell comparisons, as we describe next.

### Significant Interactions Involving Multilevel Factors

For an example involving a significant interaction with a factor that has more than two levels, we will return to the music experiment, with one modification. We subtracted 2 points from each of the five participants who lacked musical training and listened to Mozart. $MS_W$ is unaffected, but the numerator $MS$s have changed, yielding the following $F$ ratios for the data shown in Table 15.4: $F_{composer}$ (3, 32) = 2.84, $p$ = .053; $F_{background}$ (1, 32) = 4.47, $p < .05$; $F_{interaction}$ (3, 32) = 3.42, $p < .05$. Although the $F$ ratio for composers is so close to being statistically significant that it would be tempting to apply pairwise comparisons to the composer (column) means, the significant two-way interaction suggests that tests of simple main effects may be more legitimate and informative. For example, to test the simple main effect of composer *just for the musicians*, you can perform a one-way ANOVA on the four-cell means for the musicians, ignoring the cell means of the novices in finding $MS_{Bet}$, and then use $MS_W$ from the original ("omnibus") two-way ANOVA as your error term. Given the equal cell sizes, we can use the following simplified formula from Chapter 13:

$$MS_{bet} = ns_{\overline{X}}^2 = 5 \times \text{variance of } (17.4, 13.2, 12.3, 12.8) = 5(5.614) = 28.07$$

Therefore, $F_{composer\ (for\ musicians)} = 28.07/4.7 = 5.97$, which substantially exceeds the critical value for $F_{.05}$ (3, 32) = 2.90. (Note that our error $df$ is $df_W$ from the two-way ANOVA because we are using $MS_W$ from the two-way ANOVA as our error term.) Because we did not change any of the cell means for the musicians, this simple main effect would have been the same, and therefore statistically significant, for the data in Table 15.1. However, the interaction was not even close to significance before we changed one of the cell means for the novices. Therefore, we would not have been justified in testing simple main effects in the original data. Our follow-up tests would not have had the "protection" of having obtained a significant interaction.

Now that we know that the choice of musical composer affects the spatial ability (at least temporarily) of musicians, we still want to clarify this result further by conducting follow-up tests for each possible pair of composers. For this purpose, an appropriate multiple

**Table 15.4 Cell Means for the Music Experiment, With One Cell Modified to Produce a Significant Interaction**

| | Classical composer | | | | |
|---|---|---|---|---|---|
| **Music background** | **Mozart** | **Chopin** | **Bach** | **Beethoven** | **Row mean** |
| Novice | 12.2 | 13.2 | 12.4 | 12.0 | 12.45 |
| Musician | 17.4 | 13.2 | 12.2 | 12.8 | 13.90 |
| $\overline{X}_{musician} - \overline{X}_{novice}$ | 5.2 | 0.0 | −0.2 | 0.8 | 1.45 |

comparison procedure is Tukey's HSD test. In fact, we can use the same formula and the same values we used earlier to test pairs of column means, but with one important change:

$$\text{HSD} = q_{\text{crit}}\sqrt{\frac{MS_W}{n}} = 3.84\sqrt{\frac{4.7}{5}} = 3.84(0.97) = 3.72$$

Note that when we were comparing column means, we used $r\,n$ (number of rows times size of each cell) as the number of subjects per group. In this latest test, we are comparing cell means, so it is appropriate to use just $n$ as the number per group. The HSD is higher than it was for the column means (smaller groups can vary more easily by chance), but the Mozart condition is nonetheless significantly higher than the other three conditions, and none of the other composers differs significantly from each other. The test of the simple main effect of composer for novices is nowhere near statistical significance ($F < 1$). So, no pairwise tests would be conducted among those cells.

Given the significant interaction in Table 15.4, there is another way that simple main effects could be tested. We would be justified in comparing the musicians to the novices for each composer. Because these simple main effects involve only two levels each, they are the same as cell-to-cell comparisons and would yield the same results whether conducted as one-way ANOVAs or $t$ tests. As the number of levels increases for each factor, so does the number of possible follow-up tests. At some point, it becomes helpful if the researcher can plan to test just some of the comparisons and use the appropriate Bonferroni correction.

## Measuring Effect Size in a Factorial ANOVA

In Chapter 13, we presented two very different formulas for finding the most common measure of effect size for a one-way ANOVA, eta squared ($\eta^2$), which is the proportion of variance in the dependent variable accounted for by the independent variable. The two different formulas, which always yield the same answer in a one-way ANOVA, are:

$$\eta^2 = \frac{SS_{\text{Bet}}}{SS_{\text{Total}}} \qquad \text{Formula 13.12A}$$

and

$$\eta^2 = \frac{df_{\text{Bet}}F}{df_{\text{Bet}}F + df_W} \qquad \text{Formula 13.12B}$$

If applied to any one of the three effects in a two-way ANOVA (i.e., the two main effects and the interaction), however, the two formulas will give different answers—sometimes *very* different answers. In order to apply Formula 13.12A to just one effect in a two-way ANOVA, the notation has to be modified slightly, as follows:

$$\eta^2_{\text{ord}} = \frac{SS_{\text{effect}}}{SS_{\text{Total}}} \qquad \text{Formula 15.4}$$

For example, if you want to measure the effect size corresponding to the composer variable, the *SS* for that effect (from Table 15.2) is 80.1; dividing that value by the $SS_T$ for the two-way ANOVA (the first *SS* we calculated in this chapter, 257.8) gives you what is called **ordinary $\eta^2$** (abbreviated as $\eta^2_{\text{ord}}$ in Formula 15.4), which equals 80.1/257.8 = .311—a large effect size, but nonetheless an underestimate, as we demonstrate next.

Unfortunately, a serious problem arises with ordinary $\eta^2$ when both factors involve experimental manipulations. Imagine that the music background factor was really a music training factor and that participants had been assigned to either intensive musical training or a control group. Because the denominator of $\eta^2_{\text{ord}}$ contains $SS_T$, which includes the between-group *SS*s for *both* factors and the interaction, the more effective the music training, the larger $SS_T$ would get (because the *SS* for music training would get larger)—and the larger $SS_T$ gets, the smaller $\eta^2_{ord}$ gets for the composer factor. (The same thing happens to the effect size of Music Training as the composer effect increases, of course.)

The measure we usually want in a two-way ANOVA, especially when both factors involve experimental manipulations (rather than individual differences that are selected for), compares the *SS* for an effect just to that same effect combined with the amount of error, and not to a denominator inflated by the inclusion of variation from the other factor (or the interaction). Therefore, this measure is called a *partial eta squared* $\eta^2_p$ and can be found by the following formula:

$$\eta^2_p = \frac{SS_{eff}}{SS_{eff} + SS_{error}} \qquad \text{Formula 15.5A}$$

Applied to the composer effect, $\eta^2_p$ equals 80.1/(80.1 + 150.4), where 150.4 is the *SS* for error from Table 15.2; therefore, $\eta^2_p = 80.1/230.5 = .348$. Note that this proportion is larger than ordinary $\eta^2$; $\eta^2_p$ can never be smaller than $\eta^2_{ord}$ and can be a good deal larger, depending on the size of the *SS*'s for the other factor and the interaction.

However, we need to point out that when the second factor is selected, such as music background (participants already had a certain degree of background at the outset of the study, and that amount was not altered by the experimenter), the ordinary $\eta^2$, when applied to the experimental (i.e., composer) factor, does make some sense. Because the experimenter did not create the variation due to different music backgrounds, and such variation is likely to exist in any future experiment, it is not necessary to remove the *SS* for music background (or its interaction with the composer factor) from the bottom of the eta-squared proportion. In terms of reporting an effect size that will be useful for the next researcher, an argument can be made for using $\eta^2_{ord}$ in such situations. Nonetheless, partial $\eta^2$ has become so popular as an effect-size measure for factorial ANOVAs that we doubt that the ordinary $\eta^2$ is often reported, even when appropriate.

Although Formula 15.4A is easy enough to use when you have all of the *SS*s from your two-way ANOVA calculations, it can be even more convenient to obtain $\eta^2_p$ directly from your *F* ratios, as was shown in Formula 13.12B. In fact, you have no choice but to use the *F* ratio and the appropriate *df*s when you are trying to measure the effect size from results that were already published and did not include any effect-size measures. Using the *F* ratios gives you partial $\eta^2$'s automatically because the denominators of the *F* ratios are based on the error term only and do not include variation from other factors. Therefore, we just need to change the notation in Formula 13.12B to create a formula that yields the partial eta squared for a factorial ANOVA:

$$\eta^2_p = \frac{df_{eff}F_{eff}}{df_{eff}F_{eff} + df_W} \qquad \text{Formula 15.5B}$$

where $df_{eff}$ and $F_{eff}$ are the *df* and *F* ratio, respectively, for any one effect in a factorial ANOVA. Use Formula 15.5B on the composer effect to see that you get the same answer as we obtained from Formula 15.5A (within rounding error).

## Summary

A two-way ANOVA is so named because *two* independent variables are being combined. The statistical analysis of a two-way factorial design makes possible a significance test (using the *F* ratio) of the effect of *each* independent variable and of the effect of the *interaction* of the two variables—that is, the *joint* effect of the two variables over and above the separate effects of each one. Interaction is an important concept for behavioral science research.

**Sums of Squares for a Balanced (i.e., Equal-$n$) Design**

1. Total sum of squares ($SS_T$)

$$\text{Computing formula}: SS_T = \sum X^2 - N_T\overline{X}_G^2 \qquad \text{Formula 13.1B}$$

where $\overline{X}_G$ is the grand mean

2. Between-groups sum of squares ($SS_{Bet}$)

$$\text{Computing formula}: SS_{Bet} = n\sum_{i=1}^{k}\overline{X}_i^2 - N_T\overline{X}_G^2 \qquad \text{Formula 15.1}$$

where
$k$ = number of groups (i.e., cells),
$n$ = number of scores in each cell,
$\overline{X}_i$ = any of the cell means

3. Within-groups sum of squares ($SS_W$)

$$\text{Shortcut computing formula}: SS_W = SS_T - SS_{Bet}$$

4. Sum of squares for Factor 1, the "column" factor ($SS_1$):

$$\text{Computing formula}: SS_1 = rn\sum_{i=1}^{c}\overline{X}_i^2 - N_T\overline{X}_G^2 \qquad \text{Formula 15.2A}$$

where
$r$ = number of rows in the data table,
$c$ = number of columns,
$\overline{X}_i$ = any of the *column* means

5. Sum of squares for Factor 2, the "row" factor ($SS_2$):

$$\text{Computing formula}: SS_2 = cn\sum_{i=1}^{r}\overline{X}_i^2 - N_T\overline{X}_G^2 \qquad \text{Formula 15.2B}$$

where
$r$, $c$, and $n$ are defined as before, and $\overline{X}_i$ can be any of the *row* means

6. Sum of squares for interaction ($SS_{1\times 2}$):

$$\text{Computing formula}: SS_{1\times 2} = SS_{Bet} - SS_1 - SS_2 \qquad \text{Formula 15.3}$$

**Degrees of Freedom**

Total degrees of freedom:

$$df_T = N_T - 1$$

where
$N_T$ = total number of observations

Degrees of freedom within groups:

$$df_W = N_T - k$$

where
$k$ = number of cells

Degrees of freedom for the column factor (Factor 1):

$$df_1 = c - 1$$

where
$c$ = number of columns

Degrees of freedom for the row factor (Factor 2):

$$df_2 = r - 1$$

where
$r$ = number of rows

Degrees of freedom for interaction:

$$df_{1\times 2} = df_1 \times df_2 = (c - 1)(r - 1)$$

Also, note that $k = r \times c$ and $N_T = r \times c \times n$.

### Mean Squares

Mean square within groups:

$$MS_W = \frac{SS_W}{df_W}$$

Mean square for Factor 1:

$$MS_1 = \frac{SS_1}{df_1}$$

Mean square for Factor 2:

$$MS_2 = \frac{SS_2}{df_2}$$

Mean square for interaction:

$$MS_{1\times 2} = \frac{SS_{1\times 2}}{df_{1\times 2}}$$

### *F* Ratios and Tests of Significance

Main effect of Factor 1:

$$F_1 = \frac{MS_1}{MS_W}$$

Main effect of Factor 2:

$$F_2 = \frac{MS_2}{MS_W}$$

Effect of interaction:

$$F_{1\times 2} = \frac{MS_{1\times 2}}{MS_W}$$

Each computed $F$ value is compared to the critical value from the $F$ table for the degrees of freedom associated with the numerator and denominator *of that test*. If the computed $F$ is less than the critical $F$, $H_0$ is retained; otherwise, $H_0$ is rejected in favor of $H_A$ (i.e., the effect is labeled as statistically significant).

### Types of Interactions

A graph of the cell means for a two-way ANOVA allows you to see at a glance whether there is very little interaction (the lines on the graph are nearly parallel to each other) or a good deal of interaction. However, even an interaction that looks large (the lines sharply converge or diverge) could fail to attain statistical significance. If the *direction* of one factor's effect changes at different levels of the other factor, the interaction is said to be *disordinal*. If the lines converge or diverge while sloping in the same direction, the interaction is an *ordinal* one. A large interaction, especially a disordinal one, can lead to misleading main effects.

### Follow-Up Tests (Multiple Comparisons) for a Two-Way ANOVA

1. **If the interaction is *not* significant:** In this case, the focus is on the main effects of each factor. If the $F$ ratio for a factor is statistically significant and that factor has more than two levels, pairwise comparisons among the column or row means may be conducted to ascertain which levels differ from which others. Fisher's protected $t$ tests are acceptable when the factor has only three levels, whereas Tukey's test (or the modified LSD test) is recommended when there are four or more levels. Remember that when column or row means are being compared, the value for $n$ in the basic LSD or HSD formula should be the number of scores in each column (i.e., $rn$) or each row (i.e., $cn$), as appropriate.
2. **If the interaction *is* significant:** In this case, the focus usually turns to the *simple main effects*—that is, the effects of one factor at *each level* of the other factor. If the $F$ ratio for a simple main effect is significant and more than two levels are involved, cell-to-cell comparisons may be conducted to further specify the effect. LSD can be used when there are only three levels, and HSD can be used for more than three levels, where $n$ in these formulas is the size of each cell.
3. It is considered acceptable to ignore the experimentwise alpha ($\alpha_{EW}$) and instead use procedures to keep the overall alpha at .05 for the *family* of comparisons that are used to follow up each significant $F$ ratio, thus keeping the familywise alpha ($\alpha_{FW}$) at .05.

### Measuring Effect Size in a Factorial ANOVA

Especially when the levels of both factors in a two-way ANOVA have been created/manipulated by the researcher, the preferred measure of effect size for each of the three effects is the **partial eta squared** $\left(\eta_p^2\right)$, given by the following formula:

$$\eta_p^2 = \frac{SS_{\text{eff}}}{SS_{\text{eff}} + SS_{\text{error}}} \qquad \text{Formula 15.5A}$$

The exact same values can be obtained conveniently from the $F$ ratios corresponding to each effect, by using the following formula, instead:

$$\eta_p^2 = \frac{df_{\text{eff}}F_{\text{eff}}}{df_{\text{eff}}F_{\text{eff}} + df_W} \qquad \text{Formula 15.5B}$$

If you are measuring the effect size of an experimental factor when the other factor is a grouping (i.e., pre-existing) factor, it may be more appropriate to use the formula for **ordinary eta squared:**

$$\eta_{\text{ord}}^2 = \frac{SS_{\text{effect}}}{SS_{\text{Total}}} \qquad \text{Formula 15.4}$$

## Exercises

1. The following table contains the statistics quiz scores for 18 students as a function of their procrastination level and exercise status.

| | Low procrastination | Moderate procrastination | High procrastination |
|---|---|---|---|
| Regular exercise | 5 | 8 | 5 |
| | 4 | 5 | 6 |
| | 4 | 7 | 9 |
| Irregular exercise | 3 | 7 | 4 |
| | 4 | 5 | 9 |
| | 2 | 6 | 7 |

   (a) Compute the two-way ANOVA for these data and present your results in the form of an ANOVA summary table (see Table 15.2).
   (b) Conduct the appropriate follow-up tests to determine which procrastination levels differ significantly from other levels. Are these follow-up tests justified by your results in part (a)? Explain.

2. An industrial psychologist wishes to determine the effects of satisfaction with pay and satisfaction with job security on overall job satisfaction. She obtains measures of each variable for a total group of 20 employees, and the results are shown in the following table. (Cell entries represent overall job satisfaction, where 7 = very satisfied and 1 = very dissatisfied.)

| | Satisfaction with pay | |
|---|---|---|
| **Satisfaction with job security** | **High** | **Low** |
| High | 7 | 3 |
| | 7 | 1 |
| | 6 | 2 |
| | 4 | 2 |
| | 6 | 2 |
| Low | 1 | 2 |
| | 2 | 1 |
| | 5 | 3 |
| | 2 | 1 |
| | 2 | 1 |

   (a) Perform a two-way ANOVA on these data. Using an alpha of .05, what can the psychologist conclude?
   (b) Compute partial eta squared for each of the main effects. Comment on the size of each effect.
   (c) Graph the cell means for these data. What type of interaction do you see: ordinal or disordinal?
   (d) Test all of the simple main effects. What specific conclusions can you draw from these tests?

3. Suppose that a 2 × 2 factorial design is conducted to determine the effects of caffeine and stress on scores on a 20-item English test. The cell means are given in the following table:

| Stress | Caffeine factor | |
|---|---|---|
| | Caffeine | Placebo |
| High Stress | $\overline{X} = 17.3$ | $\overline{X} = 12.0$ |
| Low Stress | $\overline{X} = 12.3$ | $\overline{X} = 16.4$ |

(a) Given that $n = 7$ and $MS_W = 2.0$, compute the appropriate $F$ ratios, and test each for significance at the .05 level.
(b) Graph the cell means for these data. What type of interaction do you see: ordinal or disordinal?
(c) Test all of the simple main effects. What specific conclusions can you draw from these tests?

4. For each of the following experiments, perform a two-way ANOVA and then the follow-up tests that are appropriate for your results. Use a graph of the cell means to explain the results you obtained.

(a) Experiment 1

| Factor 2 | Factor 1 | | |
|---|---|---|---|
| | 1 | 2 | 3 |
| 1 | 8<br>17<br>2 | 14<br>10<br>3 | 1<br>7<br>15 |
| 2 | 18<br>10<br>19 | 6<br>2<br>2 | 3<br>3<br>7 |
| 3 | 5<br>17<br>6 | 16<br>15<br>9 | 4<br>16<br>1 |

(b) Experiment 2

| Factor 2 | Factor 1 | | |
|---|---|---|---|
| | 1 | 2 | 3 |
| 1 | 2<br>3<br>1 | 17<br>14<br>15 | 8<br>10<br>7 |
| 2 | 4<br>2<br>3 | 18<br>16<br>17 | 6<br>5<br>9 |
| 3 | 3<br>2<br>1 | 19<br>15<br>16 | 10<br>6<br>7 |

(c) Experiment 3

| | Factor 1 | | |
|---|---|---|---|
| **Factor 2** | **1** | **2** | **3** |
| | 18 | 8 | 2 |
| 1 | 16 | 10 | 3 |
| | 17 | 7 | 1 |
| | 4 | 19 | 6 |
| 2 | 2 | 15 | 5 |
| | 3 | 16 | 9 |
| | 10 | 3 | 17 |
| 3 | 6 | 2 | 14 |
| | 7 | 1 | 15 |

## Thought Questions

1. (a) What is a factorial design? (b) What is the difference in purpose between a two-way ANOVA and a one-way ANOVA?
2. Complete the following table by stating which SSs, MSs, and $F$ tests are computed in *both* one-way ANOVA and two-way ANOVA, and what SSs, MSs, and $F$ tests are computed in a two-way ANOVA but *not* in a one-way ANOVA.

| | Computed in *both* one-way and two-way ANOVA | Computed in two-way ANOVA but *not* in one-way ANOVA |
|---|---|---|
| Sums of squares | | |
| Mean squares | | |
| $F$ tests | | |

3. A researcher conducts a two-way ANOVA. One factor is age of onset: childhood or adulthood. The other factor is drug level: whether participants receive a pill designed to treat obsessive-compulsive disorder or a placebo that looks and tastes the same but contains no medicine. Describe some of the interactions that might occur in this experiment.
4. What is the difference between a disordinal interaction and an ordinal interaction?
5. Give an example of how a large interaction in a two-way ANOVA could cause the main effects to be misleading.
6. (a) What is a simple main effect? (b) When should a researcher test simple main effects, and what procedure should the researcher use?

## Computer Exercises

1. Using college major and exercise as your independent variables, perform a two-way ANOVA on the academic interest questionnaire. Request descriptive statistics. Use the cell means to explain your results. (Advanced exercise: Redo the two-way ANOVA without the psychology majors.)

2. Redo Exercise 1, replacing the college-major factor with the grouping variable you created in Computer Exercise 3 of Chapter 13 (from the number of previous outstanding teachers).
3. Create a grouping variable based on an approximate "median split" of the procrastination scores so that one group contains the students with the 55 lowest procrastination scores, and the other has the 45 highest procrastination scores. Then perform a two-way ANOVA on postquiz heart rate, using the new procrastination grouping variable and the quiz conditions (last question easy, moderate, difficult, or impossible) as your two factors.
   (a) Request a graph of the cell means and refer to the graph to explain the results of the two-way ANOVA.
   (b) Test all of the possible pairs of levels for the quiz-condition factor.
4. Redo the two-way ANOVA in Exercise 3 using postquiz anxiety as your dependent variable.
   (a) Request a graph of the cell means, and refer to the graph to explain the results of the two-way ANOVA.
   (b) Regardless of your ANOVA results, test the simple main effect of quiz condition for each procrastination group, and test all of the possible pairwise (cell-to-cell) comparisons for each of those simple main effects.

## Bridge to SPSS

To perform a two-way ANOVA in SPSS, select **Univariate ...** from the **Analyze/General Linear Model** menu. Then move your two IVs into the *Fixed Factor(s):* space (we will not be discussing Random Factors in this text) and your DV to the *Dependent Variable:* space (univariate means there is only one DV). There are eight buttons down the right side of the **Univariate** dialog box (see Figure 15.5A), but we will be explaining only half of them. First, if you want to create a plot of the cell means (and you probably will), click on the **Plots ...** button. If one factor has more levels than the other, the graph will look neater if you move that factor to the *Horizontal Axis:* space and the factor with fewer levels to *Separate Lines*. Do not forget to click the **Add** button before you click on **Continue**.

Under **Options ...**, you will likely want to check *Descriptive statistics* and *Estimates of effect size*. Note that SPSS gives you the *partial* eta-squared values for each effect, by default, and adds them as an additional column to the ANOVA summary table. However, for a balanced (or nearly balanced) design, it is not considered important to request *Homogeneity tests*. If you have a factor with more than two levels and you know you will want to see pairwise comparisons among its levels (e.g., you expect the main effect of that factor to be significant), click on **Post Hoc ...** and, after moving over the factor(s) of interest, check the appropriate post hoc test (e.g., LSD for a three-level factor; Tukey for more than three levels). However, if the interaction of the two factors is significant, these pairwise tests may not be very informative.

The final button that is relevant to this chapter is **EM Means ...** If you expect the interaction to be significant, you will want to see simple main effects, and you can request them from this dialog box by moving the interaction term (e.g., Training*Composer) from the list on the left to the *Display Means for:* space on the right and then checking *Compare simple main effects*. Checking this box will open a drop-down menu labeled Confidence interval adjustment; the first choice is LSD, but if either of your factors has more than three levels, we recommend choosing Bonferroni (see Figure 15.5B).

The default output of the **Univariate** procedure consists of two tables: the first contains the sample sizes for each level of each IV (not shown here) and the second is an ANOVA

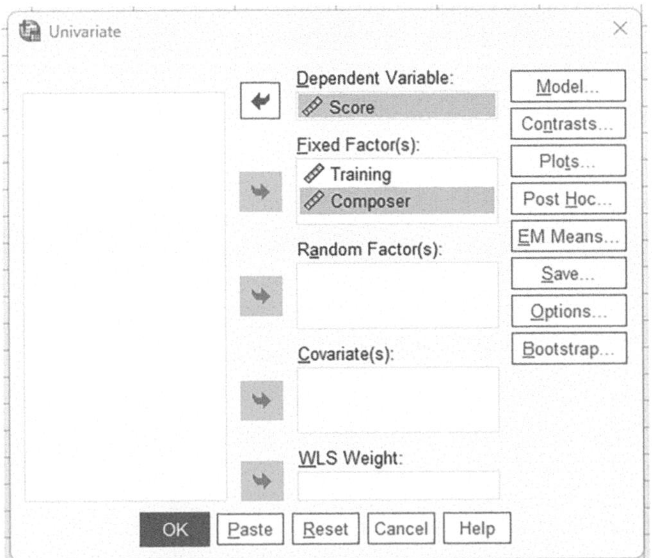


# Figure 15.5A

**The Dialog box for Univariate ANOVA**

*Source*: Courtesy of International Business Machines Corporation, © SPSS, Inc., an IBM company

summary table. However, although we used the data from Table 15.1 to create the output shown in Figure 15.6A, the summary table created by this procedure contains several more rows than the simpler one we showed you in Table 15.2, and these extra rows require some explanation. The first new row is labeled "Corrected Model." The *SS* in this row corresponds to what we have been calling $SS_{\text{Bet}}$—the between-group *SS* that you would get if you treated all the cells of the two-way design as the groups of a one-way ANOVA. (This quantity was calculated in Step 2 of the Sums of Squares calculations near the beginning of this chapter.) The corresponding *df* is 7, one less than the number of cells in this design. The *SS* in the row labeled "Intercept" is equal to $N_T \times \overline{X}_G^2$; this quantity is sometimes called the "correction factor" (CF) in computational ANOVA formulas because it is the quantity that gets subtracted in the formulas for $SS_{\text{Bet}}$ and $SS_T$. (The CF, which rounds off to 7,209.2, was the amount subtracted in Steps 1, 2, 4, and 5 of the *SS* calculations.) The test of the Intercept is really a test of the grand mean against a null hypothesis of zero, so it usually leads to a very large *F* and is rarely of any interest.

The next three rows represent the main effects of your two factors and their interaction and are clearly labeled in terms of the variable names you assigned to your factors. (The sum of the *SS*s in these three rows equals the *SS* for the "Corrected Model" in a balanced design.) The following row, labeled "Error," contains what we have been referring to as $SS_W$, $df_W$, and $MS_W$. The row labeled "Total" is the sum of the Corrected Model, Intercept, and Error, and will generally be of no interest to you (because it contains the Intercept), but the last row, "Corrected Total," contains the sum of the *SS*'s for the Corrected Model (i.e., $SS_{\text{Bet}}$) and for Error (i.e., $SS_W$) and is therefore equal to what we have been calling $SS_T$.

There is also an extra column on the right end of the summary table, which we mentioned earlier when describing that one of the choices under **Options ...** produces effect sizes. You may also notice that the heading of the SS column is *Type III Sum of Squares*. There is more than one way to calculate the sum of squares for the effects of a two-way ANOVA, but the most common way by far, and the SPSS default, is called Type III. Lastly, note that under the summary table, you will see a value for *R* Squared (.417) and Adjusted R Squared. This is the same as the statistic more commonly referred

# Figure 15.5B

**The Dialog Box for Univariate: Estimated Marginal Means**

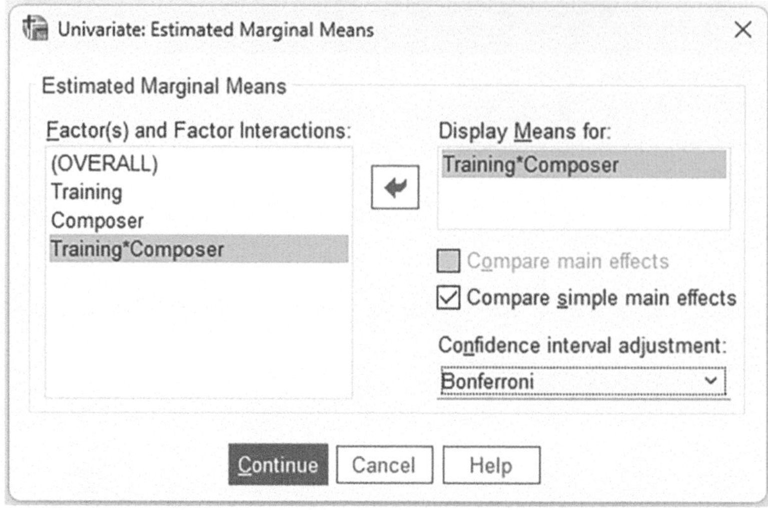


*Source*: Courtesy of International Business Machines Corporation, © SPSS, Inc., an IBM company

# Figure 15.6A

**The Summary Table for a Two-Way ANOVA**

**Tests of Between-Subjects Effects**

Dependent Variable: Score

| Source | Type III Sum of Squares | df | Mean Square | F | Sig. | Partial Eta Squared |
|---|---|---|---|---|---|---|
| Corrected Model | 107.375[a] | 7 | 15.339 | 3.264 | .010 | .417 |
| Intercept | 7209.225 | 1 | 7209 .225 | 1533.878 | <.001 | .980 |
| Training | 9.025 | 1 | 9.025 | 1.920 | .175 | .057 |
| Composer | 80.075 | 3 | 26 .692 | 5.679 | .003 | .347 |
| Training * Composer | 18.275 | 3 | 6.092 | 1.296 | .293 | .108 |
| Error | 150.400 | 32 | 4.700 | | | |
| Total | 7467.000 | 40 | | | | |
| Corrected Total | 257.775 | 39 | | | | |

a. R Squared = .417 (Adjusted R Squared = 289)

to as eta squared ($\eta^2$) in the context of one-way ANOVA. It is equal to the *SS* for the Corrected Model divided by the *SS* for the Corrected Total—in other words, it is $SS_{Bet}/SS_T$. (The adjusted *R* squared is related to the unbiased estimate of omega squared, as described in Chapter 13.) This tells you the proportion of variance accounted for by the cell means, but for a two-way ANOVA, you are much more likely to be interested in the variance proportions for each factor separately, as well as for the interaction, as given in the Partial Eta Squared column.

The EM Means selection as shown in Figure 15.5B produces an additional six output tables, but in Figure 15.6B, we will show you only the two tables of simple main effect results, both of which are labeled Univariate Tests. The first of these tests the difference in means between the two levels of musical training for each composer. It is obvious that training makes a significant difference only when Mozart's music is playing. In the second table, you can see that the composer makes a significant difference only for musicians. Finally, looking at the plot of the cell means (bottom panel of Figure 15.6B) makes it easy to see the large difference between training levels for Mozart's music and that the

**Figure 15.6B**

**Selected Additional Results from the Univariate ANOVA Procedure**

**Univariate Tests**

Dependent Variable: Score

| Composer | | Sum of Squares | df | Mean Square | F | Sig. |
|---|---|---|---|---|---|---|
| Mozart | Contrast | 25.600 | 1 | 25.600 | 5.447 | .026 |
| | Error | 150.400 | 32 | 4.700 | | |
| Chopin | Contrast | .000 | 1 | .000 | 000 | 1.000 |
| | Error | 150.400 | 32 | 4.700 | | |
| Bach | Contrast | .100 | 1 | .100 | .021 | .885 |
| | Error | 150.400 | 32 | 4.700 | | |
| Beethoven | Contrast | 1.600 | 1 | 1.600 | .340 | .564 |
| | Error | 150.400 | 32 | 4.700 | | |

Each F tests the simple effects of Training within each level combination of the other effects shown. These tests are based on the linearly independent pairwise comparisons among the estimated marginal means.

**Univariate tests**

Dependent Variable: Score

| Training | | Sum of Squares | df | Mean square | F | Sig. |
|---|---|---|---|---|---|---|
| No Training | Contrast | 14.1 50 | 3 | 4.717 | 1.004 | .404 |
| | Error | 150.400 | 32 | 4.700 | | |
| Musician | Contrast | 84.200 | 3 | 28.067 | 5.972 | .002 |
| | Error | 150.400 | 32 | 4.700 | | |

Each F tests the simple effects of Composer within each level combination of the other effects shown. These tests are based on the linearly independent pairwise comparisons among the estimated marginal means.

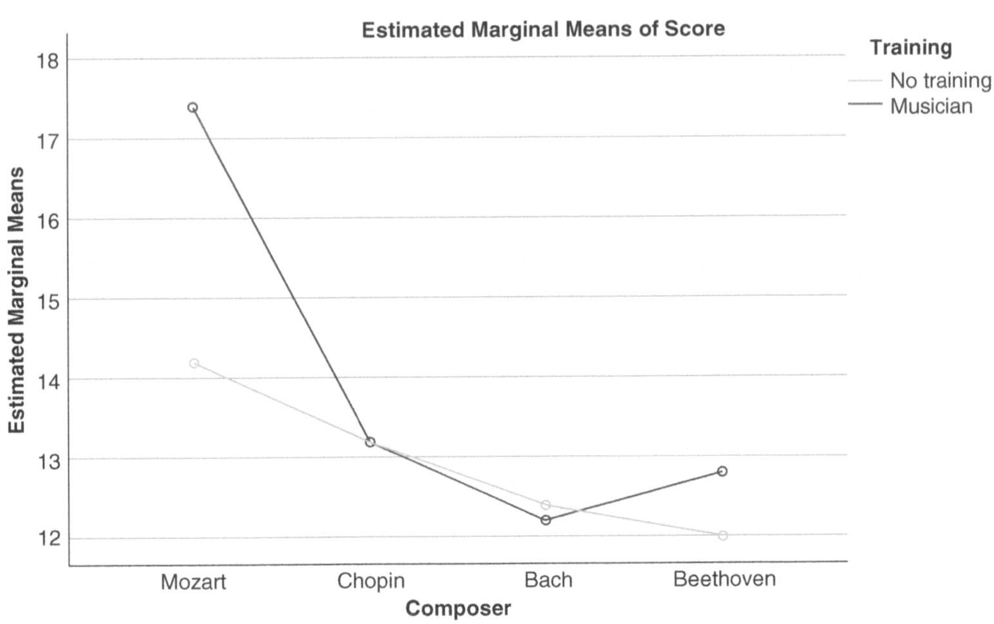


differences among the composers are more dramatic for the musicians (due mostly to Mozart). Although some tendency toward an interaction can be seen in the graph, the lines are certainly not parallel, the summary table shows us that the interaction is not even near statistical significance.

# Chapter 16
# Repeated-Measures ANOVA

## PREVIEW

### Introduction

How can the advantage of the matched-pairs $t$ test be extended to more than two conditions?

### Calculating the One-Way RM ANOVA

How can the formulas for a two-way ANOVA be used to compute a one-way repeated-measures (RM) ANOVA?

How does the critical value for a one-way RM ANOVA compare to the critical value for the corresponding independent-samples ANOVA?

### Rationale for the RM ANOVA Error Term

Why does it make sense to use the subject-by-treatment interaction as the error term for the RM ANOVA? How does this interaction compare to the variance of the difference scores when the RM ANOVA has only two levels?

### Assumptions and Other Considerations Involving the RM ANOVA

How are the usual ANOVA assumptions modified in the case of RM ANOVA? What is the sphericity assumption, and why is it important?

What precautions are recommended when conducting pairwise comparisons following a significant RM ANOVA?

How do repeated measures or matching subjects affect the power and effect size of an ANOVA?

### The RM Versus RB Design: An Introduction to the Issues of Experimental Design

Under what conditions is the RM design highly recommended, and when is it problematic?

When is it advantageous to add an individual-difference factor to an ordinary one-way ANOVA?

What are the advantages and disadvantages of the treatment-by-block and the randomized-blocks (RB) design?

### The Two-Way Mixed Design

What steps must be added to the two-way ANOVA procedure to obtain the two different error terms needed to compute a two-way mixed-design ANOVA?

What additional assumption is needed for the two-way mixed ANOVA as compared to the one-way RM ANOVA?

When a one-way RM ANOVA involves counterbalancing, why is it important to transform it into a mixed design by including order as a between-groups factor?

### Summary

### Exercises

### Thought Questions

### Computer Exercises

### Bridge to SPSS

*Introductory Statistics for the Behavioral Sciences*, Eighth Edition. R. Brooke Lea and Barry H. Cohen.

Companion website: www.wiley.com/go/ISBS8e

## Introduction

In Chapter 7, we demonstrated the advantage of within-subject designs, in which subjects are matched into pairs or subjects are measured twice as compared to giving two different treatments (or conditions) to two independent samples. Our goal in this chapter is to show how analysis of variance (ANOVA) can be used to extend the same advantage to a situation in which the independent variable (IV) has three or more levels. The procedure we describe in this chapter is commonly called the **repeated-measures (RM) ANOVA**, but it can be applied equally well to an experimental design in which subjects are matched across several conditions. The latter design is referred to as a **randomized-blocks (RB) design**; when there are only two conditions, it is identical to the matched-pairs design discussed in Chapter 7.

The trick that made it easy to compute a matched-pairs *t* test was to first calculate a difference score for each pair and then compute a one-sample *t* test to compare these difference scores to the mean for the null hypothesis (namely, zero). There is also a simple trick to calculating an RM ANOVA; it makes use of the two-way ANOVA formulas presented in the previous chapter. We discuss the calculation procedure first and then explain why it makes sense that the RM ANOVA is computed this way.

## Calculating the One-Way RM ANOVA

To demonstrate the power of the RM ANOVA, we will take a set of data for which we have already computed an independent-samples ANOVA and show how much larger the *F* ratio would be if the same data were analyzed according to an RM design. For our example, we will use the data from Table 13.1A, for which the *F* ratio was quite small ($F = 3.60/4.28 = 0.84$). These data are reprinted in Table 16.1.

Note that if each row of data contains five scores *from the same person* (each obtained while he or she listened to a different one of the five classical composers included in this experiment), it is meaningful to calculate the means for the rows, as well as the columns, in the data set. The row means are therefore subject means. To view the data in Table 16.1 as coming from a two-way factorial design, you have to view the *second* factor (composer is the first factor) as the "subject" factor. Each particular subject is automatically at a different level of the subject factor by virtue of being a different person. Let us now apply the procedure of the previous chapter to begin the computation of a two-way ANOVA for the data in Table 16.1.

The first step, as in any ANOVA in which you have all of the raw data, is to calculate the total sum of squares ($SS_T$). The same formula is used in both the one- and two-way ANOVAs, and in fact, $SS_T$ was already calculated for these data in Chapter 13:

$$SS_T = \sum X^2 - N_T\overline{X}_G^2 = 4{,}325 - 25(13.0)^2 = 4{,}325 - 4{,}225 = 100$$

**Table 16.1** **Music Experiment Data From Table 13.1A Presented as Repeated Measures**

| Subject number | Mozart | Chopin | Bach | Schubert | Beethoven | Subject $\overline{X}$ |
|---|---|---|---|---|---|---|
| 1 | 16 | 16 | 16 | 16 | 14 | 15.6 |
| 2 | 16 | 14 | 14 | 14 | 13 | 14.2 |
| 3 | 14 | 13 | 12 | 12 | 13 | 12.8 |
| 4 | 13 | 10 | 10 | 12 | 10 | 11.6 |
| 5 | 12 | 10 | 10 | 12 | 10 | 10.8 |
| Column $\overline{X}$ | 14.2 | 13.2 | 12.4 | 13.2 | 12.0 | 13.0 |

In an ordinary two-way ANOVA, the next step would be to calculate $SS_{\text{between-cells}}$. However, when dealing with a one-way RM ANOVA design, there is only one score per cell (i.e., $n = 1$), so $SS_{\text{between-cells}}$ is the same as $SS_T$ (i.e., there is no $SS_W$ component). As it turns out, this is not a problem; it simply saves us a step. Therefore, we can proceed to calculate $SS_1$ and $SS_2$ for the main effects of the two-way ANOVA. However, we can relabel these components as $SS_{\text{treatment}}$ (or $SS_{\text{treat}}$, for short) and $SS_{\text{subject}}$ (or $SS_{\text{sub}}$, for short), respectively, keeping in mind that the usual way to view the data from an RM ANOVA is with the subjects in the different rows and the treatments in the different columns. Using the formulas from the previous chapter (keeping in mind that in this application, $n = 1$) and relabeling appropriately, we obtain

$$SS_{\text{treat}} = r\sum \overline{X}_i^2 - N_T\overline{X}_G^2$$

$$= 5\left(14.2^2 + 13.2^2 + 12.4^2 + 13.2^2 + 12.0^2\right) - 4{,}225$$

$$= 5(201.64 + 174.24 + 153.76 + 174.24 + 144) - 4{,}225$$

$$= 5(847.88) - 4{,}225 = 4{,}239.4 - 4{,}225 = 14.4$$

$$SS_{\text{sub}} = c\sum \overline{X}_j^2 - N_T\overline{X}_G^2$$

$$= 5\left(15.6^2 + 14.2^2 + 12.8^2 + 11.6^2 + 10.8^2\right) - 4{,}225$$

$$= 5(860.04) - 4{,}225 = 4{,}300.2 - 4{,}225 = 75.2$$

(The index $j$ is used to represent the means for the latter summation, as a reminder that these means are different from the ones in the previous summation. The $j$ means are found across rows or subjects rather than columns or treatments.)

When applied to the one-way RM ANOVA, $r$ (the number of rows) always equals the number of different subjects (or blocks), and $c$ (the number of columns) always equals the number of levels of the IV. Also note that $SS_{\text{treat}}$ agrees with the value that was calculated in Chapter 13 (albeit by a more tedious version of the same formula) when it was called just $SS_{\text{Bet}}$. The $SS$ for the interaction, which in this case is the interaction of the subject factor with the different treatment levels, is found by subtraction, as it was in Formula 15.3. However, although this $SS$ is based on an interaction, it will be used as the basis of our error term, so we will label it $SS_{\text{error}}$ for simplicity.

$$SS_{\text{error}} = SS_T - SS_{\text{treat}} - SS_{\text{sub}} = 100 - 14.4 - 75.2 = 10.4$$

The total $df$ $(rc - 1)$ break down into three components in this design:

$$df_{\text{treat}} = c - 1;\ df_{\text{sub}} = r - 1;\ \text{and}\ df_{\text{error}} = (c - 1)(r - 1)$$

Thus,

$$MS_{\text{treat}} = \frac{SS_{\text{treat}}}{df_{\text{treat}}} = \frac{14.4}{4} = 3.6 \text{ (compare to } MS_{\text{bet}} \text{ in Chapter 13)}$$

and $$MS_{\text{error}} = \frac{SS_{\text{error}}}{df_{\text{error}}} = \frac{10.4}{16} = 0.65$$

We will not bother to calculate $MS_{\text{sub}}$ because it is not used in this design, for reasons that will be described shortly. Instead, it is $MS_{\text{error}}$ that is used as the denominator of the $F$ ratio. Therefore,

$$F_{\text{treat}} = \frac{MS_{\text{treat}}}{MS_{\text{error}}} = \frac{3.6}{0.65} = 5.54$$

### The Critical *F* for the RM ANOVA

The appropriate critical *F* at the .05 level is $F_{.05}\ (df_{\text{treat}}, df_{\text{error}}) = F_{.05}\ (4, 16) = 3.01$. In fact, our calculated *F* is significant even at the .01 level because $F_{\text{treat}} = 5.54 > F_{.01}\ (4, 16) = 4.77$. Notice how much larger the *F* for the RM ANOVA is (5.54) compared to the *F* for the independent-samples ANOVA calculated on the same data (0.84). This difference is not due to the numerators of the two *F* ratios, which are the same ($MS_{\text{treat}}$ and $MS_{\text{bet}}$ are based on the same column means). It is due to the difference in the denominators. Recall that the error term in the Chapter 13 ANOVA was based on $SS_W$, which was equal to 85.6. It is not a coincidence that the sum of $SS_{\text{error}}$ and $SS_{\text{sub}}$ in the RM ANOVA (i.e., 10.4 + 75.2) also equals 85.6. In fact, put another way, it will always be true that $SS_{\text{error}} = SS_W - SS_{\text{sub}}$. Why does it make sense to use the subject-by-treatment interaction as the error term for the RM ANOVA instead of $MS_W$, or even $MS_{\text{sub}}$? We answer this question in the section after the next one.

$SS_{\text{error}}$ is almost always smaller than $SS_W$ (it can never be larger), which works to the advantage of the RM ANOVA. It is also true that $df_{\text{error}}$ will *always* be smaller than $df_W$ (16 versus 20, for this example), which means that the critical value for the RM ANOVA will always be *larger* than the critical value for the corresponding independent-samples ANOVA. This will make it more difficult to reject $H_0$, so the loss in degrees of freedom and a corresponding increase in the critical value is clearly a disadvantage of the RM ANOVA. In practice, however, it is usually more than compensated for by the increase in the calculated *F* ratio. You may recall an analogous situation involving the matched-pairs *t* test. The connections between the RM ANOVA and the matched-pairs *t* test will become more obvious next when we apply both statistical procedures to the same data set that has only two treatment levels.

### Comparing the Matched-Pairs *t* Test With the One-Way RM ANOVA for Two Conditions

To make the comparison with the matched-pairs *t* test, let us look at only the last (i.e., rightmost) two composers in Figure 16.1. The difference scores (i.e., Schubert–Beethoven) for the five subjects are 2, 1, −1, 2, and 2. The amount of subject-by-treatment interaction for the last two composers is directly related to the variability of these difference scores. If all the difference scores were the same, the variability of the difference scores would be zero, and so would $MS_{\text{error}}$ for that pairwise comparison because the lines on the graph for the last two composers would all be parallel to each other. The mean for the Schubert–Beethoven differences is $\Sigma D/N = 6/5 = 1.2$, and the (unbiased) standard deviation (SD) is 1.304. The matched *t* value

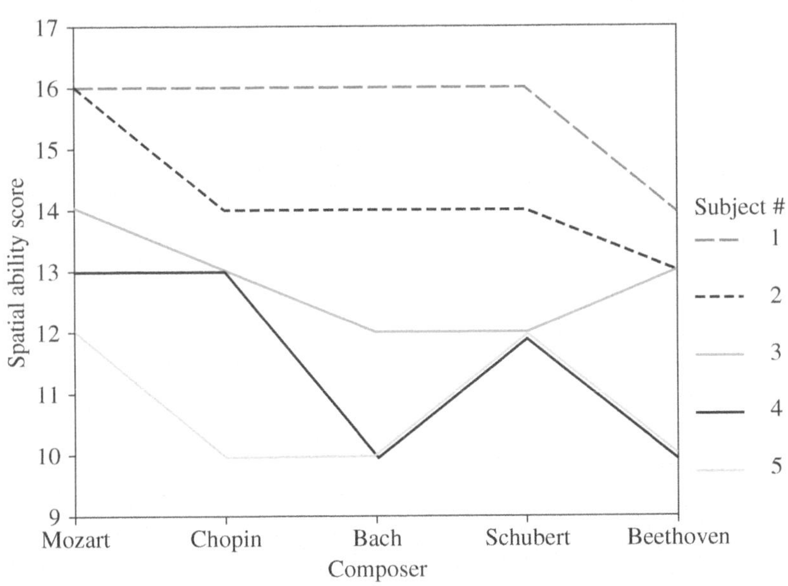


**Figure 16.1**

**Graph of Data in Table 16.1**

**Table 16.2** **Data from Table 16.1 for Schubert and Beethoven Only**

| Subject number | Schubert | Beethoven | Subject $\overline{X}$ |
|---|---|---|---|
| 1 | 16 | 14 | 15.0 |
| 2 | 14 | 13 | 13.5 |
| 3 | 12 | 13 | 12.5 |
| 4 | 12 | 10 | 11.0 |
| 5 | 12 | 10 | 11.0 |
| Column $\overline{X}$ | 13.2 | 12.0 | 12.6 |

for these two composers is therefore $1.2/\left(1.304/\sqrt{5}\right) = 1.2/0.583 = 2.06$, which is not significant with only 4 degrees of freedom. It can be instructive to compare this result to the RM ANOVA for just the last two composers.

As an exercise, the reader should verify that based on the Schubert–Beethoven data in Table 16.2:

$$SS_{\mathrm{T}} = 30.4; \; SS_{\mathrm{sub}} = 23.4; \; SS_{\mathrm{treat}} = 3.6; \; \text{and } SS_{\mathrm{error}} = 3.4$$

$$MS_{\mathrm{treat}} = \frac{3.6}{1} = 3.6; \; MS_{\mathrm{error}} = \frac{3.4}{4} = 0.85; \; F_{\mathrm{RM}} = \frac{3.6}{0.85} = 4.235$$

As in the case of independent samples, squaring the matched $t$ value yields the same value (within rounding error) as calculating the RM ANOVA on the same data: matched $t^2 = 2.06^2 = 4.24 = F_{\mathrm{RM}}$. There is a simple relationship between the variance of the difference scores for two matched sets of scores (for the data in Table 16.2, $s_D^2$ is $1.304^2 = 1.7$) and $MS_{\mathrm{error}}$ for the same data. The latter value is 0.85, which is exactly half of $s_D^2$, as will always be the case.

## Rationale for the RM ANOVA Error Term

Now let us return to the question of why it makes sense to use the subject-by-treatment interaction as the error term for the RM ANOVA. First, note that in the two-condition case, $MS_{\mathrm{error}}$ is very closely related to the variance of the difference scores, as we just demonstrated, so the rationale for using the interaction as the error term for the simplest RM ANOVA is exactly the same as the rationale for using $s_D^2$ as the basis of the error term in the matched $t$ test. Both $MS_{\mathrm{error}}$ and $s_D^2$ are measures of how closely subjects follow the same pattern over the different levels of the treatment, and therefore both can be thought of as indices of the reliability of the difference in treatment effects. However, the subject-by-treatment interaction has the advantage that it can be quantified over any number of treatments, and not just two. The amount of interaction for a multilevel study can be inspected by creating a graph of the data from an RM ANOVA, as you would for the cell means of a two-way ANOVA, except that in the RM case, the cell means are really just the individual scores. That is how we created the graph in Figure 16.1.

If all of the subjects in Figure 16.1 were to exhibit almost exactly the same profile over all of the treatment levels, the lines would be nearly parallel, and the amount of subject-by-treatment interaction for the RM ANOVA would be quite small. As $MS_{\mathrm{error}}$ decreases, the $F$ ratio for the RM ANOVA increases, appropriately reflecting the increasing reliability of the treatment effects across the different subjects. Of course, with several repeated levels, the interaction for one pair of conditions can be much larger or smaller than for another pair. The size of $MS_{\mathrm{error}}$ is an average of the different pieces of the interaction, and therefore the use of the subject-by-treatment interaction as the error term of the overall analysis requires making a homogeneity assumption, as we explain in the next section.

## Assumptions and Other Considerations Involving the RM ANOVA

The RM ANOVA makes the usual three assumptions demanded by inferential parametric statistics:

1. The dependent variable has been measured on an interval or ratio scale.
2. The dependent variable follows a normal distribution in each population.
3. The observations are all mutually independent *within each sample*.

This last assumption has had to be modified because the key element of an RM ANOVA is that the scores are *not* independent *across* samples. In fact, you are hoping that there is a fairly high positive correlation between the scores for every pair of samples.

Moreover, several statistical procedures, such as the pooled-variance $t$ test and the ordinary independent-samples ANOVA (as presented in Chapter 13), require the assumption of homogeneity of variance (HOV). The one-way RM ANOVA described in this chapter also requires an appropriate homogeneity assumption, as described next.

### The Sphericity Assumption

In the one-way independent-samples ANOVA, $MS_W$ is found by computing a weighted average of all of the sample variances. However, averaging all of the sample variances to create a single-error term leads to a straightforward statistical test only if you can assume that the variances of the populations represented by those samples are all the same. As we mentioned in Chapter 13, the HOV assumption can be tested for a one-way ANOVA, and if the HOV test is significant (indicating heterogeneity of variance) *and* the samples differ considerably in size, statisticians recommend that the ordinary one-way ANOVA procedure not be used.

Similarly, the error term in the RM ANOVA is based on an average of the variances *of the difference scores* for every different pair of levels of the RM factor. (In the case of five composers, there are 10 different pairs for which the variance of the difference scores can be calculated: 1 versus 2, 1 versus 3, 1 versus 4, 1 versus 5, 2 versus 3, etc.) The assumption that these variances of difference scores are all equal in the population is commonly referred to as the **sphericity assumption**. We could have referred to the error term of the RM ANOVA as $MS_{inter}$, based on the analogy with the two-way ANOVA, or as the $MS_{residual}$ (for reasons we will not discuss here), but the least confusing label is, undoubtedly, $MS_{error}$. The latter term will help us to avoid confusion when we discuss the two-way mixed-design ANOVA, in which there is an interaction between the two IVs, as well as an interaction between the subjects and one of the IVs.

A test of the sphericity assumption has been devised by Mauchly (1940), and is available from some statistical packages (see Bridge to SPSS at the end of this chapter). However, because a lack of population sphericity generally leads to a higher Type I error rate than the alpha used to look up the critical value, it is becoming increasingly common to adjust the degrees of freedom downward in order to increase the critical value and make the test more conservative, even without testing the sphericity assumption. The most popular *df* adjustment is the one devised by Greenhouse and Geisser (1959), but its calculation is best left to statistical software that will also give you the corresponding adjusted $p$ value.

Without statistical software, you can easily test your RM ANOVA by assuming the worst-case scenario (i.e., maximum violation of sphericity). Just use *df* of 1 for the numerator and $r - 1$ for the denominator to find the worst-case critical value. On one hand, if your calculated $F$ exceeds this conservative criterion, you can declare your results statistically significant without worrying that your Type I error rate might be higher than you think it is. On the other hand, if your RM ANOVA is not significant using the usual (unadjusted) critical $F$, you cannot be making a Type I error at all because you cannot reject the null hypothesis. (There is no adjustment that lowers your critical value to give you more power.) If your calculated $F$ falls between the unadjusted and the worst-case critical $F$ values, we advise the use of statistical software.

Returning to our example, in which the calculated $F$ (5.54) was greater than the ordinary critical $F$ (3.01), the worst-case critical $F$ would be $F_{.05}(1, 4) = 7.71$. Because the calculated $F$ does not exceed the maximally adjusted critical $F$, you would need to obtain a more exact adjustment of your critical $F$ from statistical software before you could decide whether to reject the null hypothesis.

### Follow-Up Tests

Due to concerns about sphericity, some statisticians favor abandoning the RM ANOVA entirely in favor of a multivariate approach, but that topic is well beyond the scope of an introductory text. However, even those who want to retain the RM ANOVA approach in general may be concerned about the effects of a lack of sphericity on the multiple comparisons that follow a significant ANOVA to locate the significant results more specifically.

In our example involving the music of five different composers, the overall RM ANOVA was statistically significant. To determine which pairs of composers differ significantly, we could compute Tukey's Honestly Significant Difference (HSD), substituting $MS_{error}$ for $MS_W$ in the formula given in Chapter 14. However, suppose that the subject-by-treatment interaction involving just Mozart and Chopin is considerably larger than it is for any other pair of composers. Using Tukey's HSD essentially compares those two composers with an error term that has been reduced by averaging in the much smaller interactions involving other pairs of composers.

Unless it is clear that the variability of the difference scores is quite similar for every pair of conditions in your RM ANOVA, the recommended (cautious) approach to follow-up tests is to use an error term based only on the conditions being compared. When conducting pairwise comparisons, this amounts to performing separate matched-pairs $t$ tests for each pair of levels of your RM factor. When dealing with five RM levels, as in our music example, there are a total of 10 matched $t$ tests to conduct. Because of the danger of an increased experimentwise alpha when conducting so many individual follow-up tests, it is often recommended that the researcher use a Bonferroni adjustment to determine the alpha for each comparison (see Chapter 14). In the music example, this would mean testing each pair of composers at the .005 level (i.e., .05/10).

### Power and Effect Size

In Chapter 9, we showed how the power of a matched-pairs $t$ test increases as the correlation between the two sets of scores increases. Similarly, the RM ANOVA gains power as the correlation between each pair of conditions increases. (If the correlations are not similar from one pair of factor levels to another, this is an indication that the RM ANOVA may not be appropriate for that data set.) As the correlations increase, the subject-by-treatment interaction—the error term of the RM ANOVA—decreases, so the $F$ ratio tends to get larger, thus increasing power. Thus, when applicable, repeated measures or matching participants can serve as an economical alternative to increasing your sample size in order to increase power. Generally, the highest correlations (i.e., the best matching) are obtained when measuring the same participant at all levels of the RM factor, but this experimental approach can have its drawbacks, which we will point out in the next section.

The effect size in your RM ANOVA data can be measured as eta squared by applying Formula 13.11B to your obtained $F$ ratio. This is what is typically done, but note that measuring $\eta^2$ in this way can be misleading. A large $F$ ratio that arises from modest sample sizes will be associated with a large proportion of variance accounted for, but bear in mind that highly consistent responses across participants can enlarge an $F$ ratio just as easily as a larger separation in sample means. Therefore, a large $\eta^2$ does not tell you whether your effect is large in a practical sense (i.e., the different RM levels produce markedly different average magnitudes of subject responses) or merely consistent (i.e., the levels do not make a large difference on average, but each subject exhibits, almost exactly, the same small changes). Moreover, when predicting the effect size in a future, similar experiment, you have to take into consideration whether the same degree of matching can be expected in a replication. It is not difficult to adjust your $\eta^2$ measure to indicate the effect size you would have without the benefit of matching (see B. H. Cohen, 2008), but this is not commonly done, so we will not add confusion to this matter by showing you how to do that.

## The RM Versus RB Design: An Introduction to the Issues of Experimental Design

You may recall from Chapter 7 that the same matched-pairs *t* test formula is used whether the two measurements come from the same person or two participants who were matched together. The matched-subjects version of the RM design is referred to as the *RB* design, for reasons we will soon make clear.

### Problems With the RM Design

As we just pointed out, the power of the RM ANOVA increases as the correlations between pairs of treatments increase. Generally, the highest correlations (i.e., the best matching) are obtained when measuring the same subject under all of the conditions. Although an RM design works well when the different conditions can be randomly mixed together (e.g., when five different types of words are mixed together in one long list to be memorized), this design can be problematic when the conditions must be presented successively.

For example, the only reasonable way to conduct the music experiment in terms of repeated measures would be to test a particular subject's spatial ability only once for each composer. However, if the same order of composers were used for each subject, you would not be able to separate practice and/or fatigue effects (i.e., simple order effects) from the specific effects of each composer's music (e.g., think about the difference in spatial ability measured for the composer who is always presented first, as compared to the one who is always presented last). A possible solution is counterbalancing, as described near the end of Chapter 7. But the problems inherent in this approach increase rapidly with the number of conditions.

In the case of five conditions, **complete counterbalancing** is not feasible. It would require 120 subjects ($5 \times 4 \times 3 \times 2 \times 1 = 120$) just to have one subject assigned to each possible order. Fortunately, a form of counterbalancing can be achieved using the same number of orders as there are conditions, according to a design called the **Latin square**. However, any form of counterbalancing leaves your experimental design susceptible to complex carryover effects—that is, the effect of a particular composer could depend on which composer had been presented just before. Carryover effects can sometimes be avoided simply by allowing enough time or by inserting a distracting task between each pair of conditions. When serious carryover effects cannot be avoided (as described in Chapter 7), a design that involves the matching of participants on some relevant variable should be considered, as described next.

### Advantages and Disadvantages of Alternatives to the RM Design

When appropriate, the RM design is the most economical way to increase the power of an ANOVA test. It increases power by essentially increasing the effect size you are dealing with, which avoids the expense of increasing the size of the samples. Theoretically, the effect size of an ANOVA could be increased by producing greater differences among the factor levels (as by increasing the strength of your experimental manipulations). However, this approach is often not feasible or desirable (or even ethical, if a very strong manipulation could become harmful). The alternative is to increase power by reducing variability (i.e., the denominator of any standardized measure of effect size), and the RM ANOVA does this by focusing on the *consistency* of subjects' responses (as represented by $SS_{error}$), while ignoring the subjects' overall differences on the dependent variable (as represented by $SS_{sub}$).

When the RM design is not feasible, there are alternative (albeit less effective) ways to reduce the error term of an ANOVA (and therefore increase power). We discuss three of these designs in this section.

### Adding a Grouping Factor to a One-Way Independent-Samples *ANOVA*

Suppose that you are studying three methods for increasing vocabulary size in children by assigning a different method to each of three separate groups. After six months of applying these methods to your samples, your results could come out as depicted in Figure 16.2A. Notice that the subject-to-subject variability is large compared to any average differences among the groups. You then realize that your groups are mixed with respect to gender and that vocabulary size may differ by gender, especially for the age group you are studying. By adding gender as a second between-groups factor, thus creating a two-way ANOVA, it is proper to calculate $MS_W$ separately for each *cell* of the two-way design.

Looking at Figure 16.2B, you can see that the average of the six within-*cell* $MS_W$s is much smaller than the average of the three within-*group* $MS_W$'s ignoring gender. This greatly reduces the denominator of the *F* ratio you will use to test the effect of the different methods. (The numerator of that *F* ratio would *not* change, because the means of the three methods

## Figure 16.2

**Comparison of Error Variance Without and With a Grouping Factor**

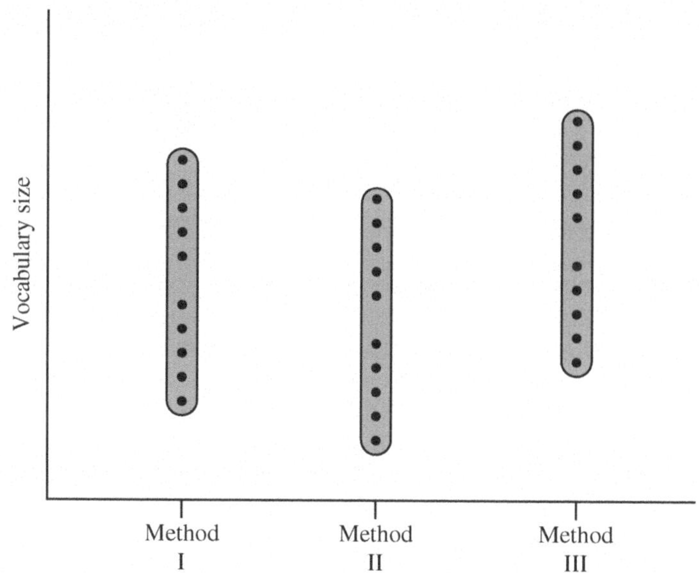


(A) Large within-group variability due partly to gender differences in Each method group

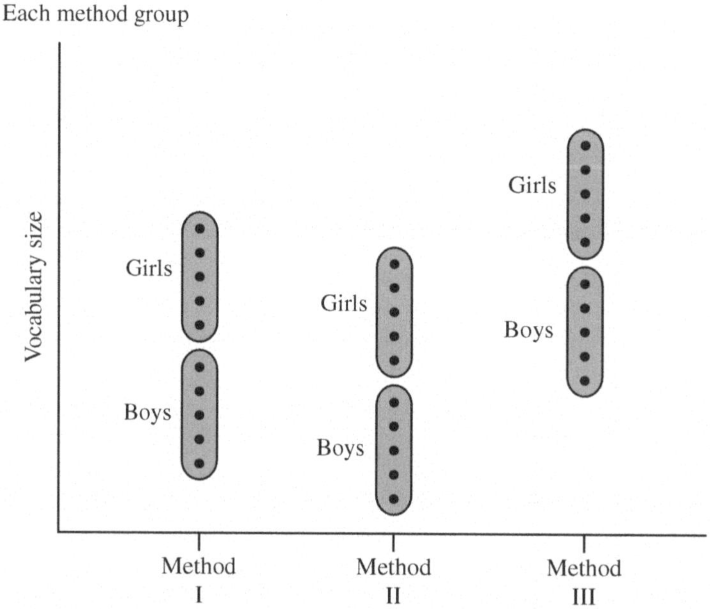


(B) Reduced within-group variability as a result of adding gender as a factor

would remain the same.) Adding a grouping factor based on pre-existing individual differences (e.g., gender) allows you to convert *within-group* variability to *between-cell* variability.

Of course, we created an unusually dramatic example to illustrate this point. Adding a grouping factor to your ANOVA is not likely to have such a large effect. But as long as you can identify a pre-existing individual-difference factor that is relevant to your dependent variable, it can be worth adding it to your ANOVA design. Moreover, you gain the opportunity to test whether the grouping variable interacts with your treatment factor (e.g., are the differences among the methods the same for both genders?).

### *The Treatment-by-Block Design*

In the preceding example, the added factor (i.e., gender) was a categorical one. However, it is also quite possible to find an individual difference factor that is related to your dependent variable and is at least crudely quantitative. In the case of children at a particular school, it should not be difficult to use existing information to classify the students into several levels of scholastic ability that can be expected to translate, on average, into differences in vocabulary size.

For convenience, you will want to make sure that the number of students included in each scholastic group (called a *block* in this context) is some multiple of the number of levels of your treatment factor. For example, if you are comparing three methods of instruction, you will want the number of children in each scholastic block to be some multiple of 3. Then, separately for each block, you would be able to assign, at random, exactly one-third of the students to each of the three methods.

The data can be collected and then analyzed as they would be for any two-way ANOVA. Of course, the main effect of the blocking factor will not yield any new information because you chose that factor for its obvious relationship to your dependent variable. However, the test of the main effect of the method factor should involve a larger $F$ ratio due to the reduction in the error term (as occurred when adding gender in the previous example), and the interaction could show whether the relative superiority of one or another method depends on the initial scholastic level of the student.

### *The Randomized-Blocks (RB) Design*

The most precise—and potentially the most powerful—form of the treatment-by-block design is the one in which the size of the block is equal to the number of different treatment conditions. If testing three vocabulary-enhancing methods, students would be matched into groups of three so that, within each block, the three students would be as similar as possible on whichever relevant criteria were available. The term *randomized- blocks design* usually refers to this special case of the treatment-by-block design, in which there is only one subject in each block/treatment combination (i.e., cell) and therefore no way to calculate $MS_W$. The data are analyzed as though the different subjects in a particular block were really different measurements of the same subject. The data can be arranged as in Table 16.1 (with block number replacing subject number) and then analyzed using the RM ANOVA procedure described earlier in this chapter. When the treatment factor has only two levels, the RB design becomes the matched-pairs design discussed in Chapter 7.

The responses of the subjects in the same block are generally not as similar as would be the responses of the same subject, and the RB design lacks the economy of measuring the same subject multiple times. But the RB design has the desirable feature of usually increasing power, while completely eliminating the possibility of carryover effects as well as the need for counterbalancing. Unfortunately, the RB design is difficult to arrange if your subjects are not all known at the beginning of the study. For instance, if patients are being randomly assigned to one of three treatment methods as they arrive at a clinic seeking help (and meet the screening criteria of the study), they can be classified as belonging to one of several crudely defined blocks (in terms of gender, age, duration, intensity of symptoms, etc.), but they cannot be matched closely to patients who have not yet sought help at that clinic. When precise matching over your entire subject pool is not possible, it may still be helpful to create blocks along the way, in preparation for a treatment-by-block ANOVA.

## The Two-Way Mixed Design

Now that you have learned how to conduct a one-way RM ANOVA as well as a two-way ANOVA, you have the tools you will need to analyze one of the most useful experimental designs in behavioral research: the two-way mixed design. There are a number of common situations that lead to this design. For example, subjects are assigned to one of two therapy conditions or to a control group, and they are then measured at the beginning of the experiment, at the midpoint, and again at the end. Or subjects are selected for being either high or low in visual imagery ability, and each is asked to memorize several lists of words that differ in how concrete the words are. Or subjects must read and respond to some opinions that they are told have been written by students and others that ostensibly come from professors. For half of these subjects, the person conducting their experimental session is obviously a student, whereas for the other half, the experimenter is clearly a professor. Each of the three two-way factorial designs just described is best designed, and then analyzed, as a mixed-design ANOVA, in which one IV is a between-groups factor and the other is a *within-subjects* (i.e., RM) factor.

### Calculating the Mixed-Design ANOVA

Just as we illustrated the calculation of the one-way RM ANOVA by taking a previous example of a one-way ANOVA and recasting it in terms of repeated measures, we will create a mixed design by taking the two-way ANOVA example from the previous chapter and imagining that the participants were matched across the four musical composers. (The analysis is the same whether the participants had been matched in blocks of four and then randomly assigned to the different composers or each participant was measured four times, once for each composer.) The matching makes it meaningful to add subject means to Table 15.1, as shown in Table 16.3. Recall that the first five subjects have no musical training, whereas the second five are professional musicians. (In case you are wondering why Schubert does not appear in Table 16.3, recall that we dropped him when we introduced two-way ANOVA, in order to simplify the analysis. Sorry, Schubert fans.)

When computing a two-way mixed ANOVA by hand, there are several reasonable ways to proceed. One instructive approach is to first calculate the numerators of the three $F$ ratios, which we already did in the previous chapter. (Matching does not affect any of the numerators.) As a reminder, we display a table of the cell means for this example (see Table 16.4), and briefly review the calculations of the numerator *MS*s.

$$SS_{\text{between-cell}} = 107.4;\ SS_{\text{composer}} = 80.1;\ SS_{\text{background}} = 9.0;$$

$$SS_{\text{inter}} = 107.4 - 80.1 - 9.0 = 18.3;$$

$$MS_{\text{composer}} = \frac{80.1}{3} = 26.7;\ MS_{\text{background}} = \frac{9.0}{1} = 9.0;\ MS_{\text{inter}} = \frac{18.3}{3} = 6.1$$

**Table 16.3 Music Experiment Data from Table 15.1 Presented as Repeated Measures**

| Subject number | Mozart | Chopin | Bach | Beethoven | Subject $\overline{X}$ |
|---|---|---|---|---|---|
| 1 | 16 | 16 | 16 | 14 | 15.5 |
| 2 | 16 | 14 | 14 | 13 | 14.25 |
| 3 | 14 | 13 | 12 | 13 | 13.0 |
| 4 | 13 | 13 | 10 | 10 | 11.5 |
| 5 | 12 | 10 | 10 | 10 | 10.5 |
| 6 | 16 | 17 | 14 | 16 | 15.75 |
| 7 | 18 | 10 | 14 | 10 | 13.0 |
| 8 | 20 | 13 | 11 | 13 | 14.25 |
| 9 | 15 | 12 | 10 | 14 | 12.75 |
| 10 | 18 | 18 | 12 | 11 | 13.75 |

## Table 16.4 Cell Means by Composer and Subject Group for the Data in Table 16.3

| Music background | Mozart | Chopin | Bach | Beethoven | Row means |
|---|---|---|---|---|---|
| No training | 14.2 | 13.2 | 12.4 | 12.0 | 12.95 |
| Professional | 17.4 | 13.2 | 12.2 | 12.8 | 13.90 |
| Column means | 15.8 | 13.2 | 12.3 | 12.4 | 13.425 |

It is when we calculate the denominators of our $F$ ratios that we must account for the fact that the participants are now matched across composers.

In the mixed-design ANOVA, the $F$ ratio for the main effect of the between-groups factor can be calculated separately from the rest of the design. You can simply perform an ordinary one-way ANOVA on the subject means in Table 16.3, completely ignoring the particular scores on which these means are based, except that you need to multiply your $SS$ components by the number of scores that contribute to each subject mean (in this example, 4):

$$\begin{aligned} SS_T &= \sum X^2 - N_T\overline{X}_G^2 \\ &= \left(15.5^2 + 14.25^2 + \ldots + 12.75^2 + 13.75^2\right) - 10(13.425)^2 \\ &= 1{,}826.56 - 1{,}802.31 = 24.25. \end{aligned}$$

After multiplying by 4,

$$SS_T = 4 \times 24.25 = 97$$

$SS_{bet}$ for this one-way between-groups ANOVA is based on the means of the two groups of subjects (the row means in Table 16.4). It was already calculated as $SS_{background}$, which is equal to 9.0 (because, to save effort, we are taking $SS_{background}$ from the original two-way ANOVA, it should *not* be multiplied by 4).

Then

$$SS_W = SS_T - SS_{bet} = 97 - 9 = 88$$

$$MS_W = \frac{SS_W}{df_W} = \frac{88}{(10-2)} = \frac{88}{8} = 11.0$$

and

$$MS_{background} = \frac{SS_{background}}{1} = \frac{9}{1} = 9.0$$

Therefore,

$$F_{background} = \frac{9.0}{11.0} = 0.82$$

Given that $F$ for the between-groups factor is less than 1, it cannot be statistically significant. Note that, although we can refer to $SS_{background}$ and $SS_{composer}$ in this example, more generally, we will refer to the $SS$ for the effect of the between-groups factor as $SS_{groups}$ and the $SS$ for the RM factor as $SS_{RM}$.

Before we complete the analysis of our example of a two-way mixed-design ANOVA, we think it is important, as with any complex design, to show how the total degrees of freedom are partitioned into the different components of the analysis. To reduce the potential for confusion, we will use symbols for this design in a way that is as consistent as possible with the use of those symbols in previous chapters, as well as the earlier part of this chapter. First, we will use $k$ to represent the number of different levels of the between-groups factor because that is consistent with our use of $k$ to represent the number of different groups in a one-way ANOVA. Next, we will use the letter $r$ to represent the number of subjects (or blocks) in each group. This only makes sense for a balanced design, in which all of the groups have the same number of subjects, but that is the only type of mixed design we will consider in this text. Therefore, the total number of subjects (or blocks) in a balanced mixed design can be expressed as $kr$. However, it is critical to note that this is not the total number of observations or scores in the analysis because each subject contributes a score for each level of the RM (also called *within-subjects*) factor. We will continue to use the letter $c$ (think columns or conditions) to represent the number of levels of the RM factor, so the total number of observations, $N_T$, is equal to $krc$. For our example in Table 16.3, you can see that the total number of scores is $(2)(5)(4) = 40$. The total $df$ is just one less than $N_T$, which leads to the following formula:

$$df_{\text{total}} = krc - 1 \qquad \text{Formula 16.1}$$

Given the symbols we just defined, the following formulas for $df$ components should be obvious:

$$df_{\text{groups}} = k - 1 \qquad \text{Formula 16.2}$$

$$df_{\text{RM}} = c - 1 \qquad \text{Formula 16.3}$$

$$df_{\text{inter}} = df_{\text{groups}} \times df_{\text{RM}} = (k - 1)(c - 1) \qquad \text{Formula 16.4}$$

These are the degrees of freedom for the numerators of the three $F$ ratios in a two-way mixed design. Note that $df_{\text{groups}}$ could be called $df_{\text{between-groups}}$, but we prefer the shorter version. For the present example, there are two between-groups levels (no training, professional training), so $k = 2$, five subjects per group, so $r = 5$, and four repeated (or matched) conditions, so $c = 4$. Therefore, $df_{\text{total}} = 39$, $df_{\text{groups}} = 1$, $df_{\text{RM}} = 3$, and $df_{\text{inter}} = 3$.

The degrees of freedom for the denominators (i.e., error terms) of the three $F$ ratios are not so obvious but should make sense after we have described how to calculate the $SS$s for the two error terms. We have already described the calculation of $SS_W$ in terms of computing a one-way ANOVA on the subject means, so we just need to change the notation a bit to incorporate this part of the analysis into the larger picture. To be consistent with the one-way RM ANOVA, the $SS$ based on the subject means (e.g., there are 10 of them in Table 16.3) should be called $SS_{\text{sub}}$, rather than $SS_{\text{total}}$. The $SS$ based on the group means (there are two, in the present example) can be called $SS_{\text{groups}}$, in which case, $SS_W = SS_{\text{sub}} - SS_{\text{groups}}$.

All that remains is to calculate the error term for the within-subjects part of the ANOVA. Both the main effect of the RM factor and the interaction of the two factors use the same error term, which is based on the amount of subject-by (RM)-treatment interaction in each group. One straightforward way to obtain this error term is to subtract the $SS_W$ just calculated for the between-groups factor from the $SS_{\text{within-cell}}$ from the original (i.e., unmatched) two-way ANOVA. $SS_{\text{within-cell}}$, which was just labeled $SS_W$ in the previous chapter (when there was only one error term to talk about), was found to be 150.4. As in the one-way RM ANOVA, we will label the $SS$ for the within-subject error term simply as $SS_{\text{error}}$. So $SS_{\text{error}} = SS_{\text{within-cell}} - SS_W = 150.4 - 88 = 62.4$. The degrees of freedom can be subtracted in the same way: $df_{\text{error}} = df_{\text{within-cell}} - df_W = 32 - 8 = 24$ ($df_{\text{within-cell}}$ is the same as $df_W$ in the previous chapter). Therefore, $MS_{\text{error}} = 62.4/24 = 2.6$.

Let us look at how we can determine the degrees of freedom for the two error terms by formulas, using the symbols we have just defined for the numerator $df$s. First, $df_{\text{within-cell}}$ is defined as in any two-way ANOVA, as the total number of observations minus the number of cells. Using our current notation:

$$df_{\text{within-cell}} = krc - kc = kc(r - 1)$$

Then, $df_W$, for the between-groups error term, is defined *not* in terms of total observations but rather in terms of the total number of different subjects (or blocks), which is $kr$. One $df$ is lost for each group, so:

$$df_W = kr - k = k(r - 1) \qquad \text{Formula 16.5}$$

Finally, because $df_{\text{error}}$ equals $df_{\text{within-cell}}$ minus $df_W$, $df_{\text{error}}$ equals $kc\ (r - 1) - k\ (r - 1)$, which equals $(kc - k)(r - 1)$. This leads to our last $df$ formula for this design:

$$df_{\text{error}} = k(c - 1)(r - 1) \qquad \text{Formula 16.6}$$

For the present example, $df_W = 2(5 - 1) = 2 \times 4 = 8$, and $df_{\text{error}} = 2(4 - 1)(5 - 1) = 2 \times 3 \times 4 = 24$.

Finally, we can find the $F$ ratios for the RM factor (composer) and the interaction of the two factors from the previously computed numerator $MS$s and the just-calculated error term:

$$F_{\text{composer}} = \frac{MS_{\text{composer}}}{MS_{\text{error}}} = \frac{26.7}{2.6} = 10.3$$

$$F_{\text{inter}} = \frac{MS_{\text{inter}}}{MS_{\text{error}}} = \frac{6.1}{2.6} = 2.35$$

The critical value for both of these $F$ ratios is $F_{.05}\ (3, 24) = 3.01$. Therefore, the statistical conclusions from Chapter 15 have not changed: The main effect of the composer is significant at the .05 level, and the interaction effect is not. However, it is important to note that the $F$s for these two effects have nearly doubled because the error term based on the matching of subjects ($MS_{\text{error}}$) is only a bit more than half as large as the original two-way ANOVA error term (2.6 versus 4.7). This will often be the case when subjects are matched or repeatedly measured.

The results of the mixed-design ANOVA just computed can be neatly summarized in a table much like the one for an independent-groups two-way ANOVA (see Table 15.2), except that now there are two error terms included (see Table 16.5). Note that the total $SS$ and total $df$ in Table 16.5 are the same as in Table 15.2, because we were working here with the same 40 data points as were contained in Table 15.1.

### Assumptions

In addition to all of the assumptions of the RM ANOVA, including sphericity, the mixed-design ANOVA also requires you to assume that the amount of interaction (in the population) between any two treatment levels is the same for each subgroup (i.e., level of the between-groups factor) in the design. This is called **multi-sample sphericity**. For example, when there are only two repeated measurements (e.g., before and after some intervention) and two different groups (e.g., drug and placebo), the usual sphericity assumption does not apply (there is only one pair of treatment levels), but the 2 × 2 mixed-design ANOVA assumes that the amount of before–after interaction is the same for the placebo subjects as it is for those given the real drug. There is a test for multi-sample sphericity, but it is well beyond the scope of an introductory text.

Just as in the one-way RM ANOVA, you can retest the $F$ ratio for the RM factor of a mixed-design ANOVA against a worst-case critical $F$: $F[1, k(r - 1)]$. For this example, the worst-case $F_{\text{crit}}$ is $F\ (1, 8)$, which equals 5.32, for an .05 test. Because $F_{\text{composer}}$ for this example is 10.3, which is larger than 5.32, you can reject the null hypothesis at the .05 level

**Table 16.5** **Summary of Two-Way Mixed-Design ANOVA of Music Experiment**

| Source | SS | df | MS | F |
|---|---|---|---|---|
| Music background | 9.0 | 1 | 9.0 | 0.82 |
| Error (within-group) | 88.0 | 8 | 11.0 | |
| Composer | 80.1 | 3 | 26.7 | 10.3 |
| Composer × background | 18.3 | 3 | 6.1 | 2.35 |
| Error (subject × composer) | 62.4 | 24 | 2.6 | |
| Total | 257.8 | 39 | | |

without having to worry about whether your data meet the sphericity assumption. The $F$ ratio for the interaction of the two factors also requires the sphericity assumption because it uses the same error term as $F_{RM}$; the worst-case $F_{crit}$ for the interaction is $F[(k - 1), k(r - 1)]$, which in this example is again $F(1, 8)$ because $k$ is only 2. However, in this example, $F_{inter}$ was not significant even when assuming sphericity (i.e., using the ordinary *dfs*), so there's no need to retest it. It will fail to be significant regardless of what is assumed about sphericity.

### Follow-Up Tests

If the interaction in a mixed design fails to approach significance, you can turn your attention to the main effects. Post hoc comparisons for a significant between-groups factor can use the $MS_W$ error term from the original mixed-design ANOVA unless there is a strong reason to be concerned about HOV among the different groups. (This is one reason it is a good idea to have the same number of participants or blocks at each level of the between-groups factor.) Following up a significant RM main effect evokes the same concern as expressed previously for the one-way design. Unless you can be quite confident in assuming sphericity in the population, it is recommended that each pair of levels of the RM factor be compared with an error term based on only those two levels.

If the interaction in a mixed design *is* significant, the follow-up procedures are the same as for an independent-samples two-way ANOVA with a significant interaction (as described in the previous chapter), except for the error terms. The simplest and most conservative approach is not to use either error term from the original mixed-design ANOVA but rather to base the error terms for your follow-up tests on only the data involved in those tests. For example, if, based on a significant interaction, you were to test the simple main effect of composer just for the group with no musical training, the recommendation is to use for your error term the subject-by-treatment interaction for only the untrained participants rather than $MS_{error}$ from the mixed-design ANOVA. Similarly, if comparing musicians to untrained participants only for the Mozart condition, base $MS_W$ on the Mozart data only, instead of using $MS_W$ from the original design.

### The Before–After Case

A very common form of mixed-design ANOVA is one in which the RM factor is time, and it has only two levels (e.g., the dependent variable is measured at the beginning and end of some experimental treatment). Perhaps half the participants receive some new form of therapy, while the other half of them receive sham therapy. In such a case, there would be little interest in the *main* effect of time because this effect would involve averaging the real with the sham therapy participants. Similarly, the main effect of the group would not be of interest because it would involve an average of the before scores (when no group difference is expected) with the after scores. The chief interest in this design is the group-by-time interaction. It is likely that you would want the before–after difference to be much larger in magnitude for the real, as compared to the sham, therapy recipients.

You do not need to conduct a mixed-design ANOVA to assess the amount of interaction in the design just described. If you calculate the before–after difference for each participant and then compute the ordinary (independent-samples) one-way ANOVA on these difference scores, the $F$ ratio you obtain will be exactly the same as the $F$ ratio for the interaction effect of the mixed-design ANOVA. When there are only two groups in the study, you also have the option of conducting an independent-samples $t$ test on the difference scores. Of course, the $p$ value for the $t$ test will be the same as for the $F$ ratio from the corresponding ANOVA. If there are more than two levels of the between-groups factor, you can still compute a one-way ANOVA on the difference scores, and this $F$ ratio will be the same as the $F$ ratio for the interaction in the corresponding mixed-design ANOVA. However, if you have a between-groups factor and there are more than two time points included (e.g., before, after, and follow-up), there will be more than one set of difference scores, so, in that case, you will need the mixed-design ANOVA.

### The One-Way RM ANOVA With Counterbalancing

The cautions described in this subsection apply to mixed designs in which the RM factor literally involves repeated measures, and not just the matching of subjects into blocks. By now, it should be clear that if you are going to present the same participants with two or more conditions, one after another, you should not present these different conditions in the same order to all participants. What is less obvious is that even though simple order (e.g., practice) effects will be eliminated from the numerator of your *F* ratio by counterbalancing, order effects will increase the size of your denominator (i.e., error term) unless you add order as a between-groups factor to your analysis. Consider the simplest case: Participants perform the same task under two different distraction conditions. If the performance of most participants is enhanced a bit by practice on whichever condition they receive second, this will cause participants receiving one order of the conditions to differ somewhat in the relative effects of the two conditions from participants receiving the conditions in the reverse order.

For instance, if the individual participants are graphed as in Figure 16.1, the lines representing participants receiving one of the orders will tend to slant a bit differently between the two conditions when compared to the lines of participants receiving the other order. Thus, a simple order effect would cause the lines to be less parallel. This translates to a larger amount of subject-by-condition interaction, which forms the basis of the error term in an RM ANOVA. The solution to this problem involves creating a mixed design in which participants receiving different orders of the conditions are considered to be in different groups—that is, they are at different levels of a between-groups factor that we will refer to as *order*. If a mixed-design ANOVA is then computed, the subject-by-treatment interaction will be measured separately for each order group and then averaged. Participants from different levels of the order factor are separated in the analysis and therefore cannot interact with each other.

When adding order to create a mixed design, the amount of interaction caused by the simple order effects appears in the *SS* for the interaction of the two factors (order and the RM factor), rather than in the subject-by-treatment interaction. This reduces the error term when testing the main effect of the RM factor. A significant interaction between order and treatment just affirms that you have sizable order effects. A main effect of order, however, is worrisome because there is no good reason for participants from one order group to be performing better *overall* than the participants in another order group. This result suggests that you may have asymmetric carryover effects and need to look more closely at your data. As discussed earlier, an RB design would eliminate the possibility of such carryover effects, while providing much of the increased power associated with the RM design. A mixed design with one between-groups factor is analyzed in the same way, of course, whether the second factor is based on repeated measures or RBs.

## Summary

The repeated-measures (RM) ANOVA extends the advantages of the matched-pairs *t* test to studies in which more than two treatments or conditions are being compared. The significance test for the RM ANOVA uses the interaction of subjects with treatments as its error term and is therefore able to ignore overall subject-to-subject differences on the dependent variable. When the RM ANOVA is not feasible (e.g., comparing three methods for teaching young children to read), much of the power of the RM ANOVA can be gained by matching subjects into blocks and then randomly assigning the members of the blocks to the different levels of the IV. This is called the randomized-blocks (RB) design. When each block contains the same number of subjects as there are different conditions, the data are analyzed in the same way as the RM ANOVA.

In the two-way mixed-design ANOVA, there are matching or repeated measures for one of the factors but not the other. This necessitates the calculation of two separate error terms. One of these, $MS_W$, is based on the subject means within each group and is used as the denominator for testing the main effect of the between-groups factor. The other, $MS_{error}$, is based on the subject-by-treatment interaction within each group and is used as the denominator for testing the main effect of the RM factor, as well as its interaction with the between-groups factor. One important application of the mixed-design ANOVA involves the one-way RM design with counterbalancing. Adding order as a between-groups factor takes variability away from the error term of an RM ANOVA and puts it in the numerator of a new $F$ ratio, which tests the interaction of the order by treatment effect.

### One-Way RM ANOVA

The computational procedure outlined next can be applied in exactly the same way to data from either an RM or an RB design. The formulas for the $SS$ components are the same as those for a two-way ANOVA, in which $n = 1$ (there is only one score in each cell).

#### *1. Sums of Squares*

1. Total sum of squares ($SS_T$)

$$\text{Computing formula}: \quad SS_T = \sum X^2 - N_T \overline{X}_G^2$$

where

$\overline{X}_G$ = (grand) mean of all the scores

2. Between-treatments sum of squares ($SS_{treat}$)

$$\text{Computing formula}: \quad SS_{treat} = r \sum \overline{X}_i^2 - N_T \overline{X}_G^2$$

where

$r$ = number of different subjects or blocks

$\overline{X}_i s$ = means of the different treatments

3. Between-subjects sum of squares ($SS_{sub}$)

$$\text{Computing formula}: \quad SS_{sub} = c \sum \overline{X}_j^2 - N_T \overline{X}_G^2$$

where

$c$ = number of different treatments or conditions

$\overline{X}_j s$ = means of the different subjects or blocks

4. Interaction (i.e., error) sum of squares ($SS_{error}$)

$$\text{Computing formula}: \quad SS_{error} = SS_T - SS_{treat} - SS_{sub}$$

#### *2. Degrees of Freedom*

Total degrees of freedom:

$$df_T = N_T - 1 = rc - 1$$

Degrees of freedom between treatments:

$$df_{treat} = c - 1$$

Degrees of freedom between subjects:

$$df_{sub} = r - 1$$

Degrees of freedom for (subject-by-treatment) interaction:

$$df_{error} = (c - 1)(r - 1)$$

***3. Mean Squares***

Mean square between treatments:

$$MS_{\text{treat}} = \frac{SS_{\text{treat}}}{df_{\text{treat}}}$$

Mean square for interaction:

$$MS_{\text{error}} = \frac{SS_{\text{error}}}{df_{\text{error}}}$$

***4. F Ratio***

$$F_{\text{RM}}(\text{i.e.}, F_{\text{treat}}) = \frac{MS_{\text{treat}}}{MS_{\text{error}}}$$

$$df = (c-1), (c-1)(r-1)$$

***5. The Sphericity Assumption***

An important assumption that applies when an RM ANOVA has more than two levels is that the variability of the difference scores for any pair of conditions is the same as it is for any other pair of conditions in the population. A significant $F$ ratio can be retested without assuming *sphericity* in the population by comparing it to a worst-case critical $F$ based on 1 and $r - 1$ $df$. Because a lack of population sphericity tends only to increase the $F$ ratio when the null hypothesis is true, there is no need to retest an RM ANOVA that does not attain significance. However, if the $F$ from an RM ANOVA is significant with the usual critical $F$, but not with the conservatively adjusted (worst-case $F_{\text{crit}}$), it is recommended that statistical software be used to adjust the critical $F$ more precisely.

***6. Alternative Experimental Designs***

Carryover effects can invalidate the results from an RM ANOVA conducted on an RM design. In such cases, the RB design may be preferable. Consult Table 16.6 for several error-reducing alternatives to the RM design.

**Table 16.6** **Recommended Experimental Designs that Reduce the Error Term of a One-Way ANOVA in Various Experimental Situations**

| | Experimental situation | Recommended experimental design |
|---|---|---|
| 1. | You have several different conditions, but trials of each can be mixed together randomly. | Repeated-measures ANOVA (each participant experiences all of the conditions) |
| 2. | You have several conditions that must be presented in order, but the effects of one condition are *not* likely to carry over into the next one. | Repeated-measures ANOVA with counterbalancing (equal number of participants receive the conditions in different orders), analyzed as a two-way mixed-design ANOVA |
| 3. | You have several conditions that must be presented in order, but the effects of one condition *are* likely to carry over into the next, and you have a basis for matching participants into blocks. | Randomized-blocks ANOVA (match $k$ participants in each block, where $k$ is the number of different conditions, and then assign the members of each block randomly to the different conditions) |
| 4. | You have several conditions that must be presented in order, and the effects of one condition are likely to carry over into the next. You have a basis for matching participants but do not have all of your participants at the outset of the experiment. | Treatment-by-blocks (two-way) ANOVA (categorize participants as belonging to one or another level of some relevant blocking factor, and make sure that the same number of participants from each block is assigned to each condition) |
| 5. | You have several conditions that must be presented in order, and the effects of one condition are likely to carry over into the next. It is *not* feasible to match your participants, but you can categorize your participants into groups that differ on your dependent variable. | Two-way ANOVA with grouping factor (start with a one-way independent-samples ANOVA and then add the grouping factor to create a two-way ANOVA, thereby reducing the error term for your factor of interest) |

## Two-Way Mixed-Design ANOVA

There are several ways to begin the partitioning of the *SS* total from a mixed design, but all result in the same final *SS* components. Probably, the easiest way to begin the analysis is by following the procedure for an ordinary two-way ANOVA, as shown next.

### *1. Sums of Squares*

1. Create a table of the cell means, and find the row and column means. Then calculate $SS_{\text{between-cell}}$, the *SS* for each of the two factors, and the *SS* for their interaction as described in Chapter 15. We can label the *SS* for the between-groups factor as $SS_{\text{bet-groups}}$, or just $SS_{\text{groups}}$ and the *SS* for the repeated-measures factor as $SS_{\text{RM}}$; their interaction will be labeled $SS_{\text{inter}}$. Also, as in Chapter 15, calculate $SS_{\text{T}}$, and subtract $SS_{\text{between-cell}}$ to get what we have been calling $SS_{\text{within-cell}}$.
2. Calculate the between-subjects sum of squares ($SS_{\text{sub}}$), based on the subject means, as shown for the one-way RM ANOVA (i.e., ignore the fact that subjects belong to one group or another). Subtract $SS_{\text{groups}}$ from $SS_{\text{sub}}$ to get $SS_{\text{W}}$ for the mixed design.
3. Subtract $SS_{\text{W}}$ from $SS_{\text{within-cell}}$ to obtain $SS_{\text{error}}$.

### *2. Degrees of Freedom*

Total degrees of freedom:

$$df_{\text{T}} = N_{\text{T}} - 1 = krc - 1 \qquad \text{Formula 16.1}$$

where *k* is the number of groups (i.e., levels of the between-groups factor), *r* is the number of subjects in each of the groups (we are considering only balanced designs), and *c* is the number of levels of the RM factor.

Numerator degrees of freedom:

$$df_{\text{group}} = k - 1 \qquad \text{Formula 16.2}$$

$$df_{\text{RM}} = c - 1 \qquad \text{Formula 16.3}$$

$$df_{\text{inter}} = (k - 1)(c - 1) \qquad \text{Formula 16.4}$$

Denominator degrees of freedom:

$$df_{\text{W}} = k(r - 1) \qquad \text{Formula 16.5}$$

$$df_{\text{error}} = k(c - 1)(r - 1) \qquad \text{Formula 16.6}$$

### *3. Mean Squares*

Numerator mean squares:

$$MS_{\text{groups}} = \frac{SS_{\text{groups}}}{df_{\text{groups}}}; \; MS_{\text{RM}} = \frac{SS_{\text{RM}}}{df_{\text{RM}}}; \; MS_{\text{inter}} = \frac{SS_{\text{inter}}}{df_{\text{inter}}}$$

Denominator mean squares:

$$MS_{\text{W}} = \frac{SS_{\text{W}}}{df_{\text{W}}}; \; MS_{\text{error}} = \frac{SS_{\text{error}}}{df_{\text{error}}}$$

### *4. F Ratios*

$$F_{\text{groups}} = \frac{MS_{\text{groups}}}{MS_{\text{W}}}; \; F_{\text{RM}} = \frac{MS_{\text{RM}}}{MS_{\text{error}}}; \; F_{\text{inter}} = \frac{MS_{\text{inter}}}{MS_{\text{error}}}$$

***5. Additional Assumption***

The mixed-design ANOVA requires all of the assumptions of the RM ANOVA, including sphericity, but in addition, it must be assumed that the variability of the difference scores for any pair of conditions is the same in each population that is represented by a different level of the between-groups factor.

### Common Situations for Which the Two-Way Mixed Design Is Useful

1. *Measuring two or more independent groups over several points in time.* The simplest version of this situation, the before–after design, usually requires only a one-way ANOVA on the difference scores to yield all the information of interest. However, when more than two points in time are involved, a mixed-design ANOVA is called for.
2. *A one-way repeated-measures ANOVA with counterbalancing.* A mixed-design ANOVA is created by adding order (of treatments) as a between-groups factor to reduce the error term.
3. *Adding a grouping factor to a one-way repeated-measures or randomized- blocks ANOVA.* Each participant serves in several conditions, or participants are matched in blocks, but each participant is also categorized as belonging to one level or another of a pre-existing individual-differences factor (e.g., psychiatric diagnosis).
4. *One experimental factor has levels that are easily repeated, but the other does not.* For example, the RM factor might involve several versions of the same task, while the levels of the between-groups factor consist of different general mindsets for approaching the tasks (e.g., your performance on each of these tasks reflects your overall level of intelligence versus some individuals perform better on some versions of the task according to their individual skills and aptitudes).

## Exercises

1. Redo Exercise 7 from Chapter 7 as a one-way RM ANOVA. What is the relationship between the *F* ratio you calculated for this exercise and the *t* value you calculated for that exercise?
2. The following group means come from Exercise 4 in Chapter 13. Compute the one-way RM ANOVA for these data, assuming that there are a total of 15 blocks of (matched) participants and that $SS_{error} = 5{,}040$. Is the *F* ratio significant at the .05 level, based on the unadjusted *df* (i.e., assuming sphericity)? Would this *F* ratio be significant at the .05 level, given the worst-case adjustment of the *df* (i.e., assuming a maximum violation of sphericity)?

| | Group | | | | |
|---|---|---|---|---|---|
| | 1 | 2 | 3 | 4 | 5 |
| $\overline{X}$ | 23 | 30 | 34 | 29 | 26 |

3. The following data come from an experiment in which each participant has been measured under three different levels of distraction (the DV is the number of errors committed on a clerical task during a 5-minute period). Compute the one-way RM ANOVA for these data. Is the *F* ratio significant at the .01 level, assuming sphericity? Would this *F* ratio be significant at the .05 level, assuming a total lack of sphericity?

| Sub. No. | Mild | Moderate | Strong |
|---|---|---|---|
| 1 | 1 | 3 | 3 |
| 2 | 0 | 0 | 1 |
| 3 | 3 | 2 | 4 |
| 4 | 0 | 2 | 2 |
| 5 | 2 | 3 | 2 |
| 6 | 1 | 1 | 0 |
| 7 | 1 | 1 | 3 |
| 8 | 2 | 4 | 5 |
| 9 | 1 | 2 | 4 |
| 10 | 0 | 3 | 6 |
| 11 | 4 | 4 | 5 |
| 12 | 2 | 1 | 6 |

4. Suppose that, prior to performing the clerical tasks in the experiment of Exercise 3, the first six participants took pills they thought to be caffeine but that were actually placebos; the remaining six participants ingested real caffeine pills.
   (a) Reanalyze the data in Exercise 3 as a mixed-design ANOVA, adding drug condition (placebo versus caffeine) as the between-groups variable. Test each *F* ratio for significance at the .05 level.
   (b) Compare the *SS* components you found as part of your analysis in part (a) to the *SS* components you found in Exercise 3. Which *SS* components are the same, and which combinations of *SS*s in part (a) add up to one of the *SS* components in Exercise 3?
   (c) Compute separate one-way RM ANOVAs for the placebo and the caffeine participants and test these simple main effects for significance at the .05 level. How do these tests relate to your analysis in part (a)?
   (d) If this design was completely counterbalanced for each of the two drug groups, how many participants would have been assigned to each possible treatment order within each group?
5. Suppose that participants are asked to memorize a list of words that range from very abstract to very concrete. The number of words recalled of each type for each participant are shown in the following table:

| Participant number | Very abstract | Mildly abstract | Mildly concrete | Very concrete |
|---|---|---|---|---|
| 1 | 2 | 5 | 11 | 14 |
| 2 | 0 | 8 | 4 | 11 |
| 3 | 7 | 9 | 13 | 20 |
| 4 | 3 | 12 | 10 | 15 |
| 5 | 3 | 7 | 8 | 6 |
| 6 | 2 | 6 | 6 | 9 |
| 7 | 5 | 8 | 7 | 7 |
| 8 | 4 | 9 | 5 | 8 |

   (a) Compute the one-way RM ANOVA for these data. Test the *F* ratio for significance at the .05 level. How many pairwise comparisons could be tested if your ANOVA were significant? What alpha would you use to test each of these comparisons, if you used a Bonferroni adjustment to keep the experimentwise alpha at .05?
   (b) Reanalyze the data as a mixed-design ANOVA, assuming that the first four participants were selected for their high scores on a spatial ability test, whereas the second group of four participants were chosen for their poor performance in spatial ability. Test each *F* ratio for significance at the .05 level.

(c) Graph the cell means of the two-way mixed-design ANOVA described in part (b). Explain how the pattern of the cell means is consistent with the results you obtained in part (b).

## Thought Questions

1. (a) What advantage does repeated-measures ANOVA have over ANOVA without repeated measures (where every score is derived from a different participant)? (b) Why might a researcher decide *not* to use repeated-measures ANOVA in spite of this advantage, and what alternatives could be considered?
2. What is the sphericity assumption in behavioral science research? Give an example of when it might be violated.
3. Give an example for which each of the following is appropriate: (a) Repeated-measures ANOVA. (b) The randomized-blocks design. (c) The two-way mixed design. (d) The before-and-after form of the two-way mixed design. (e) One-way repeated-measures design with counterbalancing.

## Computer Exercises

1. Perform an RM ANOVA to test for a significant change in anxiety level in all subjects over time (baseline, prequiz, and postquiz). Regardless of whether or not the ANOVA is significant, follow it with matched *t* tests (or two-level RM ANOVAs) for each pair of levels of the time factor.
2. Redo Exercise 1 for heart rate.
3. Add *exercise* to the analysis in Exercise 1 and compute the mixed-design ANOVA. Request a plot of the cell means and explain your ANOVA results in terms of the pattern in your graph. Regardless of whether the interaction is significant, follow the mixed-design ANOVA with separate one-way RM ANOVAs (i.e., simple main effects) for each exercise group.
4. Add the experimental factor (i.e., difficulty of the 11th quiz question) to the analysis in Exercise 1 and compute the mixed-design ANOVA. Request a plot of the cell means and explain your ANOVA results in terms of the pattern in your graph. Follow the mixed-design ANOVA with separate one-way ANOVAs for each measurement period (i.e., baseline, prequiz, and postquiz).
5. Redo Exercise 4 using heart rate as the dependent variable instead of anxiety.

## Bridge to SPSS

There is a fundamental difference between the ways in which data are entered for a repeated-measures design and for an independent-groups design. When the same subjects are measured at all the levels of your IV, there is *no one* column in your spreadsheet that contains your DV. Each level of your factor is represented by another column (i.e., variable) in your spreadsheet. You will have to tell SPSS which columns in your spreadsheet are the levels of your RM factor. If you are using an RB design, a single row does not literally represent one subject, but it does represent one *block,* which is treated as though it contained repeated measures of the same subject.

To perform an RM ANOVA in SPSS, select **Repeated Measures …** from the **Analyze/ General Linear Model** menu. The dialog box that opens will ask you to create a name for your RM factor. For this example, we will use the scores for anxiety as measured at three time points in Jackson's data set. Therefore, an appropriate name for our repeated-measures factor would be *Time.* After replacing the default *Within-Subject Factor Name* ("factor1") with *Time,* type in 3 as the *Number of Levels*:, click **Add**, and then click **Define** in the lower-left corner of the dialog box. This will open the main dialog box for this procedure, as shown in Figure 16.3.

# Figure 16.3

**The Dialog Box for Repeated-Measures ANOVA**

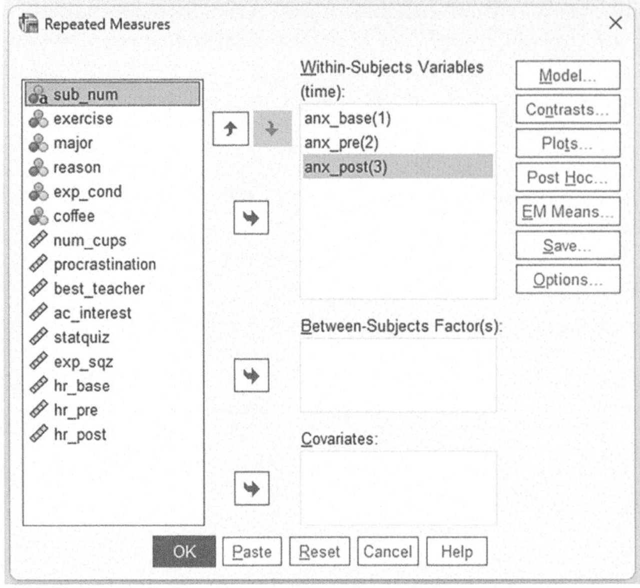


*Source*: Courtesy of International Business Machines Corporation, © SPSS, Inc., an IBM company

In the *Within-Subjects Variables* space of the main dialog box, the number of slots that appear corresponds to the number of levels you chose in the first dialog box. You must move variables from the list on the left to fill in all of those slots. The seven buttons on the right side of this dialog box are useful mostly after adding a *Between-Subjects Factor*, so we will discuss the functions of several of those buttons when we describe SPSS's approach to the Mixed-Design ANOVA.

Clicking **OK** in the dialog box shown in Figure 16.3 produces six tables of output by default but not all of those tables are relevant to our coverage in this chapter, or when there is no between-subjects factor. The first table just lists the variables that serve as the levels of your RM factor and the second one, labeled **Multivariate Tests**, contains an alternative approach to RM ANOVA that is not commonly used. The next table tests an important assumption of the RM ANOVA by means of **Mauchly's Test of Sphericity**; the results in this table are needed to interpret the results in the table that follows it: **Tests of Within-Subjects Effects**. SPSS always presents four variations of the RM ANOVA in this table, whether you want to see them or not (see Figure 16.4). The first variation, labeled **Sphericity Assumed**, is the ordinary RM ANOVA result, as described in this chapter. The next two rows of this box contain the results of two different formulas for reducing both the numerator and denominator *df* to accommodate possible violations of the sphericity assumption. The first of these, known as the **Greenhouse–Geisser (G–G) correction**, mentioned earlier in this chapter is by far the more conservative and more popular of the two; the Huynh–Feldt (H–F) correction is rarely used.

The fourth variation, "Lower-bound" epsilon, uses what we referred to as the worst-case *df* adjustment earlier in this chapter. There is no need to use this overly conservative adjustment if the G–G solution is available. If your data seem very consistent with the sphericity assumption, it is reasonable to report the $p$ value from the row labeled Sphericity Assumed. (The $p$ values change from row to row, because the *df*s change, but the calculated $F$ values do not.) However, if Mauchly's test is significant (i.e., Sig. $< .05$), it is recommended that you report the $p$ value according to the G–G correction, along with its (usually fractional) adjusted *df*.

In our example of anxiety measured at three time points, you can see that Mauchly's test is significant at the .001 level, so the $p$ value for Sphericity Assumed should be ignored, and the value reported should correspond to **Greenhouse_Geisser**: $F\,(1.64, 162.6) = 3.74$, $p = .034$. Fortunately, we can also ignore the worst-case (**Lower-bound**), which falls just short of significance at the .05 level.

## Figure 16.4

**The Main Results for the One-Way Repeated-Measures ANOVA**

**Mauchly's Test of Sphericity[a]**

Measure: MEASURE_1

| | | | | | Epsilon[b] | | |
|---|---|---|---|---|---|---|---|
| Within subjects effect | Mauchly's W | Approx. Chi-square | df | Sig. | Greenhouse-geisser | Huynh-Feldt | Lower-bound |
| time | .782 | 24.054 | 2 | <.001 | .821 | .833 | .500 |

Tests the null hypothesis that the error covariance matrix of the orthonormalized transformed dependent variables is proportional to an identity matrix.

a. Design: Intercept
Within Subjects Design: time

b. May be used to adjust the degrees of freedom for the averaged tests of significance. Corrected tests are displayed in the Tests of Within-Subjects Effects table.

**Tests of Within-Subjects Effects**

Measure: MEASURE_1

| Source | | Type III Sum of Squares | df | Mean Square | F | Sig. |
|---|---|---|---|---|---|---|
| time | Sphericity Assumed | 76.527 | 2 | 38.263 | 3.744 | .025 |
| | Greenhouse-Geisser | 76.527 | 1.643 | 46.591 | 3.744 | .034 |
| | Huynh-Feldt | 76.527 | 1.667 | 45.919 | 3.744 | .033 |
| | Lower-bound | 76.527 | 1.000 | 76.527 | 3.744 | .056 |
| Error(time) | Sphericity Assumed | 2023.473 | 198 | 10.220 | | |
| | Greenhouse-Geisser | 2023.473 | 162.608 | 12.444 | | |
| | Huynh-Feldt | 2023.473 | 164.988 | 12.264 | | |
| | Lower-bound | 2023.473 | 99.000 | 20.439 | | |

The fifth table, **Tests of Within-Subjects Contrasts Effects**, is interpretable only if the levels of your RM factor are quantitative rather than qualitative. This requirement does apply to the IV in our current example (time), but we will postpone discussion of this table until the next section (Mixed-Design ANOVA). The final table, **Tests of Between-Subjects Effects**, is relevant only for the Mixed-Design ANOVA, so discussion of this table will be postponed to the next section, as well.

### Mixed-Design ANOVA

Requesting an analysis of a mixed-design ANOVA in SPSS begins the same way as for the one-way RM ANOVA. However, after you have defined the levels of your RM factor in the **Repeated Measures** dialog box, you can move one or more between-groups factors (e.g., *exercise, major*) into the space labeled *Between-Subjects Factor(s):*. For our mixed-design example, we moved the variable *exercise* to that space (not shown in Figure 16.3). Once there is a between-subject (i.e., grouping) factor with more than two levels, the **Post Hoc** button becomes relevant. However, the exercise factor has only two levels, so we did not use that button in this example. Unfortunately, pairwise comparisons for your RM levels cannot be requested from the **Post Hoc** dialog box. Instead, you can request matched-pairs $t$ tests for each pair of RM levels and then use the Bonferroni correction to lower the alpha for each pairwise test.

Because the mixed design we have been describing has two factors and an interaction, a graph of the cell means can be informative. The **Plots ...** button opens the same dialog box as in the two-way ANOVA procedure described in the preceding chapter. For this example, we moved *Time* to the *Horizontal Axis:* and *Exercise* to the *Separate Lines:* space. Clicking on **EM Means ...**, moving the *exercise∗time* interaction to the *Display Means for:* space, and then checking the *Compare simple main effects* box will produce tables of simple main effects.

The default output for the mixed-design ANOVA consists of the same tables as the one-way RM ANOVA, except for a table with the sample size for each level of the between-subjects factor. Compared to the **Tests of Within-Subjects Effects** table for the RM ANOVA, however, the mixed-design includes a section for the interaction, inserted between the RM factor and the error term; the latter is used as the denominator for both the RM main effect and the interaction (see Figure 16.5). The **Tests of Between-Subjects Effects** box, which contains only a seldom-needed test of the intercept for a purely RM ANOVA, includes a test of the main effect of the between-groups factor in a two-way mixed design.

Looking at both the Within and Between tables in Figure 16.5, you can see that the main effect of time is significant (just barely) at the .05 level with the G–G correction (adding the *exercise* factor decreased the time effect slightly compared to the RM ANOVA) and that the main effect of *exercise* is easily significant ($p$ = .006). You can also see that the interaction effect is not nearly significant. However, the table for **Tests of Within-Subjects Contrasts**, which we didn't show for the RM ANOVA, tells a more nuanced story concerning the interaction. You can see that there is a significant interaction for the linear trend. This can be understood by looking at the graph in Figure 16.5 (bottom panel). Anxiety for the regular exercisers falls back to near baseline after the experimental quiz, creating a trend that is almost entirely quadratic, without a linear component. The anxiety score for the non-exercisers continues to rise after the quiz, though not as sharply as before the quiz, exhibiting a mostly linear trend with a small quadratic component.

Although the **EM Means ...** selection described earlier produces two tables for simple main effects, the one for the effects of time for each exercise level contains only multivariate

## Figure 16.5

**Selected Results for the Two-Way Mixed-Design ANOVA**

**Tests of Within-Subjects Effects**

Measure: MEASURE_1

| Source | | Type III Sum of Squares | df | Mean Square | F | Sig. |
|---|---|---|---|---|---|---|
| time | Sphericity Assumed | 67.365 | 2 | 33.683 | 3.316 | .038 |
| | Greenhouse-Geisser | 67.365 | 1.612 | 41.790 | 3.316 | .049 |
| | Huynh-Feldt | 67.365 | 1.652 | 40.787 | 3.316 | .048 |
| | Lower-bound | 67.365 | 1.000 | 67.365 | 3.316 | .072 |
| time * exercise | Sphericity Assumed | 32.725 | 2 | 16.363 | 1.611 | .202 |
| | Greenhouse-Geisser | 32.725 | 1.612 | 20.301 | 1.611 | .207 |
| | Huynh-Feldt | 32.725 | 1.652 | 19.814 | 1.611 | .206 |
| | Lower-bound | 32.725 | 1.000 | 32.725 | 1.611 | .207 |
| Error(time) | Sphericity Assumed | 1990.748 | 196 | 10.157 | | |
| | Greenhouse-Geisser | 1990.748 | 157.974 | 12.602 | | |
| | Huynh-Feldt | 1990.748 | 161.860 | 12.299 | | |
| | Lower-bound | 1990.748 | 98.000 | 20.314 | | |

**Tests of Within-Subjects Contrasts**

Measure: MEASURE_1

| Source | time | Type III Sum of Squares | df | Mean Square | F | Sig. |
|---|---|---|---|---|---|---|
| time | Linear | 36.208 | 1 | 36.208 | 6.098 | .015 |
| | Quadratic | 31.157 | 1 | 31.157 | 2.167 | .144 |
| time * exercise | Linear | 30.568 | 1 | 30.568 | 5.148 | .025 |
| | Quadratic | 2.157 | 1 | 2.157 | .150 | .699 |
| Error(time) | Linear | 581.887 | 98 | 5.938 | | |
| | Quadratic | 1408.861 | 98 | 14.376 | | |

Figure 16.5

(Continued)

**Tests of Between-Subjects Effects**

Measure: MEASURE_1

Transformed Variable: Average

| Source | Type III Sum of Squares | df | Mean Square | F | Sig. |
|---|---|---|---|---|---|
| Intercept | 105730.152 | 1 | 105730.152 | 1813.755 | <.001 |
| exercise | 468.632 | 1 | 468.632 | 8.039 | .006 |
| Error | 5712.765 | 98 | 58.294 | | |

**Univariate Tests**

Measure: MEASURE_1

| time | | Sum of Squares | df | Mean Square | F | Sig. |
|---|---|---|---|---|---|---|
| 1 | Contrast | 84.428 | 1 | 84.428 | 4.669 | .033 |
| | Error | 1772.082 | 98 | 18.082 | | |
| 2 | Contrast | 127.674 | 1 | 127.674 | 3.153 | .079 |
| | Error | 3968.686 | 98 | 40.497 | | |
| 3 | Contrast | 289.255 | 1 | 289.255 | 14.443 | <.001 |
| | Error | 1962.745 | 98 | 20.028 | | |

Each F tests the simple effects of exercise within each level combination of the other effects shown. These tests are based on the linearly independent pairwise comparisons among the estimated marginal means.

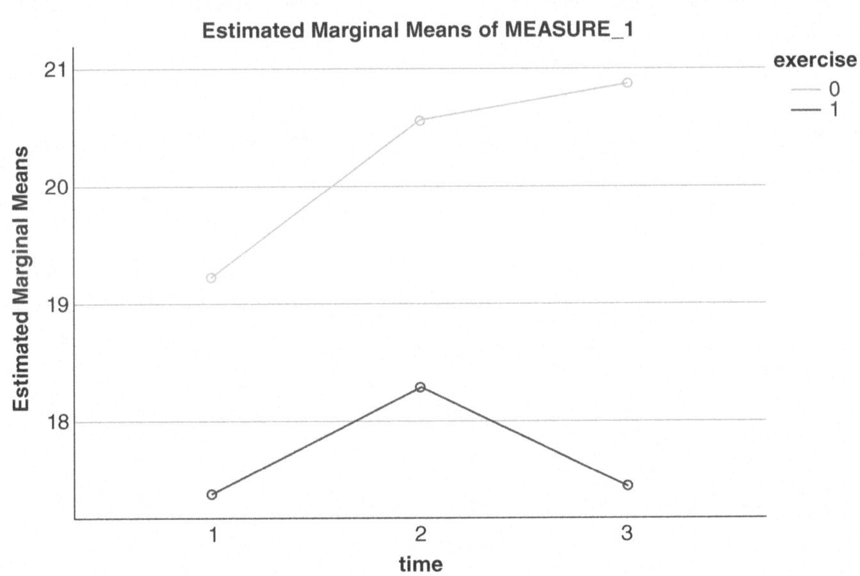


tests due to SPSS's caution concerning the sphericity assumption (not shown in Figure 16.5). However, the other table contains ordinary univariate tests for the effects of exercise level at each time point. It is clear from looking at the graph of the cell means that you would expect the largest contrast to occur at time 3. It is also easy to see why the main effect of exercise is significant.

# Part IV

# Nonparametric Statistics for Categorical Data

# Chapter 17

# Probability of Discrete Events and the Binomial Distribution

## PREVIEW



## Introduction

In the analysis of variance, as well as in the *t* test, the dependent variable (DV) is measured on a precise, quantitative (i.e., interval or ratio) scale. However, there are many possible DVs in behavioral research that cannot be measured precisely (at least at the present time). For example, if we want to test some new form of imagery exercise that is designed to increase creativity, we may be able to assemble a panel of judges that can reliably determine whether one work of art or writing is more creative than another, without being able to accurately quantify the *amount* of creativity separately for each work. Indeed, they may not even be


*Introductory Statistics for the Behavioral Sciences*, Eighth Edition. R. Brooke Lea and Barry H. Cohen.



Companion website: www.wiley.com/go/ISBS8e

able to rank the amounts by which pairs of works differ in creativity, thus ruling out the possibility of using Wilcoxon's matched-pairs test (see Chapter 8).

In another example, a behavioral outcome may have only two possible levels, to begin with. For example, an (unknowing) participant may either return the wallet they found on the street (where the experimenter left it) or not. The independent variable (IV) in this study might also consist of just a few categories, such as the type of location where the wallet was left (residential, commercial, or a tourist district).

When both the IVs and DVs in a study can only be categorized, the statistical methods discussed in previous chapters are not appropriate. Methods for dealing with purely categorical data, which, like the ordinal tests of Chapter 8, are referred to as *nonparametric statistics*, will be described in this chapter and the next. We begin by considering the simplest possible situation in which you would use a nonparametric test: You are observing an outcome that is **dichotomous**—it can fall into only one of two possible categories.

### Dichotomous Data

If you want to make a statistical decision based on how many outcomes fall into one or another of two categories, you need an appropriate model to represent the distribution of frequencies that occur when the null hypothesis is true. However, when the outcome is dichotomous, the distribution corresponding to the null hypothesis will not be a smooth mathematical curve like the $t$ or $F$ distributions. It will tend to have a rather boxy shape, which is known as the *binomial* distribution. For a small number of outcomes, it is not difficult to determine the entire binomial distribution exactly. However, to fully understand how this is done, you will need to learn the basic rules of probability as they apply to discrete (i.e., countable) events, rather than smooth distributions based on infinite populations of precise measurements, such as those discussed in Chapter 4.

Flipping a coin, which can land only as heads or tails, is the example that we will use to introduce the laws of probability. An *unbiased* coin is one that is just as likely to land heads as it is to land tails.

## Probability

After defining probability and odds, we cover the rules of probability.

### Definition of Probability

The **probability (P)** of a given event can be defined by the following fraction:

$$P(\text{event}) = \frac{\text{number of ways the specified event can occur}}{\text{total number of equally likely possible events}} \qquad \text{Formula 17.1}$$

For example, if an unbiased coin is flipped and you wish to determine the probability that heads will come up, the solution is

$$P(\text{H}) = \frac{1}{2} = \frac{\text{1 head}}{\text{2 possible events: head or tail}}$$

Similarly, the probability of tails coming up is also 1/2. Decimals are often used to express probabilities, so we can also write $P(\text{H})$ (that is, the probability of obtaining a head on one toss of the coin), or $P(\text{T})$, as .50. These numbers express the fact that if the coin is fair, in the long run, heads will occur half the time and tails will occur half the time.

To provide some further illustrations, let us assume that you have a standard 52-card deck of playing cards (any jokers or other special cards have been removed). The 52 cards are divided into four suits (spades, hearts, diamonds, and clubs), each of which contains 13 cards (2 through 10, jack, queen, king, and ace). The deck is thoroughly shuffled, and one card is drawn. The probability that it is the king of diamonds is 1/52 (or .019); there is only one way for that event to occur (only one king of diamonds) and 52 possible equally likely events (cards that could be drawn). In fact, the probability of drawing any one specific card from the deck is 1/52. The probability of drawing a 9 is 4/52 or 1/13 (or .077) since there are four ways for this event to occur (9 of spades, 9 of hearts, 9 of diamonds, and 9 of clubs) out of the total possible 52 events. If we define a *spot card* as a card with any number from 2 to 9, the probability of drawing a spot card in clubs is 8/52, or 2/13 (or .15), for there are eight specific equally likely events (the 2, 3, 4, 5, 6, 7, 8, and 9 of clubs) in the 52-card deck that fall under the category of this event.

Probability cannot be less than 0 or greater than 1. If an event has a probability of 1, that means that it must happen. The probability of drawing either a red or a black card from a standard deck is 1.00, if you draw a card at all, because all of the cards in the deck are either red or black (i.e., $P = 52/52 = 1.00$). The probability of drawing a purple card with yellow polka dots is zero ($P = 0/52 = 0$).

## Odds

The *odds against an event* are defined as the ratio of the number of unfavorable outcomes to the number of favorable outcomes, all outcomes being equally likely. See, for example, the following table.

| Outcome | Total events (T) | Number of ways of obtaining the event (W) | Number of ways of not obtaining the event (T − W) | Probability of event (W/T) | Odds against event (T − W to W) |
|---|---|---|---|---|---|
| King of diamonds | 52 | 1 | 51 | 1/52 | 51 to 1 |
| A nine | 52 | 4 | 48 | 4/52 or 1/13 | 48 to 4 or 12 to 1 |
| Spot card in clubs | 52 | 8 | 44 | 8/52 or 2/13 | 44 to 8 or 11 to 2 or 5.5 to 1 |

If the odds against you winning a game are (say) 12 to 1, you need to collect 12 times your wager when you win for the game to be fair. If you stand to win more than 12 times your bet, it is to your advantage to play the game. If you collect less than 12 times your bet when you win, you should refuse to play because you can expect to lose money in the long run; this is true of any game at a gambling casino, for instance.

## The Probability of *A* or *B*

The probability of *either of two events* taking place is computed by using the following formula:

$$P(A \text{ or } B) = P(A) + P(B) - P(A \text{ and } B) \qquad \text{Formula 17.2}$$

Example: Suppose that you will win a prize if you draw *either* a king *or* a club from a standard 52-card deck in a single try. The probability of winning is equal to

$$\begin{aligned} &P(\text{king}) + P(\text{club}) - P(\text{king and club}) \\ &= 4/52 + 13/52 - 1/52 \\ &= 16/52 (\textit{or } .31) \end{aligned}$$

To verify these calculations, let us tackle the problem the long way. There are the usual 52 total possible events, and the number of favorable events may be counted as follows:

| King (4 favorable events) | Club (13 favorable events) |
|---|---|
| King of spades | Ace of clubs |
| King of hearts | *King of clubs* |
| King of diamonds | Queen of clubs |
| *King of clubs* | Jack of clubs |
| | Ten of clubs |
| | Nine of clubs |
| | Eight of clubs |
| | Seven of clubs |
| | Six of clubs |
| | Five of clubs |
| | Four of clubs |
| | Three of clubs |
| | Two of clubs |

There appear to be 17 favorable events: 13 clubs and 4 kings. But the king of clubs has been counted twice, and there is only one king of clubs in the deck. Therefore, the probability of king *and* club (1/52) must be subtracted, and the resulting probability (16/52) shows that there are 16 separate and distinct favorable events (cards) out of the total of 52 in the deck.

If events $A$ and $B$ cannot happen simultaneously (in mathematical terminology, they *are* **mutually exclusive**), then $P(A \text{ and } B) = 0$. For example, the probability of drawing a king or a queen from the deck is equal to $4/52 + 4/52 - 0/52 = 8/52$. The events are mutually exclusive because no card in the deck is *both* a king and a queen.

### The Probability of *A* and Then *B*

Suppose that a card is drawn from a standard deck, looked at, and replaced in the deck. The deck is shuffled thoroughly, and a second card is drawn. We wish to know the probability of obtaining a king on the first draw *and then* the ace of spades on the second draw. In this situation, the two draws are *independent*—that is, what happens on the first draw does not affect the probabilities on the second draw (because of the replacement and shuffling). The probability of both A *and* B occurring is given by this formula:

$$P(A \text{ and then } B) = P(A) \times P(B) \qquad \text{Formula 17.3}$$

In the present example, the answer is

$$\begin{aligned} P(\text{king and then ace of spades}) &= P(\text{king}) \times P(\text{ace of spades}) \\ &= 4/52 \times 1/52 \\ &= 4/2704 (= .00148) \end{aligned}$$

### Conditional Probability

If the card drawn on the first try is *not* returned to the deck, the probability becomes $4/52 \times 1/51$ (or $4/2652 = .00151$). Under this procedure, only 51 cards remain in the deck after the first draw, and the probability of success on the second draw is therefore 1/51.

Drawing two cards from a deck *without* replacement results in two events that are *not* independent. That is, the probability of the second card drawn is changed by the first draw. In statistical language, the probability of the second event is *conditional* on the outcome of the first event. For example, the probability of the second card being a club is certainly less if the first card drawn happens to also be a club.

Alpha is another example of a *conditional probability*. Alpha is the probability of rejecting the null hypothesis *under the condition* that the null hypothesis is true. With alpha fixed at .05, we know that 5% of all experiments *for which the null hypothesis is actually true* will be statistically significant and therefore Type I errors, but we do *not* know what percentage of *all* experiments will be statistically significant at the .05 level.

To create a binomial distribution, as we are about to do, all of the events involved must be independent of all of the others. Therefore, we will not deal further with conditional probabilities in this chapter.

## The Binomial Distribution

Using only the simple rules of probability we have just described, we can now explain how the binomial distribution arises and how it can be used.

### Constructing the Binomial Distribution for $N = 2$

To return to the coin-tossing example, the probability of obtaining a head on each of two consecutive flips of a coin is $P(\text{H}) \times P(\text{H})$, which is equal to $1/2 \times 1/2$ or $1/4$. This can be verified by listing all possible outcomes of two flips of the coin:

| First flip | Second flip | Overall result | Probability |
|---|---|---|---|
| H | H | 2 heads | 1/4 (or .25) |
| H | T | 1 head, 1 tail | 1/4 |
| T | H | 1 head, 1 tail | 1/4 |
| T | T | 2 tails | 1/4 |

Of the four possible outcomes (all of equal probability), only one yields two heads, so, using Formula 17.1, this probability is 1/4. Similarly, the probability of two tails is 1/4, and the probability of obtaining one head and one tail irrespective of order is 2/4 or 1/2. Since the events are independent (i.e., the results of the first flip do not affect the probabilities on the second flip), the probabilities are exactly the same if we flip two coins once. Notice also that the sum of the probabilities of the four possible outcomes is equal to 1.0.

The shape of a binomial distribution depends on two values: $N$, the number of events; and $P$, the probability of one of the two categories for each event. For coin tossing, the two categories are heads (H) and tails (T), and $P$ is the probability of one of these. For a fair coin, $P$ equals .5 for both heads and tails. This will *not* be the case for a biased coin, which will be more likely to fall heads up or tails up. In the general case, we choose one category to focus on and refer to it as $X$. Then we say that $P$ is the probability of the $X$ category occurring on any one event. (For flips of a coin, it is customary to choose heads as the $X$ category.) The probability of the other category (in this case, tails) is sometimes labeled $Q$. But when there are only two categories, it is simpler to refer to the probability of the other category as $1 - P$, because $P$ plus $Q$ (the sums of the probabilities of the two possible outcomes) must equal 1.

To construct a graph of a binomial distribution, place all of the possible values for $X$ along the horizontal axis. The height of the bar for each value of $X$ depends on the relative frequency of that value occurring. For $N = 2$, there are three possible values for $X$: 0 (no heads), 1 head, and 2 heads. From the preceding table and discussion, it should be easy to imagine the binomial distribution for $N = 2$ and $P = .5$. It would consist of three bars: The bars over 0 and 2 would reach a height of .25, while the bar in the middle ($X = 1$) would reach .5.

It is particularly easy to construct a binomial distribution when $P$ (and therefore $Q$) equals .5. In that case, all of the possible sequences of the two categories have the same probability (e.g., HTTT is no more or less likely than THTH), and the task becomes one of counting how many sequences contain a particular number of $X$s (e.g., heads). If, instead, $X$ represents drawing a club from a standard deck of cards (with replacement), $P$ equals .25. If we use $Y$ to represent a card that is not a club ($Q = .75$), then a sequence like $XYYY$ is *more* likely than a sequence like $YXYX$, not because of the order but because one has more $Y$s (which are three times more likely) than the other. When $P = Q = .5$, the binomial distribution is conveniently symmetric, as you will see in the next section. Fortunately, $P = .5$ is often a reasonable value for the null hypothesis when dealing with dichotomous events, so the kind of binomial distribution that we will construct next has many uses.

### Constructing the Binomial Distribution for $N = 6 (P = .5)$

If you toss a coin six times, the number of possible different sequences is $2 \times 2 \times 2 \times 2 \times 2 \times 2$ (or $2^6$), which equals 64. When $P = .5$, the probability is the same for all 64 sequences, so $p = 1/64$ for each sequence. However, the probability for a particular value of $X$ (e.g., 5) depends on how many of the 64 sequences have that value for $X$ (e.g., how many sequences contain exactly five heads). So, how many sequences have exactly five heads? Let us count the ways, which are labeled A through F in the following table:

| | Possible outcomes | | | | | |
|---|---|---|---|---|---|---|
| | **Flip 1** | **Flip 2** | **Flip 3** | **Flip 4** | **Flip 5** | **Flip 6** |
| A | H | H | H | H | H | T |
| B | H | H | H | H | T | H |
| C | H | H | H | T | H | H |
| D | H | H | T | H | H | H |
| E | H | T | H | H | H | H |
| F | T | H | H | H | H | H |

As you can see, there are six different sequences that contain exactly five heads. Using the "or" rule of probability, the probability that you will get five heads from a fair coin that is flipped six times is equal to

$$P(\text{A or B or C or D or E or F})$$

Since these events are mutually exclusive, this equals $1/64 + 1/64 + 1/64 + 1/64 + 1/64 + 1/64 = 6/64$, or .094.

Using the same procedures, we can compute the probability of any number of heads if a coin is flipped six times. However, writing out all of the possible sequences can be quite tedious,[1] so we will just present the end product in the following table:

| Number of heads | Probability |
|---|---|
| 0 | 1/64 = .016 |
| 1 | 6/64 = .094 |
| 2 | 15/64 = .234 |
| 3 | 20/64 = .312 |
| 4 | 15/64 = .234 |
| 5 | 6/64 = .094 |
| 6 | 1/64 = .016 |
| | 64/64 = 1.000 |

### Testing Null Hypotheses With the Binomial Distribution

The binomial distribution shown can be used to test some simple null hypotheses. Suppose a friend claims that he can taste the difference between two popular brands of root beer, and you decide to test his claim.

You present the two brands of root beer in identical cups, and he has to tell you which cup contains which brand. Suppose your friend answers correctly six times in a row. What should you decide?

The appropriate null hypothesis is that your friend was just guessing, in which case $P$ equals .5 for each test. Because being correct six times out of six tests is represented by only

[1] Mathematics makes use of special techniques, called *permutations* and *combinations*, to speed the calculation of the number of sequences for each $X$. However, we do not need such fancy methods for the purposes of this chapter.

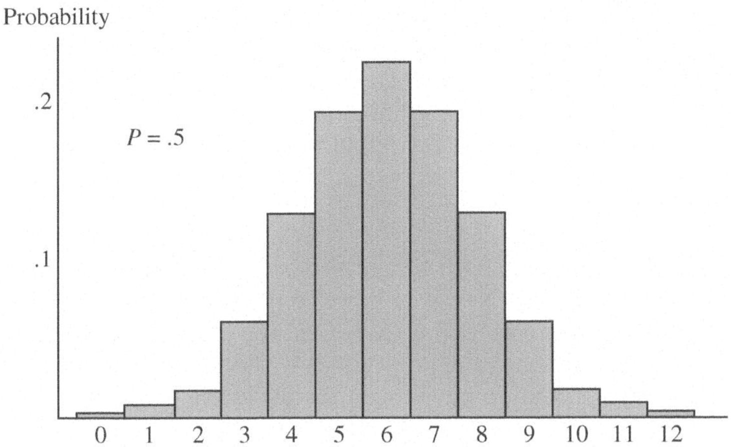


## Figure 17.1

**Binomial Distribution for $N = 12$, $P = .5$**

one sequence out of 64, the $p$ value is .016. If you want to be conservative and perform a two-tailed test (your friend being *wrong* six times in a row would also suggest that he does detect some difference), the $p$ value would be .032. In either case, using the .05 decision rule, you would reject the null hypothesis and conclude that your friend has a greater than .5 chance of being correct on each trial.

If your friend was correct on only five of the six trials, the one-tailed $p$ value would be .11. This is not significant at the .05 level, even for a one-tailed test. Notice that to get the $p$ value in this case, you must add the probabilities of both one wrong, or .094, and zero wrong, or .016. Because zero wrong is farther out in the tail of the binomial distribution, it contributes to the possibility of committing a Type I error.

If your friend is correct on 10 out of 12 repetitions of the taste test, you need the binomial distribution with $N = 12$, $P = .5$, to find his $p$ value. This distribution is shown in Figure 17.1. (You cannot read this accurately from the graph, but for $X = 10$, the one-tailed $p = .0161 + .0029 + .0002 = .0192$. So $H_0$ is rejected at the .05 level with either a one- or a two-tailed test.) We show the distribution for $N = 12$ so you can see its resemblance to the normal distribution. As $N$ increases, the binomial distribution becomes smoother and resembles the normal distribution more closely. (Theoretically, the two distributions become identical when $N$ is infinitely large.) When $P$ is higher or lower than .5, the binomial distribution is skewed. The farther $P$ is from .5, the larger $N$ needs to be before the binomial begins to resemble the normal distribution.

### Using the Normal Distribution as an Approximation

It is not common to find tables of the binomial distribution for $N$ greater than 20. Fortunately, if $P$ is .5 (or even just close to .5), the resemblance to the normal distribution becomes so close with $N > 20$ that accurate results can be obtained by calculating a $z$ score and referring it to a table of the standard normal distribution (like Table A in the Appendix).

It should be evident that the mean of a binomial distribution is $NP$ (e.g., for $N = 12$ and $P = .5$, the mean is 6). It is not at all obvious, but the standard deviation (SD) for a binomial distribution is $\sqrt{NPQ}$. This leads to the following $z$ score formula that can be applied to the binomial distribution when $N$ is fairly large:

$$z = \frac{X - NP}{\sqrt{NPQ}} \qquad \text{Formula 17.4}$$

For the example in which your friend identified the root beer brands correctly 10 out of 12 times, the $z$ score would be

$$z = \frac{10 - 6}{\sqrt{12(.5)(.5)}} = \frac{4}{\sqrt{3}} = 2.31$$

We can see from Table A that the $p$ value corresponding to a $z$ score of 2.31 is .0104. Although this $p$ value leads to the same conclusion with respect to $H_0$ that we reached by calculating $p$ exactly for the binomial distribution (.0192), it is noticeably different. With $P = .5$, $N$ needs to be at least about 20 before the normal approximation to the binomial distribution becomes sufficiently accurate to use in null hypothesis testing (NHT).

*The Formula for Proportions*

Rather than dealing with the actual number of trials on which your friend was correct (i.e., $X$) or the number correct expected by chance ($NP$), it may be more convenient to deal with the *proportion* of trials on which your friend was correct (i.e., $X/N$) and the proportion expected by chance (i.e., $NP/N = P$). Because we obtained the proportions by dividing by $N$, we also have to divide the SD of the binomial distribution by $N$ to get the SD of the proportions:

$$\frac{\sqrt{NPQ}}{N} = \frac{\sqrt{NPQ}}{\sqrt{N^2}} = \sqrt{\frac{NPQ}{N^2}} = \sqrt{\frac{PQ}{N}}$$

Therefore, the $z$ score for proportions is given by the following formula:

$$z = \frac{\frac{X}{N} - P}{\sqrt{\frac{PQ}{N}}}$$

This formula should seem familiar. To see why, let us change the notation. We will use a lowercase $p$ to represent $X/N$, the proportion you observed in your sample. (For the example of 10 correct trials out of 12, $p = 10/12 = .833$.) Instead of using an uppercase $P$ to represent the proportion under the null hypothesis, we will use the (lowercase) Greek letter pi ($\pi$) because this is the proportion of $X$ in the entire population. Finally, because $Q$ equals $1 - P$, it can now be written as $1 - \pi$. Using the modified notation, the previous formula can be written as

$$z = \frac{p - \pi}{\sqrt{\frac{\pi(1 - \pi)}{N}}} \qquad \text{Formula 6.5}$$

This is the formula we called the $z$ score for proportions in Chapter 6. If we apply this version of the formula to the root beer taste test example, it yields the same $z$ score:

$$z = \frac{.833 - .5}{\sqrt{\frac{.5(1 - .5)}{12}}} = \frac{.333}{\sqrt{.02083}} = \frac{.333}{.1443} = 2.31$$

You can also obtain a confidence interval for your friend's taste test proportion by using Formula 6.6.

**Applications of the Binomial Test**

The use of $z$ scores in conjunction with the normal distribution yields only an approximation to the exact $p$ value you would get from the binomial distribution. As we mentioned in Chapter 6, this approximation is not sufficiently accurate for NHT when *either* $N\pi$ or $N(1 - \pi)$ is less than 5.

In Chapter 6, we applied the $z$-score proportion formula to the results of an election poll. This test can also be applied to experimental situations more central to behavioral research. For example, a developmental researcher may want to test whether 10-month-old infants spend more time looking at photographs of adult faces rated as attractive by other adults than faces rated as relatively unattractive. In this case, the null hypothesis is that the infants have no preference. (Although the researcher has a hypothesis concerning which type of faces will be preferred, the more cautious procedure is to conduct a two-tailed test.) By exposing the infants to equal numbers of attractive and unattractive faces and measuring their gaze time for each type of face, each infant can be categorized as preferring attractive faces or preferring unattractive faces. The proportion of infants out of, say, 20 or more who prefer attractive faces can be tested against the null hypothesis of $\pi = .5$ using the preceding formula. (Using Formula 6.5, you can demonstrate for yourself that 15 out of 20 infants preferring attractive faces would allow rejection of $H_0$ at the .05 level with a two-tailed test.)

You can also test null hypotheses for which $\pi$ is not .5. Imagine an observational study in which you find that out of 50 leading architects, 12 are left-handed. (There is some evidence that left-handers on average have greater spatial ability than right-handers because of the way the left and right sides of their brains are organized.) In this case, $\pi$ is not .5. It is the actual proportion of left-handers in the population, which is closer to .15. The observed proportion is $p = 12/50 = .24$, and $z$ is therefore as follows:

$$z = \frac{.24 - .15}{\sqrt{\frac{.15(.85)}{50}}} = \frac{.09}{.0505} = 1.78$$

This result would be significant at the .05 level only if a one-tailed test could be justified (which is not likely). Note that because $N\pi = 50 \times .15 = 7.5$, which is greater than 5, and $N(1 - \pi)$ is also greater than 5, the use of the normal approximation would be considered reasonably accurate in this case.

### Simplified Formula for $P = .5$

In many common experimental situations, the null hypothesis will involve equal probabilities for the two alternatives. For this null hypothesis ($P = Q = .5$), the $z$-score formula that we started out with, $z = (X - NP) / \sqrt{NPQ}$, can be simplified to

$$z = (2X - N)/\sqrt{N} \qquad \text{Formula 17.5}$$

where $X$ is the number of events that fall in (either) one of the two categories.

We will have the opportunity to use this specialized version of the $z$ score formula in the next section.

## The Sign Test for Matched Samples

A particularly useful application of the binomial distribution is to serve as the null hypothesis distribution for a simple nonparametric alternative to the matched $t$ test. At the beginning of this chapter, we mentioned the example of testing a new form of imagery exercise that is designed to increase creativity. Suppose that we match young art students into pairs based primarily on the creativity they have already shown in their paintings. (The matching will not be precise, of course, but even an approximate matching can be quite helpful.) The members of each pair are assigned randomly to either the imagery training condition or a control condition that provides some plausible but ineffective version of the training. At the end of the training period, each art student produces a painting. A panel of art instructors is asked to determine, for each pair of participants, whose painting exhibited the greater amount of creativity. Let us further suppose that, of the 30 pairs of participants in the study, the experimental participant is judged to be more creative in 18 cases, the control participant is found to be more creative in 10 cases, and no determination is made in 2 cases (each of these two comparisons is judged a tie). The simplest approach is to delete the two ties from the analysis and use 28 as $N$ in the $z$ score for proportions.

Because we do not have actual measurements of creativity, we cannot calculate difference scores and compute a matched-pairs $t$ test. We will further assume that it would not be reasonable to ask the judges to compare the creativity difference for one pair to the creativity differences of other pairs, so the Wilcoxon matched-pairs test (see Chapter 8) is not a viable alternative. However, we can, at least, count how many of the creativity differences are positive and how many are negative. (We will define a difference as positive when the experimental participant is judged as more creative than his or her pair in the control condition.) The appropriate null hypothesis in such situations is that there will be equal numbers of positive and negative differences in the population (i.e., $P$, or $\pi$, = .5). We can use the binomial distribution to decide whether it is reasonable to reject the hypothesis that the signs of the difference scores in the population are just as likely to be positive as negative, and this is called **the sign test for matched samples**. Given that $H_0$ for the sign test is always that $P = .5$, as long as $N$ is at least about 15 or 20, we can use Formula 17.5, and look up our $p$ value in Table A in the Appendix. For this example, $X$ is 18, so

$$z = \frac{2X - N}{\sqrt{N}} = \frac{36 - 28}{\sqrt{28}} = \frac{8}{5.29} = 1.51$$

From Table A, the one-tailed $p$ value corresponding to this $z$ score is .0655. Even with a one-tailed test, this $z$ score would not allow rejection of the null hypothesis at the .05 level: There is not sufficient reason to believe that the imagery exercise increases creativity.

The sign test is particularly appropriate when you cannot make quantitative measurements, but you can determine the direction in which a variable changes (e.g., mental health has improved, stayed the same, or worsened). However, as we noted in Chapter 8, there are times when a nonparametric test is preferred even when the DV has been measured precisely and a matched $t$ test can be computed. Although the matched $t$ test is quite robust with respect to its assumption concerning normal distributions, when you are dealing with fairly small samples and the distribution of the difference scores bears little resemblance to a bell-shaped curve (e.g., a few of the differences are far from the others, or the difference scores form two distinct clumps), a nonparametric, distribution-free test is recommended. If the pairwise differences can be rank ordered, then the Wilcoxon matched-pairs test is much preferred to the sign test, as it usually has considerably more power than a sign test performed on the same data. When applied to quantitative data, the sign test makes absolutely no use of the amounts of the differences, which is why it usually has much less power to detect a true population effect than a matched $t$ test, or even the Wilcoxon test, applied to the same data. Nonetheless, if nearly all the differences are in the same direction, you may obtain statistical significance with the sign test, regardless of how the difference scores are distributed, or even if they cannot be compared to each other at all.

## Summary

The rules of probability for discrete events can be described in terms of counting the number of events that fall into one category or another. When a fixed number of dichotomous events are counted in terms of how many events fell into either one of the two possible categories, it is easy to construct a binomial distribution, which can then be used to draw statistical inferences.

### Probability

**Definition:**

$$P = \frac{\text{number of ways the specified event can occur}}{\text{total number of equally likely possible events}} \qquad \text{Formula 17.1}$$

**Odds against an event** = the number of unfavorable events divided by the number of favorable events

**Additive law:**

$$P(A \text{ or } B) = P(A) + P(B) - P(A \text{ and } B) \qquad \text{Formula 17.2}$$

**Multiplicative law:**

If $A$ and $B$ are independent events:

$$P(A \text{ and then } B) = P(A) \times P(B) \qquad \text{Formula 17.3}$$

### Using the Normal Distribution as an Approximation to the Binomial Distribution

First, calculate

$$z = \frac{X - NP}{\sqrt{NPQ}} \qquad \text{Formula 17.4}$$

where

$N$ = total number of events,
$X$ = number of events that fall into the category whose probability is $P$,
$Q = 1 - P$.

Then look up the $p$ value for that $z$ score in Table A. (Double this value for a two-tailed test.)

If $P = Q = .5$, the formula can be simplified to

$$z = \frac{(2X - N)}{\sqrt{N}} \qquad \text{Formula 17.5}$$

If we divide all of the terms of Formula 17.4 by $N$ (and change our notation), we obtain the same $z$-score formula for proportions that we used in Chapter 6:

$$z = \frac{p - \pi}{\sqrt{\frac{\pi(1 - \pi)}{N}}} \qquad \text{Formula 6.5}$$

### The Sign Test for Matched Samples

For each matched pair of participants, determine the *sign* of the difference between them, using some arbitrary but consistent rule (e.g., if the participant in the experimental condition performs better than the paired participant in the control group, the difference is positive). If your DV has been measured precisely, you can calculate the difference as you would for a matched-pairs $t$ test, but for this test, you would note only the sign (positive or negative) of the difference score. Then apply one of the preceding formulas, wherein $N$ is the number of pairs (after subtracting any pairs with zero differences). Generally, the null hypothesis will be such that $P = Q = .5$.

## Exercises

1. In the following problems, cards are drawn from a standard 52-card deck. Before a second draw, the first card drawn is replaced and the deck is thoroughly shuffled. Compute each of the following probabilities:
   (a) The probability of drawing a 10 on the first draw.
   (b) The probability of drawing either a deuce, a 3, a 4, a 5, a heart, or a diamond on the first draw.
   (c) The probability of drawing the ace of spades twice in a row.
   (d) The probability of drawing either a jack, a 10, a 7 of clubs, or a spade on the first draw *and then* drawing either the ace of diamonds or 9 of hearts on the second draw.
   (e) Recompute the probabilities asked for in parts (c) and (d) assuming that the first card drawn is *not* replaced.

2. The following questions refer to the throw of one fair, six-sided die.
   (a) What is the probability of obtaining an odd number on one throw?
   (b) What is the probability of obtaining seven odd numbers in seven throws?
3. One hundred slips of paper bearing the numbers from 1 to 100, inclusive, are placed in a large hat and thoroughly mixed. What is the probability of drawing ...
   (a) The number 17?
   (b) The number 92?
   (c) Either a 2 or a 4?
   (d) A number from 7 to 11, inclusive?
   (e) A number in the 20s?
   (f) An even number?
   (g) An even number, and then a number from 3 to 19, inclusive? (The number drawn first is replaced prior to the second draw.)
   (h) A number from 96 to 100, inclusive, or a number from 70 to 97, inclusive?
4. Slips of paper are placed in a large hat and thoroughly mixed. 10 slips bear the number 1, 20 slips bear the number 2, 30 slips bear the number 3, and 5 slips bear the number 4. What is the probability of drawing ...
   (a) A 1?
   (b) A 2?
   (c) A 3?
   (d) A 4?
   (e) A 1 or a 4?
   (f) A 1 or a 2 or a 3 or a 4?
   (g) A 5?
   (h) A 2 and then a 3? (The number drawn first is replaced prior to the second draw.)
5. Imagine that you want to test whether a particular coin is biased or fair by flipping the coin four times and counting the number of times it comes up heads.
   (a) How many different sequences can be produced by flipping the coin four times? How many different values can $X$ (the number of heads) take on?
   (b) Graph the binomial distribution for $N = 4$ and $P = .5$ so that the height of each bar represents the probability that corresponds to each value of $X$.
   (c) If the coin were to land on tails four times in a row, could you reject the null hypothesis that the coin is fair? Explain.
6. Apply the sign test to the data from Exercise 8 in Chapter 7, using one of the normal approximation formulas. Explain the difference in results between this exercise and the one in Chapter 7.
7. Suppose that after six months of a new form of treatment for chronic schizophrenia, 18 patients exhibited some improvement, 4 did not change, and 6 patients actually got worse.
   (a) Using the sign test, can you reject the null hypothesis (that the new treatment has no effect) at the .05 level? (Show the $z$ score and $p$ value that you used to answer this question.)
   (b) A more conservative way to conduct the binomial test is to assign half of the tied cases arbitrarily to each category (one pair is discarded if there is an odd number of ties). Redo part (a) after applying this approach to tied cases.
8. At a small rural college, incoming students are arranged into 320 *pairs* by the administration based on their high school credentials. (The students do not know this). One member of each pair is chosen randomly to participate in an enriched orientation program. At the end of the first semester, it is found that 55% of enriched-program students are getting higher grades than the (control) students they have been matched with. Can you reject the null hypothesis at the.05 level with a two-tailed test? With a one-tailed test? Explain the basis for your statistical decisions.

## Thought Questions

1. Give an example of a psychological research study where the variables can be expressed only as categories, and not in any quantitative way.
2. You want to determine if a coin is biased in favor of falling heads up or tails up. (a) You flip the coin three times and obtain three heads. What should you decide? (b) Why is the informal experiment in part (a) badly designed? (c) Which error is more likely to result from the experiment in part (a), Type I or Type II? (d) Suppose you flip the coin six times and obtain six heads. What should you decide? (e) What distribution is used to reach a decision in part (a) and part (d)?
3. When should the normal distribution be used as an approximation to the binomial distribution?
4. A researcher wishes to determine if a particular form of psychotherapy will improve the mental health of patients who suffer from depression. If the researcher cannot obtain a quantitative measure of mental health but can conclude whether it has improved, stayed the same, or become worse for each patient, what application of the binomial distribution should the researcher use?

## Computer Exercises

1. For Jackson's data, perform a binomial test on the exercise variable, with $P = .5$. Can you reject the null hypothesis that Jackson's statistics class is a random sample from a population in which both individuals that exercise regularly and those who do not are equally represented? Use the binomial test to compare the proportion of people who *do not* exercise regularly in the class against the null hypothesis of $P = .55$. Can you reject the null hypothesis in this case?
2. Redo the binomial test on exercise (with $P = .5$, only) separately for each college major.
3. In this exercise, you will be testing the (two-valued) grouping variable you created from a median split of the procrastination scores for the computer exercises in Chapter 15. Perform a binomial test for $P = .5$ separately for each exercise group. Explain the relationship between the tests of the two exercise groups.
4. If you have not already done this for a previous exercise, create a variable that is the difference between baseline and prequiz heart rate. Then recode this new variable so that negative difference scores are assigned a numeric score of 1, positive difference scores are assigned a score of 2, and zero differences are assigned as missing. Perform a binomial test on this new variable with $P = .5$. Compare the $p$ value from this test with the $p$ value you obtained for the matched-pairs $t$ test on the same two variables in Chapter 7. Explain the relationship between these two alternative tests of the same difference scores.
5. If your software has the capability, request a sign test of the baseline versus the prequiz heart rates, and compare the results to those from the binomial test in the previous exercise.
6. Repeat Exercise 5 for the baseline and prequiz anxiety scores. (If your software does not have the sign test, repeat Exercise 4 for the anxiety variable.)

## Bridge to SPSS

SPSS has a number of options for creating and deriving probabilities from a binomial distribution, and it makes some arbitrary choices without informing you. So, you will need to read this section carefully. Note that the procedures we discuss in this section appear as "Legacy Dialogs" in the Nonparametric Tests menu.

### The Binomial Test

If one of your variables has only two different numerical values, you can perform a binomial test on that variable by selecting **Binomial ...** from the **Analyze/Nonparametric/Legacy Dialogs** menu. All of your numeric variables will appear in the list on the left side of the dialog box (see Figure 17.2), from which they can be moved to the *Test Variable List*. If all of the variables in your *Test Variable List* have no more than two values each, you can leave Define Dichotomy at its default setting: *Get from data*. SPSS will not perform a binomial test on any variable that has more than two values if *Get from data* has been chosen. The default value for the Test Proportion is .50, but it can be changed to any value between .001 and .999. For instance, consider the variable *coffee* from Jackson's data set. If you want to test whether you have a significantly greater proportion of regular coffee drinkers in your sample than in the larger population, you can type your best estimate of the relevant population proportion (e.g., .30) in the *Test Proportion:* box.

Note, however, that when the *Test Proportion* is changed to a value other than .5, you will be dealing with a binomial distribution that is not symmetrical, and you will get a one-tailed $p$ value. If that value is labeled as "Exact Sig. (one-tailed)" then you know that the actual binomial distribution is being used, rather than a normal approximation (if a normal approximation is used, SPSS uses the label "Asymp." for asymptotic, rather than "Exact"). SPSS then compares the proportion in the smaller of the two groups to your test proportion (or the group associated with the *lower* numerical value if the two groups are the same size). For example, because there are fewer regular coffee drinkers ($N$ = 42) than nondrinkers ($N$ = 58) in Jackson's data set, your test proportion should be based on the population proportion of coffee drinkers. If those sample sizes were reversed, you would enter an estimate of the population proportion of those who are not regular coffee drinkers.

Also note that SPSS always gives you the $p$ value for the *tail* of the binomial distribution—that is, the portion of the binomial distribution that begins at your actual proportion and heads away from the test proportion. SPSS assumes that you would not want to know the $p$ value if you picked the wrong tail. Of course, you can always subtract the one-tailed Sig. value that SPSS gives you from 1.0 to get the other portion of the binomial distribution. For our coffee drinker example, we set the *Test Proportion* to .30 to obtain the results table shown in Figure 17.3. The table shows that the proportion of coffee drinkers in Jackson's study (.42) is significantly *higher* than the hypothesized proportion of .30 ($p$ = .007). If, instead, we had hypothesized that .54 of the population are regular coffee drinkers, the Binomial test would have returned a $p$ value of .011, showing that the proportion of coffee drinkers in the study is significantly *lower* than .54.

Figure 17.2

**The Dialog Box for the Binomial Test**

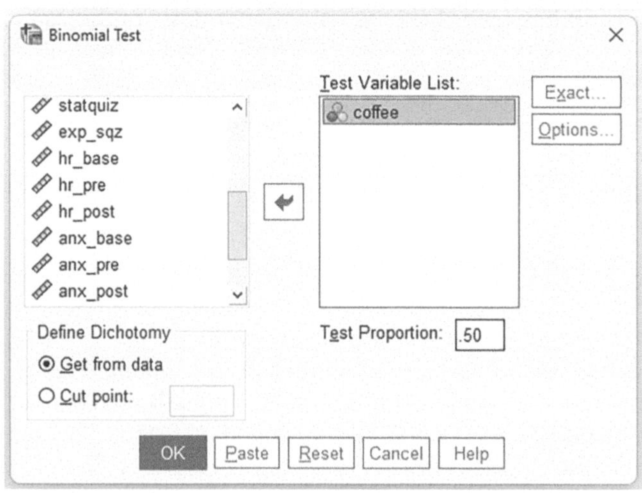


*Source*: Courtesy of International Business Machines Corporation, © SPSS, Inc., an IBM company

**Binomial Test**

| | | Category | N | Observed Prop. | Test Prop. | Exact Sig. (1-tailed) |
|---|---|---|---|---|---|---|
| Coffee | Group 1 | 1 | 42 | .4 | .3 | .007 |
| | Group 2 | 0 | 58 | .6 | | |
| | Total | | 100 | 1.0 | | |

**Figure 17.3**

**The Main Results Table for a Binomial Test on a Variable with Two Values**

**Binomial Test**

| | | Category | N | Observed Prop. | Test Prop. | Exact Sig. (2-tailed) |
|---|---|---|---|---|---|---|
| anx_base | Group 1 | <= 19 | 67 | .67 | .50 | <.001 |
| | Group 2 | > 19 | 33 | .33 | | |
| | Total | | 100 | 1.00 | | |

**Figure 17.4**

**The Main Results Table for a Binomial Test Using a Cut Point on a Variable with Many Values**

When analyzing the data from an experiment with a dichotomous outcome, the null hypothesis is usually .5 (e.g., two choices are equally preferred by the participants). In that case, the *Test Proportion* would be kept at .50, and the output table for the Binomial test will give you a *two-tailed p* value. You can also run the Binomial test if your test variable has more than two numeric values, by selecting **Cut point** to define your dichotomy (see Figure 17.2), and then typing in a value somewhere between your highest and lowest values for that variable. SPSS will find the proportions above and below that cut point and then compare the smaller of the two proportions to the *Test Proportion* that you set. For example, suppose that, in a national study of the anxiety measure Jackson is using, the median population score is 19. If Jackson wants to know if his students are significantly less anxious than the larger population, they can set the cut point to 19 and the Test Proportion to .5. The results, showing that Jackson's students are significantly less anxious (two-thirds of Jackson's students are below the national median) can be seen in Figure 17.4.

### The Sign Test

To perform a Sign test, select **2 Related Samples** from the **Analyze/Nonparametric Tests/Legacy Dialogs** menu. Just move two related numeric variables from the list at the left of the dialog box to the *Test Pair(s):* space, as you would for a matched-pairs *t* test (see Figure 17.5). Then, under *Test Type*, click on the check box for **Sign** and unselect **Wilcoxon**

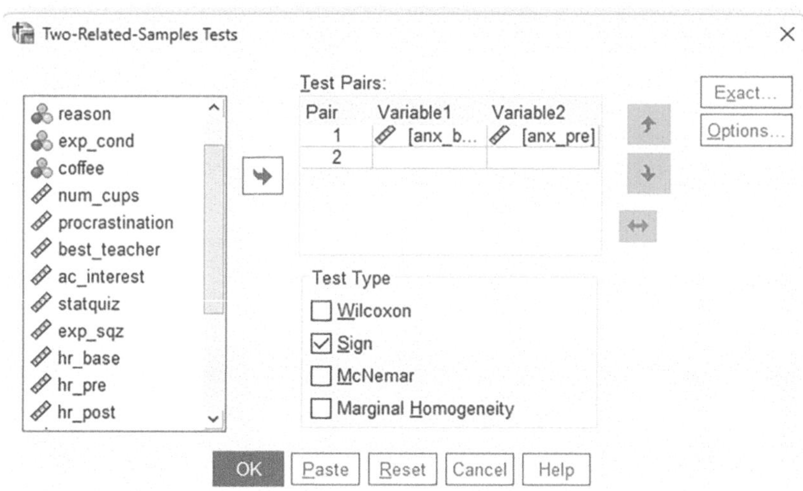


*Source*: Courtesy of International Business Machines Corporation, © SPSS, Inc., an IBM company

**Figure 17.5**

**The Dialog Box for the Sign Test**

## Figure 17.6

**The Results of a Sign Test Comparing Baseline Anxiety to Pre-Quiz Anxiety**

**Frequencies**

| | | N |
|---|---|---|
| anx_pre - anx_base | Negative Differences[a] | 35 |
| | Positive Differences[b] | 64 |
| | Ties [c] | 1 |
| | Total | 100 |

a. anx_pre < anx_base
b. anx_pre > anx_base
c. anx_pre = anx_base

**Test Statistics[a]**

| | anx_pre - anx_base |
|---|---|
| Z | −2.814 |
| Asymp. Sig. (2-tailed) | .005 |

a. Sign Test

(we already used this same dialog box for the Wilcoxon test in Chapter 8). After you click **OK**, SPSS will perform a binomial test by comparing the proportion of positive (or negative—the results are the same) difference scores (zero differences are excluded) to a test proportion (i.e., null hypothesis) of .5. (This test proportion is fixed for this procedure.)

The default output consists of two tables, as shown in Figure 17.6. The first table tells us that for 35 students, their anxiety actually decreased after the quiz was announced (*Negative Differences*); for 64 students, anxiety increased (*Positive Differences*); and for just one student, there was no change in anxiety (a "Tie"). The results in the second table are based on a normal approximation to the binomial distribution (which is quite accurate for a sample of 100), hence the $Z$ score ($-2.814$) and the asymptotic $p$ value (.005). The Sign test can be even more powerful than the matched $t$ test and Wilcoxon test (as in this case) on the same data if there is wide variation in the sizes of the difference scores, but they tend to go in the same direction. We don't know why the $Z$ score was reported as negative, but rather than relying on the sign of a test statistic, you should always look at the descriptive statistics to determine the direction of your results.

# Chapter 18
# Chi-Square Tests

## PREVIEW

### Introduction

### Chi-Square and Goodness of Fit: One-Variable Problems

What are the procedures for testing the hypothesis that more than two population frequencies are distributed in a specified way?

What is the observed frequency? The expected frequency? The chi-square statistic?

What is the correct statistical model to use with frequency data?

How do we test chi-square for statistical significance?

How does the chi-square test compare to the binomial test in the two-category case?

When should chi-square not be used?

### Chi-Square as a Test of Independence: Two-Variable Problems

What are the procedures for testing the significance of the relationship between two variables when data are expressed in terms of frequencies?

How do we compute chi-square from a 2 × 2 table?

### Measures of Strength of Association in Two-Variable Tables

Why is it desirable to convert a statistically significant value of chi-square into a measure that shows the strength of the relationship between the two variables?

What are the two most common categorical measures of strength of relationship? When should each of these measures be used?

### Summary

### Exercises

### Thought Questions

### Computer Exercises

### Bridge to SPSS

## Introduction

Consider the following problem:

A publisher introduces three titles for a new literary magazine for women—*Today*, *Choice*, and *New Alternatives*. She wonders whether these titles will be equally popular among female consumers. To test the hypothesis of equal preference, she obtains a random sample of 177 women and asks them which *one* of the three titles they like best. She finds that 65 women prefer *Today*, 60 prefer *Choice*, and 52 prefer *New Alternatives*. Are these results sufficient reason to reject the null hypothesis that equal numbers of women *in the population* prefer each of the three titles? Or are these results likely to occur as a result of

*Introductory Statistics for the Behavioral Sciences*, Eighth Edition. R. Brooke Lea and Barry H. Cohen.

Companion website: www.wiley.com/go/ISBS8e

sampling error if the preferences in the population are in fact equal, in which case the null hypothesis of equality in the population should be retained?

In the previous chapter, you saw how to test null hypotheses when events are constrained to only two possible categories. In this chapter, you will learn how to deal with data that consist of the number of objects (usually persons) falling in any one of a number of categories. That is, the data are in terms of empirically observed *frequencies* for *more than two* categories that can then be compared to frequencies expected on the basis of some hypothesis.

## Chi-Square and the Goodness of Fit: One-Variable Problems

The data for the example just given consist of frequencies or head counts—that is, *how many* prefer a particular alternative—and the objective is to decide whether it is reasonable to conclude that the corresponding population frequencies are distributed in a specified way. Thus, in the example, the publisher now knows how many women in the sample prefer each title to the other two. However, what she would like to know is whether it is reasonable to conclude that the population from which her sample was randomly drawn is equally divided with regard to preference for the three titles.

Testing hypotheses about frequency data that fall in more than two category-sets is similar in strategy to the tests we performed on dichotomous data in the previous chapter. However, to find probabilities associated with observed frequencies involving three or more outcomes for each event, we cannot use the binomial distribution. With three categories, we would need a trinomial distribution. In the general case of any number of possible categories, the appropriate statistical model would be the *multinomial distribution*.

Just as the normal distribution can be used to approximate the more exact results from the binomial distribution, a convenient statistic can be calculated from multicategory frequency data. The distribution of this statistic can be reasonably approximated by what is known as the **chi-square distribution**. (An exception does occur if you are dealing with very small frequencies, a problem we discuss later in this chapter.) This statistic is named after the distribution it tends to follow, so it is called the **chi-square statistic** and is symbolized by the lowercase Greek letter chi (pronounced *kie* to rhyme with *pie*) squared: $\chi^2$.

Because we may be dealing with any number of categories, we can no longer use $X$ to represent the frequency of one category and $N - X$ as the number in the other category. Instead, we will need to keep track of the **observed frequency ($f_o$)** in each category. Similarly, rather than dealing with a single value that is expected for $X$ according to the null hypothesis (i.e., $NP$), each category will have its own **expected frequency ($f_e$)**. The chi-square statistic is based on the discrepancy between the observed and expected frequency for each category, as expressed in the following formula:

$$\chi^2 = \sum \frac{(f_o - f_e)^2}{f_e} \qquad \text{Formula 18.1}$$

where

$f_o$ = observed frequency
$f_e$ = expected (null-hypothetical) frequency
$\sum$ is taken over all the categories

If the differences between the observed frequencies and the expected frequencies are small, $\chi^2$ will be small. The greater the difference between the observed frequencies and those expected under the null hypothesis, the larger $\chi^2$ will be. If the differences between the observed and expected values are so large collectively as to occur by chance only .05 or less of the time when the null hypothesis is true, the null hypothesis is rejected.

The problem given at the beginning of this chapter is concerned with differences in choice among different categories (or levels) of a *single* variable, so the appropriate statistical analysis is often referred to as a **one-way chi-square test**. In that example, the one variable is magazine title preference, and each respondent selects one of three possible choices.

## Table 18.1 Chi-Square Test for Magazine Title Choices

| Title | Observed frequency ($f_o$) | Expected frequency ($f_e$) | $f_o - f_e$ | $(f_o - f_e)^2$ | $\frac{(f_o - f_e)^2}{f_e}$ |
|---|---|---|---|---|---|
| *Today* | 65 | 59 | 6 | 36 | 0.610 |
| *Choice* | 60 | 59 | 1 | 1 | 0.017 |
| *New Alternatives* | 52 | 59 | −7 | 49 | 0.831 |

$$\chi^2 = \sum \frac{(f_o - f_e)^2}{f_e} = 1.458$$

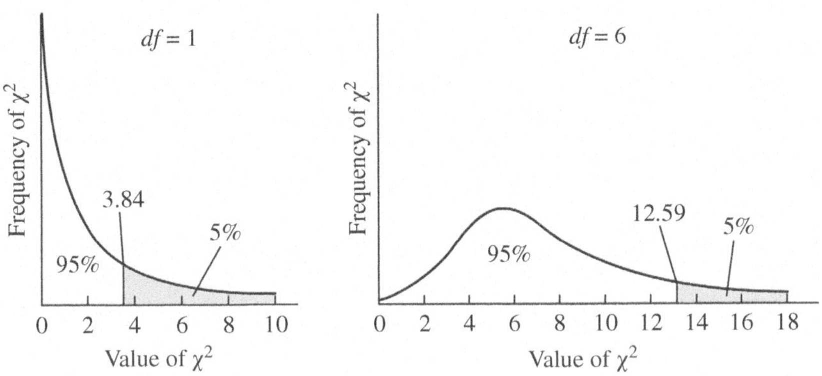


## Figure 18.1

**Chi-Square Distribution for $df = 1$ and $df = 6$**

To analyze such a problem, the first step is to set up a table with the three alternatives and tabulate the observed frequencies (see Table 18.1). Next, the expected frequencies must be stated before $\chi^2$ can be determined. The null hypothesis states that the three titles are equally preferred, so *if* it is true, you would expect the sample of 177 women to be equally divided among the three categories. Thus, the expected frequencies under the null hypothesis are 177/3, or 59, 59, and 59. (Note that the sum of the expected frequencies must be equal to the sum of the observed frequencies.) Is it likely or unlikely that observed frequencies of 65, 60, and 52 would occur if the population frequencies are exactly equal? In other words, are the observed frequencies of 65, 60, and 52 significantly different from the expected frequencies of 59, 59, and 59? To answer this question, $\chi^2$ has been computed in Table 18.1. Note that *each* value of $(f_o - f_e)^2$ is divided by its own expected frequency. (All the expected frequencies happen to be equal in this particular problem, but this will not always be true as it depends on the particular null hypothesis.) The resulting three values are then summed to obtain $\chi^2$, which is equal to 1.46.

In order to test the significance of $\chi^2$ using a specified criterion of significance, the obtained value is referred to Table H in the Appendix with the appropriate degrees of freedom (*df*). Note that in this table, there is a different value of $\chi^2$ for every *df*. Chi-square, like *t*, yields a family of curves, with the shape of a particular curve depending on the *df* (See Figure 18.1). For the $\chi^2$ distributions, however, the *df* is based on the number of *categories* rather than on the sample size as in the case of the *t* distributions. In the one-variable case, the *df* are equal to $k - 1$, where $k$ is equal to the number of categories of the variable. Thus, in the present problem, $df = 3 - 1 = 2$. For $\chi^2$ to be significant at the .05 level, therefore, the obtained value must be equal to or greater than 5.99. Since the obtained value is only 1.46, the null hypothesis is retained; the evidence is not sufficient to reject the null hypothesis that the frequencies in the population are equal. Therefore, we do not have a sufficient basis for concluding that any particular title or titles is (are) preferred.

### Expected Frequencies from a Preexisting Population

In experimental applications, the null hypothesis for a one-way chi-square test is, most commonly, that all of the expected frequencies are equal. However, if you are dealing with preexisting groups in the population, this need not be the case. Suppose it is claimed that a local newspaper has a political slant, and as proof of this claim, it is offered that its readership does not reflect the political composition of the area it serves. For instance, if a survey of 210 readers found that 120 were registered Democrats, 60 were registered Republicans, and 30 were Independents, the expected frequencies to test against would not necessarily be 70 (i.e., 210/3) for each category. If the null hypothesis is that the readership *does* reflect the political affiliations of the community, it would make sense to derive the expected frequencies from voter registration records for that community. For example, the voter registration percentages might turn out to be: 40% Democrat, 35% Republican, and 25% Independents. In that case, the expected frequencies you would want to use for your one-way chi-square test would be: 84 for Democrats (i.e., .4 × 210), 73.5 for Republicans (i.e., .35 × 210), and 52.5 for Independents (i.e., .25 × 210), which sum to 210. You can see from the survey results that the newspaper does indeed have more Democratic (120) and Independent (30) readers and fewer Republican (60) ones than you would expect from the proportions of political affiliations in that community. However, performing the chi-square test would allow you to determine whether the observed discrepancies are large enough to safely reject the null hypothesis and declare that the newspaper's readership is not a simple reflection of its community's political affiliations. (As an exercise, use Formula 18.1 to verify that the value of the $\chi^2$ statistic equals 27.55, in this case.)

The preceding example demonstrates why the one-way chi-square test is often called a **goodness-of-fit** test: In that example, we were testing the fit between the political affiliations of the newspaper's readership and the political affiliations of the community served by that newspaper. In such cases, it is likely that the purpose of the test would be to reject the null hypothesis and declare a *lack of fit*. However, the one-way chi-square can just as easily be used to test the fit between the proportions of particular responses in your data and the specific proportions that can be predicted from a theory you want to test. In this type of example, the expected frequencies are not coming from a null hypothesis but rather from an alternative hypothesis based on your theory. Therefore, you do not want to reject the hypothesis being tested; instead, you would be hoping for a very small value of the $\chi^2$ statistic (ideally, close to zero) that would demonstrate a good fit between your data and your predictions.

As an example of a case in which you would be hoping for the $\chi^2$ value to be as far from statistical significance as possible, imagine that you are observing whether participants choose to sit on the right, on the left, or in the middle, when offered a choice of three chairs facing a small stage. Suppose that your theory predicts that people who are "purely" right handed will choose the left seat; those who are "purely" left handed will choose the right seat; and all others will choose to sit in the middle. In this example, the expected frequencies would again depend on proportions in the population from which your participants are drawn—specifically, the proportions of people who are purely right handed, purely left handed, or somewhere between. If these proportions were known to be .6, .1, and .3, respectively, and you plan to record the seating choice for 60 different participants, your expected frequencies would be: 36 (i.e., .6 × 60) for the right chair; 6 (i.e., .1 × 60) for the left chair; and 18 (i.e., .3 × 60) for the middle chair. If the actual preferences were to come out roughly equal for the three chairs (about 20 participants choosing each), you would not be happy at all. Even if the chi-square test were to fall short of statistical significance, the fact that the $\chi^2$ statistic would not be very small in this case would be discouraging and should cause you to rethink your theory concerning the connection between handedness and seat preference.

### The Relationship Between the Binomial Test and the Chi-Square Test With Two Categories

Based on comments that some of her students have made, a teacher believes that statistics students in general do not want their final examinations to be cumulative (i.e., encompassing all of the material from the beginning to the end of the course). Rather, the students seem

**Table 18.2** **Chi-Square Test for Attitude Toward Cumulative Final Examinations**

| Attitude | $f_o$ | $f_e$ | $f_o - f_e$ | $(f_o - f_e)^2$ | $\frac{(f_o - f_e)^2}{f_e}$ |
|---|---|---|---|---|---|
| For | 15 | 37.5 | −22.5 | 506.25 | 13.5 |
| Against | 60 | 37.5 | 22.5 | 506.25 | 13.5 |
| | 75 | 75.0 | | | |

$$\chi^2 = \sum\frac{(f_o - f_e)^2}{f_e} = 27$$

to want the final exam to cover only the second half of the course. She decides to use her own class as though it were a random sample of all statistics students, and she asks her students to turn in anonymous notes stating whether they want the final exam to be cumulative or not. Of the 75 students who hand in notes, She finds that 15 students are in favor of cumulative final exams and 60 students are against them. If the null hypothesis is that the population of statistics students is equally divided on this issue, the expected frequencies are 75/2 = 37.5 for each category. ($f_e$s frequently come out to fractional values, even though it is not possible for half a student to vote one way or the other.) The computation of the chi-square statistic for this problem is shown in Table 18.2.

Because the calculated chi-square statistic, 27, is larger than the critical value of 3.84 from Table H for 2 − 1 = 1 *df* using the .05 criterion of significance, the null hypothesis is rejected; it is *not* reasonable to assume that the student population is evenly divided on this issue. By inspecting the $f_o$, it may be concluded that a majority is opposed to final examinations that are cumulative. The $\chi^2$ value of 27 would have been significant even if we used an alpha of .001 (e.g., we are performing 50 such tests and use the Bonferroni correction to adjust the alpha for each test) because 27 exceeds $\chi^2_{.001}(1) = 10.83$.

Because the situation just described involves only two categories (for and against), the null hypothesis can also be tested by using the binomial distribution. Given that $P = .5$ and $N$ is rather large, we can use Formula 17.5 from the previous chapter to create a $z$ score:

$$z = \frac{2X - N}{\sqrt{N}} = \frac{2(60) - 75}{\sqrt{75}} = \frac{45}{8.66} = 5.196$$

If you square this $z$ score, you will get 27, which is the value we calculated for chi-square. (If in this formula you used 15, the number of students who voted "for," instead of 60, the $z$ score would have the same magnitude but the opposite sign, which would yield the same $p$ value and make no difference at all after squaring.) Squaring the normal distribution yields a chi-square distribution with 1 degree of freedom. For example, the .05 critical value for $\chi^2(1)$ is 3.84, which equals 1.96 squared.

### Some Precautions Involving the Use of $\chi^2$

The $\chi^2$ tests just described can be validly performed only when the observations are *independent*. That is, no response should be related to or dependent upon any other response. For example, it would be incorrect to apply $\chi^2$ with $N = 100$ to the true-false responses of five patients with schizophrenia to 20 questionnaire items. The 100 responses are not independent of each other because the 20 responses given by each patient must be assumed to be mutually related. This assumption of independence is also critical to the validity of the binomial test.

Second, any case, participant, and so on must fall into *one and only one* category. Thus, in the problem involving the preference for magazine titles, each participant was asked to choose the one title she most preferred.

Third, the computations must be based on all the cases in the sample. In the previous example, a category of "for" as well as "against" had to be included so that $\chi^2$ would be based on the total frequency of 75, the total size of the sample. As a check, the sum of the observed frequencies *must* be equal to the sum of the expected frequencies.

One final precaution is concerned with the size of the expected frequencies. Like the *z*-score formula for the binomial test, chi-square is actually an approximate test for obtaining the probability values for the observed frequencies (i.e., the probability of getting the observed frequencies if the null hypothesis is true). This test is based on the expectation that within *any category*, sample frequencies are normally distributed about the population or expected value. Since frequencies cannot be negative, the distribution cannot be normal when the *expected* population values are *close to zero*. The sample frequencies cannot be much below the expected frequency, while they can be much above it—an asymmetric distribution.

Under certain conditions, therefore, you should *not* compute $\chi^2$. For 1 *df*, all of the *expected* frequencies should be at least 5; this is consistent with the rule for the binomial test—that both $NP$ and $N(1 - P)$ must be at least 5. For 2 *df*, all expected frequencies should exceed 2. With 3 or more *df*, if all expected frequencies but one are greater than or equal to 5, and if the one that is not is at least equal to 1, $\chi^2$ is still a good approximation. In general, the greater the *df*, the more lenient the requirement for minimum expected frequencies.

## Chi-Square as a Test of Independence: Two-Variable Problems

In the preceding section, $\chi^2$ was used to test an a priori hypothesis about expected frequencies—that is, a hypothesis about $f_e$ values formulated prior to the experiment—which involved frequency data concerning a single variable. Chi-square can also be used, however, to test the significance of *the relationship between two variables when data are expressed in terms of frequencies of joint occurrence*.

Suppose you want to find out if students majoring in psychology differ in their career preferences from students majoring in economics. If the two variables of major and career choice are *not* related (are *independent*), you would expect the proportion of psychology majors who prefer, say, an academic to an applied career to be the same as the proportion of economics majors who prefer academia. A numerical example in which the proportions are equal for the two majors is shown in Table 18.3. Here, in contrast to one-variable problems, no a priori expected frequencies are involved. You are dealing with the relationship between two variables, each of which may have any number of categories or levels. (In this example, each variable has two levels.) The null hypothesis is that the two variables are independent, which (as you will see) implies a set of expected frequencies. Because there

**Table 18.3 Hypothetical Illustration of Perfectly Independent Relationship Between Major and Career Preference, Using Frequency Data**

| | | Undergraduate major | | | | |
|---|---|---|---|---|---|---|
| | | **Economics** | | **Psychology** | | **Total students** |
| Career preference | Applied | 40 | | 60 | | 100 |
| | Academic | 32 | | 48 | | 80 |
| | Total | 72 | + | 108 | = | 180 |

55.56% of the econ majors (40/72) prefer applied careers; 44.44% (32/72) prefer academia.
55.56% of the psych majors (60/108) prefer applied careers; 44.44% (48/108) prefer academia.
40.00% of those preferring applied careers are econ majors (40/100); 60.00% (60/100) are psych majors.
40.00% of those preferring academia are econ majors (32/80); 60.00% (48/80) are psych majors.
Thus, there is no relationship between major and career preference.

**Table 18.4** Table of Frequencies Relating Undergraduate Major to Career Preference (Hypothetical Data)[a]

| | | Undergraduate major | | | | |
|---|---|---|---|---|---|---|
| | | Economics | | Psychology | | Total students |
| | Applied | 45 (40) | | 55 (60) | | 100 |
| Career preference | Academic | 27 (32) | | 53 (48) | | 80 |
| | Total | 72 | + | 108 | = | 180 |

Calculations

| $f_o$ | $f_e$ | $f_o - f_e$ | $(f_o - f_e)^2$ | $\frac{(f_o - f_e)^2}{f_e}$ |
|---|---|---|---|---|
| 45 | 40 | 5 | 25 | 0.625 |
| 55 | 60 | −5 | 25 | 0.417 |
| 27 | 32 | −5 | 25 | 0.781 |
| 53 | 48 | 5 | 25 | 0.521 |
| | | | | $\chi^2$ = 2.344 |

[a]Within each cell, $f_e$ is in parentheses.

are two variables being jointly tested, the appropriate analysis is usually called a **two-way chi-square test**.

As an illustration of the psychology/economics example just mentioned, in which major is *not* independent of career choice, suppose you have a random sample of 108 psychology majors and a random sample of 72 economics majors. You ask each individual to state which of the two types of careers he or she prefers. With two categories of career type and two undergraduate majors, there are four possible combined categories, or *cells*: econ-applied; econ-academic; psych-applied; and psych-academic. The entire analysis is shown in Table 18.4.

As you can see from the main entries in the top half of Table 18.4, 45 econ majors preferred applied careers, 55 psych majors preferred applied careers, 27 econ majors preferred academia, and 53 psych majors preferred academia. The total for the first column (called the *marginal frequency* for that column) indicates that the sample contains 72 (i.e., 45 + 27) economics majors. The marginal frequency for the second column shows that the sample contains 108 psychology majors. Similarly, the row marginal frequencies indicate that 100 students across both majors preferred an applied career while 80 preferred academia. The total sample size, *N*, is equal to 180.

The null hypothesis is stated in terms of the independence of the two variables, undergraduate major and career preference.

$H_0$: major and career preference are independent (*not* related)

$H_A$: major and career preference *are* related

The expected frequencies, as shown in parentheses in the top half of Table 18.4, are the same as the frequencies used in Table 18.3 to illustrate the independence of the two variables because independence is what is expected under the null hypothesis for the two-way chi-square test. The procedure for computing the expected frequencies can be summarized as follows: For any cell, the expected value is equal to the product of the two marginal frequencies common to the cell (the row total times the column total) divided by the total *N*. That is,

$$f_e = \frac{(\text{row total})(\text{column total})}{N} \qquad \text{Formula 18.2}$$

For example, in Table 18.4, the expected frequency for the econ-applied cell is equal to

$$\frac{(100)(72)}{180} = 40$$

Precisely 72/180, or 40.00%, of the sample are economics majors, and 100/180, or 55.56%, of the sample prefer applied careers. So, 40.00% of 55.56%, or 22.22%, of the 180 cases, or (72/180)(100/180)(180) = 40 of them, would be expected to be economics majors preferring applied careers if there were no relationship between major and career preference in the sample. The other $f_e$ values may be found by simply using Formula 18.2 in each case, based on the same rationale.

The computation of $\chi^2$ is shown in the bottom half of Table 18.4. The resulting value of 2.34 is tested for statistical significance by referring to Table H in the Appendix with the proper *df*. For a two-variable problem, the *df* are equal to

$$df = (r - 1)(c - 1) \qquad \text{Formula 18.3}$$

where

$r$ = number of rows
$c$ = number of columns

For the 2 × 2 table, the degrees of freedom are equal to (2 − 1)(2 − 1) = 1. The reason why there is 1 *df* in a 2 × 2 table is as follows: Consider the expected frequency of 40 for the econ-applied cell. Having computed this value, the expected frequency for the psych-applied cell is fixed (not free to vary). This is because the total of the expected frequencies in the first row of the table, like the total of the observed frequencies, must add up to 100—the marginal frequency for that row. Thus, the expected frequency for the psych-applied cell is equal to

$$100 - 40 = 60$$

Similarly, the expected values in the first column must add up to the marginal frequency of 72. Having computed the expected value of 40 for the econ-applied cell, the expected frequency for the econ-academic cell is fixed; it must be

$$72 - 40 = 32$$

Finally, the expected frequency for the psych-academic cell is equal to

$$108 - 60 \text{ (or } 80 - 32) = 48$$

Thus, once any one $f_e$ is known, all the others are automatically determined. In other words, only one $f_e$ is free to vary; the values of the others depend on its value. Therefore, the 2 × 2 table has only 1 *df*.

In a larger table, all but one of the $f_e$ values in a given row or column are free to vary. Once they are specified, the last one is fixed by virtue of the fact that the expected frequencies must add up to the marginal frequency. An illustration of the *df* for a 4 × 3 table is shown in Figure 18.2.

Returning to the problem in Table 18.4, the minimum value of $\chi^2$ required to reject $H_0$ for $df = 1$ and $\alpha = .05$ is 3.84. Because the obtained value of 2.34 is less than the tabled value, you retain the null hypothesis and conclude that your data do not provide a sufficient reason to believe that the variables of undergraduate major and career preference are related.

## Figure 18.2

**Illustration of Degrees of Freedom for a 4 × 3 Table**

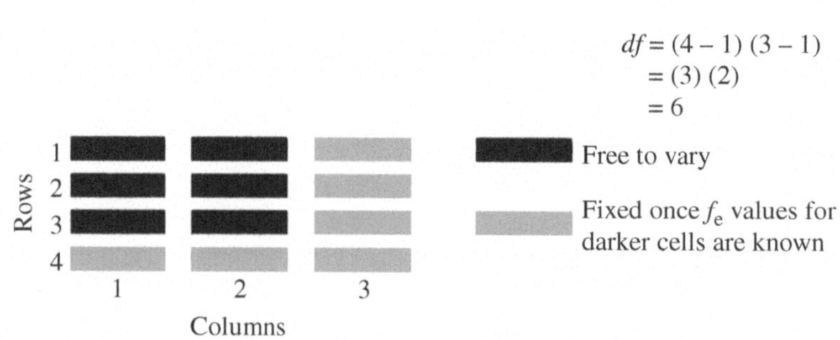

### Computing Formula for 2 × 2 Tables

For a 2 × 2 table, the following formula for $\chi^2$ is equivalent to the one given previously and requires somewhat less work computationally:

Observed frequencies:

| | |
|---|---|
| A | B |
| C | D |

$$\chi^2 = \frac{N(AD - BC)^2}{(A + B)(C + D)(A + C)(B + D)} \qquad \text{Formula 18.4}$$

As an illustration, consider the following 2 × 2 table:

| | | |
|---|---|---|
| 50 | 70 | 120 |
| 19 | 41 | 60 |
| 69 | 111 | 180 = *N* |

$$\chi^2 = \frac{180(50 \times 41 - 70 \times 19)^2}{(120)(60)(69)(111)}$$

$$= \frac{(180)(720)^2}{(120)(60)(69)(111)}$$

$$= 1.692$$

We leave it as an exercise for the reader to apply the 2 × 2 shortcut formula to the data in Table 18.4 and verify that it yields the same chi-square value as the longer method shown in that table.

### Two-Way Chi-Square Example With Multiple Levels for Both Variables

The chi-square statistical procedure discussed in this section can be used for two-variable problems with any number of levels for each variable. To illustrate, consider a new environmental treatment that is designed to improve the general mental health of all inpatient psychiatric patients and facilitate the effectiveness of whatever other treatments they are receiving. (The environmental treatment could involve redecorating the ward in cheerful colors, playing soothing music, or even releasing pleasant aromas.) Suppose that this environmental manipulation is implemented in a large psychiatric facility in which 300 patients reside: 138 have been diagnosed with schizophrenia, 88 with bipolar disorder, 42 with major depressive disorder, and 32 with borderline personality disorder. After three months of living in the new environment, each patient is assessed by the hospital staff as having either improved, remained the same, or worsened with respect to his or her state of mental health. The (hypothetical) data for this experiment, along with the calculations needed to obtain the chi-square statistic, are displayed in Table 18.5.

As in the 2 × 2 chi-square design, the expected frequency for a given cell is obtained by multiplying the row total for the cell by the column total and dividing by the total *N*. For example, $f_e$ for the Schizophrenic/No change cell is:

$$\frac{(200)(138)}{300} = 92.0$$

The computed value of $\chi^2$ of 27.55 (see bottom of Table 18.5) exceeds the tabled value for $(r - 1)(c - 1) = (3 - 1)(4 - 1)$ or 6 *df* of 12.59 at the .05 level of significance. Though you might have expected amounts of improvement to be independent of diagnosis, the analysis

**Table 18.5** Table of Frequencies Relating Treatment Response to Psychiatric Diagnosis

| Response | Psychiatric diagnosis | | | | |
|---|---|---|---|---|---|
| | Schizophrenic | Bipolar | Depressed | Borderline | Row total |
| Improved | 36 | 8 | 14 | 2 | 60 |
| No change | 84 | 72 | 18 | 26 | 200 |
| Worse | 18 | 8 | 10 | 4 | 40 |
| Column Total | 138 | 88 | 42 | 32 | 300 |

Calculations

| $f_o$ | $f_e$ | $f_o - f_e$ | $(f_o - f_e)^2$ | $\frac{(f_o - f_e)^2}{f_e}$ |
|---|---|---|---|---|
| 36 | 27.60 | 8.40 | 70.56 | 2.56 |
| 8 | 17.60 | −9.60 | 92.16 | 5.24 |
| 14 | 8.40 | 5.60 | 31.36 | 3.73 |
| 2 | 6.40 | −4.40 | 19.36 | 3.02 |
| 84 | 92.00 | −8.00 | 64.00 | 0.70 |
| 72 | 58.67 | 13.33 | 177.69 | 3.03 |
| 18 | 28.00 | −10.00 | 100.00 | 3.57 |
| 26 | 21.33 | 4.67 | 21.81 | 1.02 |
| 18 | 18.40 | −0.40 | 0.16 | 0.01 |
| 8 | 11.73 | −3.73 | 13.91 | 1.19 |
| 10 | 5.60 | 4.40 | 19.36 | 3.46 |
| 4 | 4.27 | −0.27 | 0.07 | 0.02 |
| | | | | $\chi^2 = 27.55$ |

indicates that you should reject $H_0$ and conclude that psychiatric diagnosis and response to the new environment *are* related.

Note that if you ignore the middle row of the data ("No change"), you could perform a sign test on the row totals to determine whether, across all diagnoses, patients were more likely to improve than to worsen. You could also perform sign tests for each diagnosis separately, but if you perform a two-way (4 × 2) chi-square test (without the middle row), you accomplish something that the sign test cannot do. You are testing the differences in the proportions of improvement among the four diagnoses against the null hypothesis that all four proportions are equal to each other (not that they are equal to .5). Adding the middle category of "No change" complicates the interpretation of the results, but it also gives you additional data and greater power to detect differences among the diagnostic groups.

If you perform smaller chi-square tests for follow-up purposes, you should use the Bonferroni correction. That is, the alpha you use for each of these tests should be .05 divided by the number of follow-up tests you plan to conduct.

One important application of the chi-square test is as an alternative to a parametric test when your quantitative data have a very unusual distribution. (For example, the data have multiple modes or extreme outliers on one or both ends of the distribution.) For the psychiatric environment experiment just described, the researcher may have collected clinical ratings for each patient at the beginning and end of the three-month period, with plans to perform a one-way analysis of variance (ANOVA) to compare the patient groups on the difference scores. However, the difference scores may be distributed so strangely that the researcher feels more comfortable simply categorizing those differences as exhibiting some improvement, some decline, or virtually no change at all. Considerable statistical power can be lost when converting precise measurements to just a few categories, but this conversion is justified (and even encouraged) if your data seriously violate the assumptions that underlie the planned parametric test. By just ignoring such violations, it is quite possible that you would be committing a higher rate of Type I errors than the alpha you are using would indicate.

## Measures of Strength of Association in Two-Variable Tables

The $\chi^2$ test of independence allows you to make decisions about *whether* there is a relationship between two variables, using frequency data. Thus, if $H_0$ is rejected, you can conclude that there *is* a statistically significant relationship between the two variables. As we have pointed out frequently in this text, however, statistical significance does not indicate the *strength* of the relationship; a significant result suggests only that the relationship in the population is unlikely to be zero. Here again, it is desirable to have a measure of the strength of the relationship—that is, an index of the degree of correlation. We next discuss such a measure.

### The Phi Coefficient (ϕ) for 2 × 2 Tables

Recall that the chi-square statistic was not statistically significant for the relationship between undergraduate major and type of career preference. Suppose, however, that the researcher, undeterred, collects data for twice as many students, and the results come out proportionally the same as they did in Table 18.4. That is, the frequency in every individual cell is doubled as well. The data for these 360 students are shown in Table 18.6.

Using Formula 18.4 (for 2 × 2 tables), we find that not only have $N$ and all of the observed frequencies doubled, but the value for $\chi^2$ is also exactly twice as large as it was in Table 18.4. This new value for $\chi^2$ is now statistically significant: $4.688 > 3.84$. Does it seem reasonable that increasing $N$, without changing the proportions, should turn a nonsignificant result into one that *is* significant?

Just like $t$ or $F$, $\chi^2$ depends on sample size; it tells you about statistical significance but says nothing about the size of the effect. To resolve this shortcoming, you need a measure of the strength of the relationship between two dichotomous variables that depends only on the proportions and does not change just because every frequency is, for instance, doubled. Such a measure does exist, and it is easy to calculate. For a 2 × 2 table, $\chi^2$ can be converted to a correlation coefficient called the **phi coefficient** (symbolized by **ϕ**) by the following formula:

$$\phi = \sqrt{\frac{\chi^2}{N}} \qquad \text{Formula 18.5}$$

Thus, for the data in Table 18.6,

$$\begin{aligned} \phi &= \sqrt{\frac{4.688}{360}} \\ &= \sqrt{0.013} \\ &= 0.114 \end{aligned}$$

**Table 18.6** **Table of Frequencies Relating Undergrad Major to Career Preference**

| | | Undergraduate major | | |
|---|---|---|---|---|
| | | **Economics** | **Psychology** | **Total students** |
| Career preference | Applied | 90 | 110 | 200 |
| | Academic | 54 | 106 | 160 |
| | Total | 144 | 216 | 360 = $N$ |

$$\begin{aligned} \chi^2 &= \frac{360[(90 \times 106) - (110 \times 54)]^2}{200 \times 160 \times 144 \times 216} \\ &= \frac{4{,}665{,}600.00}{995{,}328{,}000} \\ &= 4.688 \end{aligned}$$

If you compare Table 18.4 to Table 18.6, you will see that $N$ gets doubled (from 180 to 360), but so does $\chi^2$ (from 2.344 to 4.688). Because 4.688/360 equals 2.344/180, the ratio under the square root sign in the formula for $\phi$ does not change (it remains .013), and therefore $\phi$ is also .114 for the original data in Table 18.4.

The $\phi$ is interpreted as a Pearson $r$. In fact, it *is* a Pearson $r$; you would get the same answer of .114 if you assigned scores of 1 and 2 to major (e.g., 1 = economics, 2 = psychology) and to career preference (1 = prefers academia, 2 = prefers an applied career) and then computed the correlation between the two variables using one of the usual formulas for the Pearson $r$ given in Chapter 9. Thus the absolute value of $\phi$ varies between 0 and 1. The larger the value of $\phi$, the stronger is the relationship between the two variables. When the assignment of score values is arbitrary (as it is in the present case; it would be just as acceptable to score 2 for economics majors and 1 for psychology majors), the sign of $\phi$ is irrelevant, and $\phi$ is therefore reported without sign. Just as with the point-biserial $r$, you must look at the data to determine the direction of the findings (which, in our example, is that the proportion of economics majors who prefer applied careers to academic careers is greater than that proportion for psychology majors).

In the present example, you would conclude from the value of $\phi$ that the relationship between major and career preference in the population, while likely to be greater than zero, is also likely to be fairly weak. In other words, there are likely to be numerous exceptions to the conclusion. This is evidenced, for instance, by the 54 out of 144 economics majors who expressed a preference for an academic career (see Table 18.6). The phi coefficient sums up this information conveniently in a readily interpretable correlation.

Note that there is no need to carry out a special significance test for $\phi$. It has already been done with $\chi^2$. In fact, the $\chi^2$ test on a 2 × 2 table can just as well be viewed as testing the null hypothesis that the population $\phi$ value equals zero. When $\chi^2$ is sufficiently large, this hypothesis is rejected, which then suggests that there is some degree of relationship other than zero.

### Cramér's Phi Coefficient ($\phi_c$)

When one (or both) of the two categorical variables whose relationship you are exploring has more than two levels, you cannot simply assign arbitrary numbers to their levels and calculate a Pearson $r$. The magnitude of the $r$ will depend on how you assign the arbitrary numbers to the levels of any variable with more than two categories, so it will be uninterpretable. However, there is a strength-of-relationship index that is closely related to the phi coefficient and can be used with tables larger than 2 × 2. It is known as **Cramér's $\phi$**, sometimes symbolized as $\boldsymbol{\phi_c}$ (but also sometimes referred to as *Cramer's V*). It is applicable to any $r \times c$ contingency table, and always varies between 0 and 1 regardless of the size of the table. It is simply

$$\text{Cramer's } \phi = \sqrt{\frac{\chi^2}{N(k-1)}} \qquad \text{Formula 18.6}$$

Where

$k$ = smaller of $r$ or $c$ (or, when $r = c$, $k = r = c$)

For the example in Table 18.5, which is a 3 × 4 table,

$$\phi = \sqrt{\frac{27.55}{300(3-1)}}$$
$$= \sqrt{0.0459}$$
$$= 0.214$$

This degree of relationship is somewhere between small and moderate in size. For $\phi_C$, like ordinary $\phi$, and other correlational measures of effect size, a value of .1 is considered to be small, .3 is considered to be of medium size, and .5 is large.

## Summary

Chi-square tests are designed to be used with *frequency* data.

$$\chi^2 = \sum \frac{(f_o - f_e)^2}{f_e} \qquad \text{Formula 18.1}$$

where
$f_o$ = observed frequency
$f_e$ = expected frequency

### One-Variable Problems

$$df = k - 1$$

where
$k$ = number of categories of the variable

Expected frequencies are readily determined from the null hypothesis. For example, if $H_0$ specifies that people in the population are equally divided among the $k$ categories, $f_e$ for each category is equal to $N/k$ (where $N$ = number of people in the sample).

### Two-Variable Problems: Test of Association

$$df = (r - 1)(c - 1) \qquad \text{Formula 18.3}$$

where
$r$ = number of rows
$c$ = number of columns

For a cell in a given row and column, the expected frequency is equal to

$$f_e = \frac{(\text{row total})(\text{column total})}{N} \qquad \text{Formula 18.2}$$

For a 2 × 2 table, the following formula for $\chi^2$ is equivalent to determining expected frequencies and using the usual chi-square formula, but does so in one step and therefore requires somewhat less work computationally:

| | |
|---|---|
| A | B |
| C | D |

$$\chi^2 = \frac{N(AD - BC)^2}{(A + B)(C + D)(A + C)(B + D)} \qquad \text{Formula 18.4}$$

### Measures of Strength of Association in Two-Variable Tables

For 2 × 2 tables, compute the phi coefficient:

$$\phi = \sqrt{\frac{\chi^2}{N}} \qquad \text{Formula 18.5}$$

$\phi$ is interpreted like any other Pearson $r$.

For *larger tables*, compute Cramér's $\phi$:

$$\text{Cramer's } \phi = \sqrt{\frac{\chi^2}{N(k - 1)}} \qquad \text{Formula 18.6}$$

where

$k$ = the *smaller* of $r$ (number of rows) or $c$ (number of columns);
$k$ = either one if $r = c$

These two effect-size measures are statistically significant if $\chi^2$ is statistically significant. For both $\phi$ and $\phi_C$, the following are the conventional guidelines: .1 is small; .3 is moderate; and .5 is large.

### Some Precautions on the Use of $\chi^2$

$\chi^2$ should be used only when the observations are independent—that is, when no observation is related to or dependent on any other observation. The significance test of $\chi^2$ is not sufficiently accurate when either of the following conditions applies:

(a) $df = 1$, and any *expected* frequency is less than 5.

(b) $df = 2$, and any *expected* frequency is less than 3.

## Exercises

1. Out of 100 psychiatric patients given a new form of treatment, 60 improved, while 40 got worse.
   (a) Use a chi-square test to decide whether you can reject the null hypothesis that the treatment is totally ineffective (i.e., that patients are equally likely to improve or get worse).
   (b) Redo the significance test in part (a) using the $z$ score formula (i.e., normal approximation) for the binomial test. Explain the relationship between the $z$ score you calculated in this part and the $\chi^2$ value you calculated in part (a).
2. A developmental researcher has observed that in a random sample of 60 toddlers, 27 preferred blue toys, 19 preferred red toys, and 14 preferred green toys. Perform a chi-square test of the null hypothesis that, in the entire population of toddlers, the preference for these three colors is equally divided.
3. Repeat the previous exercise, except that this time there are 79 toddlers choosing among toys that come in a total of four different colors. The number of toddlers preferring each color is as follows:
   (a) 21 chose red, 14 chose blue, 26 chose yellow, 18 chose green.
   (b) The same toddlers were tested a year later with the following results: 28 chose red, 20 chose blue, 9 chose yellow, 22 chose green.
4. At Bigbrain University, the typical grade distribution is A, 15%; B, 25%; C, 45%; D, 10%; F, 5%. The grades given by two professors are shown here. For each one (separately), test the null hypothesis that the professor is a typical grader, using a one-variable chi-square analysis.

| | (a) Professor 1 | | | | | (b) Professor 2 | | | | |
|---|---|---|---|---|---|---|---|---|---|---|
| | A | B | C | D | F | A | B | C | D | F |
| $f_o$ | 7 | 13 | 22 | 4 | 6 | 13 | 12 | 8 | 3 | 3 |

5. A psychologist wants to test the hypothesis that college women will do better on a particular type of problem-solving task than will college men. He obtains the following results:

| | Result on problem-solving task | |
|---|---|---|
| **Gender** | **Succeed** | **Fail** |
| Male | 12 | 18 |
| Female | 10 | 10 |

(a) Test the null hypothesis that gender and success on the problem-solving task are independent. Does this mean the same as a statement about whether the percent success differs between males and females?

(b) Calculate the phi coefficient for these data. Does it look like the psychologist is dealing with a small, medium, or large effect?

6. A bond issue is to be put before the voters in a forthcoming election. An opinion poll company obtains a random sample of 200 registered voters and asks them what party they belong to and how they intend to vote on the bond issue. The results are as follows:

| | Prospective vote on bond issue | | |
|---|---|---|---|
| **Political party** | **Yes** | **No** | **Undecided** |
| Democratic | 20 | 30 | 10 |
| Republican | 30 | 30 | 20 |
| Independent | 10 | 40 | 10 |

(a) Test the null hypothesis that political party and prospective vote on the bond issue are independent. What is your conclusion?

(b) Calculate Cramér's $\phi$ for these data. Does it look like there's a strong association between party affiliation and attitude toward this particular bond issue?

7. Suppose that students majoring in the natural sciences, social sciences, and humanities were asked whether they were in favor of having greater student participation in academic decisions at their college. The data from the survey appear in the following table:

| | **Natural sciences** | **Social sciences** | **Humanities** |
|---|---|---|---|
| In favor | 20 | 18 | 16 |
| Neutral | 8 | 8 | 30 |
| Against | 6 | 19 | 18 |

(a) Compute the two-way chi-square test for these data. Can you reject the null hypothesis that students' attitudes on this question are not at all related to their academic areas? What critical value did you use to make your decision?

(b) Calculate the appropriate measure for the strength of relationship between the two variables, regardless of your decision in part (a). How strong does the relationship appear to be?

8. Repeat the previous exercise with the data in the following table. Note that the marginal frequencies have not changed, so you do not have to recompute the expected frequencies.

| | **Natural sciences** | **Social sciences** | **Humanities** |
|---|---|---|---|
| In favor | 11 | 17 | 26 |
| Neutral | 10 | 15 | 21 |
| Against | 13 | 13 | 17 |

(a) Now can you reject the null hypothesis that students' attitudes on this question are not at all related to their academic areas?

(b) Calculate the appropriate measure for the strength of relationship between the two variables, regardless of your decision in part (a). How strong does the relationship appear to be?

## Thought Questions

1. A researcher wants to test the null hypothesis that three types of television programming are preferred equally by college students. She obtains a sample of 139 college students and finds that 57 prefer sitcoms, 44 prefer sports, and 38 prefer reality shows. What statistical procedure should be used to test this null hypothesis? Why?
2. The same researcher wants to test the relationship between education level and preference for the three television shows. For the data in Question 1, she finds the following: Of the 57 students who prefer sitcoms, 29 are seniors and 28 are first-year. Of the 44 who prefer sports, 9 are seniors and 35 are first-year. Of the 38 who prefer reality shows, 30 are seniors and 8 are first-year. What statistical procedure should be used to test this null hypothesis? Why?
3. Suppose that the same researcher asks the 139 students to vote yes or no as to whether they like each of the three types of television shows and obtains the following results: for sitcoms, 92 yes and 47 no; for sports, 61 yes and 78 no; for reality shows, 74 yes and 65 no. What type of chi-square analysis should be used to analyze these data, if any? What assumption of chi-square tests is being violated in this example? Explain.
4. (a) When using chi-square for a two-variable problem, why is it desirable to compute either the $\phi$ or Cramér's $\phi$ coefficient when statistically significant results are obtained? (b) When should each of the coefficients in part (a) be used?

## Computer Exercises

1. (a) Perform a one-way chi-square test to determine whether you can reject the null hypothesis that, at Jackson's college, there are the same number of students majoring in each of the five areas represented in Jackson's class, if you assume that their students represent a random sample with respect to major area.

   (b) Perform the test in part (a) separately for both exercise groups in Jackson's class.
2. Suppose that Jackson obtains registration information from their college and finds that the numbers of undergraduates who have declared each of the five majors are as follows: psychology, 400; geology, 310; biology, 270; sociology, 140; economics, 240. Can you reject the null hypothesis that Jackson's statistics class is a random sample from the undergraduate population of their university?
3. Conduct a two-way chi-square analysis of Jackson's data to test the null hypothesis that the proportion of nonregular exercisers is the same for each of the five represented majors in the entire university population. Request a statistic to describe the strength of the relationship between exercise and major.
4. Conduct a two-way chi-square analysis of Jackson's data to test the null hypothesis that the (two-valued) grouping variable you previously created from a median split of the procrastination scores is independent of exercise status. Request the phi coefficient for this relationship. Use your software to compute the correlation directly between the high/low procrastination variable and exercise, and compare this value to the phi coefficient.

5. Create a grouping variable that equals 1 if a student has had no more than one previous teacher they would identify as outstanding prior to registering for statistics and 2 if the student has had two or more outstanding teachers. Test whether having more than one previous outstanding teacher is independent of a student's major. Request a statistic to describe the strength of the relationship between these two variables.

## Bridge to SPSS

The Analyze menu selection that SPSS uses to access one-way chi-square tests bears no resemblance to the menu selection that leads to a two-way chi-square test. It is not obvious why this is so, but both types of the test are easily obtained, as described next.

### One-Way Chi-Square Tests

If one of your variables has been coded with at least two different numerical values, you can perform a one-way chi-square test on that variable by selecting **Chi-Square** from the **Analyze/Nonparametric Tests/Legacy Dialogs** menu. The chi-square test dialog box is similar to the one for the binomial test, but by necessity, there are two major differences (see Figure 18.3). First, replacing the *Define Dichotomy* choice is the selection of the *Expected Range*, but its default setting, **Get from data**, should be familiar from the previous chapter. In the chi-square test, this setting means that the number of categories for your one-way chi-square test (i.e., $k$) will be determined by the number of different values that SPSS reads for a variable that you selected (i.e., moved to the *Test Variable:* list).

The values of your test variable can be numeric codes that have been arbitrarily assigned to your categories (e.g., 1 = Psych major; 2 = Pre-Med; 3 = Bio; etc.). However, the values can also be real integer quantities, such as the loudness of ambient noise measured in decibels (db) as participants perform a task. Equal numbers of participants would be assigned to each loudness condition (e.g., 60, 80, and 100 db), and the number of participants who give up and refuse to continue is counted as the observed frequency. If there are quite a few possible values for your IV but you do not want to analyze all of the conditions, you can accomplish that by selecting **Use specified range** instead of **Get from data** and entering the desired upper and lower limits for your IV.

The second major difference from the binomial test dialog box is that *Test Proportion* is replaced by *Expected Values*. The default choice is **All categories equal**, and in the two-category case, this is equivalent to the binomial test's default *Test Proportion* of .5. This choice is often appropriate for experimental null hypotheses in which you want to see if some categorical distinction makes a difference to, or is even noticed by, your participants—for example, are people who find a wallet on the street (placed by a hidden experimental observer) more likely to return the wallet if it appears that the owner is poor, middle class, or wealthy? The other choice for *Expected Values* is to actually type in your expected frequencies for all of the possible values of your test variable. You must add your $f_e$s to the Values box starting with the category with the lowest numerical value and working your way up to the highest. For example, if you have coded right-handers as 1, left-handers as 2, and truly ambidextrous persons as 3, and the corresponding population proportions are 80%, 15%, and 5%, and your total $N$ is 60, you would enter the following expected values in this order: 48, 9, and 3 (i.e., $.8 \times 60$, $.15 \times 60$, and $.05 \times 60$).

For a simple example of the one-way chi-square test, we will test whether the five categories for undergrad major in Jackson's data set are consistent with the null hypothesis that, for all students at the same college level in the population, the five majors are equally preferred (i.e., the proportion of students in the population choosing each major is .2, assuming, of course, that these majors are their only choices).

The basic output from the chi-square test consists of two tables: one containing the observed and expected frequencies for each category (plus the residual, which is just $f_o - f_e$), and one containing the chi-square statistic, along with its $df$ $(k - 1)$, and the $p$ value

## Figure 18.3

**The Dialog Box for the One-Way Chi-Square Test**

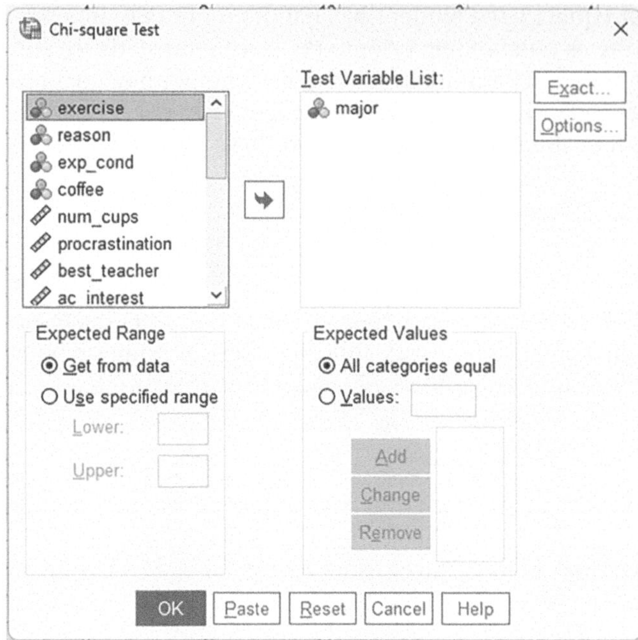


*Source*: Courtesy of International Business Machines Corporation, © SPSS, Inc., an IBM company

for your test. This *p* value is labeled as **Asymp. Sig.** (asymptotic significance) to remind you that this *p* is not an exact value from the appropriate multinomial distribution but an approximation analogous to the normal approximation of the binomial distribution (see Figure 18.4). That is why SPSS puts a note under this table, telling you how many cells have $f_e$s less than 5. The chi-square approximation is not very accurate when several cells have $f_e$s less than 5 (or any cells have $f_e$s less than 1 or 2). Because the *p* value is greater than .05, we cannot conclude from Jackson's sample that, in the population, these majors are *not* equally preferred. Stating this conclusion in terms of a double negative is important because, as usual, failing to reject the null hypothesis does not allow us to support the null and therefore conclude, in this case, that these majors *are* equally preferred in the population.

## Figure 18.4

**The Results of a One-Way Chi-Square Test on Undergrad Major with respect to a Null Hypothesis of All Frequencies Being Equal in the Population**

**major**

| | Observed N | Expected N | Residual |
|---|---|---|---|
| Psych | 29 | 20.0 | 9.0 |
| Pre-Med | 25 | 20.0 | 5.0 |
| Bio | 21 | 20.0 | 1.0 |
| Socio | 15 | 20.0 | –5.0 |
| Econ | 10 | 20.0 | –10.0 |
| Total | 100 | | |

**Test Statistics**

| | major |
|---|---|
| Chi-Square | 11.600[a] |
| df | 4 |
| Asymp. Sig. | .021 |

a. 0 cells (0.0%) have expected frequencies less than 5. The minimum expected cell frequency is 20.0.

### Two-Way Chi-Square Tests

Somewhat surprisingly, the **Chi-Square Test** dialog box from the **Analyze/Nonparametric Tests/Legacy Dialogs** menu does not accommodate the two-way chi-square test. Even more surprising, perhaps, is that the two-way test is accessed from the **Analyze/Descriptive Statistics** menu; choose **Crosstabs** (short for *Cross-tabulations*) from that list. The descriptive function of **Crosstabs** is that it produces a matrix of rows and columns that makes it easy to see the joint frequencies of any two categorical variables—that is, how frequently each possible combination of levels of the two variables is represented in the data set. For example, if you were to move *reason* to the *Row(s):* space and *major* to the *Column(s):* space and then click **OK**, you would get a matrix with three rows (one for each reason for taking stats) and five columns (one for each of the five majors represented in Jackson's classes). In each of the 15 cells of the matrix would be the number of students who fell into the two particular categories represented by that cell (e.g., one of the cells would have the number of psychology majors who took stats as a program requirement).

To test the null hypothesis that the two variables are independent of each other (i.e., not associated at all), click on the **Statistics ...** button. The two checkboxes you will then want to select because those statistics were covered in this chapter, are **Chi-square** and **Phi and Cramer's V** (that value which SPSS calls *Cramer's V* is the effect-size measure that we have been calling *Cramer's phi*, or, as a symbol, $\phi_C$).

The output that is produced by clicking OK in the dialog box shown in Figure 18.5, given the Statistics selections we mentioned, consists of four tables. The first table only addresses the number of missing values. The second table is the matrix that shows you the frequency in your sample for each possible combination of the levels of your two variables (see Figure 18.6). You can see by just a glance at this table that there are some wide discrepancies (e.g., apparently, in this hypothetical college, stats is a program requirement for Psychology, but not for Sociology).

The first row of the third table gives you the chi-square statistic for testing the association (or conversely, the independence) between the two variables (this statistic is called *Pearson chi-square*, after its originator, Karl Pearson), and its degrees of freedom (*df*). Recall that for a two-way chi-square, the *df* equal $(r - 1)(c - 1)$; in this case, $df = 2 \times 4 = 8$. The *p* value is labeled **Asymp. Sig.** for reasons we have already explained. (We cannot explain

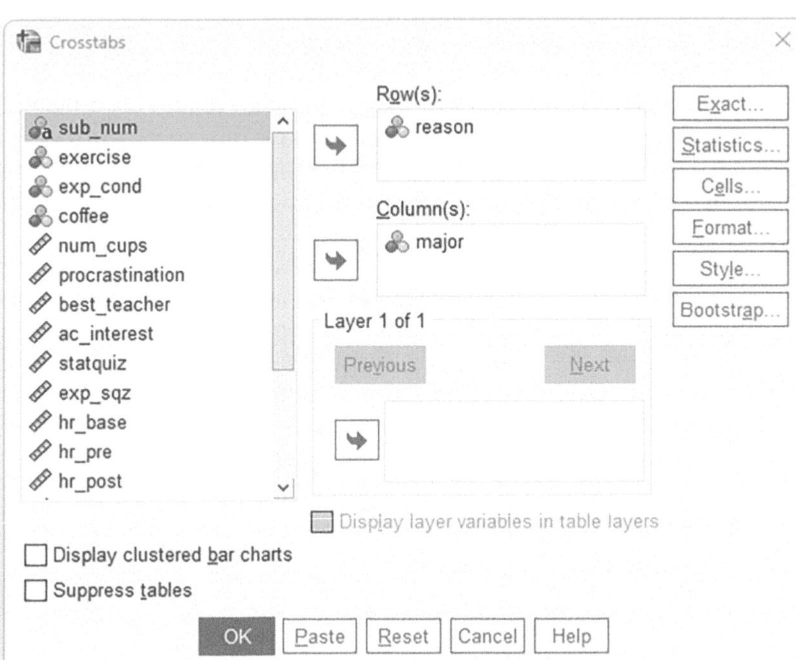


## Figure 18.5

**The Dialog Box for the Two-Way Chi-Square Test**

*Source*: Courtesy of International Business Machines Corporation, © SPSS, Inc., an IBM company

## Figure 18.6

**The Results of a Two-Way Chi-Square Test to Determine Whether there is an Association between Undergrad Major and Reason for Taking Statistics**

**reason * major Crosstabulation**

Count

| | | major | | | | | |
|---|---|---|---|---|---|---|---|
| | | Psych | Pre-Med | Bio | Socio | Econ | Total |
| reason | Program | 21 | 10 | 7 | 0 | 3 | 41 |
| | Personal | 3 | 8 | 6 | 7 | 5 | 29 |
| | Advisor | 5 | 7 | 8 | 8 | 2 | 30 |
| Total | | 29 | 25 | 21 | 15 | 10 | 100 |

**Chi-Square Tests**

| | Value | df | Asymptotic Significance (2-sided) |
|---|---|---|---|
| Pearson Chi-Square | 25.548[a] | 8 | .001 |
| Likelihood Ratio | 30.845 | 8 | <.001 |
| Linear-by-Linear Association | 10.465 | 1 | .001 |
| N of Valid Cases | 100 | | |

a. 5 cells (33.3%) have expected count less than 5. The minimum expected count is 2.90.

**Symmetric Measures**

| | | Value | Approximate Significance |
|---|---|---|---|
| Nominal by Nominal | Phi | .505 | .001 |
| | Cramer's V | .357 | .001 |
| N of Valid Cases | | 100 | |

why it is also labeled as "2-sided," as this does not make sense in this context.) Because this $p$ equals .001, we can reject the null hypothesis that these two variables are independent. That is, students have significantly different reasons for taking stats, depending on their major. There are two alternative test statistics included in this output table, but explaining them goes beyond the scope of this textbook.

The fourth table, titled **Symmetric Measures**, gives you two choices for effect size. If both of your variables are dichotomous, thus creating a 2 × 2 matrix, the appropriate measure of effect size is the one in the first row labeled **Phi**. Phi is a true correlation coefficient, akin to the point-biserial $r$, except that in this case both variables have only two values each. Because the output in this example is based on a 3 × 5 matrix, the appropriate effect-size measure is **Cramér's V** (which we prefer to call Cramer's phi to emphasize its similarity to Phi).

When both of your variables are dichotomous, two more rows are added to the third table, labeled **Chi-Square Tests**; they are **Continuity Correction** and **Fisher's Exact Test**. The continuity correction reduces the chi-square somewhat to improve the approximation of the chi-square statistic. Fisher's Exact Test is based on the multinomial distribution rather than the chi-square approximation to that distribution. It is analogous to obtaining results from the binomial test, rather than the normal approximation to that test. The only drawback is that for very large matrices, the exact test can be very computer time consuming.

# Appendix

Statistical Tables

Answer Key

Data from Jackson's Experiment

*Introductory Statistics for the Behavioral Sciences*, Eighth Edition. R. Brooke Lea and Barry H. Cohen.

Companion website: www.wiley.com/go/ISBS8e

## Statistical Tables

### Table A Percent Area Under the Normal Curve Between the Mean and $z$

| $z$ | 0.00 | 0.01 | 0.02 | 0.03 | 0.04 | 0.05 | 0.06 | 0.07 | 0.08 | 0.09 |
|---|---|---|---|---|---|---|---|---|---|---|
| 0.0 | 00.00 | 00.40 | 00.80 | 01.20 | 01.60 | 01.99 | 02.39 | 02.79 | 03.19 | 03.59 |
| 0.1 | 03.98 | 04.38 | 04.78 | 05.17 | 05.57 | 05.96 | 06.36 | 06.75 | 07.14 | 07.53 |
| 0.2 | 07.93 | 08.32 | 08.71 | 09.10 | 09.48 | 09.87 | 10.26 | 10.64 | 11.03 | 11.41 |
| 0.3 | 11.79 | 12.17 | 12.55 | 12.93 | 13.31 | 13.68 | 14.06 | 14.43 | 14.80 | 15.17 |
| 0.4 | 15.54 | 15.91 | 16.28 | 16.64 | 17.00 | 17.36 | 17.72 | 18.08 | 18.44 | 18.79 |
| 0.5 | 19.15 | 19.50 | 19.85 | 20.19 | 20.54 | 20.88 | 21.23 | 21.57 | 21.90 | 22.24 |
| 0.6 | 22.57 | 22.91 | 23.24 | 23.57 | 23.89 | 24.22 | 24.54 | 24.86 | 25.17 | 25.49 |
| 0.7 | 25.80 | 26.11 | 26.42 | 26.73 | 27.04 | 27.34 | 27.64 | 27.94 | 28.23 | 28.52 |
| 0.8 | 28.81 | 29.10 | 29.39 | 29.67 | 29.95 | 30.23 | 30.51 | 30.78 | 31.06 | 31.33 |
| 0.9 | 31.59 | 31.86 | 32.12 | 32.38 | 32.64 | 32.89 | 33.15 | 33.40 | 33.65 | 33.89 |
| 1.0 | 34.13 | 34.38 | 34.61 | 34.85 | 35.08 | 35.31 | 35.54 | 35.77 | 35.99 | 36.21 |
| 1.1 | 36.43 | 36.65 | 36.86 | 37.08 | 37.29 | 37.49 | 37.70 | 37.90 | 38.10 | 38.30 |
| 1.2 | 38.49 | 38.69 | 38.88 | 39.07 | 39.25 | 39.44 | 39.62 | 39.80 | 39.97 | 40.15 |
| 1.3 | 40.32 | 40.49 | 40.66 | 40.82 | 40.99 | 41.15 | 41.31 | 41.47 | 41.62 | 41.77 |
| 1.4 | 41.92 | 42.07 | 42.22 | 42.36 | 42.51 | 42.65 | 42.79 | 42.92 | 43.06 | 43.19 |
| 1.5 | 43.32 | 43.45 | 43.57 | 43.70 | 43.82 | 43.94 | 44.06 | 44.18 | 44.29 | 44.41 |
| 1.6 | 44.52 | 44.63 | 44.74 | 44.84 | 44.95 | 45.05 | 45.15 | 45.25 | 45.35 | 45.45 |
| 1.7 | 45.54 | 45.64 | 45.73 | 45.82 | 45.91 | 45.99 | 46.08 | 46.16 | 46.25 | 46.33 |
| 1.8 | 46.41 | 46.49 | 46.56 | 46.64 | 46.71 | 46.78 | 46.86 | 46.93 | 46.99 | 47.06 |
| 1.9 | 47.13 | 47.19 | 47.26 | 47.32 | 47.38 | 47.44 | 47.50 | 47.56 | 47.61 | 47.67 |
| 2.0 | 47.72 | 47.78 | 47.83 | 47.88 | 47.93 | 47.98 | 48.03 | 48.08 | 48.12 | 48.17 |
| 2.1 | 48.21 | 48.26 | 48.30 | 48.34 | 48.38 | 48.42 | 48.46 | 48.50 | 48.54 | 48.57 |
| 2.2 | 48.61 | 48.64 | 48.68 | 48.71 | 48.75 | 48.78 | 48.81 | 48.84 | 48.87 | 48.90 |
| 2.3 | 48.93 | 48.96 | 48.98 | 49.01 | 49.04 | 49.06 | 49.09 | 49.11 | 49.13 | 49.16 |
| 2.4 | 49.18 | 49.20 | 49.22 | 49.25 | 49.27 | 49.29 | 49.31 | 49.32 | 49.34 | 49.36 |
| 2.5 | 49.38 | 49.40 | 49.41 | 49.43 | 49.45 | 49.46 | 49.48 | 49.49 | 49.51 | 49.52 |
| 2.6 | 49.53 | 49.55 | 49.56 | 49.57 | 49.59 | 49.60 | 49.61 | 49.62 | 49.63 | 49.64 |
| 2.7 | 49.65 | 49.66 | 49.67 | 49.68 | 49.69 | 49.70 | 49.71 | 49.72 | 49.73 | 49.74 |
| 2.8 | 49.74 | 49.75 | 49.76 | 49.77 | 49.77 | 49.78 | 49.79 | 49.79 | 49.80 | 49.81 |
| 2.9 | 49.81 | 49.82 | 49.82 | 49.83 | 49.84 | 49.84 | 49.85 | 49.85 | 49.86 | 49.86 |
| 3.0 | 49.87 | | | | | | | | | |
| 3.5 | 49.98 | | | | | | | | | |
| 4.0 | 49.997 | | | | | | | | | |
| 5.0 | 49.99997 | | | | | | | | | |

# Table B Critical Values of *t*

| | Level of significance for one-tailed test | | | | | |
|---|---|---|---|---|---|---|
| | 0.10 | 0.05 | 0.025 | 0.01 | 0.005 | 0.0005 |
| | Level of significance for two-tailed test | | | | | |
| *df* | 0.20 | 0.10 | 0.05 | 0.02 | 0.01 | 0.001 |
| 1 | 3.078 | 6.314 | 12.706 | 31.821 | 63.657 | 636.619 |
| 2 | 1.886 | 2.920 | 4.303 | 6.965 | 9.925 | 31.598 |
| 3 | 1.638 | 2.343 | 3.182 | 4.541 | 5.841 | 12.941 |
| 4 | 1.533 | 2.132 | 2.776 | 3.747 | 4.604 | 8.610 |
| 5 | 1.476 | 2.015 | 2.571 | 3.365 | 4.032 | 6.859 |
| 6 | 1.440 | 1.943 | 2.447 | 3.143 | 3.707 | 5.959 |
| 7 | 1.415 | 1.895 | 2.365 | 2.998 | 3.449 | 5.405 |
| 8 | 1.397 | 1.860 | 2.306 | 2.896 | 3.355 | 5.041 |
| 9 | 1.383 | 1.833 | 2.262 | 2.821 | 3.250 | 4.781 |
| 10 | 1.372 | 1.812 | 2.228 | 2.764 | 3.169 | 4.587 |
| 11 | 1.363 | 1.796 | 2.201 | 2.718 | 3.106 | 4.437 |
| 12 | 1.356 | 1.782 | 2.179 | 2.681 | 3.055 | 4.318 |
| 13 | 1.350 | 1.771 | 2.160 | 2.650 | 3.012 | 4.221 |
| 14 | 1.345 | 1.761 | 2.145 | 2.624 | 2.977 | 4.140 |
| 15 | 1.341 | 1.753 | 2.131 | 2.602 | 2.947 | 4.073 |
| 16 | 1.337 | 1.746 | 2.120 | 2.583 | 2.921 | 4.015 |
| 17 | 1.333 | 1.740 | 2.110 | 2.567 | 2.898 | 3.965 |
| 18 | 1.330 | 1.734 | 2.101 | 2.552 | 2.878 | 3.922 |
| 19 | 1.328 | 1.729 | 2.093 | 2.539 | 2.861 | 3.883 |
| 20 | 1.325 | 1.725 | 2.086 | 2.528 | 2.845 | 3.850 |
| 21 | 1.323 | 1.721 | 2.080 | 2.518 | 2.831 | 3.819 |
| 22 | 1.321 | 1.717 | 2.074 | 2.508 | 2.819 | 3.792 |
| 23 | 1.319 | 1.714 | 2.069 | 2.500 | 2.807 | 3.767 |
| 24 | 1.318 | 1.711 | 2.064 | 2.492 | 2.797 | 3.745 |
| 25 | 1.316 | 1.708 | 2.060 | 2.485 | 2.787 | 3.725 |
| 26 | 1.315 | 1.706 | 2.056 | 2.479 | 2.779 | 3.707 |
| 27 | 1.314 | 1.703 | 2.052 | 2.473 | 2.771 | 3.690 |
| 28 | 1.313 | 1.701 | 2.048 | 2.467 | 2.763 | 3.674 |
| 29 | 1.311 | 1.699 | 2.045 | 2.462 | 2.756 | 3.659 |
| 30 | 1.310 | 1.697 | 2.042 | 2.457 | 2.750 | 3.646 |
| 40 | 1.303 | 1.684 | 2.021 | 2.423 | 2.704 | 3.551 |
| 60 | 1.296 | 1.671 | 2.000 | 2.390 | 2.660 | 3.460 |
| 120 | 1.289 | 1.658 | 1.980 | 2.358 | 2.617 | 3.373 |
| ∞ | 1.282 | 1.645 | 1.960 | 2.326 | 2.576 | 3.291 |

# Table C Critical Values of the Pearson *r*

| *df* (= *N* − 2; *N* = number of pairs) | Level of significance for one-tailed test | | | |
|---|---|---|---|---|
| | 0.05 | 0.025 | 0.01 | 0.005 |
| | Level of significance for two-tailed test | | | |
| | 0.10 | 0.05 | 0.02 | 0.01 |
| 1 | 0.988 | 0.997 | 0.9995 | 0.9999 |
| 2 | 0.900 | 0.950 | 0.980 | 0.990 |
| 3 | 0.805 | 0.878 | 0.934 | 0.959 |
| 4 | 0.729 | 0.811 | 0.882 | 0.917 |
| 5 | 0.669 | 0.754 | 0.833 | 0.874 |
| 6 | 0.622 | 0.707 | 0.789 | 0.834 |
| 7 | 0.582 | 0.666 | 0.750 | 0.798 |
| 8 | 0.549 | 0.632 | 0.716 | 0.765 |
| 9 | 0.521 | 0.602 | 0.685 | 0.735 |
| 10 | 0.497 | 0.576 | 0.658 | 0.708 |
| 11 | 0.476 | 0.553 | 0.634 | 0.684 |
| 12 | 0.458 | 0.532 | 0.612 | 0.661 |
| 13 | 0.441 | 0.514 | 0.592 | 0.641 |
| 14 | 0.426 | 0.497 | 0.574 | 0.623 |
| 15 | 0.412 | 0.482 | 0.558 | 0.606 |
| 16 | 0.400 | 0.468 | 0.542 | 0.590 |
| 17 | 0.389 | 0.456 | 0.528 | 0.575 |
| 18 | 0.378 | 0.444 | 0.516 | 0.561 |
| 19 | 0.369 | 0.433 | 0.503 | 0.549 |
| 20 | 0.360 | 0.423 | 0.492 | 0.537 |
| 21 | 0.352 | 0.413 | 0.482 | 0.526 |
| 22 | 0.344 | 0.404 | 0.472 | 0.515 |
| 23 | 0.337 | 0.396 | 0.462 | 0.505 |
| 24 | 0.330 | 0.388 | 0.453 | 0.496 |
| 25 | 0.323 | 0.381 | 0.445 | 0.487 |
| 26 | 0.317 | 0.374 | 0.437 | 0.479 |
| 27 | 0.311 | 0.367 | 0.430 | 0.471 |
| 28 | 0.306 | 0.361 | 0.423 | 0.463 |
| 29 | 0.301 | 0.355 | 0.416 | 0.456 |
| 30 | 0.296 | 0.349 | 0.409 | 0.449 |
| 35 | 0.275 | 0.325 | 0.381 | 0.418 |
| 40 | 0.257 | 0.304 | 0.358 | 0.393 |
| 45 | 0.243 | 0.288 | 0.338 | 0.372 |
| 50 | 0.231 | 0.273 | 0.322 | 0.354 |
| 60 | 0.211 | 0.250 | 0.295 | 0.325 |
| 70 | 0.195 | 0.232 | 0.274 | 0.302 |
| 80 | 0.183 | 0.217 | 0.256 | 0.283 |
| 90 | 0.173 | 0.205 | 0.242 | 0.267 |
| 100 | 0.164 | 0.195 | 0.230 | 0.254 |
| 200 | 0.116 | 0.138 | 0.164 | 0.181 |
| 500 | 0.073 | 0.088 | 0.104 | 0.115 |
| 1,000 | 0.052 | 0.062 | 0.073 | 0.081 |

# Table D Power as a Function of δ and Significance Criterion (α)

| | One-tailed test (α) | | | | | One-tailed test (α) | | | |
|---|---|---|---|---|---|---|---|---|---|
| | 0.05 | 0.025 | 0.01 | 0.005 | | 0.05 | 0.025 | 0.01 | 0.005 |
| | Two-tailed test (α) | | | | | Two-tailed test (α) | | | |
| δ | 0.10 | 0.05 | 0.02 | 0.01 | δ | 0.10 | 0.05 | 0.02 | 0.01 |
| 0.0 | 0.10[a] | 0.05[a] | 0.02 | 0.01 | 2.5 | 0.80 | 0.71 | 0.57 | 0.47 |
| 0.1 | 0.10[a] | 0.05[a] | 0.02 | 0.01 | 2.6 | 0.83 | 0.74 | 0.61 | 0.51 |
| 0.2 | 0.11[a] | 0.05 | 0.02 | 0.01 | 2.7 | 0.85 | 0.77 | 0.65 | 0.55 |
| 0.3 | 0.12[a] | 0.06 | 0.03 | 0.01 | 2.8 | 0.88 | 0.80 | 0.68 | 0.59 |
| 0.4 | 0.13[a] | 0.07 | 0.03 | 0.01 | 2.9 | 0.90 | 0.83 | 0.72 | 0.63 |
| 0.5 | 0.14 | 0.08 | 0.03 | 0.02 | 3.0 | 0.91 | 0.85 | 0.75 | 0.66 |
| 0.6 | 0.16 | 0.09 | 0.04 | 0.02 | 3.1 | 0.93 | 0.87 | 0.78 | 0.70 |
| 0.7 | 0.18 | 0.11 | 0.05 | 0.03 | 3.2 | 0.94 | 0.89 | 0.81 | 0.73 |
| 0.8 | 0.21 | 0.13 | 0.06 | 0.04 | 3.3 | 0.96 | 0.91 | 0.83 | 0.77 |
| 0.9 | 0.23 | 0.15 | 0.08 | 0.05 | 3.4 | 0.96 | 0.93 | 0.86 | 0.80 |
| 1.0 | 0.26 | 0.17 | 0.09 | 0.06 | 3.5 | 0.97 | 0.94 | 0.88 | 0.82 |
| 1.1 | 0.30 | 0.20 | 0.11 | 0.07 | 3.6 | 0.97 | 0.95 | 0.90 | 0.85 |
| 1.2 | 0.33 | 0.22 | 0.13 | 0.08 | 3.7 | 0.98 | 0.96 | 0.92 | 0.87 |
| 1.3 | 0.37 | 0.26 | 0.15 | 0.10 | 3.8 | 0.98 | 0.97 | 0.93 | 0.89 |
| 1.4 | 0.40 | 0.29 | 0.18 | 0.12 | 3.9 | 0.99 | 0.97 | 0.94 | 0.91 |
| 1.5 | 0.44 | 0.32 | 0.20 | 0.14 | 4.0 | 0.99 | 0.98 | 0.95 | 0.92 |
| 1.6 | 0.48 | 0.36 | 0.23 | 0.16 | 4.1 | 0.99 | 0.98 | 0.96 | 0.94 |
| 1.7 | 0.52 | 0.40 | 0.27 | 0.19 | 4.2 | 0.99 | 0.99 | 0.97 | 0.95 |
| 1.8 | 0.56 | 0.44 | 0.30 | 0.22 | 4.3 | [b] | 0.99 | 0.98 | 0.96 |
| 1.9 | 0.60 | 0.48 | 0.33 | 0.25 | 4.4 | | 0.99 | 0.98 | 0.97 |
| 2.0 | 0.64 | 0.52 | 0.37 | 0.28 | 4.5 | | 0.99 | 0.99 | 0.97 |
| 2.1 | 0.68 | 0.56 | 0.41 | 0.32 | 4.6 | | [b] | 0.99 | 0.98 |
| 2.2 | 0.71 | 0.59 | 0.45 | 0.35 | 4.7 | | | 0.99 | 0.98 |
| 2.3 | 0.74 | 0.63 | 0.49 | 0.39 | 4.8 | | | 0.99 | 0.99 |
| 2.4 | 0.77 | 0.67 | 0.53 | 0.43 | 4.9 | | | 0.99 | 0.99 |
| | | | | | 5.0 | | | [b] | 0.99 |
| | | | | | 5.1 | | | | 0.99 |
| | | | | | 5.2 | | | | [b] |

[a]Values inaccurate for *one-tailed* test by more than 0.01.
[b]The power at and below this point is greater than 0.995.

# Table E δ as a Function of Significance Criterion (α) and Power

| | One-tailed test (α) | | | |
|---|---|---|---|---|
| | 0.05 | 0.025 | 0.01 | 0.005 |
| | Two-tailed test (α) | | | |
| Power | 0.10 | 0.05 | 0.02 | 0.01 |
| 0.25 | 0.97 | 1.29 | 1.65 | 1.90 |
| 0.50 | 1.64 | 1.96 | 2.33 | 2.58 |
| 0.60 | 1.90 | 2.21 | 2.58 | 2.83 |
| 0.67 | 2.08 | 2.39 | 2.76 | 3.01 |
| 0.70 | 2.17 | 2.48 | 2.85 | 3.10 |
| 0.75 | 2.32 | 2.63 | 3.00 | 3.25 |
| 0.80 | 2.49 | 2.80 | 3.17 | 3.42 |
| 0.85 | 2.68 | 3.00 | 3.36 | 3.61 |
| 0.90 | 2.93 | 3.24 | 3.61 | 3.86 |
| 0.95 | 3.29 | 3.60 | 3.97 | 4.22 |
| 0.99 | 3.97 | 4.29 | 4.65 | 4.90 |
| 0.999 | 4.37 | 5.05 | 5.42 | 5.67 |

# Table F **Critical Values of *F* (α = 0.05 in Standard Type, α = 0.01 in Boldface)**

| | $n_1$ degrees of freedom (for numerator mean square) | | | | | | | | | | | |
|---|---|---|---|---|---|---|---|---|---|---|---|---|
| $n_2$ | 1 | 2 | 3 | 4 | 5 | 6 | 7 | 8 | 9 | 10 | 11 | 12 |
| 1 | 161 | 200 | 216 | 225 | 230 | 234 | 237 | 239 | 241 | 242 | 243 | 244 |
| | **4,052** | **4,999** | **5,403** | **5,625** | **5,764** | **5,859** | **5,928** | **5,981** | **6,022** | **6,056** | **6,082** | **6,106** |
| 2 | 18.51 | 19.00 | 19.16 | 19.25 | 19.30 | 19.33 | 19.36 | 19.37 | 19.38 | 19.39 | 19.40 | 19.41 |
| | **98.49** | **99.00** | **99.17** | **99.25** | **99.30** | **99.33** | **99.34** | **99.36** | **99.38** | **99.40** | **99.41** | **99.42** |
| 3 | 10.13 | 9.55 | 9.28 | 9.12 | 9.01 | 8.94 | 8.88 | 8.84 | 8.81 | 8.78 | 8.76 | 8.74 |
| | **34.12** | **30.82** | **29.46** | **28.71** | **28.24** | **27.91** | **27.67** | **27.49** | **27.34** | **27.23** | **27.13** | **27.05** |
| 4 | 7.71 | 6.94 | 6.59 | 6.39 | 6.26 | 6.16 | 6.09 | 6.04 | 6.00 | 5.96 | 5.93 | 5.91 |
| | **21.20** | **18.00** | **16.69** | **15.98** | **15.52** | **15.21** | **14.98** | **14.80** | **14.66** | **14.54** | **14.45** | **14.37** |
| 5 | 6.61 | 5.79 | 5.41 | 5.19 | 5.05 | 4.95 | 4.88 | 4.82 | 4.78 | 4.74 | 4.70 | 4.68 |
| | **16.26** | **13.27** | **12.06** | **11.39** | **10.97** | **10.67** | **10.45** | **10.27** | **10.15** | **10.05** | **9.96** | **9.89** |
| 6 | 5.99 | 5.14 | 4.76 | 4.53 | 4.39 | 4.28 | 4.21 | 4.15 | 4.10 | 4.06 | 4.03 | 4.00 |
| | **13.74** | **10.92** | **9.78** | **9.15** | **8.75** | **8.47** | **8.26** | **8.10** | **7.98** | **7.87** | **7.79** | **7.72** |
| 7 | 5.59 | 4.74 | 4.35 | 4.12 | 3.97 | 3.87 | 3.79 | 3.73 | 3.68 | 3.63 | 3.60 | 3.57 |
| | **12.25** | **9.55** | **8.45** | **7.85** | **7.46** | **7.19** | **7.00** | **6.84** | **6.71** | **6.62** | **6.54** | **6.47** |
| 8 | 5.32 | 4.46 | 4.07 | 3.84 | 3.69 | 3.58 | 3.50 | 3.44 | 3.39 | 3.34 | 3.31 | 3.28 |
| | **11.26** | **8.65** | **7.59** | **7.01** | **6.63** | **6.37** | **6.19** | **6.03** | **5.91** | **5.82** | **5.74** | **5.67** |
| 9 | 5.12 | 4.26 | 3.86 | 3.63 | 3.48 | 3.37 | 3.29 | 3.23 | 3.18 | 3.13 | 3.10 | 3.07 |
| | **10.56** | **8.02** | **6.99** | **6.42** | **6.06** | **5.80** | **5.62** | **5.47** | **5.35** | **5.26** | **5.18** | **5.11** |
| 10 | 4.96 | 4.10 | 3.71 | 3.48 | 3.33 | 3.22 | 3.14 | 3.07 | 3.02 | 2.97 | 2.94 | 2.91 |
| | **10.04** | **7.56** | **6.55** | **5.99** | **5.64** | **5.39** | **5.21** | **5.06** | **4.95** | **4.85** | **4.78** | **4.71** |
| 11 | 4.84 | 3.98 | 3.59 | 3.36 | 3.20 | 3.09 | 3.01 | 2.95 | 2.90 | 2.86 | 2.82 | 2.79 |
| | **9.65** | **7.20** | **6.22** | **5.67** | **5.32** | **5.07** | **4.88** | **4.74** | **4.63** | **4.54** | **4.46** | **4.40** |
| 12 | 4.75 | 3.88 | 3.49 | 3.26 | 3.11 | 3.00 | 2.92 | 2.85 | 2.80 | 2.76 | 2.72 | 2.69 |
| | **9.33** | **6.93** | **5.95** | **5.41** | **5.06** | **4.82** | **4.65** | **4.50** | **4.39** | **4.30** | **4.22** | **4.16** |
| 13 | 4.67 | 3.80 | 3.41 | 3.18 | 3.02 | 2.92 | 2.84 | 2.77 | 2.72 | 2.67 | 2.63 | 2.60 |
| | **9.07** | **6.70** | **5.74** | **5.20** | **4.86** | **4.62** | **4.44** | **4.30** | **4.19** | **4.10** | **4.02** | **3.96** |

# Table F (*continued*)

| $n_1$ degrees of freedom (for numerator mean square) | | | | | | | | | | | |
|---|---|---|---|---|---|---|---|---|---|---|---|
| 14 | 16 | 20 | 24 | 30 | 40 | 50 | 75 | 100 | 200 | 500 | ∞ |
| 245 | 246 | 247 | 248 | 259 | 251 | 252 | 253 | 253 | 254 | 254 | 254 |
| **6,142** | **6,169** | **6,208** | **6,234** | **6,258** | **6,286** | **6,302** | **6,323** | **6,334** | **6,352** | **6,361** | **6,366** |
| 19.42 | 19.43 | 19.44 | 19.45 | 19.46 | 19.47 | 19.47 | 19.48 | 19.49 | 19.49 | 19.50 | 19.50 |
| **99.43** | **99.44** | **99.45** | **99.46** | **99.47** | **99.48** | **99.48** | **99.49** | **99.49** | **99.49** | **99.50** | **99.50** |
| 8.71 | 8.69 | 8.66 | 8.64 | 8.62 | 8.60 | 8.58 | 8.57 | 8.56 | 8.54 | 8.54 | 8.53 |
| **26.92** | **26.83** | **26.69** | **26.60** | **26.50** | **26.41** | **26.35** | **26.27** | **26.23** | **26.18** | **26.14** | **26.12** |
| 5.87 | 5.84 | 5.80 | 5.77 | 5.74 | 5.71 | 5.70 | 5.68 | 5.66 | 5.65 | 5.64 | 5.63 |
| **14.24** | **14.15** | **14.02** | **13.93** | **13.83** | **13.74** | **13.69** | **13.61** | **13.57** | **13.52** | **13.48** | **13.46** |
| 4.64 | 4.60 | 4.56 | 4.53 | 4.50 | 4.46 | 4.44 | 4.42 | 4.40 | 4.38 | 4.37 | 4.36 |
| **9.77** | **9.68** | **9.55** | **9.47** | **9.38** | **9.29** | **9.24** | **9.17** | **9.13** | **9.07** | **9.04** | **9.02** |
| 3.96 | 3.92 | 3.87 | 3.84 | 3.81 | 3.77 | 3.75 | 3.72 | 3.71 | 3.69 | 3.68 | 3.67 |
| **7.60** | **7.52** | **7.39** | **7.31** | **7.23** | **7.14** | **7.09** | **7.02** | **6.99** | **6.94** | **6.90** | **6.88** |
| 3.52 | 3.49 | 3.44 | 3.41 | 3.38 | 3.34 | 3.32 | 3.29 | 3.28 | 3.25 | 3.24 | 3.23 |
| **6.35** | **6.27** | **6.15** | **6.07** | **5.98** | **5.90** | **5.85** | **5.78** | **5.75** | **5.70** | **5.67** | **5.65** |
| 3.23 | 3.20 | 3.15 | 3.12 | 3.08 | 3.05 | 3.03 | 3.00 | 2.98 | 2.96 | 2.94 | 2.93 |
| **5.56** | **5.48** | **5.36** | **5.28** | **5.20** | **5.11** | **5.06** | **5.00** | **4.96** | **4.91** | **4.88** | **4.86** |
| 3.02 | 2.98 | 2.93 | 2.90 | 2.86 | 2.82 | 2.80 | 2.77 | 2.76 | 2.73 | 2.72 | 2.71 |
| **5.00** | **4.92** | **4.80** | **4.73** | **4.64** | **4.56** | **4.51** | **4.45** | **4.41** | **4.36** | **4.33** | **4.31** |
| 2.86 | 2.82 | 2.77 | 2.74 | 2.70 | 2.67 | 2.64 | 2.61 | 2.59 | 2.56 | 2.55 | 2.54 |
| **4.60** | **4.52** | **4.41** | **4.33** | **4.25** | **4.17** | **4.12** | **4.05** | **4.01** | **3.96** | **3.93** | **3.91** |
| 2.74 | 2.70 | 2.65 | 2.61 | 2.57 | 2.53 | 2.50 | 2.47 | 2.45 | 2.42 | 2.41 | 2.40 |
| **4.29** | **4.21** | **4.10** | **4.02** | **3.94** | **3.86** | **3.80** | **3.74** | **3.70** | **3.66** | **3.62** | **3.60** |
| 2.64 | 2.60 | 2.54 | 2.50 | 2.46 | 2.42 | 2.40 | 2.36 | 2.35 | 2.32 | 2.31 | 2.30 |
| **4.05** | **3.98** | **3.86** | **3.78** | **3.70** | **3.61** | **3.56** | **3.49** | **3.46** | **3.41** | **3.38** | **3.36** |
| 2.55 | 2.51 | 2.46 | 2.42 | 2.38 | 2.34 | 2.32 | 2.28 | 2.26 | 2.24 | 2.22 | 2.21 |
| **3.85** | **3.78** | **3.67** | **3.59** | **3.51** | **3.42** | **3.37** | **3.30** | **3.27** | **3.21** | **3.18** | **3.16** |

(*continued*)

# Table F (*continued*)

| | $n_1$ degrees of freedom (for numerator mean square) | | | | | | | | | | | |
|---|---|---|---|---|---|---|---|---|---|---|---|---|
| $n_2$ | 1 | 2 | 3 | 4 | 5 | 6 | 7 | 8 | 9 | 10 | 11 | 12 |
| 14 | 4.60<br>**8.86** | 3.74<br>**6.51** | 3.34<br>**5.56** | 3.11<br>**5.03** | 2.96<br>**4.69** | 2.85<br>**4.46** | 2.77<br>**4.28** | 2.70<br>**4.14** | 2.65<br>**4.03** | 2.60<br>**3.94** | 2.56<br>**3.86** | 2.53<br>**3.80** |
| 15 | 4.54<br>**8.68** | 3.68<br>**6.36** | 3.29<br>**5.42** | 3.06<br>**4.89** | 2.90<br>**4.56** | 2.79<br>**4.32** | 2.70<br>**4.14** | 2.64<br>**4.00** | 2.59<br>**3.89** | 2.55<br>**3.80** | 2.51<br>**3.73** | 2.48<br>**3.67** |
| 16 | 4.49<br>**8.53** | 3.63<br>**6.23** | 3.24<br>**5.29** | 3.01<br>**4.77** | 2.85<br>**4.44** | 2.74<br>**4.20** | 2.66<br>**4.03** | 2.59<br>**3.89** | 2.54<br>**3.78** | 2.49<br>**3.69** | 2.45<br>**3.61** | 2.42<br>**3.55** |
| 17 | 4.45<br>**8.40** | 3.59<br>**6.11** | 3.20<br>**5.18** | 2.96<br>**4.67** | 2.81<br>**4.34** | 2.70<br>**4.10** | 2.62<br>**3.93** | 2.55<br>**3.79** | 2.50<br>**3.68** | 2.45<br>**3.59** | 2.41<br>**3.52** | 2.38<br>**3.45** |
| 18 | 4.41<br>**8.28** | 3.55<br>**6.01** | 3.16<br>**5.09** | 2.93<br>**4.58** | 2.77<br>**4.25** | 2.66<br>**4.01** | 2.58<br>**3.85** | 2.51<br>**3.71** | 2.46<br>**3.60** | 2.41<br>**3.51** | 2.37<br>**3.44** | 2.34<br>**3.37** |
| 19 | 4.38<br>**8.18** | 3.52<br>**5.93** | 3.13<br>**5.01** | 2.90<br>**4.50** | 2.74<br>**4.17** | 2.63<br>**3.94** | 2.55<br>**3.77** | 2.48<br>**3.63** | 2.43<br>**3.52** | 2.38<br>**3.43** | 2.34<br>**3.36** | 2.31<br>**3.30** |
| 20 | 4.35<br>**8.10** | 3.49<br>**5.85** | 3.10<br>**4.94** | 2.87<br>**4.43** | 2.71<br>**4.10** | 2.60<br>**3.87** | 2.52<br>**3.71** | 2.45<br>**3.56** | 2.40<br>**3.45** | 2.35<br>**3.37** | 2.31<br>**3.30** | 2.28<br>**3.23** |
| 21 | 4.32<br>**8.02** | 3.47<br>**5.78** | 3.07<br>**4.87** | 2.84<br>**4.37** | 2.68<br>**4.04** | 2.57<br>**3.81** | 2.49<br>**3.65** | 2.42<br>**3.51** | 2.37<br>**3.40** | 2.32<br>**3.31** | 2.28<br>**3.24** | 2.25<br>**3.17** |
| 22 | 4.30<br>**7.9** | 3.44<br>**5.72** | 3.05<br>**4.82** | 2.82<br>**4.31** | 2.66<br>**3.99** | 2.55<br>**3.76** | 2.47<br>**3.59** | 2.40<br>**3.45** | 2.35<br>**3.35** | 2.30<br>**3.26** | 2.26<br>**3.18** | 2.23<br>**3.12** |
| 23 | 4.28<br>**7.88** | 3.42<br>**5.66** | 3.03<br>**4.76** | 2.80<br>**4.26** | 2.64<br>**3.94** | 2.53<br>**3.71** | 2.45<br>**3.54** | 2.38<br>**3.41** | 2.32<br>**3.30** | 2.28<br>**3.21** | 2.24<br>**3.14** | 2.20<br>**3.07** |
| 24 | 4.26<br>**7.82** | 3.40<br>**5.61** | 3.01<br>**4.72** | 2.78<br>**4.22** | 2.62<br>**3.90** | 2.51<br>**3.67** | 2.43<br>**3.50** | 2.36<br>**3.36** | 2.30<br>**3.25** | 2.26<br>**3.17** | 2.22<br>**3.09** | 2.18<br>**3.03** |
| 25 | 4.24<br>**7.77** | 3.38<br>**5.57** | 2.99<br>**4.68** | 2.76<br>**4.18** | 2.60<br>**3.86** | 2.49<br>**3.63** | 2.41<br>**3.46** | 2.34<br>**3.32** | 2.28<br>**3.21** | 2.24<br>**3.13** | 2.20<br>**3.05** | 2.16<br>**2.99** |
| 26 | 4.22<br>**7.72** | 3.37<br>**5.53** | 2.98<br>**4.64** | 2.74<br>**4.14** | 2.59<br>**3.82** | 2.47<br>**3.59** | 2.39<br>**3.42** | 2.32<br>**3.29** | 2.27<br>**3.17** | 2.22<br>**3.09** | 2.18<br>**3.02** | 2.15<br>**2.96** |

# Table F (*continued*)

| $n_1$ degrees of freedom (for numerator mean square) | | | | | | | | | | | |
|---|---|---|---|---|---|---|---|---|---|---|---|
| 14 | 16 | 20 | 24 | 30 | 40 | 50 | 75 | 100 | 200 | 500 | ∞ |
| 2.48 | 2.44 | 2.39 | 2.35 | 2.31 | 2.27 | 2.24 | 2.21 | 2.19 | 2.16 | 2.14 | 2.13 |
| **3.70** | **3.62** | **3.51** | **3.43** | **3.34** | **3.26** | **3.21** | **3.14** | **3.11** | **3.06** | **3.02** | **3.00** |
| 2.43 | 2.39 | 2.33 | 2.29 | 2.25 | 2.21 | 2.18 | 2.15 | 2.12 | 2.10 | 2.08 | 2.07 |
| **3.56** | **3.48** | **3.36** | **3.29** | **3.20** | **3.12** | **3.07** | **3.00** | **2.97** | **2.92** | **2.89** | **2.87** |
| 2.37 | 2.33 | 2.28 | 2.24 | 2.20 | 2.16 | 2.13 | 2.09 | 2.07 | 2.04 | 2.02 | 2.01 |
| **3.45** | **3.37** | **3.25** | **3.18** | **3.10** | **3.01** | **2.96** | **2.89** | **2.86** | **2.80** | **2.77** | **2.75** |
| 2.33 | 2.29 | 2.23 | 2.19 | 2.15 | 2.11 | 2.08 | 2.04 | 2.02 | 1.99 | 1.97 | 1.96 |
| **3.35** | **3.27** | **3.16** | **3.08** | **3.00** | **2.92** | **2.86** | **2.79** | **2.76** | **2.70** | **2.67** | **2.65** |
| 2.29 | 2.25 | 2.19 | 2.15 | 2.11 | 2.07 | 2.04 | 2.00 | 1.98 | 1.95 | 1.93 | 1.92 |
| **3.27** | **3.19** | **3.07** | **3.00** | **2.91** | **2.83** | **2.78** | **2.71** | **2.68** | **2.62** | **2.59** | **2.57** |
| 2.26 | 2.21 | 2.15 | 2.11 | 2.07 | 2.02 | 2.00 | 1.96 | 1.94 | 1.91 | 1.90 | 1.88 |
| **3.19** | **3.12** | **3.00** | **2.92** | **2.84** | **2.76** | **2.70** | **2.63** | **2.60** | **2.54** | **2.51** | **2.49** |
| 2.23 | 2.18 | 2.12 | 2.08 | 2.04 | 1.99 | 1.96 | 1.92 | 1.90 | 1.87 | 1.85 | 1.84 |
| **3.13** | **3.05** | **2.94** | **2.86** | **2.77** | **2.69** | **2.63** | **2.56** | **2.53** | **2.47** | **2.44** | **2.42** |
| 2.20 | 2.15 | 2.09 | 2.05 | 2.00 | 1.96 | 1.93 | 1.89 | 1.87 | 1.84 | 1.82 | 1.81 |
| **3.07** | **2.99** | **2.88** | **2.80** | **2.72** | **2.63** | **2.58** | **2.51** | **2.47** | **2.42** | **2.38** | **2.36** |
| 2.18 | 2.13 | 2.07 | 2.03 | 1.98 | 1.93 | 1.91 | 1.87 | 1.84 | 1.81 | 1.80 | 1.78 |
| **3.02** | **2.94** | **2.83** | **2.75** | **2.67** | **2.58** | **2.53** | **2.46** | **2.42** | **2.37** | **2.33** | **2.31** |
| 2.14 | 2.10 | 2.04 | 2.00 | 1.96 | 1.91 | 1.88 | 1.84 | 1.82 | 1.79 | 1.77 | 1.76 |
| **2.97** | **2.89** | **2.78** | **2.70** | **2.62** | **2.53** | **2.48** | **2.41** | **2.37** | **2.32** | **2.28** | **2.26** |
| 2.13 | 2.09 | 2.02 | 1.98 | 1.94 | 1.89 | 1.86 | 1.82 | 1.80 | 1.76 | 1.74 | 1.73 |
| **2.93** | **2.85** | **2.74** | **2.66** | **2.58** | **2.49** | **2.44** | **2.36** | **2.33** | **2.27** | **2.23** | **2.21** |
| 2.11 | 2.06 | 2.00 | 1.96 | 1.92 | 1.87 | 1.84 | 1.80 | 1.77 | 1.74 | 1.72 | 1.71 |
| **2.89** | **2.81** | **2.70** | **2.62** | **2.54** | **2.45** | **2.40** | **2.32** | **2.29** | **2.23** | **2.19** | **2.17** |
| 2.10 | 2.05 | 1.99 | 1.95 | 1.90 | 1.85 | 1.82 | 1.78 | 1.76 | 1.72 | 1.70 | 1.69 |
| **2.86** | **2.77** | **2.66** | **2.58** | **2.50** | **2.41** | **2.36** | **2.28** | **2.25** | **2.19** | **2.15** | **2.13** |

(*continued*)

# Table F (*continued*)

| | $n_1$ degrees of freedom (for numerator mean square) | | | | | | | | | | | |
|---|---|---|---|---|---|---|---|---|---|---|---|---|
| $n_2$ | 1 | 2 | 3 | 4 | 5 | 6 | 7 | 8 | 9 | 10 | 11 | 12 |
| 27 | 4.21 | 3.35 | 2.96 | 2.73 | 2.57 | 2.46 | 2.37 | 2.30 | 2.25 | 2.20 | 2.16 | 2.13 |
| | **7.68** | **5.49** | **4.60** | **4.11** | **3.79** | **3.56** | **3.39** | **3.26** | **3.14** | **3.06** | **2.98** | **2.93** |
| 28 | 4.20 | 3.34 | 2.95 | 2.71 | 2.56 | 2.44 | 2.36 | 2.29 | 2.24 | 2.19 | 2.15 | 2.12 |
| | **7.64** | **5.45** | **4.57** | **4.07** | **3.76** | **3.53** | **3.36** | **3.23** | **3.11** | **3.03** | **2.95** | **2.90** |
| 29 | 4.18 | 3.33 | 2.93 | 2.70 | 2.54 | 2.43 | 2.35 | 2.28 | 2.22 | 2.18 | 2.14 | 2.10 |
| | **7.60** | **5.42** | **4.54** | **4.04** | **3.73** | **3.50** | **3.33** | **3.20** | **3.08** | **3.00** | **2.92** | **2.87** |
| 30 | 4.17 | 3.32 | 2.92 | 2.69 | 2.53 | 2.42 | 2.34 | 2.27 | 2.21 | 2.16 | 2.12 | 2.09 |
| | **7.56** | **5.39** | **4.51** | **4.02** | **3.70** | **3.47** | **3.30** | **3.17** | **3.06** | **2.98** | **2.90** | **2.84** |
| 32 | 4.15 | 3.30 | 2.90 | 2.67 | 2.51 | 2.40 | 2.32 | 2.25 | 2.19 | 2.14 | 2.10 | 2.07 |
| | **7.50** | **5.34** | **4.46** | **3.97** | **3.66** | **3.42** | **3.25** | **3.12** | **3.01** | **2.94** | **2.86** | **2.80** |
| 34 | 4.13 | 3.28 | 2.88 | 2.65 | 2.49 | 2.38 | 2.30 | 2.23 | 2.17 | 2.12 | 2.08 | 2.05 |
| | **7.44** | **5.29** | **4.42** | **3.93** | **3.61** | **3.38** | **3.21** | **3.08** | **2.97** | **2.89** | **2.82** | **2.76** |
| 36 | 4.11 | 3.26 | 2.86 | 2.63 | 2.48 | 2.36 | 2.28 | 2.21 | 2.15 | 2.10 | 2.06 | 2.03 |
| | **7.39** | **5.25** | **4.38** | **3.89** | **3.58** | **3.35** | **3.18** | **3.04** | **3.04** | **2.86** | **2.78** | **2.72** |
| 38 | 4.10 | 3.25 | 2.85 | 2.62 | 2.46 | 2.35 | 2.26 | 2.19 | 2.14 | 2.09 | 2.05 | 2.02 |
| | **7.35** | **5.21** | **4.34** | **3.86** | **3.54** | **3.32** | **3.15** | **3.02** | **2.91** | **2.82** | **2.75** | **2.69** |
| 40 | 4.08 | 3.23 | 2.84 | 2.61 | 2.45 | 2.34 | 2.25 | 2.18 | 2.12 | 2.07 | 2.04 | 2.00 |
| | **7.31** | **5.18** | **4.31** | **3.83** | **3.51** | **3.29** | **3.12** | **2.99** | **2.88** | **2.80** | **2.73** | **2.66** |
| 42 | 4.07 | 3.22 | 2.83 | 2.59 | 2.44 | 2.32 | 2.24 | 2.17 | 2.11 | 2.06 | 2.02 | 1.99 |
| | **7.27** | **5.15** | **4.29** | **3.80** | **3.49** | **3.26** | **3.10** | **2.96** | **2.86** | **2.77** | **2.70** | **2.64** |
| 44 | 4.06 | 3.21 | 2.82 | 2.58 | 2.43 | 2.31 | 2.23 | 2.16 | 2.10 | 2.05 | 2.01 | 1.98 |
| | **7.24** | **5.12** | **4.26** | **3.78** | **3.46** | **3.24** | **3.07** | **2.94** | **2.84** | **2.75** | **2.68** | **2.62** |
| 46 | 4.05 | 3.20 | 2.81 | 2.57 | 2.42 | 2.30 | 2.22 | 2.14 | 2.09 | 2.04 | 2.00 | 1.97 |
| | **7.21** | **5.10** | **4.24** | **3.76** | **3.44** | **3.22** | **3.05** | **2.92** | **2.82** | **2.73** | **2.66** | **2.60** |
| 48 | 4.04 | 3.19 | 2.80 | 2.56 | 2.41 | 2.30 | 2.21 | 2.14 | 2.08 | 2.03 | 1.99 | 1.96 |
| | **7.19** | **5.08** | **4.22** | **3.74** | **3.42** | **3.20** | **3.04** | **2.90** | **2.80** | **2.71** | **2.64** | **2.58** |

# Table F (*continued*)

| $n_1$ degrees of freedom (for numerator mean square) | | | | | | | | | | | |
|---|---|---|---|---|---|---|---|---|---|---|---|
| 14 | 16 | 20 | 24 | 30 | 40 | 50 | 75 | 100 | 200 | 500 | ∞ |
| 2.08 | 2.03 | 1.97 | 1.93 | 1.88 | 1.84 | 1.80 | 1.76 | 1.74 | 1.71 | 1.68 | 1.67 |
| **2.83** | **2.74** | **2.63** | **2.55** | **2.47** | **2.38** | **2.33** | **2.25** | **2.21** | **2.16** | **2.12** | **2.10** |
| 2.06 | 2.02 | 1.96 | 1.91 | 1.87 | 1.81 | 1.78 | 1.75 | 1.72 | 1.69 | 1.67 | 1.65 |
| **2.80** | **2.71** | **2.60** | **2.52** | **2.44** | **2.35** | **2.30** | **2.22** | **2.18** | **2.13** | **2.09** | **2.06** |
| 2.05 | 2.00 | 1.94 | 1.90 | 1.85 | 1.80 | 1.77 | 1.73 | 1.71 | 1.68 | 1.65 | 1.64 |
| **2.77** | **2.68** | **2.57** | **2.49** | **2.41** | **2.32** | **2.27** | **2.19** | **2.15** | **2.10** | **2.06** | **2.03** |
| 2.04 | 1.99 | 1.93 | 1.89 | 1.84 | 1.79 | 1.76 | 1.72 | 1.69 | 1.66 | 1.64 | 1.62 |
| **2.74** | **2.66** | **2.55** | **2.47** | **2.38** | **2.29** | **2.24** | **2.16** | **2.13** | **2.07** | **2.03** | **2.01** |
| 2.02 | 1.97 | 1.91 | 1.86 | 1.82 | 1.76 | 1.74 | 1.69 | 1.67 | 1.64 | 1.61 | 1.59 |
| **2.70** | **2.62** | **2.51** | **2.42** | **2.34** | **2.25** | **2.20** | **2.12** | **2.08** | **2.02** | **1.98** | **1.96** |
| 2.00 | 1.95 | 1.89 | 1.84 | 1.80 | 1.74 | 1.71 | 1.67 | 1.64 | 1.61 | 1.59 | 1.57 |
| **2.66** | **2.58** | **2.47** | **2.38** | **2.30** | **2.21** | **2.15** | **2.08** | **2.04** | **1.98** | **1.94** | **1.91** |
| 1.98 | 1.93 | 1.87 | 1.82 | 1.78 | 1.72 | 1.69 | 1.65 | 1.62 | 1.59 | 1.56 | 1.55 |
| **2.62** | **2.54** | **2.43** | **2.35** | **2.26** | **2.17** | **2.12** | **2.04** | **2.00** | **1.94** | **1.90** | **1.87** |
| 1.96 | 1.92 | 1.85 | 1.80 | 1.76 | 1.71 | 1.67 | 1.63 | 1.60 | 1.57 | 1.54 | 1.53 |
| **2.59** | **2.51** | **2.40** | **2.32** | **2.22** | **2.14** | **2.08** | **2.00** | **1.97** | **1.90** | **1.86** | **1.84** |
| 1.95 | 1.90 | 1.84 | 1.79 | 1.74 | 1.69 | 1.66 | 1.61 | 1.59 | 1.55 | 1.53 | 1.51 |
| **2.56** | **2.49** | **2.37** | **2.29** | **2.20** | **2.11** | **2.05** | **1.97** | **1.94** | **1.88** | **1.84** | **1.81** |
| 1.91 | 1.89 | 1.82 | 1.78 | 1.73 | 1.68 | 1.64 | 1.60 | 1.57 | 1.54 | 1.51 | 1.49 |
| **2.54** | **2.46** | **2.35** | **2.26** | **2.17** | **2.08** | **2.02** | **1.94** | **1.91** | **1.85** | **1.80** | **1.78** |
| 1.92 | 1.88 | 1.81 | 1.76 | 1.72 | 1.66 | 1.63 | 1.58 | 1.56 | 1.52 | 1.50 | 1.48 |
| **2.52** | **2.44** | **2.32** | **2.24** | **2.15** | **2.06** | **2.00** | **1.92** | **1.88** | **1.82** | **1.78** | **1.75** |
| 1.91 | 1.87 | 1.80 | 1.75 | 1.71 | 1.65 | 1.62 | 1.57 | 1.54 | 1.51 | 1.48 | 1.46 |
| **2.50** | **2.42** | **2.30** | **2.22** | **2.13** | **2.04** | **1.98** | **1.90** | **1.86** | **1.80** | **1.76** | **1.72** |
| 1.90 | 1.86 | 1.79 | 1.74 | 1.70 | 1.64 | 1.61 | 1.56 | 1.53 | 1.50 | 1.47 | 1.45 |
| **2.48** | **2.40** | **2.28** | **2.20** | **2.11** | **2.02** | **1.96** | **1.88** | **1.84** | **1.78** | **1.73** | **1.70** |

(*continued*)

# Table F (*continued*)

| | $n_1$ degrees of freedom (for numerator mean square) | | | | | | | | | | | |
|---|---|---|---|---|---|---|---|---|---|---|---|---|
| $n_2$ | 1 | 2 | 3 | 4 | 5 | 6 | 7 | 8 | 9 | 10 | 11 | 12 |
| 50 | 4.03 | 3.18 | 2.79 | 2.56 | 2.40 | 2.29 | 2.20 | 2.13 | 2.07 | 2.02 | 1.98 | 1.95 |
| | **7.17** | **5.06** | **4.20** | **3.72** | **3.41** | **3.18** | **3.02** | **2.88** | **2.78** | **2.70** | **2.62** | **2.56** |
| 55 | 4.02 | 3.17 | 2.78 | 2.54 | 2.38 | 2.27 | 2.18 | 2.11 | 2.05 | 2.00 | 1.97 | 1.93 |
| | **7.12** | **5.01** | **4.16** | **3.68** | **3.37** | **3.15** | **2.98** | **2.85** | **2.75** | **2.66** | **2.59** | **2.53** |
| 60 | 4.00 | 3.15 | 2.76 | 2.52 | 2.37 | 2.25 | 2.17 | 2.10 | 2.04 | 1.99 | 1.95 | 1.92 |
| | **7.08** | **4.98** | **4.13** | **3.65** | **3.34** | **3.12** | **2.95** | **2.82** | **2.72** | **2.63** | **2.56** | **2.50** |
| 65 | 3.99 | 3.14 | 2.75 | 2.51 | 2.36 | 2.24 | 2.15 | 2.08 | 2.02 | 1.98 | 1.94 | 1.90 |
| | **7.04** | **4.95** | **4.10** | **3.62** | **3.31** | **3.09** | **2.93** | **2.79** | **2.70** | **2.61** | **2.54** | **2.47** |
| 70 | 3.98 | 3.13 | 2.74 | 2.50 | 2.35 | 2.23 | 2.14 | 2.07 | 2.01 | 1.97 | 1.93 | 1.89 |
| | **7.01** | **4.92** | **4.08** | **3.60** | **3.29** | **3.07** | **2.91** | **2.77** | **2.67** | **2.59** | **2.51** | **2.45** |
| 80 | 3.96 | 3.11 | 2.72 | 2.48 | 2.33 | 2.21 | 2.12 | 2.05 | 1.99 | 1.95 | 1.91 | 1.88 |
| | **6.96** | **4.88** | **4.04** | **3.56** | **3.25** | **3.04** | **2.87** | **2.74** | **2.64** | **2.55** | **2.48** | **2.41** |
| 100 | 3.94 | 3.09 | 2.70 | 2.46 | 2.30 | 2.19 | 2.10 | 2.03 | 1.97 | 1.92 | 1.88 | 1.85 |
| | **6.90** | **4.82** | **3.98** | **3.51** | **3.20** | **2.99** | **2.82** | **2.69** | **2.59** | **2.51** | **2.43** | **2.36** |
| 125 | 3.92 | 3.07 | 2.68 | 2.44 | 2.29 | 2.17 | 2.08 | 2.01 | 1.95 | 1.90 | 1.86 | 1.83 |
| | **6.84** | **4.78** | **3.94** | **3.47** | **3.17** | **2.95** | **2.79** | **2.65** | **2.56** | **2.47** | **2.40** | **2.33** |
| 150 | 3.91 | 3.06 | 2.57 | 2.43 | 2.27 | 2.16 | 2.07 | 2.00 | 1.94 | 1.89 | 1.85 | 1.82 |
| | **6.81** | **4.75** | **3.91** | **3.44** | **3.14** | **2.92** | **2.76** | **2.62** | **2.53** | **2.44** | **2.37** | **2.30** |
| 200 | 3.89 | 3.04 | 2.65 | 2.41 | 2.26 | 2.14 | 2.05 | 1.98 | 1.92 | 1.87 | 1.83 | 1.80 |
| | **6.76** | **4.71** | **3.88** | **3.41** | **3.11** | **2.90** | **2.73** | **2.60** | **2.50** | **2.41** | **2.34** | **2.28** |
| 400 | 3.86 | 3.02 | 2.62 | 2.39 | 2.23 | 2.12 | 2.03 | 1.96 | 1.90 | 1.85 | 1.81 | 1.78 |
| | **6.70** | **4.66** | **3.83** | **3.36** | **3.06** | **2.85** | **2.69** | **2.55** | **2.46** | **2.37** | **2.29** | **2.23** |
| 1000 | 3.85 | 3.00 | 2.61 | 2.38 | 2.22 | 2.10 | 2.02 | 1.95 | 1.89 | 1.84 | 1.80 | 1.76 |
| | **6.66** | **4.62** | **3.80** | **3.34** | **3.04** | **2.82** | **2.66** | **2.53** | **2.43** | **2.34** | **2.26** | **2.20** |
| ∞ | 3.84 | 2.99 | 2.60 | 2.37 | 2.21 | 2.09 | 2.01 | 1.94 | 1.88 | 1.83 | 1.79 | 1.75 |
| | **6.64** | **4.60** | **3.78** | **3.32** | **3.02** | **2.80** | **2.64** | **2.51** | **2.41** | **2.32** | **2.24** | **2.18** |

# Table F (*continued*)

| $n_1$ degrees of freedom (for numerator mean square) | | | | | | | | | | | |
|---|---|---|---|---|---|---|---|---|---|---|---|
| 14 | 16 | 20 | 24 | 30 | 40 | 50 | 75 | 100 | 200 | 500 | ∞ |
| 1.90 | 1.85 | 1.78 | 1.74 | 1.69 | 1.63 | 1.60 | 1.55 | 1.52 | 1.48 | 1.46 | 1.44 |
| **2.46** | **2.39** | **2.26** | **2.18** | **2.10** | **2.00** | **1.94** | **1.86** | **1.82** | **1.76** | **1.71** | **1.68** |
| 1.88 | 1.83 | 1.76 | 1.72 | 1.67 | 1.61 | 1.58 | 1.52 | 1.50 | 1.46 | 1.43 | 1.41 |
| **2.43** | **2.35** | **2.23** | **2.15** | **2.06** | **1.96** | **1.90** | **1.82** | **1.78** | **1.71** | **1.66** | **1.64** |
| 1.86 | 1.81 | 1.75 | 1.70 | 1.65 | 1.59 | 1.56 | 1.50 | 1.48 | 1.44 | 1.41 | 1.39 |
| **2.40** | **2.32** | **2.20** | **2.12** | **2.03** | **1.93** | **1.87** | **1.79** | **1.74** | **1.68** | **1.63** | **1.60** |
| 1.84 | 1.80 | 1.73 | 1.68 | 1.63 | 1.57 | 1.54 | 1.49 | 1.46 | 1.42 | 1.39 | 1.37 |
| **2.37** | **2.30** | **2.18** | **2.09** | **2.00** | **1.90** | **1.84** | **1.76** | **1.71** | **1.64** | **1.60** | **1.56** |
| 1.84 | 1.79 | 1.72 | 1.67 | 1.62 | 1.56 | 1.53 | 1.47 | 1.45 | 1.40 | 1.37 | 1.35 |
| **2.35** | **2.28** | **2.15** | **2.07** | **1.98** | **1.88** | **1.82** | **1.74** | **1.69** | **1.62** | **1.56** | **1.53** |
| 1.82 | 1.77 | 1.70 | 1.65 | 1.60 | 1.54 | 1.51 | 1.45 | 1.42 | 1.38 | 1.35 | 1.32 |
| **2.32** | **2.24** | **2.11** | **2.03** | **1.94** | **1.84** | **1.78** | **1.70** | **1.65** | **1.57** | **1.52** | **1.49** |
| 1.79 | 1.75 | 1.68 | 1.63 | 1.57 | 1.51 | 1.48 | 1.42 | 1.39 | 1.34 | 1.30 | 1.28 |
| **2.26** | **2.19** | **2.06** | **1.98** | **1.89** | **1.79** | **1.73** | **1.64** | **1.59** | **1.51** | **1.46** | **1.43** |
| 1.77 | 1.72 | 1.65 | 1.60 | 1.55 | 1.49 | 1.45 | 1.39 | 1.36 | 1.31 | 1.27 | 1.25 |
| **2.23** | **2.15** | **2.03** | **1.94** | **1.85** | **1.75** | **1.68** | **1.59** | **1.54** | **1.46** | **1.40** | **1.37** |
| 1.76 | 1.71 | 1.64 | 1.59 | 1.54 | 1.47 | 1.44 | 1.37 | 1.34 | 1.29 | 1.25 | 1.22 |
| **2.20** | **2.12** | **2.00** | **1.91** | **1.83** | **1.72** | **1.66** | **1.56** | **1.51** | **1.43** | **1.37** | **1.33** |
| 1.74 | 1.69 | 1.62 | 1.57 | 1.52 | 1.45 | 1.42 | 1.35 | 1.32 | 1.26 | 1.22 | 1.19 |
| **2.17** | **2.09** | **1.97** | **1.88** | **1.79** | **1.69** | **1.62** | **1.53** | **1.48** | **1.39** | **1.33** | **1.28** |
| 1.72 | 1.67 | 1.60 | 1.54 | 1.49 | 1.42 | 1.38 | 1.32 | 1.28 | 1.22 | 1.16 | 1.13 |
| **2.12** | **2.04** | **1.92** | **1.84** | **1.74** | **1.65** | **1.57** | **1.47** | **1.42** | **1.32** | **1.24** | **1.19** |
| 1.70 | 1.65 | 1.58 | 1.53 | 1.47 | 1.41 | 1.36 | 1.30 | 1.26 | 1.19 | 1.13 | 1.08 |
| **2.09** | **2.01** | **1.89** | **1.81** | **1.71** | **1.61** | **1.54** | **1.44** | **1.38** | **1.28** | **1.19** | **1.11** |
| 1.69 | 1.64 | 1.57 | 1.52 | 1.46 | 1.40 | 1.35 | 1.28 | 1.24 | 1.17 | 1.11 | 1.00 |
| **2.07** | **1.99** | **1.87** | **1.79** | **1.69** | **1.59** | **1.52** | **1.41** | **1.36** | **1.25** | **1.15** | **1.00** |

# Table G Critical Values of the Studentized Range Statistic ($q$) for $\alpha = 0.05$

| *df* for error term | Number of groups | | | | | | | | | | | | | | | | | | |
|---|---|---|---|---|---|---|---|---|---|---|---|---|---|---|---|---|---|---|---|
| | 2 | 3 | 4 | 5 | 6 | 7 | 8 | 9 | 10 | 11 | 12 | 13 | 14 | 15 | 16 | 17 | 18 | 19 | 20 |
| 1 | 17.97 | 26.98 | 32.82 | 37.08 | 40.41 | 43.12 | 45.40 | 47.36 | 49.07 | 50.59 | 51.96 | 53.20 | 54.33 | 55.36 | 56.32 | 57.22 | 58.04 | 58.83 | 59.56 |
| 2 | 6.08 | 8.33 | 9.80 | 10.88 | 11.74 | 12.44 | 13.03 | 13.54 | 13.99 | 14.39 | 14.75 | 15.08 | 15.38 | 15.65 | 15.91 | 16.14 | 16.37 | 16.57 | 16.77 |
| 3 | 4.50 | 5.91 | 6.82 | 7.50 | 8.04 | 8.48 | 8.85 | 9.18 | 9.46 | 9.72 | 9.95 | 10.15 | 10.35 | 10.52 | 10.69 | 10.84 | 10.98 | 11.11 | 11.24 |
| 4 | 3.93 | 5.04 | 5.76 | 6.29 | 6.71 | 7.05 | 7.35 | 7.60 | 7.83 | 8.03 | 8.21 | 8.37 | 8.52 | 8.66 | 8.79 | 8.91 | 9.03 | 9.13 | 9.23 |
| 5 | 3.64 | 4.60 | 5.22 | 5.67 | 6.03 | 6.33 | 6.58 | 6.80 | 6.99 | 7.17 | 7.32 | 7.47 | 7.60 | 7.72 | 7.83 | 7.93 | 8.03 | 8.12 | 8.21 |
| 6 | 3.46 | 4.34 | 4.90 | 5.30 | 5.63 | 5.90 | 6.12 | 6.32 | 6.49 | 6.65 | 6.79 | 6.92 | 7.03 | 7.14 | 7.24 | 7.34 | 7.43 | 7.51 | 7.59 |
| 7 | 3.34 | 4.16 | 4.68 | 5.06 | 5.36 | 5.61 | 5.82 | 6.00 | 6.16 | 6.30 | 6.43 | 6.55 | 6.66 | 6.76 | 6.85 | 6.94 | 7.02 | 7.10 | 7.17 |
| 8 | 3.26 | 4.04 | 4.53 | 4.89 | 5.17 | 5.40 | 5.60 | 5.77 | 5.92 | 6.05 | 6.18 | 6.29 | 6.39 | 6.48 | 6.57 | 6.65 | 6.73 | 6.80 | 6.87 |
| 9 | 3.20 | 3.95 | 4.41 | 4.76 | 5.02 | 5.24 | 5.43 | 5.59 | 5.74 | 5.87 | 5.98 | 6.09 | 6.19 | 6.28 | 6.36 | 6.44 | 6.51 | 6.58 | 6.64 |
| 10 | 3.15 | 3.88 | 4.33 | 4.65 | 4.91 | 5.12 | 5.30 | 5.46 | 5.60 | 5.72 | 5.83 | 5.93 | 6.03 | 6.11 | 6.19 | 6.27 | 6.34 | 6.40 | 6.47 |
| 11 | 3.11 | 3.82 | 4.26 | 4.57 | 4.82 | 5.03 | 5.20 | 5.35 | 5.49 | 5.61 | 5.71 | 5.81 | 5.90 | 5.98 | 6.06 | 6.13 | 6.20 | 6.27 | 6.33 |
| 12 | 3.08 | 3.77 | 4.20 | 4.51 | 4.75 | 4.95 | 5.12 | 5.27 | 5.39 | 5.51 | 5.61 | 5.71 | 5.80 | 5.88 | 5.95 | 6.02 | 6.09 | 6.15 | 6.21 |
| 13 | 3.06 | 3.73 | 4.15 | 4.45 | 4.69 | 4.88 | 5.05 | 5.19 | 5.32 | 5.43 | 5.53 | 5.63 | 5.71 | 5.79 | 5.86 | 5.93 | 5.99 | 6.05 | 6.11 |
| 14 | 3.03 | 3.70 | 4.11 | 4.41 | 4.64 | 4.83 | 4.99 | 5.13 | 5.25 | 5.36 | 5.46 | 5.55 | 5.64 | 5.71 | 5.79 | 5.85 | 5.91 | 5.97 | 6.03 |
| 15 | 3.01 | 3.67 | 4.08 | 4.37 | 4.59 | 4.78 | 4.94 | 5.08 | 5.20 | 5.31 | 5.40 | 5.49 | 5.57 | 5.65 | 5.72 | 5.78 | 5.85 | 5.90 | 5.96 |
| 16 | 3.00 | 3.65 | 4.05 | 4.33 | 4.56 | 4.74 | 4.90 | 5.03 | 5.15 | 5.26 | 5.35 | 5.44 | 5.52 | 5.59 | 5.66 | 5.73 | 5.79 | 5.84 | 5.90 |
| 17 | 2.98 | 3.63 | 4.02 | 4.30 | 4.52 | 4.70 | 4.86 | 4.99 | 5.11 | 5.21 | 5.31 | 5.39 | 5.47 | 5.54 | 5.61 | 5.67 | 5.73 | 5.79 | 5.84 |
| 18 | 2.97 | 3.61 | 4.00 | 4.28 | 4.49 | 4.67 | 4.82 | 4.96 | 5.07 | 5.17 | 5.27 | 5.35 | 5.43 | 5.50 | 5.57 | 5.63 | 5.69 | 5.74 | 5.79 |
| 19 | 2.96 | 3.59 | 3.98 | 4.25 | 4.47 | 4.65 | 4.79 | 4.92 | 5.04 | 5.14 | 5.23 | 5.31 | 5.39 | 5.46 | 5.53 | 5.59 | 5.65 | 5.70 | 5.75 |
| 20 | 2.95 | 3.58 | 3.96 | 4.23 | 4.45 | 4.62 | 4.77 | 4.90 | 5.01 | 5.11 | 5.20 | 5.28 | 5.36 | 5.43 | 5.49 | 5.55 | 5.61 | 5.66 | 5.71 |
| 24 | 2.92 | 3.53 | 3.90 | 4.17 | 4.37 | 4.54 | 4.68 | 4.81 | 4.92 | 5.01 | 5.10 | 5.18 | 5.25 | 5.32 | 5.38 | 5.44 | 5.49 | 5.55 | 5.59 |
| 30 | 2.89 | 3.49 | 3.85 | 4.10 | 4.30 | 4.46 | 4.60 | 4.72 | 4.82 | 4.92 | 5.00 | 5.08 | 5.15 | 5.21 | 5.27 | 5.33 | 5.38 | 5.43 | 5.47 |
| 40 | 2.86 | 3.44 | 3.79 | 4.04 | 4.23 | 4.39 | 4.52 | 4.63 | 4.73 | 4.82 | 4.90 | 4.98 | 5.04 | 5.11 | 5.16 | 5.22 | 5.27 | 5.31 | 5.36 |
| 60 | 2.83 | 3.40 | 3.74 | 3.98 | 4.16 | 4.31 | 4.44 | 4.55 | 4.65 | 4.73 | 4.81 | 4.88 | 4.94 | 5.00 | 5.06 | 5.11 | 5.15 | 5.20 | 5.24 |
| 120 | 2.80 | 3.36 | 3.68 | 3.92 | 4.10 | 4.24 | 4.36 | 4.47 | 4.56 | 4.64 | 4.71 | 4.78 | 4.84 | 4.90 | 4.95 | 5.00 | 5.04 | 5.09 | 5.13 |
| ∞ | 2.77 | 3.31 | 3.63 | 3.86 | 4.03 | 4.17 | 4.29 | 4.39 | 4.47 | 4.55 | 4.62 | 4.68 | 4.74 | 4.80 | 4.85 | 4.89 | 4.93 | 4.97 | 5.01 |

*Source:* Adapted from *Biometrika Tables for Statisticians* Vol. 1, 3rd ed., by E. Pearson and H. Hartley, Table 29. Copyright © 1966 University Press. Used with the permission of the Biometrika Trustees.

# Table H Critical Values of Chi Square

| | Level of significance | | | | | |
|---|---|---|---|---|---|---|
| *df* | 0.20 | 0.10 | 0.05 | 0.02 | 0.01 | 0.001 |
| 1 | 1.64 | 2.71 | 3.84 | 5.41 | 6.63 | 10.83 |
| 2 | 3.22 | 4.61 | 5.99 | 7.82 | 9.21 | 13.82 |
| 3 | 4.64 | 6.25 | 7.82 | 9.84 | 11.34 | 16.27 |
| 4 | 5.99 | 7.78 | 9.49 | 11.67 | 13.28 | 18.46 |
| 5 | 7.29 | 9.24 | 11.07 | 13.39 | 15.09 | 20.52 |
| 6 | 8.56 | 10.64 | 12.59 | 15.03 | 16.81 | 22.46 |
| 7 | 9.80 | 12.02 | 14.07 | 16.62 | 18.48 | 24.32 |
| 8 | 11.03 | 13.36 | 15.51 | 18.17 | 20.09 | 26.12 |
| 9 | 12.24 | 14.68 | 16.92 | 19.68 | 21.67 | 27.88 |
| 10 | 13.44 | 15.99 | 18.31 | 21.16 | 23.21 | 29.59 |
| 11 | 14.63 | 17.28 | 19.68 | 22.62 | 24.72 | 31.26 |
| 12 | 15.81 | 18.55 | 21.03 | 24.05 | 26.22 | 32.91 |
| 13 | 16.98 | 19.81 | 22.36 | 25.47 | 27.69 | 34.53 |
| 14 | 18.15 | 21.06 | 23.68 | 26.87 | 29.14 | 36.12 |
| 15 | 19.31 | 22.31 | 25.00 | 28.26 | 30.58 | 37.70 |
| 16 | 20.46 | 23.54 | 26.30 | 29.63 | 32.00 | 39.25 |
| 17 | 21.62 | 24.77 | 27.59 | 31.00 | 33.41 | 40.79 |
| 18 | 22.76 | 25.99 | 28.87 | 32.35 | 34.81 | 42.31 |
| 19 | 23.90 | 27.20 | 30.14 | 33.69 | 36.19 | 43.82 |
| 20 | 25.04 | 28.41 | 31.41 | 35.02 | 37.57 | 45.32 |
| 21 | 26.17 | 29.62 | 32.67 | 36.34 | 38.93 | 46.80 |
| 22 | 27.30 | 30.81 | 33.92 | 37.66 | 40.29 | 48.27 |
| 23 | 28.43 | 32.01 | 35.17 | 38.97 | 41.64 | 49.73 |
| 24 | 29.55 | 33.20 | 36.42 | 40.27 | 42.98 | 51.18 |
| 25 | 30.68 | 34.38 | 37.65 | 41.57 | 44.31 | 52.62 |
| 26 | 31.80 | 35.56 | 38.89 | 42.86 | 45.64 | 54.05 |
| 27 | 32.91 | 36.74 | 40.11 | 44.14 | 46.96 | 55.48 |
| 28 | 34.03 | 37.92 | 41.34 | 45.42 | 48.28 | 56.89 |
| 29 | 35.14 | 39.09 | 42.56 | 46.69 | 49.59 | 58.30 |
| 30 | 36.25 | 40.26 | 43.77 | 47.96 | 50.89 | 59.70 |

## Table I Critical Values of $r_s$ (Spearman Rank-Order Correlation Coefficient)

| | Level of significance for one-tailed test | | | |
|---|---|---|---|---|
| | 05 | 025 | 01 | 005 |
| | Level of significance for two-tailed test | | | |
| No. of pairs (*N*) | 0.10 | 0.05 | 02 | 0.01 |
| 5 | 0.900 | 1.000 | 1.000 | — |
| 6 | 0.829 | 0.886 | 0.943 | 1.000 |
| 7 | 0.714 | 0.786 | 0.893 | 0.929 |
| 8 | 0.643 | 0.738 | 0.833 | 0.881 |
| 9 | 0.600 | 0.683 | 0.783 | 0.833 |
| 10 | 0.564 | 0.648 | 0.746 | 0.794 |
| 12 | 0.506 | 0.591 | 0.712 | 0.777 |
| 14 | 0.456 | 0.544 | 0.645 | 0.715 |
| 16 | 0.425 | 0.506 | 0.601 | 0.665 |
| 18 | 0.399 | 0.475 | 0.564 | 0.625 |
| 20 | 0.377 | 0.450 | 0.534 | 0.591 |
| 22 | 0.359 | 0.428 | 0.508 | 0.562 |
| 24 | 0.343 | 0.409 | 0.485 | 0.537 |
| 26 | 0.329 | 0.392 | 0.465 | 0.515 |
| 28 | 0.317 | 0.377 | 0.448 | 0.496 |
| 30 | 0.306 | 0.364 | 0.432 | 0.478 |

*Source:* G. E. Olds, *Ann. Math. Statistics* 9 (1938); 20 (1949). Reproduced by permission of the publisher.

## Answers to Odd-Numbered Exercises

### Chapter 1

1. (a) $\sum X = 195$ $\sum X^2 = 2,801$ $(\sum X)^2 = 38,025$
   (b) $\sum X = 138$ $\sum X^2 = 1,512$ $(\sum X)^2 = 19,044$
   (c) $\sum X = 70$ $\sum X^2 = 550$ $(\sum X)^2 = 4,900$
   (d) $\sum X = 55$ $\sum X^2 = 685$ $(\sum X)^2 = 3,025$
3. (a) $\sum X = 10$ $(\sum X)^2 = 100$ $\sum(X - Y) = -20$
   $\sum Y = 30$ $(\sum Y)^2 = 900$ $\sum X - \sum Y = -20$
   $\sum X^2 = 30$ $(\sum X + Y) = 40$ $\sum XY = 73$
   $\sum Y^2 = 206$ $\sum X + \sum Y = 40$ $\sum X \sum Y = 300$
   (b) $40 = 40;$ $-20 = -20;$ $73 \neq 300;$ $30 \neq 100$
   (c) $\sum(X + k) = 10 + 20 = 30;$ $\sum X + k = 10 + 4 = 14;$ Sum of new scores = 30
   (d) $30 - 15 = 15; 30 - 3 = 27;$ Sum of new scores = 15
   (e) $2 * 10 = 20;$ Sum of new scores = 20
5. *Data set* 3: $N = 14$
   $\sum X = 1,176;$ $\sum X^2 = 100,288;$ $(\sum X)^2 = 1,382,976;$
   $\sum XY = 96,426;$ $\sum(X + Y) = 2,305;$
   $\sum Y = 1,129;$ $\sum Y^2 = 93,343;$ $(\sum Y)^2 = 1,274,641;$
   $\sum X \sum Y = 1,327,704;$ $\sum(X - Y) = 47$

### Chapter 2

1.

| | Turck | | Kirk | | Dupre | |
|---|---|---|---|---|---|---|
| **Score** | *f* | *cf* | *f* | *cf* | *f* | *cf* |
| 20 | 1 | 15 | 0 | 15 | | |
| 19 | 0 | 14 | 0 | 15 | | |
| 18 | 1 | 14 | 0 | 15 | | |
| 17 | 2 | 13 | 1 | 15 | | |
| 16 | 1 | 11 | 0 | 14 | | |
| 15 | 1 | 10 | 0 | 14 | | |
| 14 | 1 | 9 | 1 | 14 | | |
| 13 | 2 | 8 | 0 | 13 | | |
| 12 | 0 | 6 | 3 | 13 | | |
| 11 | 2 | 6 | 2 | 10 | 1 | 10 |
| 10 | 1 | 4 | 0 | 8 | 1 | 9 |
| 9 | 1 | 3 | 1 | 8 | 2 | 8 |
| 8 | 0 | 2 | 2 | 7 | 0 | 6 |
| 7 | 0 | 2 | 2 | 5 | 1 | 6 |
| 6 | 1 | 2 | 1 | 3 | 1 | 5 |
| 5 | 1 | 1 | 0 | 2 | 2 | 4 |
| 4 | | | 0 | 2 | 2 | 2 |
| 3 | | | 1 | 2 | 0 | 0 |
| 2 | | | 0 | 1 | 0 | 0 |
| 1 | | | 1 | 1 | 0 | 0 |

3.

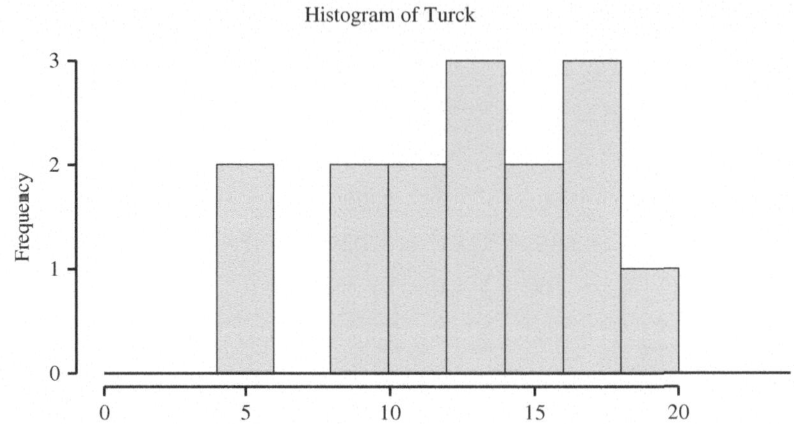


Approximately symmetrical, bimodal in shape.

5.

| Stems (intervals) | Leaves (observations) |
|---|---|
| 3–5 | 5 |
| 6–8 | 6 |
| 9–11 | 9 0 1 1 |
| 12–14 | 3 3 4 |
| 15–17 | 5 6 7 7 |
| 18–20 | 8 0 |

7. (a) The *cf* corresponding to a score of 8 is 2; PR = (2/15) × 100 = 13.33%, so the PR for 8 is about 13.
   (b) The *cf* forascoreof12is6;PR = (6/15) × 100 = 40%, so the PR for 12 is 40.
9. (a) The desired *cf* = (25/100) × 15 = 3.75, so the score at the 25th percentile is about 10.
   (b) The desired *cf* = (75/100) × 15 = 11.25, so the score at the 75th percentile is about 16.

## Chapter 3

1. The means are: Turck Hall = 13.0; Kirk Hall = 9.2; Dupre Hall = 7.0; Doty Hall = 11.0
3. The mode for Kirk Hall is 12. (It has a frequency of 3, which is more than any other score in the distribution.)
5. Subtracting the mean from each score, we obtain: −3, −3, −2, −2, −1, 0, +2, +2, +3, +4. These deviations from the mean sum to zero.
7. First, obtain the sum of squares (*SS*) for a dorm, and then divide by $N$ for $\sigma^2$, or $N - 1$ for $s^2$.The *SS* is found by first squaring the sum of the scores and dividing it by $N$; then subtract that amount from the sum of the squared scores. For Turck Hall, $SS = 2{,}801 - (195^2/15) = 2{,}801 - 2{,}535 = 266$, so $\sigma^2 = 266/15 = 17.733$, $s^2 = 266/14 = 19$, $\sigma = \sqrt{17.733} = 4.21$, and $s = \sqrt{19} = 4.36$. For Kirk Hall, $\sigma^2 = 16.16$, $s^2 = 17.314$, $\sigma = 4.02$, and $s = 4.16$. For Dupre Hall, $\sigma^2 = 6.0$, $s^2 = 6.67$, $\sigma = 2.45$, and $s = 2.58$. For Doty Hall, $\sigma^2 = 16.0$, $s^2 = 20.0$, $\sigma = 4.0$, and $s = 4.47$.
9. (a) The SD will not be changed (but the mean will increase to 62.3). (b) The SD will not be changed (but the mean will decrease to 53.3). (c) The SD will become 9.6 × 2 = 19.2. (d) The SD will become 9.6/3 = 3.2.
11. The interquartile range for Turck Hall is 16 − 10 = 6; the SIQR = 6/2 = 3. For Kirk Hall, the interquartile range is about 11.5 − 6.5 = 5; therefore the SIQR is about 2.5.
13. The MD for Doty Hall = $[|14 - 11| + |18 - 11| + |17 - 11| + |6 - 11| + |10 - 11|]/5 = (3 + 3 + 6 + 5 + 1)/5 = 18/5 = 3.6$. The MD is slightly smaller than $\sigma$ for Doty

Hall, which equals 4.0. An extreme score would affect σ more than MD, thus increasing the difference between the two measures.

## Chapter 4

1.

| | $\overline{X}$ | $\sigma$ | $\sigma^2$ |
|---|---|---|---|
| (a) | 14.8 | 4 | 16 |
| (b) | 4 | 4 | 16 |
| (c) | 25.6 | 12.8 | 163.84 |
| (d) | 2 | 1 | 1 |
| (e) | 7 | 2 | 4 |

3. Psychology $z$ score = −1.60; English $z$ score = −1.00. This student performed relatively better on the English test.

5.

| S | $z_x$ | $z_y$ |
|---|---|---|
| 1 | +1.25 | +0.65 |
| 2 | −1.54 | −1.85 |
| 3 | +0.10 | +0.50 |
| 4 | −0.97 | −0.36 |
| 5 | +0.77 | +1.28 |
| 6 | +0.77 | −0.13 |
| 7 | +1.54 | +0.81 |
| 8 | −2.03 | −2.39 |
| 9 | +0.10 | +0.34 |
| 10 | +0.29 | +0.26 |
| 11 | −0.29 | +0.81 |
| 12 | +0.87 | +0.81 |
| 13 | −0.68 | −0.44 |
| 14 | −0.19 | −0.28 |

7. $z = -0.90$, area between mean and $z$ = 31.59%; Answer = 18.41%
9. $z = -2.25$ (area = 48.78%); $z = -1.25$ (area = 39.44%); Answer = 9.34%
11. $z = +1.65$; Answer = 665
13. (a) $p = .184$; (b) $p = .376$; (c) $p = .093$; (d) $p = .050$

## Chapter 5

1. (a) $\sigma_{\overline{X}} = 0.5$
   (b) $z = -1.60$, one-tailed $p = .0548$, two-tailed $p = .1096$
   (c) $p > .05$ for both one- and two-tailed tests, so retain $H_0$ in both cases
   (d) $z = +4.80$, reject $H_0$ at .05 and .01
3. (a) $z = +1.50 < +1.96$ (two-tailed $z_{crit}$), retain $H_0$; $z = +1.50 < +1.65$ (one-tailed $z_{crit}$), not significant for a one-tailed test either.
   (b) $z = +2.40 > +1.96$; reject $H_0$ for both the one- and two-tailed tests.
5. (a) $z = (\overline{X} - \mu)/(\sigma/\sqrt{N})$; for Dupre Hall, $z = (7 - 10)/(5/\sqrt{10}) = -3/1.58 = -1.90$. Area beyond $z = -1.90$ is .0287, so the one-tailed $p$ value = .0287, and the two-tailed $p$ value = 2 × .0287 = .0574.
   (b) Reject $H_0$ at the .05 level for a one-tailed test, but not for a two-tailed test. Compared to Exercise 4, the difference between the sample mean and the population mean is the same in magnitude, though opposite in direction. This changes the sign of the $z$ score for groups, but that by itself does not affect the size of the $p$ values. The population standard deviation is also the same as in Exercise 4, but the sample size is smaller (10 instead of 15), which makes the standard error larger, which makes the $z$ score smaller, which makes the $p$ values larger.

## Chapter 6

1. (a) $s_{\overline{X}} = 1.125$; $t = -2.67$. Reject $H_0$ $(2.67 > t_{crit} = 2.145)$
   (b) $s_{\overline{X}} = 1.074$; $t = -1.675$. Retain $H_0$
3. (a) $s_{\overline{X}} = 0.63$; $t = -0.62$; $df = 8$. Retain $H_0$
   (b) $s_{\overline{X}} = 0.63$; $t = -6.17$; $df = 8$. Reject $H_0$
   (c) $ts_{\overline{X}} = 1.45$; CI = 0.66 to 3.56. When the CI includes the null hypothesis mean, retain the null; when the null hypothesis mean falls outside of the CI, reject the null.
5. $z = (0.6 - 0.5)/\sqrt{(.5)(.5)/100} = 2.0$; $p < 0.05$. Reject $H_0$; conclude that the voters will support the highland cow.

## Chapter 7

1. $s^2_{pooled} = 18.16$; $df = 28$; $t = 3.8/1.56 = 2.44$; reject $H_0$ at the .05 but not .01 level. This is a possible Type I error. On average, the scores for Turck Hall were higher ($M = 13.0$) than the scores for Kirk Hall ($M = 9.2$); this difference was found to be statistically significant, $t(28) = 2.44$, $p < .05$, $d = 0.89$.
   
   The 95% CI = 3.8 ± 2.048 (1.56), so it goes from 0.61 to 7.0. The 99% CI = 3.8 ± 2.763 (1.56), so it goes from −0.50 to 8.1. Because zero is contained in the 99% CI, but not the 95% CI, we know that the difference of means is significant at the .05 but not the .01 level.
3. $t = 2.2/1.48 = 1.49$; one-tailed $t_{.05}(23) = 1.714 > 1.49$, so $t$ is not significant at the .05 level (possible Type II error).
   
   The mean for Kirk Hall ($M = 9.2$) was higher than the mean for Dupre Hall ($M = 7.0$), but this trend did not attain significance at the .05 level, even for a one-tailed test, $t(23) = 1.49$, $p > .05$, $d = 0.61$. The 95% CI goes from −0.86 to +5.26, and the 99% CI goes from (approx.) −2.0 to +6.4. The separate variance $t = 1.63$, which is slightly larger than the pooled-variances $t$ value.
5. $s_1^2 = 9.49$; $s_2^2 = 10.00$; $s^2_{pooled} = 9.75$; $df = 8$; $t = 7/1.97 = 3.55$; $p < .05$. Reject $H_0$; the two groups differ significantly.
7. $\sum D = 10$; $\overline{D} = 1$; $s_D^2 = 4.67$; $df = 9$; $t = 1.46$; $p > .05$. Retain $H_0$; there is not sufficient reason to conclude that the two techniques differ.
9. $\sum D = 47$; $\overline{D} = 3.36$; $s_D = 18.54$; $df = 13$; $t = 0.68$; $p > .05$. Retain $H_0$.

## Chapter 8

1. (a) $\sum R = (19)(20)/2 = 190$; $T_1 = 63.5$; $T_2 = 126.5$. Based on Group 1, $T_E = 90$ and $\sigma_T = 12.25$, so $z = -26.5/12.25 = -2.16$; $p < .05$. Conclude that the new signal (Group 1) is superior.
   (b) $r_c = -.59$, indicating a strong relationship between group membership and reaction times, such that Group 1 reacts more quickly.
3. (a) Exp. 1: $\sum R = 378$; $T_1 = 159.5$; $T_2 = 218.5$. Based on Group 1, $T_E = 210$ and $\sigma_T = 20.49$, so $z = -50.5/20.49 = -2.46$; $p < .05$. Reject $H_0$.
   
   Exp. 2: $\sum R = 378$; $T_1 = 213.5$; $T_2 = 164.5$. Based on Group 1, $T_E = 210$ and $\sigma_T = 20.49$, so $z = +3.5/20.49 = 0.17$; $p > .05$. Retain $H_0$.
   (b) Exp. 1: $r_G = -.56$. Exp. 2: no need to calculate $r_G$ (= .04), because rank-sum test is not statistically significant.
5. (a) Exp. 3: $N = 14$; $z = (95.5 - 52.5)/29(52.5)/\sqrt{29(52.5)/6} = 2.70$, $p < .05$; reject $H_0$. $r_c = .82$.
   (b) Exp. 4: $N = 13$; $z = (33.5 - 45.5)/\sqrt{27(45.5)/6} = -0.84$, $p > .05$; retain $H_0$.
7. (a) $N = 10$; $z = (42.5 - 27.5)/\sqrt{21(27.5)/6} = 1.53$, $p > .05$; retain $H_0$.
   (b) $r_c = 4(42.5 - 27.5)/10(11) = .55$

## Chapter 9

1. (a) $\sum z_x z_y = 4.70$; $r_{xy} = +.67$. (This is a highly positive correlation, which tells us that there is a strong tendency for students who do well in the first semester to also do well in their sophomore year and that those who do poorly in the first semester tend to $\sum$ do poorly in their sophomore year.)
   (b) $\sum Y = 61$; $\mu_c = 3$; $\sigma_c = 1.41$; $\mu_y = 3$; $\sigma_y = 0.60$; $r_{cy} = -.34$, $df = 5$, $p > .05$. Retain $H_0$; there is not sufficient reason to believe that the correlation is different from zero in the population.
   (c) $\sum X = 1695$; $\mu_c = 3$; $\sigma_c = 1.41$; $\mu_x = 80$; $\sigma_x = 8.02$; $r_{cx} = .19$, $df = 5$, $p > .05$. Retain $H_0$; there is not sufficient reason to believe that the correlation is different from zero in the population.
3. (a) $\sum XY = 96{,}426$; $\overline{X} = 84$; $s_x = 10.76$; $\overline{Y} = 80.64$; $s_y = 13.29$; $r_{xy} = .855$, $df = 12$, $p < .01$. Reject $H_0$.
   (b) $\sum z_x z_y = 11.98$; $r_{xy} = .855$. The same value, except for possible rounding error.
   (c) $\sum XY = 94{,}502$; $\overline{X} = 84$; $s_x = 10.76$; $\overline{Y} = 80.64$; $s_y = 13.29$; $r_{xy} = -.18$, $df = 12$, $p > .05$. Retain $H_0$.
5. (a) $r_s = 1 - (6 \times 104)/2{,}730 = .77 > 0.648$ (critical $r_s$ for $N = 10$ for a .05, two-tailed test); reject $H_0$. $r_s = .77 < r_{xy} = .855$. The Spearman correlation throws away some of the quantitative information in the data and is, therefore, often smaller (and has less power) than the corresponding Pearson $r$ calculated on the raw data. However, for some data patterns, $r_s$ will be higher than $r$. Also, the critical value for $r_s$ will be higher than the corresponding critical value for $r$, although the difference becomes negligible when $N$ is large.
   (b) $r_s = 1 - (6 \times 586)/2{,}730 = |-.28| < .648$, so $p > .05$; retain $H_0$. This is a case in which $r_s$ is actually larger in magnitude than the corresponding $r_{xy}$—that is, $|-.28| > |-.18|$.

## Chapter 10

1. $b_{yc} = -0.14$; $a_{yc} = 3.42$; $Y' = -.14C + 3.42$; $Y' = 2.3$
3. Data set 3: $b_{YX} = 0.86\,(13.29/10.76) = 1.06$; $a_{YX} = 80.64 - 1.06\,(84) = -8.4$; $Y' = 1.06X - 8.4$; $r_{xy}^2 = .86^2 = .74$; $s_{est} = 13.29\sqrt{13/12(1-0.74)} = 7.05$

   Data set 3: $X$ rearranged: $b_{YX} = -0.18\,(13.29/10.76) = -0.22$; $a_{YX} = 80.64 - (-0.22)(84) = 99.12$; $Y' = -0.22X + 99.12$; $r_{xy}^2 = -.18^2 = .0324$; $s_{est} = 13.29\sqrt{13/12(1-0.0324)} = 13.61$
5. The $t$ value for these data = 1.19, so, using Formula 7.12, $d = 1.19\sqrt{\dfrac{41+31}{41 \times 31}} = 0.28$;

   using Formula 10.9, $r_{pb} = \sqrt{\dfrac{0.28^2}{0.28^2 + 4\left(\dfrac{70}{72}\right)}} = \sqrt{\dfrac{0.28^2}{0.28^2 + 3.89}} = .141$; $r_{pb}^2 = .141^2 = .02$;

   only 2% of the variance is accounted for.

7.

| $t$ | $N_1$ | $N_2$ | $r_{pb}$ |
|---|---|---|---|
| 2.11 | 12 | 8 | 0.445 |
| 2.75 | 12 | 8 | 0.544 |
| 6.00 | 12 | 8 | 0.816 |
| 2.11 | 19 | 23 | 0.316 |
| 2.75 | 19 | 23 | 0.399 |
| 6.00 | 19 | 23 | 0.688 |
| 2.11 | 51 | 51 | 0.206 |
| 2.75 | 51 | 51 | 0.265 |
| 6.00 | 51 | 51 | 0.514 |

## Chapter 11

1. (a) **d** = 20/100 = 0.20; $\delta = 0.20\sqrt{25} = 1.0$; for a two-tailed test, power = .17; for a one-tailed test, power = .26. Power is higher for the one-tailed test, because you are putting all your alpha in just one tail and therefore have a smaller critical value to beat.
   (b) **d** = 50/100 = 0.50; $\delta = 0.50\sqrt{25} = 2.5$; power = .71 (two-tailed), or power = .80 (one-tailed). These power values are higher because the effect size increased from .20 to .50. As **d** increases, power increases.
   (c) $\delta = 0.20\sqrt{100} = 2.0$; power = .52 (two-tailed), or power = .64 (one-tailed). Larger samples yield higher values for delta (all else equal) and thus increase the likelihood of correctly rejecting $H_0$. So, power increases as $N$ increases.
   (d) For .05, two-tailed: $N = (2.80/0.20)^2 = 196$; for .01, two-tailed: $N = (3.40/0.20)^2 = 289$.
   (e) For .05, two-tailed: $N = (2.80/0.50)^2 = 31$; for .01, two-tailed: $N = (3.40/0.50)^2 = 46$
3. (a) **d** = $0.10/\sqrt{(.5)(.5)} = 0.10/0.50 = 0.20$; $\delta = 0.20\sqrt{81} = 1.8$; Power = .22. She is very unlikely to reject $H_0$ even when specified $H_A$ is true, so the experimental plan is not good.
   (b) Power improves to .44, but she still has a better chance of failing to reject $H_0$ than rejecting it, when the specified $H_A$ is true.
   (c) $N = (2.63/0.20)^2 = 173$
5. $\rho = .52$; $N = (2.47/0.52)^2 + 1 = 24$
7. (a) $n = 2(41)(31)/(41 + 31) = 35.31$; $d = \sqrt{2/35.31}\,(1.19) = 0.283$; $n = 2(2.47/0.283)^2 = 152$.
   (b) **d** = $\sqrt{2/35.31}(2.8) = 0.67$
9. $\delta\ \sqrt{1/.6}\sqrt{15/2}(1.32) = 4.66$; power = .98
11. (a) $d = \sqrt{2/10}(1.45) = 0.65$; $\delta = \sqrt{10/2}(0.65) = 0.29$; power = .06
    (b) $\delta = \sqrt{1/.5}\sqrt{2/10}(0.65) = 2.01$; power = .52

## Chapter 13

1. (a) There appears to be less within-group variability in Experiment 1 and the between-group variability is equal, so the results of Experiment 1 are more likely to be statistically significant.
   (b) $\sum X^2 = 302$; $\overline{X}_G = 4$; $SS_T = 62$; $SS_{bet} = 40$; $SS_W = 22$; $MS_{bet} = 20$; $MS_W = 1.83$; $F(2, 12) = 10.93$, $p < .05$. Reject $H_0$; conclude that different audio types affect performance. $\eta^2 = 40/62 = .65$.
   (c) $\sum X^2 = 490$; $\overline{X}_G = 4$; $SS_T = 250$; $SS_{bet} = 40$; $SS_W = 210$; $MS_{bet} = 20$; $MS_W = 17.5$; $F(2, 12) = 1.14$, $p > .05$. Retain $H_0$; there is not sufficient reason to conclude that audio types affects performance. $\eta^2 = 40/250 = .16$.
   (d) There is much more within-group (error) variation in Experiment 2, thus group differences comprise a smaller part of the total variance. In Experiment 1, most of the total variation is due to between-groups differences.
   (e) Est. $\omega^2 = 36.34/63.83 = .57$
3. (a) Experiment 1:

| **Source** | ***SS*** | ***df*** | ***MS*** | ***F*** | ***p*** |
|---|---|---|---|---|---|
| Between-groups | 172.810 | 2 | 86.405 | 1.63 | .230 |
| Within-groups | 797.690 | 15 | 53.179 | | |
| Total | 970.500 | 17 | | | |

There is not sufficient evidence to reject the null hypothesis that the means of the three populations are equal, $F(2, 15) = 1.63$, $p > .05$, $\eta^2 = .217$.

(b) Experiment 2:

| Source | SS | df | MS | F | p |
|---|---|---|---|---|---|
| Between-groups | 518.843 | 2 | 259.421 | 8.616 | < .01 |
| Within-groups | 451.657 | 15 | 30.110 | | |
| Total | 970.500 | 17 | | | |

The three population means are not all equal, $F(2, 15) = 8.77$, $p < .01$, est. $\omega^2 = .46$.

5. (a) Exp. 1: $\sum R = 406$; $df = 2$; $H = 12(40.99)/812 = 0.61$, $p > .05$ (not significant).
   (b) Exp. 2: $H = 12(966.45)/812 = 14.28$, $p < .01$; reject $H_0$. $\eta^2 = 966.45/3745.25 = .258$. Group 1 versus 2, $z = 25/13.16 = 1.90$, $p > .05$ (not significant, but close). Group 2 versus 3, $z = 33.5/10.39 = 3.22$, $p < .01$. Group 1 versus 3, $z = 34.5/12.11 = 2.85$, $p < .01$.
7. (a) $\eta^2 = (2 \times 3.69)/[(2 \times 3.69) + 57] = 7.38/64.38 = .115$; this is a moderately large effect size.
   (b) $\omega^2 = (2 \times 2.69)/[2 \times 2.69) + 57] = .086$.

## Chapter 14

1. Exp. 1: LSD $= 2.179\sqrt{3.66/5} = 1.82$; Exp. 2: LSD $= 2.179\sqrt{35/5} = 5.64$. LSD is justified only for Experiment 1 because $H_0$ is rejected in that case. All pairs of means differ significantly for Experiment 1.
3. Exp. 1: LSD $= 2.131\sqrt{106.36/5.89} = 9.06$; no pairs of means differ.
   HSD $= 3.67\sqrt{53.18/5.89} = 11.03$; no pairs of means differ.
   Exp. 2: LSD $= 2.131\sqrt{60.22/5.89} = 6.81$; Group 1 differs from Groups 2 and 3.
   HSD $= 3.67\sqrt{30.11/5.89} = 8.30$; Group 1 differs from Groups 2 and 3.
   LSD values are smaller than HSD, which increases the chance of finding a significant difference between pairs of means. LSD is generally more powerful than HSD.
5. (a) Sample 1 differs significantly from Samples 2, 3, and 5; also, Samples 4 and 5 differ significantly. (b) The 95% CI for Sample 5 is 15.8 ± 5.3, so it goes from 10.5 to 21.1 [Note that the means for Samples 1 and 4 do not fall within the CI for Sample 5, which is consistent with the results in part (a)].
7. (a) $MS_W = MS_{Bet}/F = 66.467/3.69 = 18.01$; LSD $= 2.0 \times \sqrt{(36.02/20)} = 2.68$. (b) Only the means for positive and negative feedback differ significantly according to the LSD test.

## Chapter 15

1. (a)

| Source | SS | df | MS | F | p |
|---|---|---|---|---|---|
| Procrastination | 32.44 | 2 | 16.22 | 6.34 | < .05 |
| Exercise | 2.00 | 1 | 2.00 | 0.78 | > .05 |
| Procrastination × Exercise | 1.33 | 2 | 0.67 | 0.26 | > .05 |
| Error | 30.67 | 12 | 2.56 | | |
| Total | 66.44 | 17 | | | |

   (b) LSD $= 2.179\sqrt{5.12/6} = 2.01$. Therefore, the low-procrastination group differs from the moderate- and high-procrastination groups. The $F$ value for procrastination is statistically significant, so these follow-up tests comparing the marginal means of the procrastination groups are justified.
3. (a) $MS_{caffeine} = 2.52$; $MS_W = 2.0$; $F_{caffeine}\ (1, 24) = 1.26$, $p > .05$ $MS_{stress} = 0.63$; $MS_W = 2.0$; $F_{stress}\ (1, 24) = 0.32$, $p > .05$
   $MS_{inter} = 154.63$; $MS_W = 2.0$; $F_{inter}\ (1, 24) = 77.32$, $p < .01$

(b) Graph shows a disordinal interaction.

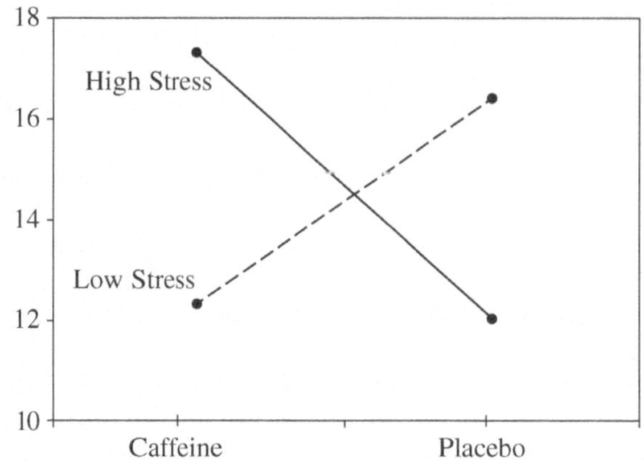


(c) The significant interaction justifies the testing of the four simple main effects. Thus, individuals with high stress given caffeine scored significantly higher than high stress individuals given placebos, $t(24) = 7.01, p < .01$, whereas individuals with low stress with caffeine had lower scores than low stress individuals given placebos, $t(24) = -5.42$, $p < .01$. Individuals with high stress with caffeine had higher scores than individuals with low stress with caffeine, $t(24) = 6.61$, $p < .01$, but individuals with high stress given placebos had lower scores than individuals with low stress given placebos, $t(24) = -5.82$, $p < .01$. In general, these results show that caffeine increases performance for individuals with high stress but decreases performance for individuals with low stress.

## Chapter 16

1. $SS_{\text{total}} = 91.0$; $SS_{\text{treat}} = 5.0$; $SS_{\text{sub}} = 65$; $SS_{\text{error}} = 21.0$; $MS_{\text{treat}} = 5.0$; $MS_{\text{error}} = 2.33$; $F(1, 9) = 2.14, p > .05$. Relationship is that $F = t^2$.
3. $SS_{\text{total}} = 100.0$; $SS_{\text{treat}} = 24.5$; $SS_{\text{sub}} = 46.1$; $SS_{\text{error}} = 29.4$; $MS_{\text{treat}} = 12.25$; $MS_{\text{error}} = 1.336$; $F(2, 22) = 9.17$, $p > .01$, sphericity assumed. Adjusted $F_{.05}(1, 11) = 4.84$; therefore, $F$ is significant at the .05 level without assuming sphericity.
5. (a) $SS_{\text{total}} = 551.5$; $SS_{\text{treat}} = 260.5$; $SS_{\text{sub}} = 155.5$; $SS_{\text{error}} = 135.5$; $MS_{\text{treat}} = 86.83$; $MS_{\text{error}} = 6.45$; $F(3, 21) = 13.46$, $p < .01$; $(4 - 1)/2 = 6$ pairwise comparisons; $\alpha_{\text{pc}} = .05/6 = .00833$.
   (b) $SS_{\text{groups}} = 33.89$; $SS_{\text{W}} = 121.61$; $SS_{\text{RM}} = 260.5$; $SS_{\text{inter}} = 99.11$; $SS_{\text{error}} = 36.39$; $F_{\text{ability}} = 33.89/20.27 = 1.67$, $p > .05$; $F_{\text{word}} = 86.83/2.02 = 42.99$, $p < .01$; $F_{\text{inter}} = 33.04/2.02 = 16.36$, $p < .01$.
   (c) Graph of cell means.

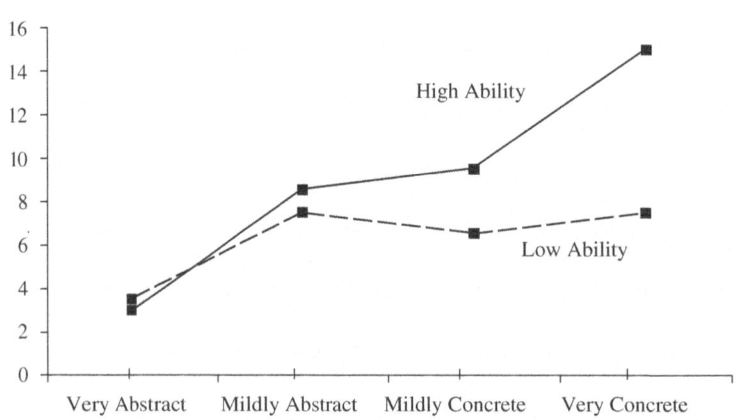

You can see from the graph that there is some interaction, as the lines increasingly diverge with greater concreteness. There is also a visible main effect for word concreteness, due chiefly to the high-ability participants. Also, the high-ability participants are somewhat higher than low-ability participants across word conditions, but the (usually) larger error term for the between-group test prevented this result from attaining significance.

## Chapter 17

1. (a) $p = 4/52 = 1/13 = 0.077$
   (b) $p = 45/52 - 8/52 = 34/52 = 0.654$
   (c) $p(A \text{ and } B) = 1/52 \times 1/52 = 0.00037$
   (d) $p(A \text{ and } B) = 20/52 \times 2/52 = 0.0148$
   (e) $p(A \text{ and } B) = 1/52 \times 0 = 0$; $p(A \text{ and } B) = 20/52 \times 2/51 = 0.0151$
3. (a) $1/100 = 0.01$
   (b) $1/100 = 0.01$
   (c) $0.01 + 0.01 = 0.02$
   (d) $5/100 = 0.05$
   (e) $10/100 = 0.10$
   (f) $50/100 = 0.50$
   (g) $0.50 \times 0.17 = 0.085$
   (h) $0.05 + 0.28 - 0.02 = 0.31$
5. (a) $2^4 = 16$; 5
   (b) Binomial distribution for $N = 4$ and $P = 0.5$
   (c) From your answer to part b, it should be clear that the $p$ for 4 tails (or 0 heads) is 0.0625. Even for a one-tailed test, $p > 0.05$, so you cannot reject the null hypothesis that the coin is fair (i.e., $P = 0.5$).
7. (a) $z = [18 - 24(.5)]/\sqrt{24(.5)(.5)} = 2.45$, Yes, $p < .05$.
   (b) $z = [20 - 28(0.5)]/\sqrt{28(0.5)(0.5)} = 2.27$; Yes, $p = .0116$, two-tailed; reject $H_0$.

## Chapter 18

1. (a) $\chi^2 = 4.0$; $df = 1$; $p < .05$. Reject $H_0$; conclude that patients are more likely to improve with new treatment.
   (b) $z = [60 - 100(0.5)]/\sqrt{100(.5)(.5)} = 2.0$; relationship is $z^2 = \chi^2$
3. (a) $\chi^2 = [(21 - 19.75)^2/19.75] + [(14 - 19.75)^2/19.75] + [(26 - 19.75)^2/19.75] + [(18 - 19.75)^2/19.75] = 3.89$. Retain $H_0$; $3.89 < \chi^2_{.05}(3) = 7.82$
   (b) $\chi^2 = 9.56$; $df = 3$, $p < .05$. Reject $H_0$; $9.56 > \chi^2_{.05}(3) = 7.82$
5. (a) $\chi^2 = [(12 - 13.2)^2/13.2] + [(10 - 8.8)^2/8.8] + [(18 - 16.8)^2/16.8] + [(10 - 11.2)^2/11.2] = 0.49$. Retain $H_0$; $0.49 < \chi^2_{0.05}(1) = 3.84$. This is equivalent to saying that the percent success is not significantly different between male and female genders.
   (b) $\phi = \sqrt{0.49/50} = .10$; small effect
7. (a) $\chi^2 = 18.84$; $df = 4$; $p < .01$. Reject $H_0$; $18.84 > \chi^2_{.10}(4) = 13.28$
   (b) $\phi_c = \sqrt{18.84/(143)(2)} = 0.26$; moderate relationship

## Data from Jackson's Experiment

| sub _num | exercise | major | reason | exp_cond | coffee | num_ cups | procrastination | teacher | acad_ interest | statquiz | exp_ sqz | hr_ base | hr_ pre | hr_ post | anx_ base | anx_ pre | anx_ post |
|---|---|---|---|---|---|---|---|---|---|---|---|---|---|---|---|---|---|
| 1 | 0 | 1 | 3 | 1 | 1 | 0 | 1 | 3 | 43 | 6 | 7 | 71 | 68 | 65 | 17 | 22 | 20 |
| 2 | 0 | 1 | 2 | 1 | 0 | 0 | 1 | 4 | 49 | 9 | 11 | 73 | 75 | 68 | 17 | 19 | 16 |
| 3 | 0 | 1 | 1 | 1 | 0 | 0 | 4 | 1 | 26 | 8 | 8 | 69 | 76 | 72 | 19 | 14 | 15 |
| 4 | 0 | 1 | 1 | 1 | 0 | 0 | 4 | 0 | 29 | 7 | 8 | 72 | 73 | 78 | 19 | 13 | 16 |
| 5 | 0 | 1 | 1 | 1 | 0 | 1 | 10 | 1 | 31 | 6 | 6 | 71 | 83 | 74 | 26 | 30 | 25 |
| 6 | 0 | 1 | 1 | 2 | 1 | 1 | 4 | 1 | 20 | 7 | 6 | 70 | 71 | 76 | 12 | 15 | 19 |
| 7 | 0 | 1 | 1 | 2 | 0 | 0 | 4 | 2 | 13 | 3 | 4 | 71 | 70 | 66 | 12 | 16 | 17 |
| 8 | 0 | 1 | 3 | 2 | 1 | 2 | 4 | 1 | 23 | 7 | 7 | 77 | 87 | 84 | 17 | 19 | 22 |
| 9 | 0 | 1 | 1 | 2 | 0 | 0 | 4 | 1 | 38 | 8 | 7 | 73 | 72 | 67 | 20 | 14 | 17 |
| 10 | 0 | 1 | 1 | 2 | 1 | 2 | 5 | 0 |  | 7 | 6 | 78 | 76 | 74 | 20 | 24 | 19 |
| 11 | 0 | 1 | 1 | 2 | 0 | 1 | 5 | 1 | 29 | 8 | 10 | 74 | 72 | 73 | 21 | 25 | 22 |
| 12 | 0 | 1 | 1 | 2 | 0 | 0 | 4 | 0 | 32 | 8 | 7 | 73 | 74 | 74 | 32 | 35 | 33 |
| 13 | 0 | 1 | 1 | 2 | 0 | 1 | 7 | 0 | 18 | 1 | 3 | 73 | 76 | 78 | 19 | 23 | 20 |
| 14 | 0 | 1 | 3 | 3 | 0 | 2 | 4 | 1 |  | 5 | 4 | 72 | 83 | 77 | 18 | 27 | 28 |
| 15 | 0 | 1 | 1 | 4 | 1 | 3 | 3 | 1 | 21 | 8 | 6 | 72 | 74 | 68 | 21 | 27 | 22 |
| 16 | 0 | 1 | 1 | 4 | 1 | 0 | 8 | 0 |  | 3 | 1 | 76 | 76 | 79 | 14 | 18 | 21 |
| 17 | 0 | 1 | 1 | 4 | 0 | 0 | 4 | 1 | 37 | 8 | 7 | 68 | 67 | 74 | 15 | 19 | 18 |
| 18 | 0 | 1 | 1 | 4 | 1 | 3 | 5 | 1 | 37 | 7 | 4 | 77 | 78 | 73 | 39 | 39 | 40 |
| 19 | 0 | 1 | 3 | 4 | 1 | 2 | 0 | 3 | 32 | 10 | 9 | 74 | 74 | 75 | 20 | 12 | 18 |
| 20 | 0 | 2 | 1 | 2 | 1 | 1 | 4 | 1 |  | 7 | 7 | 74 | 75 | 73 | 15 | 11 | 20 |
| 21 | 0 | 2 | 2 | 2 | 1 | 2 | 4 | 0 | 25 | 7 | 6 | 74 | 84 | 77 | 19 | 27 | 23 |
| 22 | 0 | 2 | 1 | 3 | 1 | 0 | 3 | 1 | 22 | 4 | 3 | 73 | 71 | 79 | 18 | 13 | 19 |
| 23 | 0 | 2 | 1 | 3 | 0 | 2 | 4 | 1 | 35 | 8 | 7 | 71 | 74 | 76 | 18 | 22 | 25 |
| 24 | 0 | 2 | 2 | 2 | 1 | 2 | 0 | 3 | 47 | 8 | 7 | 75 | 75 | 71 | 23 | 28 | 24 |
| 25 | 0 | 2 | 2 | 3 | 1 | 2 | 1 | 3 | 41 | 6 | 6 | 76 | 73 | 72 | 18 | 24 | 26 |
| 26 | 0 | 2 | 2 | 4 | 0 | 0 | 1 | 4 | 26 | 7 | 6 | 71 | 76 | 75 | 14 | 10 | 18 |
| 27 | 0 | 2 | 2 | 4 | 1 | 2 | 0 | 6 | 39 | 8 | 8 | 74 | 79 | 79 | 17 | 12 | 16 |
| 28 | 0 | 2 | 3 | 1 | 1 | 3 | 4 | 2 | 21 | 7 | 8 | 78 | 79 | 73 | 18 | 13 | 16 |
| 29 | 0 | 2 | 1 | 1 | 0 | 0 | 3 | 2 |  | 7 | 9 | 70 | 63 | 66 | 18 | 12 | 14 |
| 30 | 0 | 2 | 3 | 1 | 1 | 0 | 5 | 1 | 22 | 4 | 7 | 73 | 78 | 69 | 21 | 14 | 17 |
| 31 | 0 | 3 | 1 | 1 | 1 | 1 | 9 | 1 | 21 | 8 | 8 | 75 | 83 | 73 | 18 | 21 | 23 |
| 32 | 0 | 3 | 1 | 3 | 1 | 3 | 3 | 0 |  | 7 | 6 | 78 | 76 | 84 | 24 | 27 | 25 |
| 33 | 0 | 3 | 1 | 4 | 1 | 2 | 4 | 1 | 26 | 8 | 7 | 76 | 74 | 81 | 17 | 26 | 15 |
| 34 | 0 | 3 | 2 | 1 | 1 | 1 | 1 | 1 | 20 | 8 | 9 | 76 | 69 | 71 | 17 | 25 | 19 |
| 35 | 0 | 3 | 2 | 1 | 0 | 0 | 2 | 1 | 30 | 6 | 9 | 69 | 69 | 64 | 22 | 16 | 18 |
| 36 | 0 | 3 | 2 | 2 | 1 | 1 | 0 | 0 | 40 | 8 | 9 | 77 | 79 | 74 | 21 | 14 | 19 |
| 37 | 0 | 3 | 2 | 4 | 1 | 2 | 3 | 1 | 35 | 8 | 7 | 78 | 73 | 78 | 19 | 12 | 17 |
| 38 | 0 | 3 | 3 | 1 | 0 | 0 | 2 | 0 | 10 | 7 | 8 | 74 | 72 | 72 | 15 | 21 | 16 |
| 39 | 0 | 3 | 3 | 2 | 0 | 1 | 8 | 1 | 35 | 6 | 5 | 71 | 70 | 75 | 20 | 27 | 22 |
| 40 | 0 | 3 | 3 | 4 | 0 | 0 | 4 | 1 | 44 | 6 | 4 | 67 | 67 | 73 | 12 | 19 | 17 |
| 41 | 0 | 3 | 3 | 4 | 0 | 0 | 4 | 1 | 26 | 7 | 5 | 77 | 78 | 78 | 21 | 15 | 15 |
| 42 | 0 | 4 | 2 | 1 | 0 | 0 | 0 | 1 |  | 9 | 11 | 71 | 72 | 67 | 20 | 23 | 21 |
| 43 | 0 | 4 | 2 | 2 | 1 | 0 | 3 | 1 | 15 | 3 | 4 | 76 | 79 | 71 | 19 | 21 | 17 |
| 44 | 0 | 4 | 2 | 4 | 0 | 0 | 2 | 1 | 42 | 7 | 7 | 69 | 70 | 64 | 13 | 24 | 22 |
| 45 | 0 | 4 | 2 | 4 | 0 | 0 | 2 | 1 | 33 | 8 | 7 | 72 | 64 | 68 | 20 | 14 | 22 |
| 46 | 0 | 4 | 2 | 4 | 0 | 1 | 1 | 0 | 29 | 6 | 6 | 72 | 79 | 76 | 22 | 27 | 24 |
| 47 | 0 | 4 | 3 | 1 | 0 | 0 | 1 | 1 | 39 | 7 | 8 | 71 | 63 | 66 | 15 | 21 | 13 |
| 48 | 0 | 4 | 3 | 1 | 1 | 1 | 5 | 0 | 38 | 8 | 8 | 71 | 82 | 79 | 20 | 26 | 26 |
| 49 | 0 | 4 | 3 | 2 | 1 | 1 | 5 | 1 |  | 5 | 4 | 75 | 76 | 70 | 16 | 18 | 23 |
| 50 | 0 | 4 | 3 | 3 | 0 | 0 | 4 | 1 | 24 | 7 | 6 | 74 | 76 | 75 | 22 | 27 | 23 |
| 51 | 0 | 4 | 3 | 3 | 0 | 0 | 4 | 1 |  | 7 | 5 | 74 | 74 | 69 | 30 | 36 | 32 |
| 52 | 0 | 4 | 3 | 4 | 1 | 0 | 10 | 0 | 26 | 5 | 2 | 78 | 80 | 80 | 19 | 24 | 24 |
| 53 | 0 | 4 | 3 | 4 | 0 | 0 | 7 | 1 | 14 | 5 | 3 | 68 | 78 | 73 | 20 | 29 | 30 |
| 54 | 0 | 5 | 1 | 3 | 1 | 1 | 3 | 5 | 45 | 9 | 9 | 76 | 79 | 75 | 15 | 8 | 17 |
| 55 | 0 | 5 | 1 | 3 | 1 | 2 | 4 | 1 | 28 | 8 | 8 | 73 | 78 | 77 | 19 | 13 | 18 |
| 56 | 0 | 5 | 2 | 4 | 0 | 1 | 2 | 2 | 31 | 8 | 7 | 74 | 78 | 82 | 27 | 21 | 24 |

## Data from Jackson's Experiment (*continued*)

| sub _num | exercise | major | reason | exp_cond | coffee | num_ cups | procrastination | teacher | acad_ interest | statquiz | exp_ sqz | hr_ base | hr_ pre | hr_ post | anx_ base | anx_ pre | anx_ post |
|---|---|---|---|---|---|---|---|---|---|---|---|---|---|---|---|---|---|
| 57 | 0 | 5 | 3 | 1 | 0 | 0 | 2 | 0 | | 8 | 9 | 72 | 67 | 67 | 16 | 20 | 22 |
| 58 | 1 | 1 | 1 | 1 | 1 | 1 | 1 | 1 | 32 | 8 | 10 | 74 | 84 | 76 | 17 | 19 | 14 |
| 59 | 1 | 1 | 2 | 1 | 0 | 0 | 7 | 0 | 15 | 3 | 6 | 73 | 73 | 71 | 19 | 16 | 16 |
| 60 | 1 | 1 | 1 | 2 | 0 | 1 | 5 | 0 | 26 | 8 | 7 | 73 | 74 | 76 | 16 | 20 | 17 |
| 61 | 1 | 1 | 1 | 3 | 1 | 1 | 5 | 1 | 26 | 8 | 7 | 80 | 82 | 86 | 18 | 27 | 20 |
| 62 | 1 | 1 | 1 | 3 | 0 | 0 | 1 | 2 | 32 | 5 | 5 | 67 | 68 | 73 | 20 | 25 | 24 |
| 63 | 1 | 1 | 1 | 3 | 0 | 0 | 3 | 0 | 21 | 8 | 7 | 65 | 75 | 69 | 26 | 29 | 23 |
| 64 | 1 | 1 | 1 | 4 | 0 | 0 | 3 | 1 | 43 | 8 | 8 | 71 | 72 | 76 | 18 | 13 | 16 |
| 65 | 1 | 1 | 1 | 4 | 0 | 1 | 5 | 1 | | 8 | 5 | 72 | 75 | 68 | 21 | 26 | 27 |
| 66 | 1 | 1 | 2 | 2 | 0 | 0 | 1 | 4 | 34 | 9 | 11 | 70 | 73 | 65 | 11 | 19 | 9 |
| 67 | 1 | 1 | 3 | 1 | 0 | 0 | 0 | 3 | 33 | 10 | 11 | 72 | 67 | 68 | 16 | 25 | 15 |
| 68 | 1 | 2 | 1 | 2 | 0 | 0 | 5 | 0 | 30 | 7 | 7 | 69 | 70 | 74 | 18 | 15 | 15 |
| 69 | 1 | 2 | 1 | 3 | 0 | 0 | 6 | 2 | 46 | 8 | 6 | 68 | 74 | 71 | 16 | 10 | 17 |
| 70 | 1 | 2 | 1 | 3 | 1 | 1 | 3 | 2 | | 7 | 7 | 76 | 79 | 71 | 19 | 13 | 15 |
| 71 | 1 | 2 | 1 | 4 | 1 | 1 | 5 | 0 | 33 | 7 | 4 | 68 | 71 | 77 | 18 | 21 | 19 |
| 72 | 1 | 2 | 1 | 4 | 1 | 2 | 4 | 2 | 29 | 7 | 6 | 76 | 79 | 71 | 22 | 15 | 20 |
| 73 | 1 | 2 | 2 | 3 | 0 | 0 | 0 | 4 | 30 | 9 | 10 | 69 | 70 | 74 | 15 | 21 | 17 |
| 74 | 1 | 2 | 2 | 3 | 0 | 0 | 2 | 4 | 34 | 8 | 7 | 68 | 70 | 69 | 18 | 21 | 19 |
| 75 | 1 | 2 | 2 | 4 | 0 | 0 | 2 | 3 | 32 | 9 | 8 | 70 | 62 | 71 | 17 | 23 | 19 |
| 76 | | 2 | 3 | 1 | 0 | 0 | 3 | 1 | 37 | 6 | 8 | 70 | 73 | 77 | 15 | 23 | 14 |
| 77 | 1 | 2 | 1 | 2 | 1 | 1 | 1 | 2 | | 8 | 10 | 69 | 67 | 64 | 14 | 17 | 15 |
| 78 | 1 | 2 | 3 | 2 | 0 | 0 | 1 | 3 | 31 | 7 | 8 | 64 | 74 | 70 | 22 | 25 | 19 |
| 79 | 1 | 2 | 3 | 3 | 1 | 1 | 0 | 2 | 30 | 8 | 8 | 68 | 72 | 71 | 10 | 13 | 15 |
| 80 | 1 | 2 | 3 | 3 | 0 | 0 | 2 | 2 | 28 | 5 | 5 | 71 | 74 | 67 | 15 | 17 | 17 |
| 81 | 1 | 2 | 3 | 4 | 0 | 1 | 10 | 0 | 14 | 6 | 3 | 69 | 78 | 76 | 20 | 25 | 23 |
| 82 | 1 | 3 | 1 | 1 | 0 | 0 | 3 | 1 | 9 | 7 | 8 | 71 | 69 | 76 | 13 | 10 | 13 |
| 83 | 1 | 3 | 1 | 1 | 0 | 1 | 3 | 1 | 11 | 4 | 7 | 72 | 83 | 73 | 14 | 16 | 15 |
| 84 | 1 | 3 | 1 | 1 | 1 | 1 | 8 | 1 | 30 | 6 | 7 | 76 | 78 | 71 | 15 | 18 | 17 |
| 85 | 1 | 3 | 1 | 1 | 1 | 0 | 3 | 1 | 15 | 5 | 7 | 76 | 67 | 71 | 18 | 12 | 17 |
| 86 | 1 | 3 | 3 | 2 | 0 | 0 | 3 | 1 | 32 | 3 | 5 | 72 | 70 | 67 | 15 | 9 | 13 |
| 87 | 1 | 3 | 2 | 2 | 1 | 0 | 1 | 0 | 22 | 5 | 4 | 74 | 78 | 72 | 20 | 27 | 18 |
| 88 | 1 | 3 | 2 | 3 | 0 | 0 | 1 | 4 | | 8 | 9 | 70 | 71 | 69 | 10 | 16 | 14 |
| 89 | 1 | 3 | 3 | 2 | 0 | 0 | 2 | 0 | 25 | 6 | 7 | 68 | 70 | 73 | 16 | 20 | 19 |
| 90 | 1 | 3 | 3 | 3 | 0 | 0 | 8 | 1 | 18 | 4 | 3 | 71 | 73 | 77 | 15 | 19 | 21 |
| 91 | 1 | 3 | 3 | 3 | 1 | 1 | 3 | 1 | 11 | 4 | 3 | 72 | 75 | 76 | 24 | 17 | 21 |
| 92 | 1 | 4 | 2 | 1 | 0 | 0 | 0 | 1 | 11 | 6 | 9 | 71 | 78 | 72 | 16 | 21 | 15 |
| 93 | 1 | 4 | 2 | 4 | 0 | 0 | 0 | 2 | 37 | 9 | 9 | 72 | 71 | 73 | 17 | 11 | 17 |
| 94 | 1 | 4 | 3 | 3 | 1 | 2 | 4 | 0 | 28 | 7 | 6 | 73 | 77 | 79 | 24 | 18 | 20 |
| 95 | 1 | 5 | 3 | 3 | 0 | 0 | 7 | 2 | | 8 | 6 | 69 | 70 | 71 | 23 | 28 | 24 |
| 96 | 1 | 5 | 2 | 2 | 1 | 1 | 0 | 1 | 33 | 7 | 9 | 74 | 73 | 71 | 17 | 18 | 17 |
| 97 | 1 | 5 | 2 | 3 | 0 | 0 | 0 | 2 | 28 | 8 | 9 | 70 | 66 | 65 | 17 | 12 | 13 |
| 98 | 1 | 5 | 2 | 3 | 0 | 0 | 2 | 2 | 38 | 9 | 10 | 65 | 65 | 69 | 18 | 14 | 19 |
| 99 | 1 | 5 | 2 | 4 | 1 | 1 | 1 | 4 | 41 | 8 | 8 | 72 | 68 | 73 | 17 | 11 | 18 |
| 100 | 1 | 5 | 1 | 2 | 0 | 0 | 2 | 2 | 39 | 7 | 7 | 70 | 70 | 64 | 17 | 11 | 14 |

**Key**

Sub_num: arbitrary ID number for each participant.

exercise: 0 = non-regular exerciser; 1 = regular exerciser.

major: 1 = Psychology; 2 = Geology; 3 = Biology; 4 = Sociology; 5 = Economics.

reason: 1 = Program requirement; 2 = Personal interest; 3 = Advisor recommendation.

exp_cond: 1 = Easy; 2 = Moderate; 3 = Difficult; 4 = Impossible.

coffee: 0 = not a regular coffee drinker; 1 = regularly drinks coffee.

num_cups = number of cups of coffee drunk prior to the experiment on the same day.

procrastination: 0 = very low procrastination to 10 = very high procrastination.

teacher = Number of previous outstanding teachers the student reported having.

ac_interest = Score on Academic Interest Questionnaire: 9 = very low academic interest to 50 = very high academic interest (a blank for this value indicates that a student did not take the quiz).
statquiz = Score on 10-question stats quiz given one week before the experiment.
exp_sqz = Score on stats quiz given as part of the experiment (number correct, including the 11th question).
hr_base = Baseline heart rate (in beats per minute).
hr_pre = Prequiz heart rate.
hr_post = Postquiz heart rate.
anx_base = Baseline anxiety score.
anx_pre = Prequiz anxiety score.
anx_post = Postquiz anxiety score.

# Glossary of Terms

**Absolute values** Numerical values with the signs ignored; thus all absolute values are treated as positive.

**Alpha** The (conditional) probability of committing a Type I error (i.e., rejecting the null hypothesis when the null hypothesis is true).

**Alternative hypothesis** The hypothesis that states that the null hypothesis is *not true* and specifies some other value or set of values for the population parameter(s) in question.

**Analysis of variance (ANOVA)** Procedures for testing hypotheses about the equality of population means.

**A priori (planned) comparison** In analysis of variance, a significance test involving two or more group means that was planned before the results were in.

**Area of rejection** All numerical results of a statistical test that will cause the null hypothesis to be rejected.

**Bar chart** A graph in which the heights of the bars show how often each measure or range of measures was obtained. Spaces between adjacent bars are used to indicate that the data come from a discrete variable.

**Bayesian Statistics** A system of data analysis that uses Bayes's theorem in order to incorporate prior relevant results to more accurately assess the probabilities of both the null and alternative hypotheses. It is used as an alternative or supplement to traditional null hypothesis testing.

**Beta** The (conditional) probability of committing a Type II error (i.e., failing to reject the null hypothesis when the null hypothesis is *not* true).

**Between-group variance** In analysis of variance, variability based on differences among the means of the various groups.

**Bimodal distribution** A distribution that has two pronounced peaks when graphed as a frequency polygon.

**Bivariate normal distribution** A joint distribution of two variables, wherein scores on one variable are normally distributed for each score value of the other.

**Bonferroni correction (or adjustment)** A reduction in the alpha used for each of several planned comparisons based on dividing the desired experimentwise alpha by the number of tests planned.

**Carryover effects** In a repeated-measures design, it is the persistence of the effect of one condition that affects a participant's response to subsequent conditions; they can produce misleading results.

**Central tendency** The general location of a set of scores.

**Chi square distribution** This statistical model can be used to test hypotheses when data are in the form of frequencies.

**Class interval** One set of score values in a grouped frequency distribution.

**Cohen's *d*** When not in boldface, the standardized difference between two sample means (a measure of observed effect size).

**Complex comparison (or contrast)** In analysis of variance, a significance test involving a weighted combination of three or more group means.

*Introductory Statistics for the Behavioral Sciences*, Eighth Edition. R. Brooke Lea and Barry H. Cohen.

Companion website: www.wiley.com/go/ISBS8e

**Conditional probability** The probability of an event occurring that depends on some condition (e.g., the probability of retaining the null hypothesis *given* the condition that the null hypothesis is not true, which is the probability of committing a Type II error).

**Confidence interval** A range of values within which a specified population parameter has a given probability of falling.

**Confidence limits** The upper and lower end points of a confidence interval.

**Constant** A numerical value that is exactly the same for all cases or subjects; the opposite of a *variable*.

**Continuous variable** A variable for which it is theoretically possible for an intermediate value to occur between any specified pair of score values.

**Control group** A group that does *not* receive the treatment whose effects are being investigated by the researcher. Used as a baseline against which to evaluate the performance of the experimental group.

**Correlation coefficient** A measure of the extent to which scores on one variable are related to scores on a second variable.

**Counterbalancing** In a repeated-measures design, it is a system of assigning different orders of the RM conditions to different participants, so that no one condition will benefit from order effects more than any other condition.

**Cramér's phi (or V)** A measure of the strength of the relationship between two variables, both of which are categorical, and at least one of which has more than two levels.

**Criterion** The variable being predicted in a linear regression analysis; a dependent variable.

**Criterion of significance** A numerical value or decision rule that specifies when the null hypothesis is to be rejected.

**Cumulative frequency distribution** A listing that shows how many times a given score or less was obtained.

**d** When in boldface, the standardized difference between two population means (a measure of population effect size).

**Data Transformation** Applying a mathematical function (e.g., the square root) to each value in a data set in order to more closely satisfy the assumptions of a statistical test (e.g., to make the distribution of a set of observed data more closely resemble the normal distribution).

**Decile** One of nine transformed scores that divides the number of cases into 10 equal parts.

**Degrees of freedom** The number of quantities that are free to vary when we estimate the value of a parameter from a statistic.

**Delta** In power analysis, an index that combines the population effect size and the sample size.

**Dependent variable** A variable that may or may not change with changes in one or more independent variables.

**Descriptive statistics** Mathematical procedures for summarizing and describing the characteristics of a set of data.

**Deviation score** The difference between a score and the mean of the set of scores of which it is a member.

**Dichotomous variable** A variable with only two categories.

**Discrete variable** A variable for which it is *not* theoretically possible for any value to occur between any specified pair of score values.

**Distribution-free statistical test** A statistical test that does *not* require any assumptions about the shape of the distribution in the population.

**Effect size** How large the phenomenon we wish to investigate is in the population; the extent to which the null hypothesis is false.

**Error variance** In analysis of variance, differences among scores or group means that *cannot* be explained by the experimental treatment(s).

**Eta squared** The proportion of variance accounted for in analysis of variance.

**Experimental group** A group that receives the treatment whose effects are being investigated by the researcher.

**Experimental (or empirical) sampling distribution** A distribution whose elements are statistics (e.g., sample means) obtained by drawing repeated samples from the population and computing that statistic for each sample.

**Experimentwise error rate** The rate of occurrence of *any* (one or more) Type I errors when a *series* of individual statistical tests is conducted within the same study.

***F* distributions** The statistical model used to test hypotheses when the analysis involves the comparison of variance estimates.

**Factorial design** A procedure used to study the relationship of two or more independent variables (factors) to a dependent variable.

**Familywise error rate** The experimentwise error rate for a subset of the tests in an experiment, such that the subset can be viewed as a "family" of tests (e.g., pairwise comparisons for the main effect of one factor in a two-way factorial analysis of variance).

**File Drawer Problem** An excessive number of published results that are Type I or Type I-ish errors due to a bias towards publishing significant results. Knowing this bias causes researcher's to store their non-significant results in a "file drawer," rather than bothering to send them to a journal that will probably reject them.

**Frequency** The number of times a specified score or range of scores was obtained.

**Frequency polygon** A graph showing how often each score or range of scores was obtained.

**Glass rank biserial correlation** A measure of the strength of the relationship between membership in one of two independent samples and a set of ranked data.

**Grouped frequency distribution** A listing of sets of two or more score values (class intervals) together with the number of times that scores in each class interval were obtained.

**HARKing** An acronym for Hypothesizing After Results are Known. This is an unethical practice that involves rewriting one's initial hypotheses to match one's unexpected results, which may be accidental and therefore unlikely to be replicated.

**Histogram** A bar graph in which there is no space between adjacent bars (for use with continuous data).

**Homogeneity of variance** In a *t* test or analysis of variance, equality of the variances for all treatment populations.

**Hypothesis testing** Procedures for deciding whether to retain or reject the null hypothesis about one or more population parameters.

**Independent events** Events whose occurrence or nonoccurrence is unrelated to the occurrence (or probability of occurrence) of specified other events.

**Independent samples** Samples such that any element in one sample has no connection of any kind with any element in another sample.

**Independent variable** A variable whose variation is studied with regard to its effect on another (dependent) variable.

**Inferential statistics** Mathematical procedures for drawing inferences about characteristics of a population, based on what is observed in samples from that population.

**Interaction** The joint effect of two or more independent variables on a dependent variable, over and above their separate (main) effects.

**Interval estimate** A synonym for confidence interval.

**J-curve** A distribution that looks like the letter J or its mirror image when graphed as a frequency polygon.

**Kruskal-Wallis *H* test** A nonparametric procedure for testing hypotheses about differences among the locations of two or more independent populations, used when data are in the form of ranks.

**Least squares regression line** The regression line that minimizes the sum of squared errors in prediction (the sum of squared deviations between the predicted scores and actual scores).

**Linear regression** Procedure for determining the straight line that will enable us to predict scores on one variable (the criterion) from scores on another variable (the predictor), while minimizing the amount of (squared) error.

**Lower real limit** The lower end point of a score or class interval when dealing with continuous data.

**Marginal frequency** A column or row total in a two-way table of joint frequencies.

**Matched samples** Samples such that each element in one sample is paired with one element in the other sample(s).

**Matched *t* test** Procedure for testing a hypothesis about the difference between two population means, using matched samples, or two measurements on each participant.

**Mean (arithmetic)** A measure of the central tendency of a set of scores, obtained by summing all scores and dividing by the number of scores. When the term *mean* is used without a qualifier, it is the arithmetic mean that is meant.

**Mean (Harmonic)** A measure of central tendency that is based on averaging the reciprocals of the scores and then taking the reciprocal of that average.

**Mean square** In analysis of variance, an estimate of the population variance that is obtained by dividing a sum of squares by its associated degrees of freedom.

**Median** The score in the middle of a distribution such that exactly half the scores are higher and exactly half the scores are equal or lower; the score corresponding to the 50th percentile.

**Midpoint** The value halfway between the lower real limit and the upper real limit of a class interval.

**Mode** The score that occurs most often in a frequency distribution.

**Multiple comparisons** In analysis of variance, procedures for testing hypotheses about differences between specified pairs of means.

**Mutually exclusive events** Events that cannot happen simultaneously.

**Negatively skewed distribution** A distribution with some extremely low scores, resulting in a frequency polygon with a pronounced tail at the left.

**Nonparametric statistical test** A statistical test that does not involve the estimation of any population parameters.

**Normal distribution** A particular bell-shape, symmetric, and unimodal distribution defined by a specific mathematical equation.

**Null hypothesis** The hypothesis that specifies the value of a population parameter or a difference between two or more population parameters (usually zero) and that is assumed to be true at the outset of the statistical analysis.

**Odds against an event** The ratio of the number of unfavorable outcomes to the number of favorable outcomes.

**Omega-squared** The proportion of population variance of one variable accounted for by another.

**One-tailed test of significance** A statistical test wherein the null hypothesis can be rejected only if results are in the direction predicted by the experimenter.

**Order effects** In a repeated-measures design, simple order effects can be produced on later conditions from the practice or fatigue of having been involved in previous conditions. Simple order effects can be neutralized by counterbalancing, but carryover effects may not; see also carryover effects.

***p* hacking** The unethical practice of analyzing one's results in many ways and only reporting the few that attain statistical significance. This increases the probability of committing and then publishing a Type I or Type I-ish error. Also called *data fishing*.

***p* value** The probability of obtaining results as extreme or more extreme than a given result given that the null hypothesis is true.

**Parameter** A numerical quantity that summarizes some characteristic of a population.

**Pearson *r*** A measure of the strength of the linear relationship between two variables, both of which are either continuous or dichotomous.

**Percentile** A score at or below which a given percent of the cases fall.

**Percentile rank** The percent of cases in a specific reference group that fall at or below a given score.

**Phi coefficient** A measure of the strength of the relationship between two variables, both of which are dichotomous (that is, they have only two categories).

**Point-biserial correlation coefficient** A measure of the strength of the relationship between one dichotomous and one continuous variable.

**Point estimate** A single statistic used to estimate the value of the corresponding population parameter.

**Pooled variance** In the significance test of the difference between two means, an estimate of the population variance (assumed to be the same for both populations) obtained from a weighted average of the sample variances.

**Population** *All* of the cases in which a researcher is interested; a (usually very large) group of people, animals, objects, or responses that are alike in at least one respect.

**Population variance estimate** An estimate of the variability in a population, obtained by dividing the sum of squared deviations from the sample mean by $N - 1$.

**Positively skewed distribution** A distribution with some extremely high scores, resulting in a frequency polygon with a pronounced tail at the right.

**Post hoc comparison** In analysis of variance, a significance test involving two or more group means that was not planned prior to obtaining the results.

**Power efficiency of a nonparametric test** How many fewer cases, expressed as a percent, that are required by a parametric test to have the same power as the nonparametric test when the assumptions of the parametric test are met.

**Power of a statistical test** The probability of obtaining a statistically significant result if the null hypothesis is actually false; the probability of rejecting a false null hypothesis.

**Predicted score** In linear regression, the score on the criterion that is estimated for a case with a specified score on the predictor.

**Predictor** In linear regression, the variable from which criterion scores are estimated; an independent variable.

**Preregistration** The desirable practice of officially declaring your data analysis methods before inspecting (or even collecting) your data. This practice greatly reduces the chances of using HARKing or *p* hacking to obtain significant results by accident and thereby increases the trust that other researchers can place in your reported results.

**Probability of an event** The number of ways the specified event can occur, divided by the total number of possible occurrences.

**Protected *t* test** In analysis of variance, a procedure for conducting multiple comparisons between pairs of group means that requires the analysis of variance to be statistically significant before proceeding.

**Quartile** One of three transformed scores that divide the number of cases into four equal parts.

**Randomized-blocks design** This is an extension of the matched-pairs design to more than two conditions. Participants are matched in blocks according to the number of different levels of the factor, and then one member of each block is randomly assigned to each condition.

**Random sample** A sample drawn in such a way that each element in the population has an equal chance of being included in the sample.

**Range** The largest score minus the smallest score in a set of scores.

**Rank** A measure that shows where a given case falls with respect to others in the group but gives no indication of the distance between the cases. Thus, ranks provide less information than do scores.

**Rank-sum test** A nonparametric procedure for testing hypotheses about the difference in location between two populations; used when data are in the form of ranks.

**Raw score** A score not subjected to any statistical transformations, such as the number correct on a test.

**Rectangular distribution** A distribution wherein each score occurs with the same frequency.

**Regular frequency distribution** A listing of every score value, from the lowest to the highest, together with the number of times that each score was obtained (its frequency).

**Repeated-measures analysis of variance** Procedure for testing differences between two or more population means by measuring each participant under all of the conditions (i.e., levels) of the independent variable or by matching participants across two or more conditions; it is an extension of the matched (or RM) *t* test to any number of matched samples.

**Restriction of range** Low variability, caused by the way in which the samples were defined, which will usually underestimate the correlation between two variables.

**Robust statistical test** A statistical test that gives fairly accurate results even if the underlying assumptions are not met.

**Sample** Any subgroup of cases drawn from a clearly specified population.

**Sample size** The number of cases in a sample.

**Sampling error** Differences between the value of a statistic observed in a sample and the corresponding population parameter (thus, error), caused by the accident of which cases happened to be included in the sample.

**Scatter plot** A graph showing the relationship between two variables, wherein the score of each case on both variables is expressed as a point in two dimensions.

**Sign test** A nonparametric method for testing hypotheses about differences in location between two matched populations that is based only on the directions of the differences between the members of the various pairs; it can be used as an alternative to the matched-pairs *t* test or the Wilcoxon test.

**Significance test** A procedure for deciding whether to retain or reject a null hypothesis.

**Simple main effect** In a two-way analysis of variance, it is the effect of one factor at only one level of the other factor.

**Skewed distribution** A distribution with some extremely low scores (negatively skewed, skewed to the left) or some extremely high scores (positively skewed, skewed to the right), resulting in a frequency polygon with a pronounced tail in one direction.

**Slope** In linear regression, the average rate of change in the criterion per unit increase in the predictor.

**Spearman (rank-order) correlation coefficient** A measure of the strength of the relationship between two variables, both of which are in the form of ranks.

**Sphericity** In an repeated-measures analysis of variance, the assumption that the variance of the difference scores is the same, in the population, for every possible pair of levels of the RM factor.

**Standard deviation** A measure of the variability of the scores in a specified group, or how much the scores differ from one another; the positive square root of the variance.

**Standard error** The standard deviation of some statistic in a sampling distribution (e.g., based on means rather than individual scores), which tells us how trustworthy the single statistic we have on hand is as an estimate of the corresponding parameter. Examples of parameters that are estimated include a population mean (*standard error of the mean*), a population proportion (*standard error of a proportion*), and the difference between two population means (*standard error of the difference*).

**Standard error of estimate** In linear regression, a measure of the variability (average error) in prediction.

**Standard scores** A synonym for $z$ scores.

**Statistic** A numerical quantity that summarizes some characteristic of a sample.

**Statistical model** A summary of the probability of all possible events of a specified type (e.g., each possible value of a sample mean drawn at random from one population) under specified conditions (e.g., that the null hypothesis is true). Examples include the normal distribution, the $t$, the $F$, and the chi-square distributions.

**Statistical significance** Occurs when the results of a statistical analysis indicate that the null hypothesis can be rejected.

**Stem-and-leaf display** A pictorial description of a set of data that combines features of the frequency distribution and the histogram.

**Summation sign** A mathematical symbol that represents the sum of a set of numbers.

**Sum of squares** The sum of the squared deviations of observations from their mean.

**Symmetric distribution** A distribution wherein one half is the mirror image of the other half.

***t* distributions** This statistical model can be used to test hypotheses about: the mean of one population when the population standard deviation is *not* known; the differences between the means of two populations (independent or matched samples); the significance of certain correlation coefficients; and certain multiple comparison procedures in the analysis of variance.

***T* scores** Transformed scores that have a mean of 50 and standard deviation of 10.

**Theoretical sampling distribution** An estimate of an experimental sampling distribution that is determined mathematically (rather than by drawing repeated samples).

**Transformed score** A score that has been altered mathematically to show its standing relative to a specified group.

**Two-tailed test of significance** A statistical test of two population means wherein the null hypothesis may be rejected regardless of the direction of the results.

**Type I error** Rejecting a null hypothesis that is actually true.

**Type I-ish error** Rejecting a null hypothesis that is very close to being true (i.e., the population effect size is tiny, but the sample effect size is accidentally large enough to produce a misleading statistically significant test).

**Type II error** Retaining a null hypothesis that is actually false.

**Unimodal distribution** A distribution that has one pronounced peak when graphed as a frequency polygon.

**Upper real limit** The upper endpoint of a score or class interval when dealing with continuous data.

**Variability** The extent to which the scores in a specified group differ from one another, or how spread out or scattered the scores are.

**Variable** Any characteristic that can take on different values.

**Variance** A measure of the variability of the scores in a specified group, or how much the scores differ from one another; the square of the standard deviation.

**Wilcoxon (matched-pairs) test** A nonparametric procedure for testing hypotheses about the difference between the locations of two matched populations, used when data are in the form of ranks.

**Within-group variance** In analysis of variance, differences among the scores within the groups. Reflects variation that cannot be explained by the experimental treatment(s) (error variance).

***Y*-intercept** The point at which the regression line crosses the *Y*-axis.

***z* scores** Transformed scores that have a mean of 0 and standard deviation of 1.

# References


American Psychological Association. (2020). *Publication manual of the American Psychological Association 2020: The official guide to APA style* (7th ed.). Washington, DC: American Psychological Association.

Cohen, B. H. (2008). *Explaining psychological statistics* (3rd ed.). Hoboken, NJ: Wiley.

Cohen, B. H. (2013). *Explaining psychological statistics* (4th ed.). Hoboken, NJ: Wiley.

Cohen, B.H. (2017). Why the resistance to statistical innovations? A comment on Sharpe (2013). *Psychological Methods*, *22*, 204–210

Cohen, J. (1988). *Statistical power analysis for the behavioral sciences* (2nd ed.). Hillsdale, NJ: Erlbaum.

Delacre, M., Lakens, D., & Leys, C. (2017). Why psychologistis should by default use Welch's t-test instead of student's t-test. *International Review of Social Psychology*, *30*(1), 92–101.

Finch, S., & Cumming, G. (2009). Putting research in context: Understanding confidence intervals from one or more studies. *Journal of Pediatric Psychology*, *34*, 903–916.

Fisher, R. A. (1951). *The design of experiments* (6th ed.). Edinburgh: Oliver and Boyd.

Greenhouse, S. W., & Geisser, S. (1959). On methods in the analysis of profile data. *Psychometrika*, *24*, 95–112.

Hochberg, Y. (1988). A sharper Bonferroni procedure for multiple tests of significance. *Biometrika*, *75*, 800–802.

Kline, R. B. (2004). *Beyond significance testing: Reforming data analysis methods in behavioral research*. Washington, DC: American Psychological Association.

Mauchly, J. W. (1940). Significance test for sphericity of a normal $n$-variate distribution. *Annals of Mathematical Statistics*, *11*, 204–209.

Nelson, N. C., Chung, J., Ichikawa, K., & Malik, M. M. (2022). Psychology exceptionalism and the multiple discovery of the replication crisis. *Review of General Psychology*, *26*(2), 184–198.

Neyman, J., & Pearson, E. S. (1928). On the use and interpretation of certain test criteria for purposes of statistical inference. Part I. *Biometrika*, *20A*, 175–263.

Neyman, J., & Pearson, E. S. (1933). On the problem of the most efficient tests of statistical hypotheses. *Philosophical Transactions of the Royal Society of London, Series A*, *231*, 289–337.

Nickerson, R. S. (2000). Null hypothesis significance testing: A review of an old and continuing controversy. *Psychological Methods*, *5*, 241–301.

Nosek, B. A., & Lindsay, D. S. (2018). Preregistration becoming the norm in psychological science. *APS Observer*, *31*.

Open Science Collaboration. (2015). Estimating the reproducibility of psychological science. *Science*, *349*(6251), aac4716.

Paez, A. (2017). Gray literature: an important resource in systematic reviews. *Journal of Evidence-Based Medicine*, *10*(3), 233–240.

Rauscher, F. H., Shaw, G. L., & Ky, K. N. (1993). Music and spatial task performance. *Nature*, *365*, 611.

Siegel, S., & Castellan, N. J., Jr. (1988). *Nonparametric statistics for the behavioral sciences* (2nd ed.). New York, NY: McGraw-Hill.

Simmons J. P., Nelson L. D., & Simonsohn U. (2011). False-positive psychology: undisclosed flexibility in data collection and analysis allows presenting anything as significant. *Psychological Science*, *22*(11), 1359–1366.

Stroop, J. R. (1935). Studies of interference in serial verbal reactions. *Journal of Experimental Psychology*, *18*(6), 643–662.

Tukey, J. W. (1977). *Exploratory data analysis*. Reading, MA: Addison-Wesley.



*Introductory Statistics for the Behavioral Sciences*, Eighth Edition. R. Brooke Lea and Barry H. Cohen.



Companion website: www.wiley.com/go/ISBS8e

# Index

Note: Page numbers in *italics* refer to figures; those in **bold** to tables.




*Introductory Statistics for the Behavioral Sciences*, Eighth Edition. R. Brooke Lea and Barry H. Cohen.



Companion website: www.wiley.com/go/ISBS8e

## C

## E



## F

## G



## H



## I



## J



## K

## L



## M

## N

## O



## P

## Q



## R

## S

## T



## U

## V



## W



## Y



## Z

Wiley The manufacturer's authorized representative according to the EU General Product Safety Regulation is Wiley-VCH GmbH, Boschstr. 12, 69469 Weinheim, Germany, e-mail: Product_Safety@wiley.com.

Printed and bound by CPI Group (UK) Ltd, Croydon, CR0 4YY
17/08/2026
02199169-0001